The New Wider World

Third Edition

David Waugh

Solanger

OXFORD

UNIVERSITY PRESS

CONTENTS

Distribution and density

Distribution describes the way in which people are spread out across the Earth's surface. This distribution is uneven and changes over periods of time. It is usual to show population distribution by means of a dot map (Figure 1.1). Notice how people are concentrated into certain parts of the world making those places very crowded. At the same time, other areas have relatively few people living there. These are said to be sparsely populated.

Figure 1.1
Dot map showing world population distribution

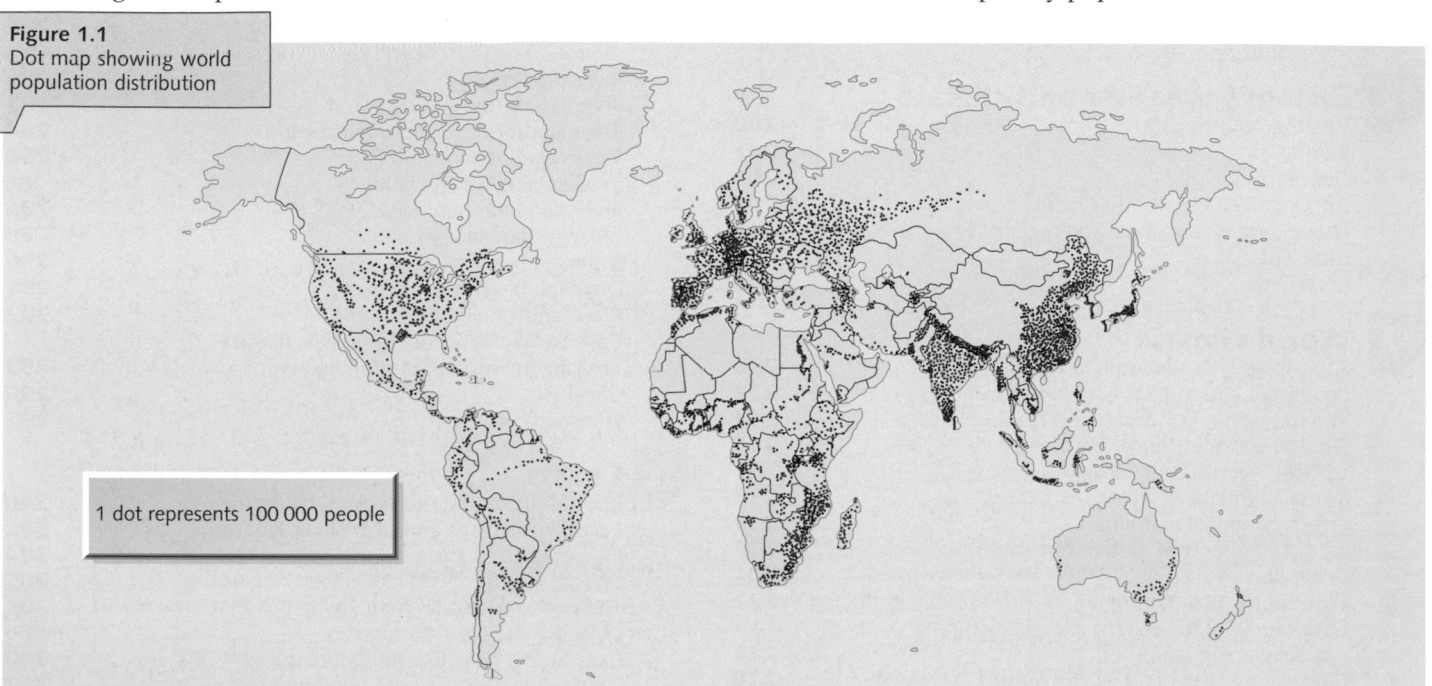

1 dot represents 100 000 people

Figure 1.3
Factors affecting distribution and density of population

	PHYSICAL			
	Densely populated	**Examples**	**Sparsely populated**	**Examples**
Relief	Flat plains and low-lying undulating areas	Bangladesh	High, rugged mountains	Andes
	Broad river valleys	Ganges Valley	Worn-down shield lands	Canadian Shield
	Foothills of active volcanoes	Etna, Pinatubo		
Climate	Evenly distributed rainfall with no temperature extremes	North-west Europe	Limited annual rainfall	Sahara Desert
	Areas with (i) high sunshine totals (ii) heavy snowfall for tourism	(i) Spanish costas	Low annual temperatures	Greenland
		(ii) Swiss alpine valleys	High annual humidity	Amazon rainforest
	Seasonal monsoon rainfall	Bangladesh	Unreliable seasonal rainfall	Sahel
Vegetation	Grasslands – easy to clear/farm	Paris Basin	Forest	Amazonia, Canadian Shield
Soil	Deep fertile silt left by rivers	Nile Valley and Delta	Thin soils in mountainous or glaciated areas	Northern Scandinavia
	Volcanic soils	Etna	(i) Lacking humus or (ii) Affected by leaching	(i) Sahel (ii) Rainforests
Natural resources	Minerals, e.g. coal, iron ore	Pennsylvania, Johannesburg	Lacking minerals	Ethiopia
	Energy supplies, e.g. HEP	Rhône Valley	Lacking energy supplies	North-east Brazil
Water supply	Reliable supplies	North-west Europe	Unreliable supplies	Afghanistan
Natural routes	Gaps through mountains, confluence of valleys	Rhine Valley, Paris	Mountain barrier	Himalayas

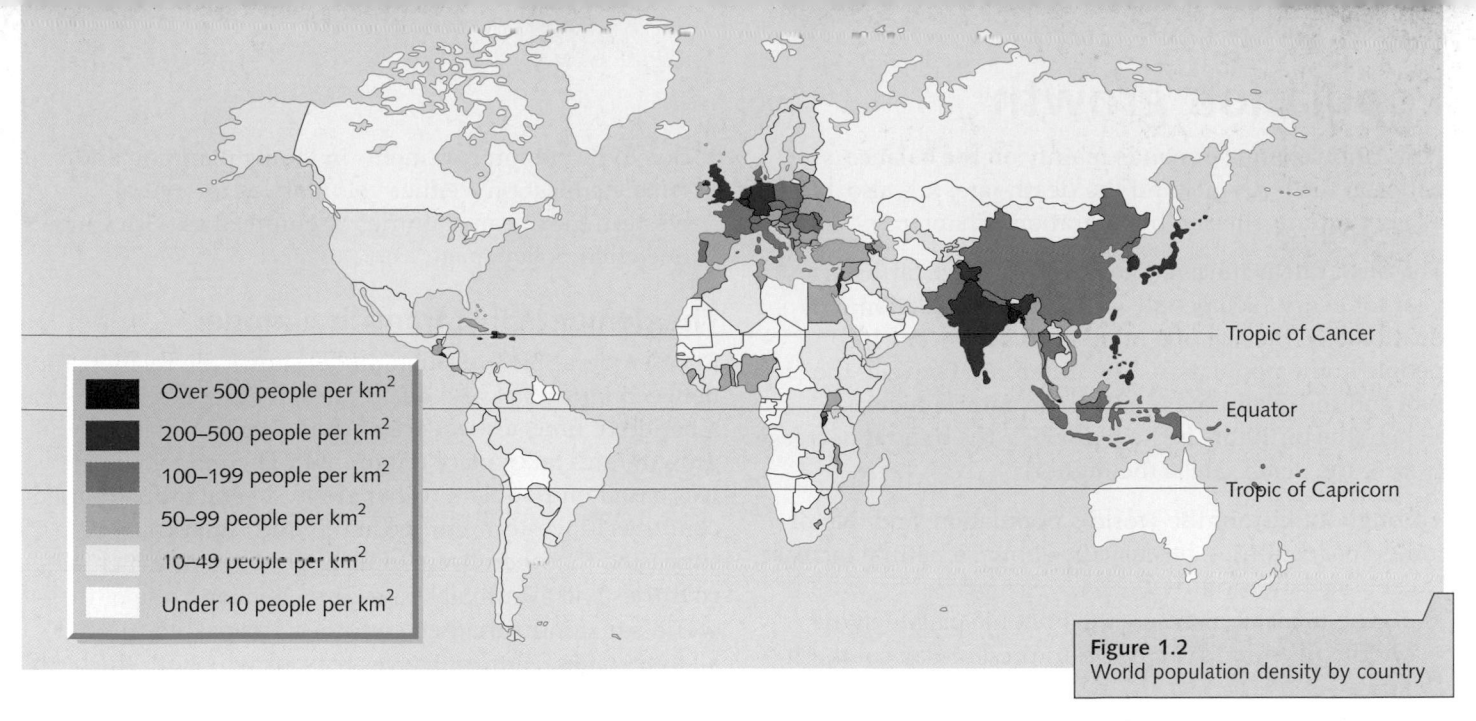

Figure 1.2
World population density by country

Legend:
- Over 500 people per km²
- 200–500 people per km²
- 100–199 people per km²
- 50–99 people per km²
- 10–49 people per km²
- Under 10 people per km²

Tropic of Cancer
Equator
Tropic of Capricorn

Density describes the number of people living in a given area, usually a square kilometre (km²). Density is found by dividing the total population of a place by its area. Population density is usually shown by a choropleth map (Figure 1.2). A choropleth map is easy to read as it shows generalisations, but it does tend to hide concentrations. For example:

- Brazil appears on the world map in Figure 1.2 to have a low population density. However, on a larger-scale map (see Figure 1.24), several parts of the country are shown to have very high densities.
- Figure 1.2 suggests that the population of Egypt is evenly spread whereas in reality it is concentrated along the Nile Valley (Figure 1.1).

On global and continental scales, patterns of distribution and density are mainly affected by **physical** factors such as relief, climate, vegetation, soils, natural resources and water supply. At regional and more local scales, patterns are more likely to be influenced by **human** factors which may be economic, political or social. Figure 1.3 gives reasons, with specific examples, why some parts of the world are densely populated while others are sparsely populated. You should be aware, however, that:

- for a given place there are usually several reasons for its dense or sparse population, e.g. Nile Valley and Sahara Desert
- even within areas there are variations in density, e.g. parts of Japan have some of the highest densities in the world, yet less than one-fifth of the country is inhabited.

			HUMAN		
	Densely populated	**Examples**	**Sparsely populated**	**Examples**
Economic	Ports	New York, Sydney	Limited facilities for ports	Bangladesh
	Good roads, railways, airports	Germany, California	Poor transport links	Himalayas
	Industrial areas (traditional)	Pittsburgh, Ruhr	Lack of industrial development	Sudan
	Development of tourism	Banff (Canada), Jamaica	Lack of tourist developments	Iraq
	Money available for new high-tech industries	California, south of France	Lack of money for new investments	Nepal, Gaza
Political	Government investment	Tokyo region, north Italy	Lack of government investment	Dem. Rep. of Congo
	New towns	Satellite towns around Cairo, Brasilia	Depopulation of rural and old industrial areas	North-east Brazil, Belgian coalfield
	Reclamation of land	Hong Kong Island, Dutch polders	Loss of land, e.g. deforestation, and soil erosion	Amazonia, Apennines, Sahel
Social	Better housing opportunities	Arizona	Poor housing opportunities	Afghanistan, Soweto
	Education, health facilities, entertainment	Sydney, Milan	Limited education, health facilities, entertainment	Rwanda
	Retirement areas	Spanish costas, Canary Islands	Poor facilities for retirement	Eritrea

Population growth

Population change depends mainly on the balance between the birth rate and the death rate. It is also, but to a lesser extent, affected by migration (Chapter 2).

The **birth rate** is the average number of live births in a year for every 1000 people in the total population. The **death rate** is the average number of deaths per 1000 people in the population. The difference between the birth rate and the death rate is either the **natural increase** (where the birth rate is the higher) or the **natural decrease** (where the death rate is the higher).

Throughout history the world's population, and that of individual countries, has usually shown a natural increase. Exceptions have occurred:
- during times of disease – the bubonic plague in the Middle Ages and **HIV/AIDS** in present-day southern Africa
- as a result of war – western Europe during the two World Wars and in present-day Afghanistan

- due to recent improvements in family planning and female education – either voluntary as in several western European countries or enforced as with China's one-child policy (page 13).

The demographic transition model

The demographic transition model shows the relationship between birth rates and death rates. It describes how, over a period of time, a sequence of change in population growth rates takes place (Figure 1.4). The model, which was based on population changes in several industrialised countries in western Europe and North America, suggested that there were four stages through which **all** countries will eventually pass. The UK, being one of the world's first industrialised countries, has passed through all four stages. Figure 1.5 gives reasons why both the birth rate and the death rate have changed over time.

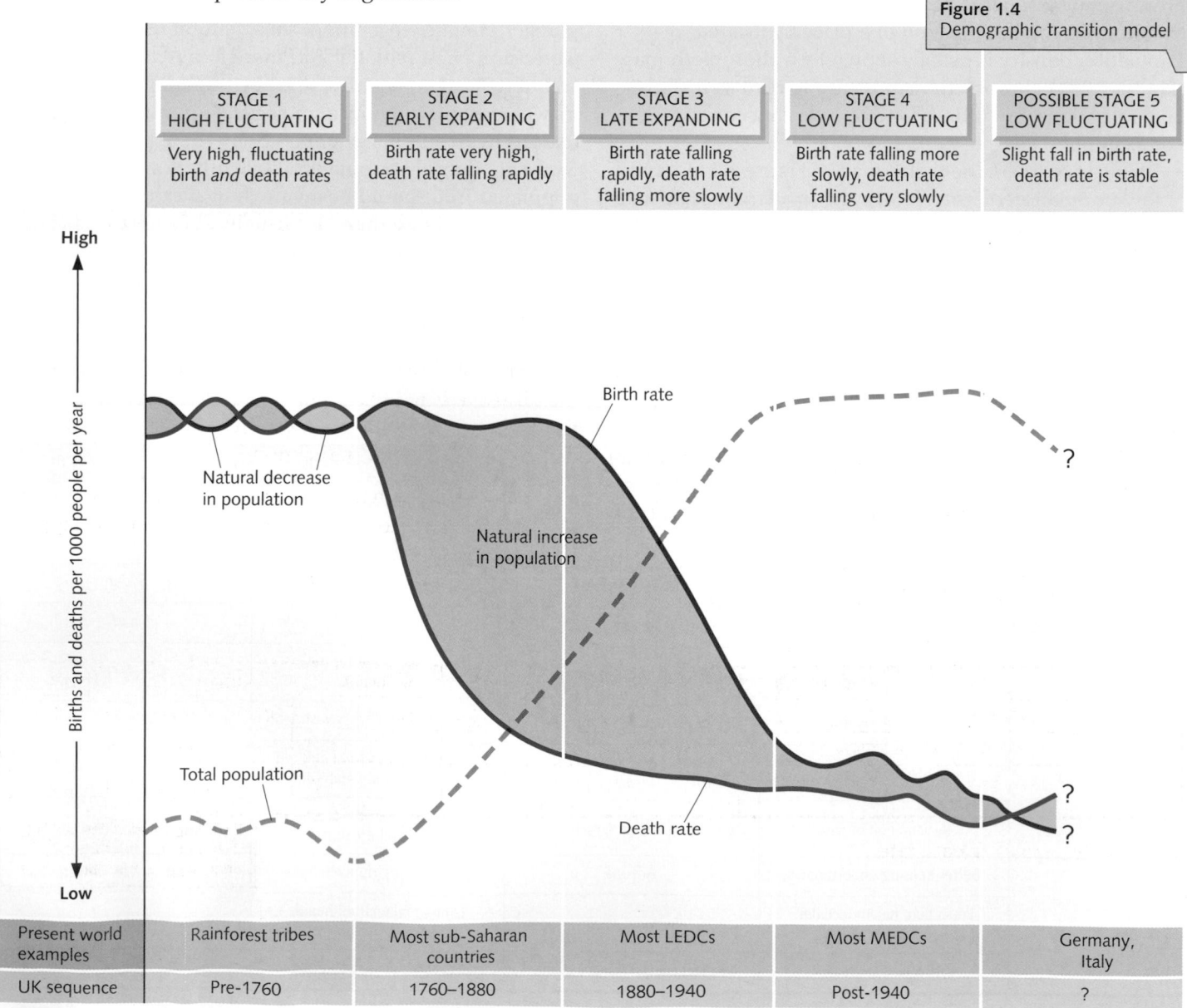

Figure 1.4
Demographic transition model

	STAGE 1 HIGH FLUCTUATING	STAGE 2 EARLY EXPANDING	STAGE 3 LATE EXPANDING	STAGE 4 LOW FLUCTUATING	POSSIBLE STAGE 5 LOW FLUCTUATING
	Very high, fluctuating birth *and* death rates	Birth rate very high, death rate falling rapidly	Birth rate falling rapidly, death rate falling more slowly	Birth rate falling more slowly, death rate falling very slowly	Slight fall in birth rate, death rate is stable
Present world examples	Rainforest tribes	Most sub-Saharan countries	Most LEDCs	Most MEDCs	Germany, Italy
UK sequence	Pre-1760	1760–1880	1880–1940	Post-1940	?

Figure 1.5
Change in birth rates and death rates in the UK

Stage 1 Here both birth rates and death rates fluctuate at a high level (about 35 per 1000) giving a small population growth.

Birth rates are high because:
- No birth control or family planning.
- So many children die in infancy that parents tend to produce more in the hope that several will live.
- Many children are needed to work on the land.
- Children are regarded as a sign of virility.
- Religious beliefs (e.g. Roman Catholics, Muslims and Hindus) encourage large families.

High death rates, especially among children, are due to:
- Disease and plague (bubonic, cholera, kwashiorkor).
- Famine, uncertain food supplies, poor diet.
- Poor hygiene – no piped, clean water and no sewage disposal.
- Little medical science – few doctors, hospitals, drugs.

Stage 2 Birth rates remain high, but death rates fall rapidly to about 20 per 1000 people giving a rapid population growth.

The fall in death rate results from:
- Improved medical care – vaccinations, hospitals, doctors, new drugs and scientific inventions.
- Improved sanitation and water supply.
- Improvements in food production (both quality and quantity).
- Improved transport to move food, doctors, etc.
- A decrease in child mortality.

Stage 3 Birth rates fall rapidly, to perhaps 20 per 1000 people, while death rates continue to fall slightly (15 per 1000 people) to give a slowly increasing population.

The fall in birth rate may be due to:
- Family planning – contraceptives, sterilisation, abortion and government incentives.
- A lower infant mortality rate therefore less need to have so many children.
- Increased industrialisation and mechanisation meaning fewer labourers are needed.
- Increased desire for material possessions (cars, holidays, bigger homes) and less for large families.
- Emancipation of women, enabling them to follow their own careers rather than being solely child bearers.

Stage 4 Both birth rates (12 per 1000) and death rates (10 per 1000) remain low, fluctuating slightly to give a steady population.

A **model** is useful when a difficult or a complex world situation is simplified to make it easier to explain or understand. Like all models, the demographic transition model has its limitations – in this case it is the assumption that the sequence of population change results from an increase in industrialisation. While it is mainly true that the **MEDCs** (the **more economically developed countries**) have reached stage 4, it is now widely accepted that many of the **LEDCs** (the **less economically developed countries**) will never become industrialised. Although the more developed LEDCs have indeed reached stage 3, most of those that are the least developed remain at stage 2 (Figures 1.4 and 1.6).

It has been noticed that since the 1990s several MEDCs appear to be entering a new and fifth stage. This stage, not shown in the original model (but indicated on Figure 1.4), is where the birth rate is beginning to fall below the death rate. It is predicted that, should this trend continue, countries entering this stage will eventually see a natural decrease in their total population.

This model shows changes in the population growth of a country over a period of time and can be used to compare these changes with those of other countries.

Stages 1–2	Stage 2		Stages 2–3		Stage 3		Stages 3–4		Stage 4		Stages 4–5		Stage 5	
	Chad	44–17	Egypt	27–6	India	24–8	Jamaica	17–6	China	12–7	Japan	9–9	Italy	9–10
	Nigeria	43–18	Ghana	32–10	Mexico	20–5	Chile	14–5	UK	13–9	Sweden	10–10	Germany	8–10
	Kenya	40–12	Bangladesh	24–7	Brazil	20–6	New Zealand	15–7	USA	14–8				

Figure 1.6
Birth and death rates at different stages of the demographic transition model

44–17 = Birth rate – Death rate

Population structures

The rate of natural increase, the birth rate, the death rate and life expectancy (**life expectancy** is the number of years that the average person born in a country can expect to live) all affect the **population structure** of a country. The population structure can be shown by a **population pyramid** or, as it is sometimes known, an **age–gender pyramid**.

A population pyramid shows:
- the total population divided into five-year age groups, e.g. 5 to 9 years, 10 to 14 years
- the percentage of the total population, subdivided into males and females, in each of those groups.

The population pyramid for the UK is taken from figures for mid-2008 (Figure 1.7). The graph shows:
- a 'rectangular' shape indicating approximately the same number in each age group, a low death rate and a steady, or even a static, population growth
- a narrow base indicating a low and falling birth rate
- that more boys are born than girls (a higher percentage of boys aged under 4 years)

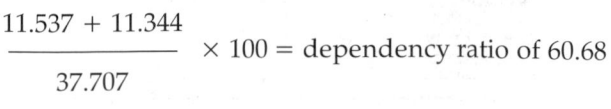

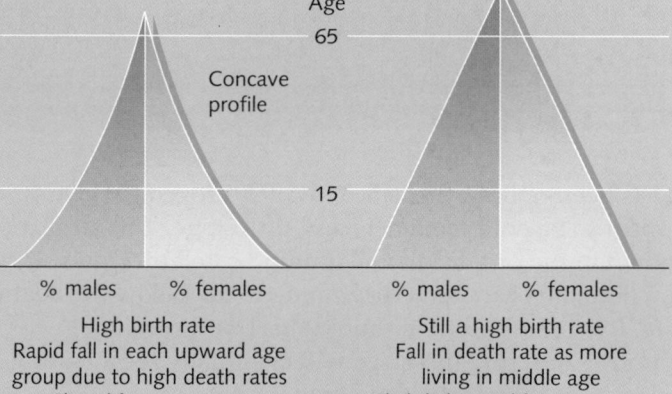

Figure 1.7
Population pyramid for the UK, 2008

- relatively large numbers aged 65 years and over, indicating a long life expectancy
- that more females than males live to over 65 years
- a relatively large proportion of the population in the pre- and post-reproductive age groups and a relatively small number in the 15–64 age group, which is the one that produces children and most of the national wealth.

This last feature can be shown as the **dependency ratio**, i.e.

$$\frac{\text{Non-economically active}}{\text{Economically active}} \quad \text{i.e.} \quad \frac{\text{Children (0–14) and elderly (65+)}}{\text{Those of working age (15–64)}}$$

e.g. UK 2008 (figures in millions):

$$\frac{11.537 + 11.344}{37.707} \times 100 = \text{dependency ratio of } 60.68$$

This means that for every 100 people of working age, there were 60.68 people dependent on them.

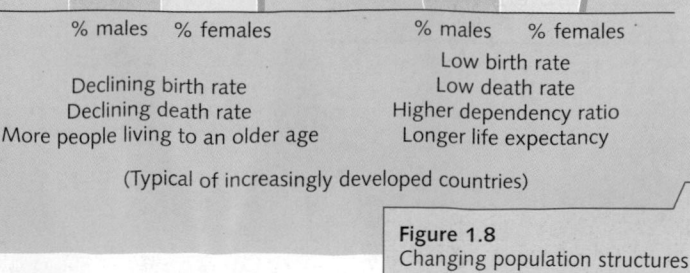

Stage 1

Age 65 — Concave profile — 15

% males % females

High birth rate
Rapid fall in each upward age group due to high death rates
Short life expectancy

Stage 2

Age 65 — 15

% males % females

Still a high birth rate
Fall in death rate as more living in middle age
Slightly longer life expectancy

(Typical of least developed countries)

Stage 3

Age 65 — 15

% males % females

Declining birth rate
Declining death rate
More people living to an older age

Stage 4

Age 65 — 15

% males % females

Low birth rate
Low death rate
Higher dependency ratio
Longer life expectancy

(Typical of increasingly developed countries)

Figure 1.8
Changing population structures

Compared with 1971, the dependency ratio had fallen slightly. This was because despite a fall in Britain's birth rate, there had been an increase in the number of economically active people and an increase in life expectancy. (The dependency ratio does not take into account those of working age who are unemployed.) Most developed countries have a dependency ratio of between 50 and 70, whereas in developing countries the ratio is often over 100 due to the large numbers of children.

Population pyramids enable comparisons to be made between countries, and can help a country to plan for future service needs such as old people's homes if it has an ageing population or fewer schools if it has a declining, younger population. Unlike the demographic transition model (Figure 1.4), population pyramids include immigrants, but like that model they can produce four idealised types of graph representing different stages of development (Figure 1.8).

Stages of development

Three countries at different stages of economic development and which will be referred to in several later chapters are Kenya, India and Japan. The most recent population pyramids for these countries are given in Figure 1.9. How well, do you think, do the three pyramids fit with the models shown in Figure 1.8? Figure 1.9 confirms the point that population structures change over periods of time. Kenya is the least wealthy of the three. Its wide base (0–4 years) confirms that it has the highest birth rate and that its population declines rapidly due to a high infant mortality rate. Birth and infant mortality rates in India, which is slightly more wealthy, are both declining. This is shown by the narrower base (0–4 years) and relatively more people reaching child-bearing age (20–24 years). Both countries, and especially Kenya, have a rapidly narrowing pyramid indicating a high death rate and a short life expectancy (there are relatively few people aged over 65 years). The pyramid for Japan shows, by contrast, a low birth rate, a low death rate, a low infant mortality rate and a long life expectancy. The graphs also show that in Japan and India, in common with most other countries,

there are slightly more male than female births; but that females have the longer life expectancy.

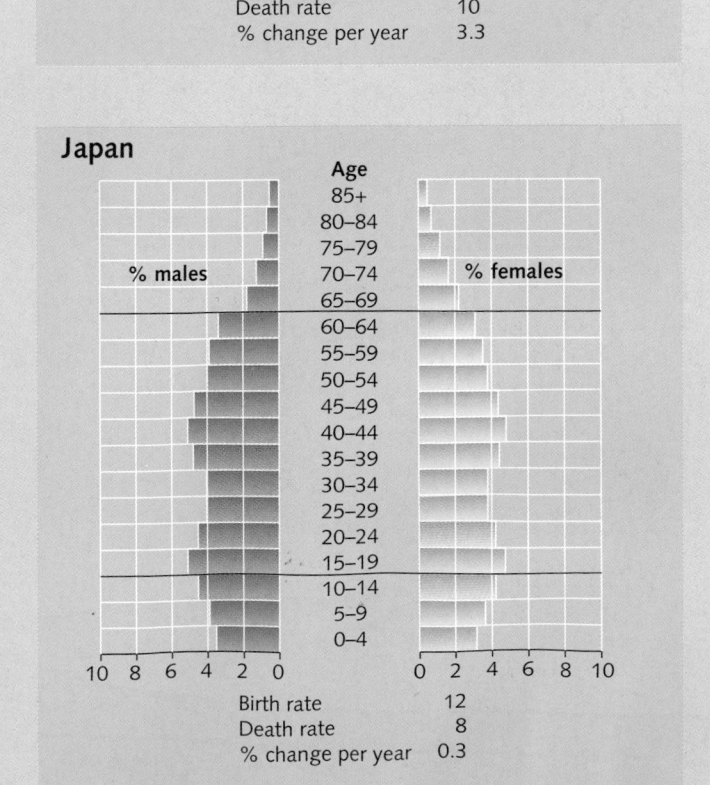

Kenya

Birth rate	47
Death rate	10
% change per year	3.3

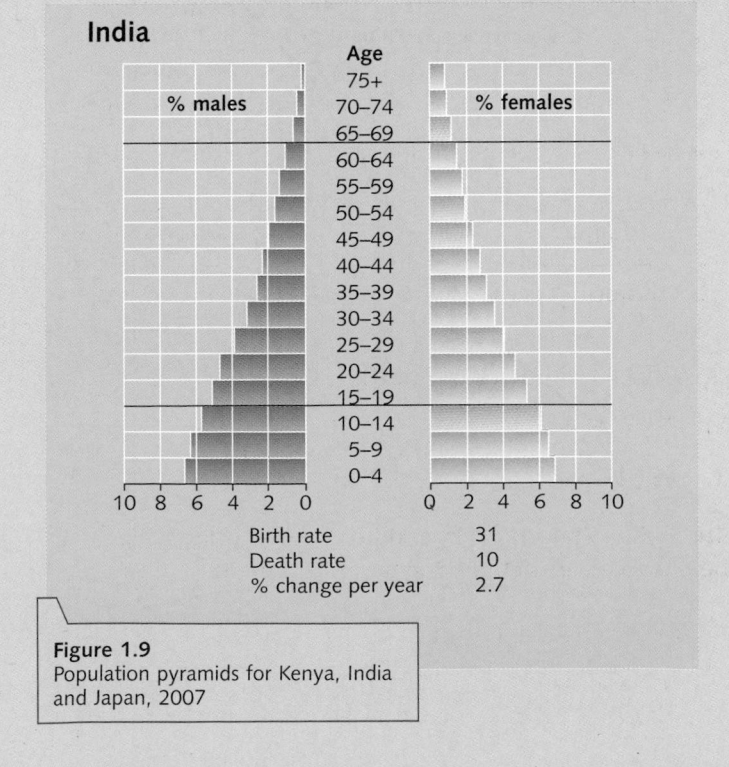

India

Birth rate	31
Death rate	10
% change per year	2.7

Japan

Birth rate	12
Death rate	8
% change per year	0.3

Figure 1.9
Population pyramids for Kenya, India and Japan, 2007

Population trends

The annual growth rate of the world's population rose slowly but steadily until the beginning of the nineteenth century. For the next century and a half it grew at an increasingly faster rate, a process that has been referred to as a **population explosion**. During the 1960s and 1970s the world's population grew, on average, by a record 2 per cent per year (Figure 1.10). This growth rate, which was even higher in the less economically developed countries, caused increasing concern. Estimates made at that time suggested that the world's population of 3.04 billion in 1960 would have reached 7.60 billion by 2000 and 11.00 billion in 2050.

In 2008, the United Nations (UN) claimed that the annual growth rate had fallen to 1.2 per cent and was likely to drop to 1.0 per cent by 2020 and to 0.5 per cent in 2050. This meant that the world's total population only reached 6 billion in 1999 instead of the 1960s prediction of 7.6 billion by the end of the century. Even so, it is still growing at a rate of 156 persons per minute, which is 82 million each year.

So why, although people across most of the world are living longer (pages 14 and 15), is the annual growth rate slowing down? This is due to a combination of factors which include:

- a faster than predicted decline in the birth rate across the world
- improvements in family planning, basic education and female literacy that have all led to a reduction in family size (i.e. a lower **fertility rate**)
- the one-child policy in China, a country with over 20 per cent of the world's population
- diseases such as HIV/AIDS and malaria which reduce life expectancies, especially in sub-Saharan countries in Africa that have, at present, the world's highest birth rates.

These population figures need to be treated with some caution as they can, at the best, only be an estimation. This is because, for example, census data for most countries is only collected every ten years and for some countries even less often. Even then, many people are unlikely either to return forms or to complete them accurately (only 94 per cent in the 2001 UK census), while groups of people like refugees, illegal immigrants and shanty town dwellers are likely to be excluded or missed.

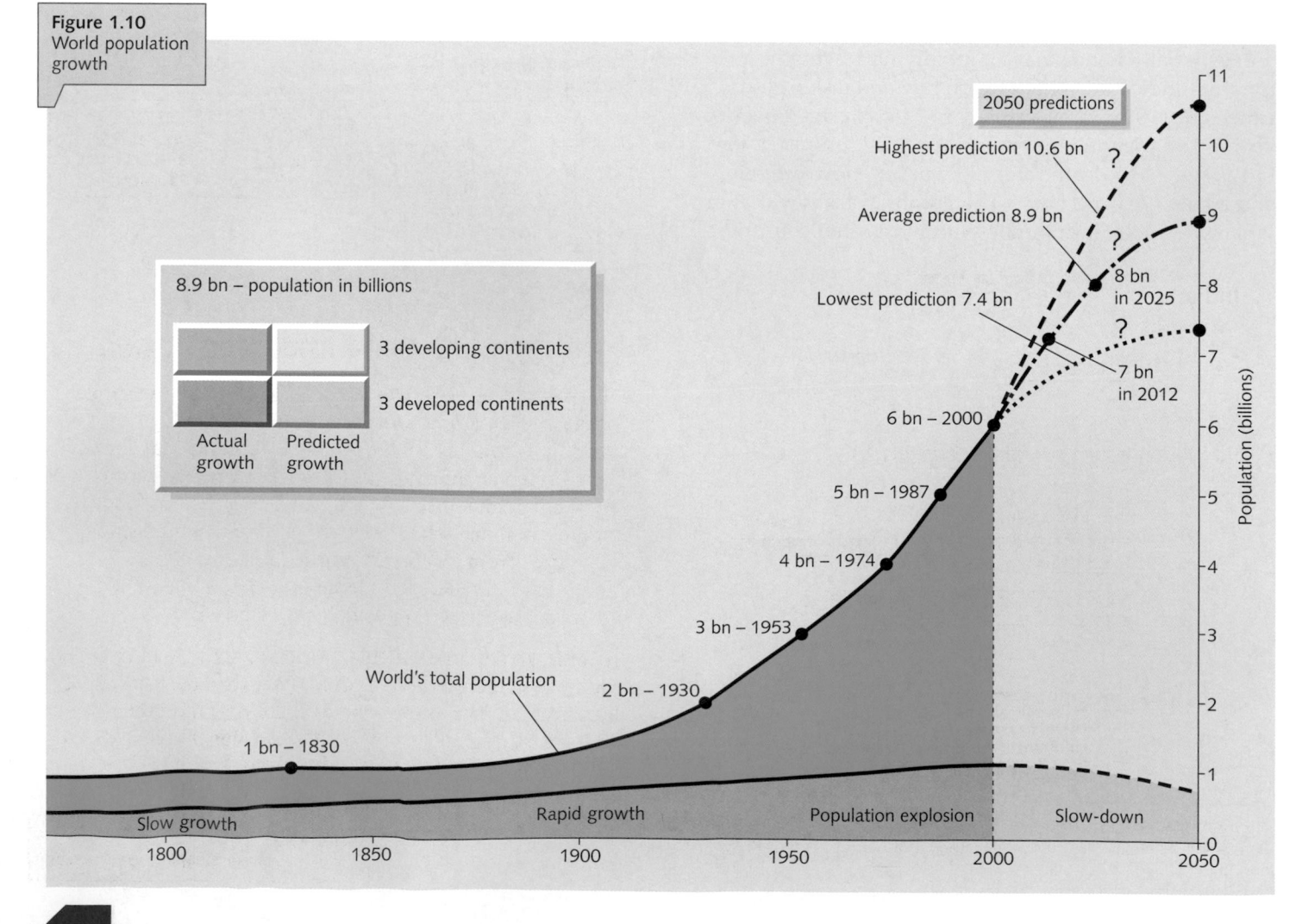

Figure 1.10
World population growth

Population change in developed and developing countries

It has already been seen on Figure 1.4 that many of the less economically developed countries (LEDCs) still fit into either stage 2 or stage 3 of the demographic transition model whereas most of the more economically developed countries (MEDCs) have long since reached stage 4. This is because the LEDCs have a higher birth rate and a greater increase in their total population (Figure 1.11).

Figure 1.12 is a bar graph that shows both the uneven growth and distribution of the world's population between the various continents. It emphasises that:

- the continents with the fastest growth in population are the three developing ones of Africa, Asia and Latin America. Africa has the greatest growth despite its high death rate caused by AIDS, crop failure and civil war
- the continents with the slowest growth rate are the three developed ones of Europe, North America and Australasia (Oceania); estimates suggest that Europe is likely to show a population decrease by the middle of the twenty-first century
- most of the world's population live in Asia (about 40 per cent live in the two countries of China and India); in 1950, 76 per cent of the total lived in the three developing continents – this proportion had risen to 82 per cent by 2000 and is estimated to reach 88 per cent by 2050
- some places have become **overpopulated** (Figure 1.13).

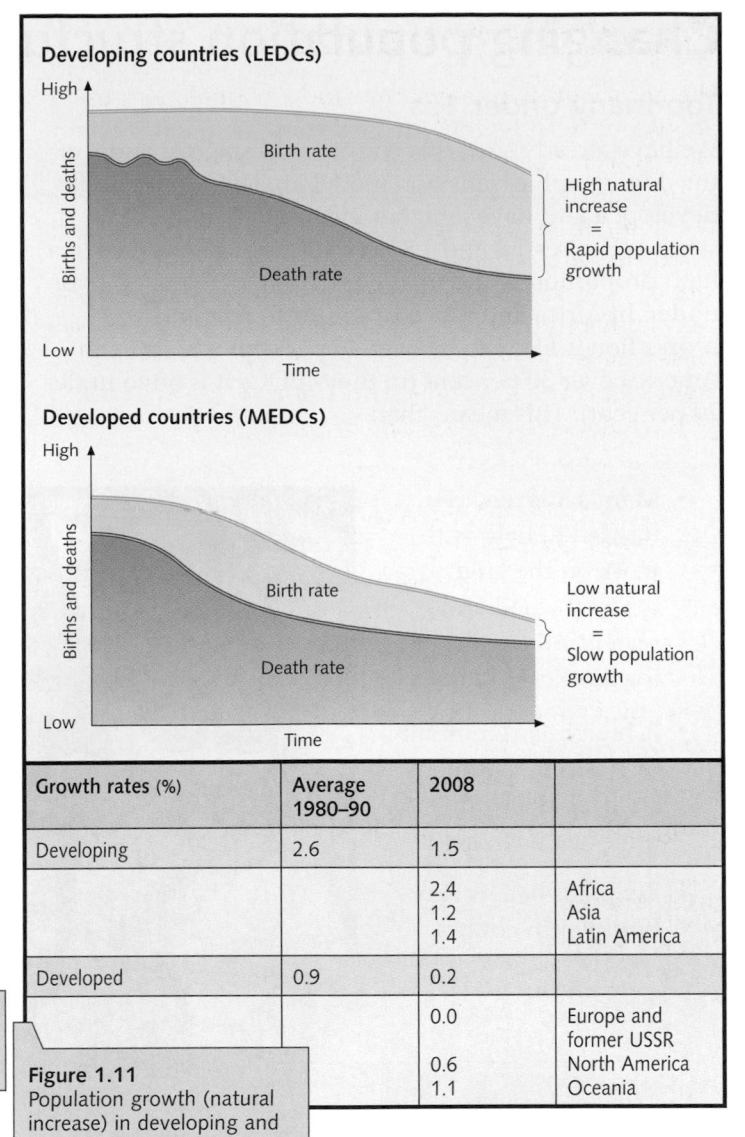

Figure 1.11
Population growth (natural increase) in developing and developed countries

Growth rates (%)	Average 1980–90	2008	
Developing	2.6	1.5	
		2.4	Africa
		1.2	Asia
		1.4	Latin America
Developed	0.9	0.2	
		0.0	Europe and former USSR
		0.6	North America
		1.1	Oceania

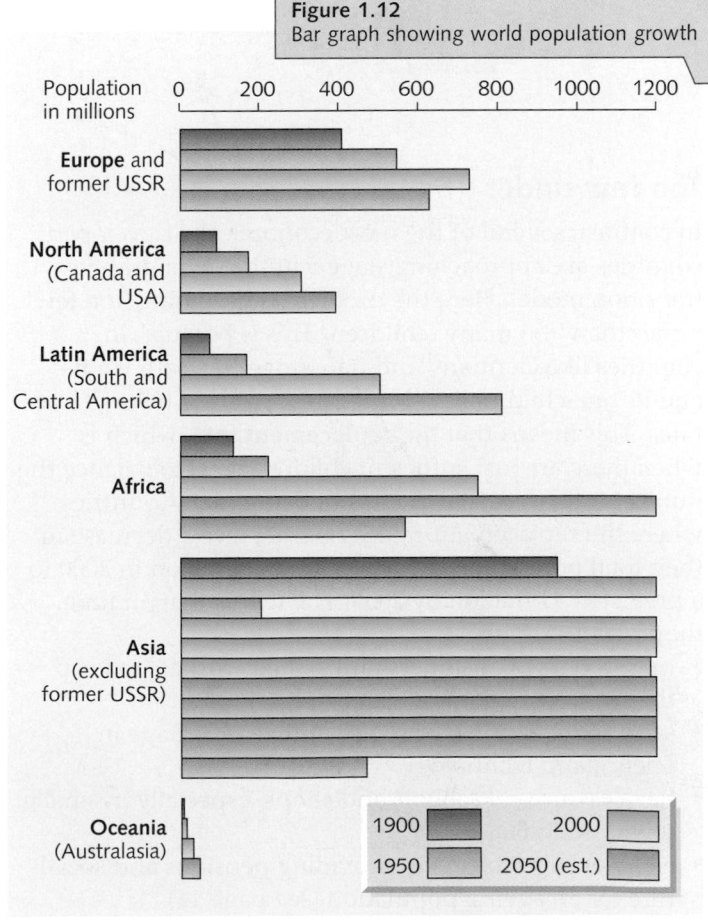

Figure 1.12
Bar graph showing world population growth

Overpopulation and underpopulation

Overpopulation occurs when there are too many people living in an area for the resources and technology available for them to maintain an acceptable standard of living. Many countries in sub-Saharan Africa are said to be overpopulated because they have insufficient food, minerals and energy resources to sustain themselves, they suffer from localised natural disasters such as drought and famine, and are characterised by low incomes, poverty and poor living conditions

In contrast, countries such as Australia and Canada are said to be underpopulated. This is because they have more food, minerals and energy resources than they need and, being able to export the surplus, they have high incomes, good living conditions and a high standard of living.

Figure 1.13
Balance between people and resources

Changing population structures

Too many under 15s

We have already seen that countries at stages 2 and 3 of the demographic transition model are less economically developed and have a higher birth rate than those at stage 4 (Figures 1.4 and 1.8). A high birth rate results in a high proportion of the total population being aged 15 or under. In Africa and much of southern Asia this proportion is likely to be over 40 per cent and in Latin America over 30 per cent (in the MEDCs it is often under 20 per cent). This means that:

- at present, the large youthful population will need child-health care and education – two services that these countries can ill-afford
- in the future, there will be more people reaching child-bearing age.

Figure 1.14 suggests reasons for the high birth and fertility rates, and the large family size, of many LEDCs.

- Many children are needed to help with work on the land, to carry wood and water, and to care for the elderly.

- A high infant mortality rate means that many children die before their first birthday.

- Religious beliefs may forbid birth control.

- A lack of education, especially for women, has an effect on family planning.

- There is a lack of access to and insufficient money for the supply of contraceptives.

- A large family can enhance a family's reputation in the local community, especially for the husband.

Figure 1.14
Why birth and fertility rates are high in LEDCs

It is now recognised that the three key factors leading to a decline in birth rate and fertility are improvements in the following areas:

Family planning by:
- increasing the knowledge of contraception methods and access to family planning services
- reducing the cost and increasing the supply of contraception commodities
- overcoming the disapproval of husbands and the opposition of certain religious groups.

Health care, which includes safer abortions and a reduction in infant mortality (the latter means fewer children need to be born as more of them will survive).

Women's education and status – this delays marriage and gives women the right to decide between having more children or using birth control.

The UN states that the freedom to choose how many children to have, and when to have them, is a woman's basic right, while family planning improves maternal health and child survival, and helps to reduce poverty.

Too few under 15s

In contrast, several of the most economically developed countries are approaching stage 5 in the demographic transition model. Here the problem is becoming 'too few' rather than 'too many' children. This is because, in countries like Germany and Italy (together with China and its one-child policy) birth rates are at, or below, death rates. This means that the **replacement rate**, which is when there are just sufficient children born to balance the number of people that die, is not being met. Countries where the replacement rate is not met face a decrease in their total population, e.g. Italy from 56 million in 2000 to a predicted 41 million by 2050. The fear is that, in time, they will:

- have too few consumers and skilled workers to keep their economy going
- see a reduction in their competitive advantage in science and technology
- have closures of schools and shops, especially in smaller towns and villages
- experience problems in providing pensions and social care for an ageing population (see page 14).

China's one-child policy

During the mid-twentieth century the Chinese, in response to the state's philosophy that 'a large population gives a strong nation', were encouraged to have large families. This resulted in a population growth of 55 million (about the same as the UK's total population) every three years. Concern over this growth rate grew during the 1970s despite a fall, between 1950 and 1975, in:

• the birth rate from 44 per 1000 to 31 per 1000
• family size from 5 children per family to 3.

In 1979, the state decided to try to control population growth. To do this, family size would have to be reduced to 1.5 children per family. To play safe, it was decided to restrict it to 1.0 – the so-called **one-child policy** – and to raise the marriageable age for men to 22 and for women to 20 (Figure 1.15). Couples had to apply to be married and again before having a child. Those who conformed were given free education, priority housing and family benefits. Those who did not were deprived of these benefits and fined heavily. Women who became pregnant a second time were forced to have an abortion and persistent offenders were 'offered' sterilisation. According to human rights activists, this resulted in thousands of forced abortions, and there is also evidence of female infanticide by parents wishing their one child to be a boy.

There were, however, notable exceptions to the one-child policy. For example:

• Couples could have a second child if the first was mentally or physically handicapped, or died.
• In most rural areas, farmers could have a second child if the first was a girl (although women do much of the farmwork, they are considered less useful as farmworkers).
• In some rural areas, local officials allowed a second child on payment of a fine (or bribe!).
• The 56 ethnic minority groups, who account for 6 per cent of China's total population, were always allowed at least two, and sometimes more, children.

Since the policy was implemented, the birth rate has fallen from 31 to 12 in less than 30 years and now too few babies are being born to maintain the population. Estimates suggest that China's total population was, in 2008, 230 million less than it would have been had the one-child policy not been introduced. Recent relaxations in the policy include:

• all families in rural areas being allowed two children
• if two 'one-only children', the first of whom had by now reached marriageable age, married they could have two children
• giving women, for the first time, an informed choice between different types of contraception
• allowing, in 300 trial districts, the option of voluntary family planning.

Figure 1.15
A one-child family in Kunming, Yunnan

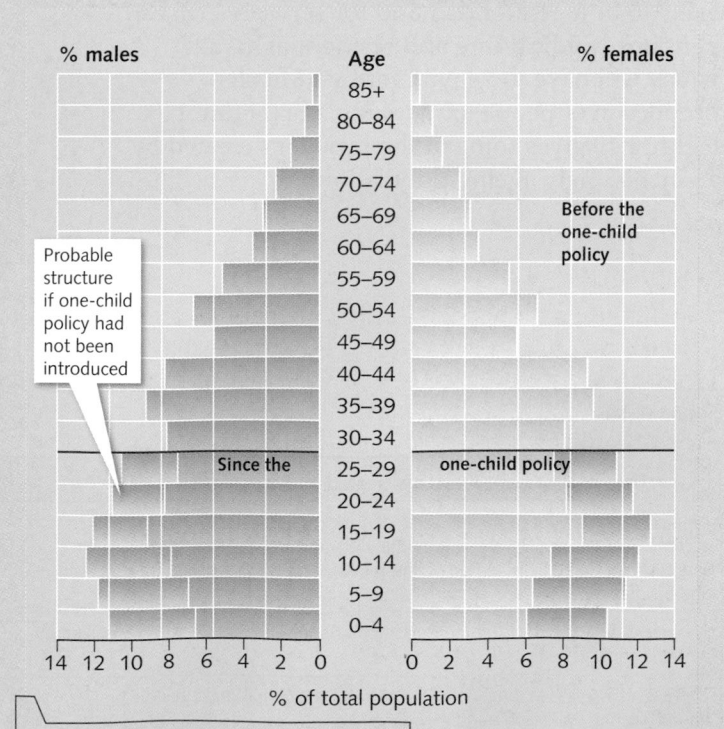

Figure 1.16
Population structure of China, 2008

People aged over 65 – an ageing population

There has been, initially in the more economically developed countries but also more recently in some of the less economically developed countries, an increase in **life expectancy** (Figure 1.17). Life expectancy is the number of years that a person born in a particular country might be expected to live. It has increased due to:

- improved standards of hygiene and health education
- improvements in primary health care
- the development of new drugs and vaccines
- a better diet, both in quantity and quality
- advances in medical knowledge and techniques by which some diseases and conditions formerly considered untreatable are now routine.

The increase in life expectancy, together with a falling birth rate, means that an increasingly higher proportion of a country's population live beyond 65, and even beyond 80. This process is referred to as **ageing**. By 2008, several of the most economically developed countries (mainly Japan and several in western Europe) had over 20 per cent of their population aged over 65 (Figure 1.18). They had, for the first time in history, more people aged over 65 than children aged under 15. The proportion of over 65s is predicted to continue to rise and, in some countries such as Japan and Italy, to reach 35 per cent by 2050. The consequences are likely to be a greater demand for services such as pensions, medical care and residential homes which will have to be paid for by a smaller proportion of people of working age (Figure 1.7). Figure 1.19 gives some of the problems created by an increasingly ageing population.

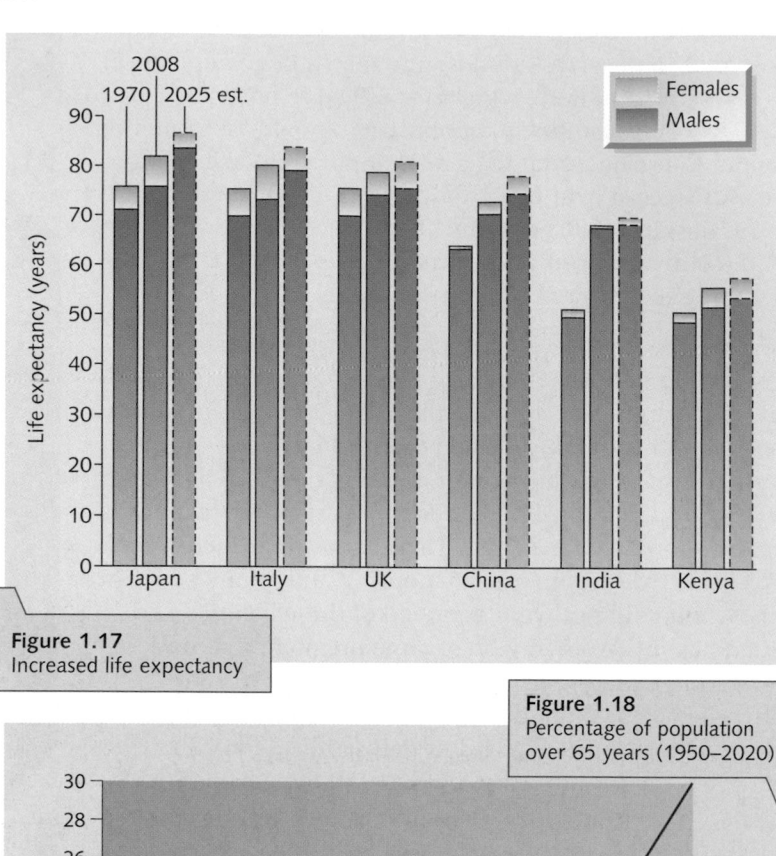

Figure 1.17
Increased life expectancy

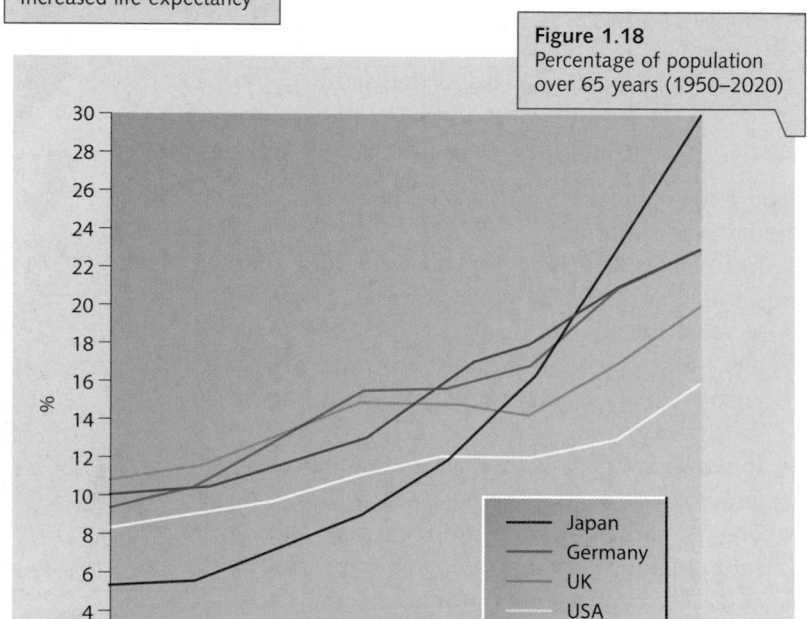

Figure 1.18
Percentage of population over 65 years (1950–2020)

Figure 1.19
Problems of an ageing population

A Problems created in a country ...	B Problems facing the elderly ...
An increasing amount of money is needed for residential homes and sheltered accommodation; health care (e.g. home visits and free prescriptions); social services (e.g. home help and providing meals); and subsidies (e.g. free TV licences and bus passes).	Many have to live alone, especially women, when their married partner dies.
An increasing amount of the family doctor's financial budget and time is taken up by the elderly.	Most cannot afford the cost of a residential home which may be over £2000 a month, especially in the south of England. Others may only be able to afford such accommodation by selling their home or using their life savings.
There is an increase in long-term illnesses and those that make people house-bound, such as Parkinson's and Alzheimer's.	There are lengthy waits for hospital operations such as hip replacements.
There is an increasing dependence on a smaller group of economically active people to provide consumer goods and services as well as money through taxation.	Those who are still fit enough and willing to work often face bias and prejudice due to their age.
Less money is available for younger age groups, e.g. for education, improvements in transport or the provision of leisure and social amenities.	Those living in urban areas have a fear of crime and traffic; those in rural areas who cannot drive and where there is no public transport have difficulty getting to the doctor, hospital and shops.

Italy

Life expectancy in Italy has increased from 64 years for men and 68 for women in 1950 to 79 for men and 84 for women in 2008. At present, 20 per cent of the total population is aged over 65, a proportion predicted to rise to 25 by 2025 and 35 per cent by 2050 (Figure 1.20). Already the ratio of workers to pensioners is 0.7 (compared with 3.5 in the UK). This means that every Italian worker has to support his/her own family and one other pensioner. In 2008, 21 per cent of Italy's gross national product (GNP – page 180) was spent looking after the over 65s. Estimates suggest that by 2030 this figure will have risen to an unsustainable 33 per cent, with 20 per cent being spent on pensions alone (Figure 1.21).

China

At present, China is more concerned with its ageing population than it is with population growth or the number of children. Life expectancy has increased from 39 for men and 42 for women in 1950 to 72 and 75 respectively in 2008. That means a person born in 2008 can expect to live 33 years longer than did their grandparents. Predictions suggest that the proportion of Chinese aged 65 and over will have increased from 10 per cent in 2000 to 30 per cent by 2050 (Figure 1.22).

In 1998, the UN introduced a new term: *the oldest old*. This term refers to people aged 80 years and over. At that time, the Chinese had 10.5 million people in that age group – 16 per cent of the world's total. By 2050, it can expect to have 100 million (nearly twice the present-day total population of the UK) aged over 80. During the same time, the world's total 'oldest old' will rise from 66 million to 370 million.

These changes will, according to the *China Business Handbook*: 'have a massive impact on Chinese society. They will require urgent reform in pensions (there are none at present for the majority of the population), health care and other benefits.' The increase in life expectancy and the one-child policy have led to significant changes in China's population structure (Figure 1.16).

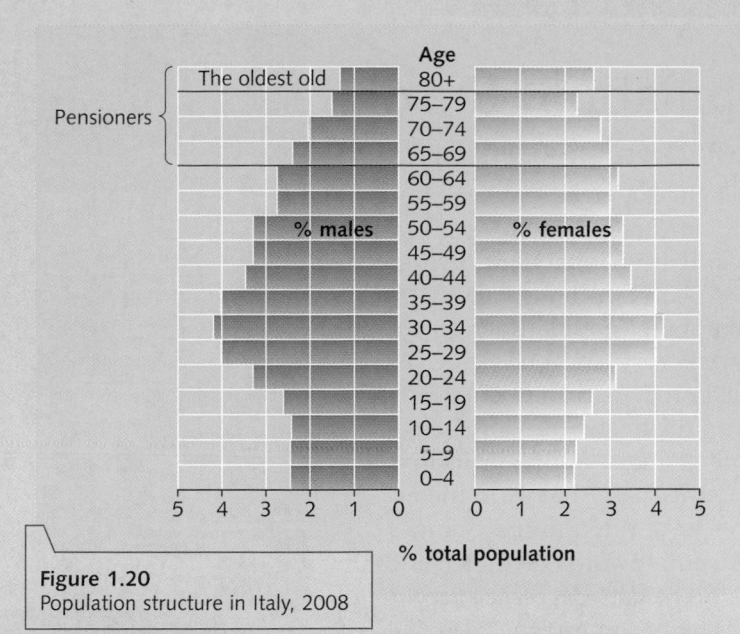

Figure 1.20
Population structure in Italy, 2008

Figure 1.21
Ageing Italians

Figure 1.22
Ageing Chinese

Population: Brazil

Distribution and density

In 2008 the average population density for Brazil was 22.9 per km² (i.e. 195 104 000 inhabitants living in 8 511 970 km²). Although the distribution of population over the country is very uneven, the population density map (Figure 1.24) shows a relatively simple pattern. Over 90 per cent of Brazilians live near to the coast, mainly in the south-east of the country. Going inland, and towards the north and west, the density decreases very rapidly, with some of the more remote areas being virtually uninhabited.

The highest population densities occur either at irregular intervals along the coast or around the cities of São Paulo (Figure 1.23) and Belo Horizonte. Although the coastal climate is hot and wet, and flat land is limited due to mountains which frequently extend down to the sea, this region has a reliable water supply and a range of natural resources. Salvador and, later, Rio de Janeiro were the country's first two capital cities. Both had good natural harbours which encouraged trade, immigration, industry and, more recently, tourism (page 149). São Paulo, the world's second largest city, and Belo Horizonte grew up on the higher, cooler, healthier plateau of the eastern Brazilian Highlands. The rich soils around São Paulo were ideal for the growing of coffee. Later the presence of nearby minerals, such as iron ore, and energy supplies allowed the city to develop into a major industrial centre. The south-east region has the best transport system in Brazil, the greatest number of services, and has benefited most from government help.

Population density decreases with distance from the coast. This is because places further inland have fewer natural resources and cannot

Figure 1.23
São Paulo

support as many people. Those areas shown as having an average density (between 1 and 5 on Figure 1.24) have a less favourable climate, a less reliable water supply, and fewer minerals and energy supplies than places to the south-east. Brasilia is an **anomaly**. An anomaly is something you would not normally expect. Brasilia exists only because in 1960 the Brazilian government chose this virtually uninhabited area as the site for the new federal capital. It was chosen in an attempt to spread out Brazil's population more evenly. Today Brasilia has a population of 2.35 million.

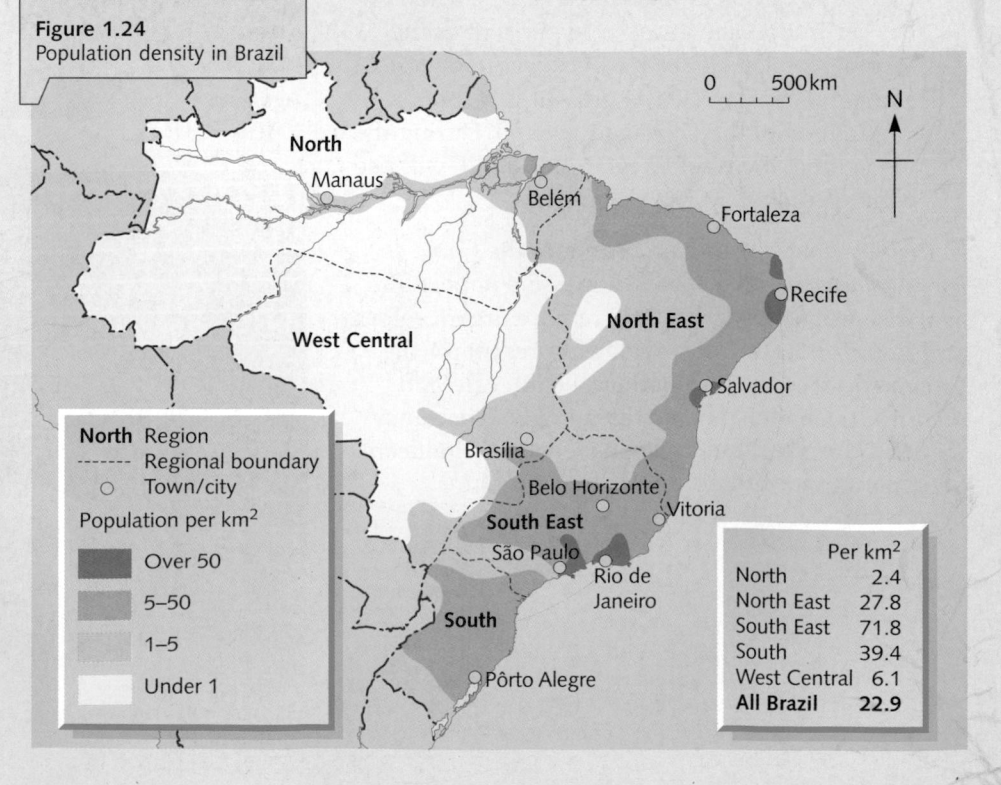

Figure 1.24
Population density in Brazil

0   500km

N

North
Manaus
Belém
Fortaleza
Recife
West Central
North East
Salvador
Brasília
Belo Horizonte
Vitoria
South East
São Paulo
Rio de Janeiro
South
Pôrto Alegre

North Region
------- Regional boundary
○ Town/city
Population per km²
Over 50
5–50
1–5
Under 1

	Per km²
North	2.4
North East	27.8
South East	71.8
South	39.4
West Central	6.1
All Brazil	**22.9**

Figure 1.25
The Amazon and tropical rainforest

Figure 1.26
Population structure for Brazil, 2007

The sparsely populated region to the north-west is drained by the River Amazon and its numerous tributaries (Figure 1.25). The area, which is covered by tropical rainforest (page 232), is hot, wet and unhealthy. Soils are poor (page 253) and most areas lack known natural resources. Transport through the forest is difficult and much of the region lacks such basic services as health care, education and electricity. Compared with south-east Brazil, birth, death and infant mortality rates are high and life expectancy is short. The region has, until recently, suffered from a lack of government investment. There are two anomalies in this region. The first is Manaus (population 1.6 million), which grew as the centre for rubber collecting and is now a large river port with a growing tourist industry. The other includes several places near to the mouth of the Amazon which have benefited either from the discovery of iron ore and bauxite or from the production of hydro-electricity (page 122).

Structure

Although figures for Brazil suggest that it has reached stage 3 of the demographic transition model, parts of the interior still show the characteristics of stage 1 (Amazon Indians) and stage 2. Brazil's population pyramid, with its relatively wide base (Figure 1.26), shows that although both birth and infant mortality rates are still high, they are falling (birth rate 26 in 1991, 20 in 2008; infant mortality rate 59 in 1991, 24 in 2008). This means that, as shown by the widening pyramid, more people are reaching child-bearing age (20–24 year age group). Beyond that, the pyramid narrows, indicating that death rates are still relatively high, although falling (10 in 1991; 6 in 2008), and life expectancy, although shorter than in a more

economically developed country, is slowly increasing (69 for males; 76 for females in 2008).

Trends

Brazil's annual average population growth rate was 1.3 in 2008 compared with 2.1 in 1991 and 3.1 in 1960. The country still has a high proportion of its population aged 15 and under although this is now showing signs of a decline (39 per cent in 1991; 28.1 per cent in 2008). In contrast the proportion aged 65 and over is small but is beginning to increase (4.3 in 1991; 7.5 in 2008). The fertility rate (the number of children per female) of 2.8 in 1991 and 2.3 in 2008 hides the difference between the better-off urban areas in the south-east, and the more rural areas (Figure 1.27) and urban *favelas* (pages 82 and 86).

"Families with children into the double figures are very common in north-east Brazil. Rural mothers in this state, Rio Grande do Norte, have an average of more than seven living children. A quarter of them have ten or more. And that is not counting the dead ones. You often meet women like Luisa Gomez, a slight, small 39-year-old. She married at 14. Since then she has been pregnant sixteen times, once every eighteen months. For half her adult life she has been pregnant, and for the other half breast-feeding the most recent addition. Only six of those sixteen are still alive. There were three stillbirths, and seven died in their first year. Ten wasted pregnancies. Seven and a half years of drain on an already weak organism for nothing. Worse than nothing, for all the anxiety, all the care, all the concern, and then the grief."

Figure 1.27
Having children in north-east Brazil

Population

Key Words and Terms

a You should know the meaning of the following terms:
- birth rate • death rate • natural increase
- infant mortality rate • life expectancy
- birth control • dependent population
- dependency ratio • family planning
- population pyramids/age–gender structures
- replacement rate • fertility rate
- ageing population.

b You should know the difference between:
- population distribution and population density
- a more economically developed country and a less economically developed country
- overpopulation and underpopulation.

Key Ideas

You should know and understand the following:
- The global distribution of population is uneven. This is due to a variety of factors, both physical and human.
- Population change depends upon birth rate, death rate and migration.
- Countries pass through several stages of growth, as shown in the demographic transition model.
- Birth rate and death rate affect population (age–gender) structure and population growth.
- Population structure and population growth vary between MEDCs and LEDCs.
- The uneven balance between population growth and resources can lead to overpopulation or underpopulation.
- Population change affects population structure. It can result in:
  – either too many or too few in the under 15 age group
  – an ageing population.
- Changes in population structure have social and economic implications. Planning is needed if the problems created by change are to be managed.
- Various strategies have been introduced to try to control population growth.

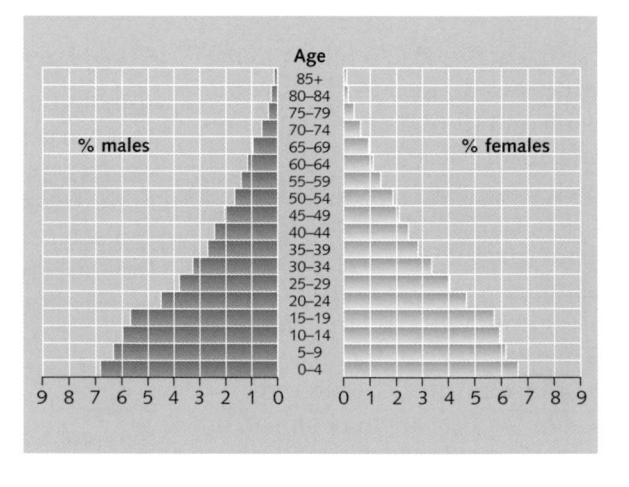

Skills, Theories and Models

- Analyse and evaluate statistical data
- Interpret and use graphs
- Interpret and use population pyramids
- Interpret dot and choropleth maps
- Use census data
- The demographic transition model
- The dependency ratio

1 *(Pages 4 and 5)*

a How is population density calculated? *(1)*

b

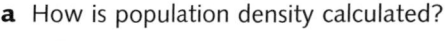

Boxes numbered 1, 2, 3, 4 on the left; 5, 6, 7, 8 on the right, with lines pointing to a world map.

Population density
Less than one person per km²
Over 100 persons per km²

♦ no extremes of climate ♦ too hot
♦ too wet ♦ too dry
♦ good water supply ♦ low-lying
♦ many resources ♦ few resources
♦ flat land ♦ fertile soil
♦ too mountainous ♦ good transport
♦ money for development

i) Make a copy of the table below.
Match the numbers on the map with the places named in the table. *(8)*

ii) Give reasons for the population density of each place.
The list (right) should help you with your answer.
Each label in the list may be used more than once. *(16)*

Densely populated	Map number	Reasons	Sparsely populated	Map number	Reasons
Eastern USA			Himalayas		
Ganges Valley (India/Bangladesh)			Sahara Desert		
North-west Europe					
Nile Valley (Egypt)			Greenland		

2 *(Pages 6 and 7)*

a This diagram is the demographic transition model. It shows, in a simplified way, how population changes are likely to take place over a period of time. Make a copy of the figure and the table beneath it.

i) On your diagram label the birth rate and the death rate. *(2)*

ii) On your diagram draw a line to show the total population. Start at X. Shade and label the natural increase. *(2)*

iii) Complete the birth rate and death rate part of the table by adding the words: **high**, **low** or **decreasing** in the correct places. *(3)*

iv) Complete the population change part of the table by adding **slow decrease**, **slow increase** or **rapid increase** in the correct places. *(3)*

b i) *Either* Give one reason for the population change in stage 2 and one reason for the population change in stage 3. *(2)*

Or Describe and give reasons for the population changes in stages 2 and 3. *(4)*

ii) When was the UK at stages 2 and 3 of population growth? *(2)*

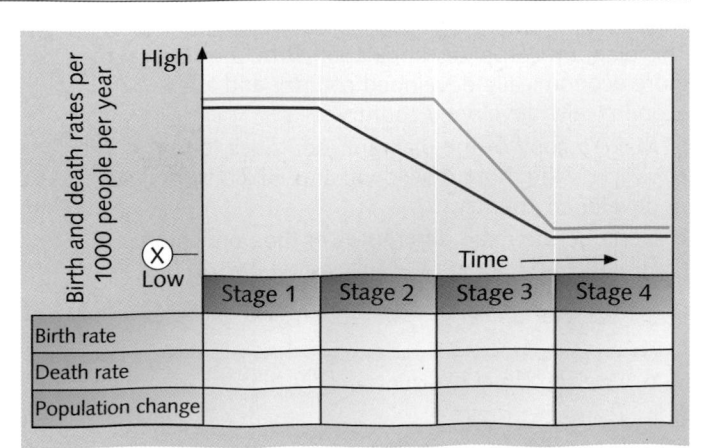

	Stage 1	Stage 2	Stage 3	Stage 4
Birth rate				
Death rate				
Population change				

3 (Pages 8 and 9)

Study these population pyramids for Bangladesh (a less economically developed country) and Australia (a more economically developed country).

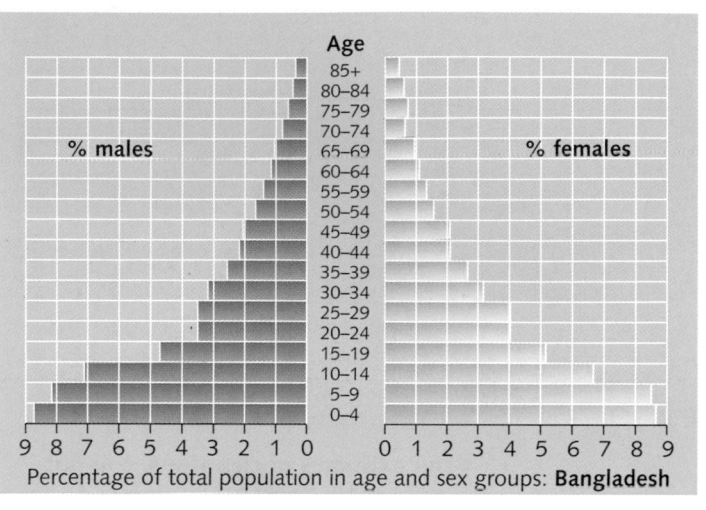

Percentage of total population in age and sex groups: **Bangladesh**

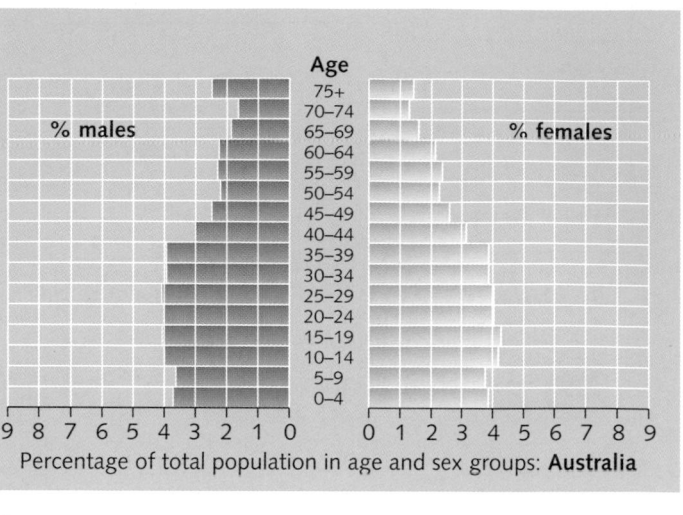

Percentage of total population in age and sex groups: **Australia**

a i) What percentage of the **total** population of Bangladesh is aged:
   - between 0 and 4   • over 65? (2)

ii) What percentage of the **total** population of Australia is aged:
   - between 0 and 4   • over 65? (2)

iii) Suggest one reason for the differences between the two countries, in:
   - children aged 0 to 4   • people aged over 65. (2)

iv) State one problem that might result from having a population structure such as that shown by:
   - Bangladesh   • Australia. (2)

b i) Describe four differences between the two population pyramids. (4)

ii) Which country has the more rapid population growth? Give reasons for your answer. (3)

iii) Which of the two graphs is most like each of the following countries?
   - UK   • Egypt   • Japan   • Kenya. (4)

c Choose *either* Bangladesh *or* Australia and explain the issues that arise in a country with such a population structure. (3)

4 (Pages 8 and 9)

This diagram compares population structures typical of a more economically developed country and a less economically developed country.

a Make a copy of the diagram. Add titles to say which is the more developed and which is the least developed country. (1)

b Describe the main differences in the population structures by adding the following labels in the correct places. (10)

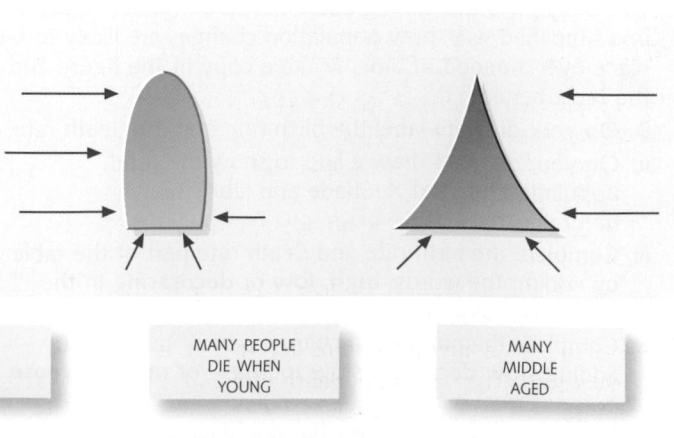

| HIGH BIRTH RATE, MANY CHILDREN | MANY ELDERLY, LONG LIFE EXPECTED | SQUAT-PYRAMID SHAPE | MANY PEOPLE DIE WHEN YOUNG | MANY MIDDLE AGED |

| NEARLY EQUAL NUMBERS AT EACH AGE | GENERALLY NARROW PYRAMID | MANY MORE YOUNG THAN OLD | LOW BIRTH RATE, FEW CHILDREN | FEW ELDERLY, SHORT LIFE EXPECTANCY |

5 *(Pages 10 and 11)*

a i) What was the world's population in 1750 and in 2000? *(2)*

ii) What is the average prediction for the world's population in 2050? *(1)*

b i) Name the three developing continents. *(1)*

ii) What, approximately, was the total population of these three continents in 1750 and in 2000? *(2)*

iii) Which continent had the largest population in 2000? *(1)*

c i) Give three reasons why the world's population growth has begun to slow down. *(3)*

ii) Which continent might see a population decrease by 2050? *(1)*

d Refer to the table below.

i) Why is Bangladesh said to be overpopulated? *(3)*

ii) Why is Canada said to be underpopulated? *(3)*

	Population density/ km²	Birth rate	Natural increase	Fertility rate	Main exports	GDP (page 180)
Bangladesh	1062	29	1.7	2.7	garments, jute	US$ 1300
Canada	3	16	0.3	1.6	oil, minerals, cereals	US$ 38 400

6 *(Pages 12 to 15)*

a i) Which graph is likely to be for a *developing* country with a high birth rate? *(1)*

ii) Give three reasons why many *developing* countries have a high birth rate. *(3)*

iii) What problems does a high birth rate create for a *developing* country? *(3)*

b i) Which graph is likely to be for a *developing* country that has been successful in reducing its birth rate? *(1)*

ii) Why have some countries felt it to be important to try to reduce the birth rate in order to control population growth? *(2)*

iii) For any *developing* country that you have studied, describe the attempts made to try to reduce the birth rate. *(3)*

iv) What problems now face some *developed* countries as a result of their success in lowering the birth rate? *(3)*

c i) Which graph is likely to be for a *developed* country with an ageing population? *(1)*

ii) Give three reasons why life expectancy has increased in *developed* countries. *(3)*

iii) What problems does an increase in life expectancy create for:
  • a government
  • elderly people in a *developed* country that you have studied? *(3)*

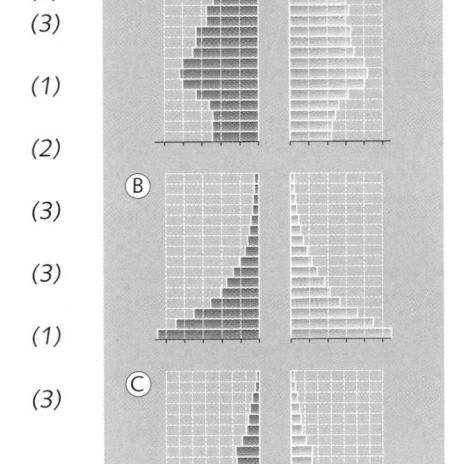

7 *(Pages 4 to 7 and 16 to 17)*

a Refer to Figure 1.24, which shows differences in population density in Brazil.

i) Describe the density of population in Brazil by locating places with:
  • a high density   • an average density
  • a low density. *(6)*

ii) For each of these places, give reasons why it has a high, average or low density. *(9)*

b Refer to the demographic transition model of Brazil.

i) What was the birth rate, death rate and total population in: • 1960   • 2000? *(6)*

ii) Predict what you think Brazil's population may be in 2050. Give three reasons for your answer. *(3)*

iii) How do you think changes in the birth and death rates will affect the shape of the population pyramid shown in Figure 1.26? *(2)*

Demographic transition: Brazil

What is migration?

Migration is a movement and in human terms usually means a change of home. However, as seen in Figure 2.1, it can be applied to temporary, seasonal and daily movements as well as to permanent changes both between countries and within a country.

Permanent international migration is the movement of people between countries. **Emigrants** are people who leave a country; **immigrants** are those who arrive in a country. The **migration balance** is the difference between the numbers of emigrants and immigrants. Countries with a **net migration loss** lose more people through emigration than they gain by immigration, and depending on the balance between their birth and death rates (page 6) they may have a declining population. Countries with a **net migration gain** receive more people by immigration than they lose through emigration, and so will have an overall population increase (assuming birth and death rates are evenly balanced).

Figure 2.1
Types of migration

	External (international)	Between countries
Permanent	i) Voluntary	West Indians to Britain
	ii) Forced (refugees)	Negro slaves to America, Kurds, Rwandans
	Internal	**Within a country**
	i) Rural depopulation	Most developing countries
	ii) Urban depopulation	British conurbations
	iii) Regional	North-west to south-east of Britain
Semi-permanent	For several years	Migrant workers (Turks into Germany)
Seasonal	For several months or several weeks	Mexican harvesters in California, holiday-makers, university students
Daily	Commuters	South-east England

International migration can be divided into two types – voluntary and forced (Figures 2.2 and 2.3).

Voluntary migration is the free movement of migrants looking for an improved quality of life and personal freedom. For example:
- Employment – either to find a job, to earn a higher salary or to avoid paying tax
- Pioneers developing new areas
- Trade and economic expansion
- Territorial expansion
- Better climate, especially on retirement
- Social amenities such as hospitals, schools and entertainment
- To be with friends and relatives

Forced migration is when the migrant has no personal choice but has to move due to natural disaster or to economic or social imposition. For example:
- Religious and/or political persecution
- Wars, creating large numbers of refugees
- Forced labour as slaves or prisoners of war
- Racial discrimination
- Lack of food due to famine
- Natural disasters caused by floods, drought, earthquakes, volcanic eruptions or hurricanes
- Overpopulation, when the number of people living in an area exceeds the resources available to them

Figure 2.2
Voluntary and forced migration

Figure 2.3
Major international migrations since 1945

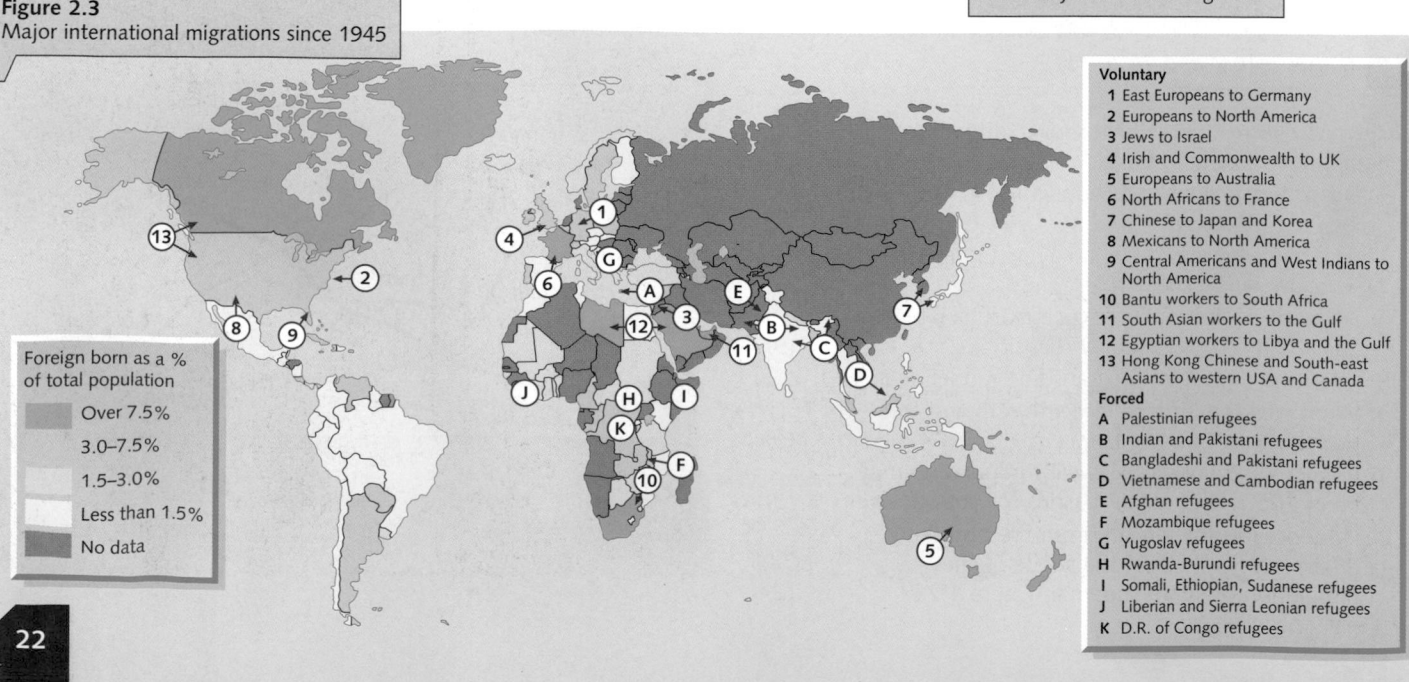

Foreign born as a % of total population
- Over 7.5%
- 3.0–7.5%
- 1.5–3.0%
- Less than 1.5%
- No data

Voluntary
1 East Europeans to Germany
2 Europeans to North America
3 Jews to Israel
4 Irish and Commonwealth to UK
5 Europeans to Australia
6 North Africans to France
7 Chinese to Japan and Korea
8 Mexicans to North America
9 Central Americans and West Indians to North America
10 Bantu workers to South Africa
11 South Asian workers to the Gulf
12 Egyptian workers to Libya and the Gulf
13 Hong Kong Chinese and South-east Asians to western USA and Canada

Forced
A Palestinian refugees
B Indian and Pakistani refugees
C Bangladeshi and Pakistani refugees
D Vietnamese and Cambodian refugees
E Afghan refugees
F Mozambique refugees
G Yugoslav refugees
H Rwanda-Burundi refugees
I Somali, Ethiopian, Sudanese refugees
J Liberian and Sierra Leonian refugees
K D.R. of Congo refugees

Refugees

Refugees are people who have been forced to leave their home country for fear of persecution for reasons of race, religion, politics, internal strife (civil war) or due to environmental disaster. They move to other countries hoping to find help and asylum. Refugees do not include displaced persons, who are people who have been forced to move within their own country. In 2008, the United Nations (UN) reported that there were just over 16 million refugees in the world but admitted that, as this did not include illegal immigrants, the figure could be very inaccurate. The UN claims that over half the world's refugees are children and most of the adults are women. About 85 per cent of refugees are in developing countries – countries that are least able to help (Figure 2.4). They live in extreme poverty, lacking adequate food, shelter, clothing, education and medical care (Figure 2.5). They have no citizenship, few (if any) rights, virtually no prospects, and are unlikely to return to their homeland.

The present problematic refugee situation began over 60 years ago when Palestinian Arabs were forced to live in camps set up following the creation of the state of Israel in 1948. Today over 4.5 million people still live this way in this part of the Middle East. More recent refugee movements, nearly always between developing countries, include the following:

- Over 3 million Afghan refugees have been forced to flee, mainly to neighbouring Pakistan and Iran, as a result of 30 years of both foreign and civil war and, more recently, 2.3 million Iraqis have sought refuge in Syria and Jordan.
- Over 1 million Ethiopians, Sudanese and Somalis have been driven from their homes in East Africa and the Horn of Africa by drought, famine and civil war. A further 1 million refugees have been created in Central Africa, initially by the Rwandan civil war of the early 1990s and now by fighting in the Democratic Republic of Congo.

Asylum seekers

These are people who have left their home country, have applied to another country for recognition as a refugee and are awaiting a decision on their application.

Illegal immigrants

These enter a country with the minimum of belongings and without personal documents such as a passport. As such, they fail to meet the legal requirements of that country and so may become part of the 'hidden economy'. They are likely to have to rely on local people to find them food, shelter and work, which leaves them vulnerable to exploitation.

Economic migrants

These are people who make a conscious choice to leave their home country knowing that they can return there if they wish without problems at a later date. This movement, which they make voluntarily, is to find a better job and, consequently, to improve their standard of living and quality of life. This group includes **migrant workers** (page 28).

Figure 2.5
A Rwandan refugee camp

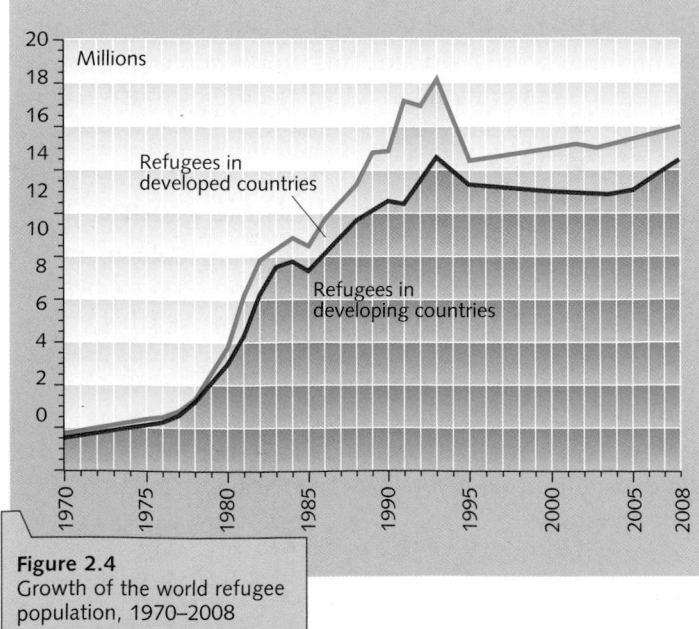

Figure 2.4
Growth of the world refugee population, 1970–2008

Migration into the UK

The UK has experienced waves of immigrants, giving it a society that has always been one of mixed races and cultures. The majority of UK residents are descended from immigrants – Romans, Vikings, Angles, Saxons and Normans. The Irish have settled in Britain for several centuries, while many other Europeans migrated here both during and after the Second World War. Usually the number of immigrants was small enough to be easily assimilated into the existing population. However, the immediate post-war years saw a much larger influx (around 3 million), many of whom had different colours of skin. It is this biological difference, rather than a social, linguistic or cultural difference, that has led to racial tensions in parts of the UK.

A small proportion of immigrants have come from the **Old Commonwealth** (Australia, New Zealand and Canada) and are descendants of earlier British migrants to those countries. The largest proportion are **New Commonwealth** immigrants from former British colonies in the Indian subcontinent (India, Pakistan and Bangladesh), Africa (Nigeria) and the West Indies (Jamaica). This group (Figure 2.6), who are mainly non-white, migrated here because:

- Britain had a labour shortage after the Second World War. Many West Indians came here partly due to the 'push' factor of overcrowding in their own islands, but mainly due to the 'pull' of jobs in Britain (page 80). The British government actively encouraged people to apply for specific jobs, for example with London Transport. Unfortunately, many immigrants were underskilled and were forced to take poorly-paid jobs (page 28).

Ethnic groups	1991 %	2001 %
White	94.5	92.1
Mixed*	–	1.1
Asian or Asian British		
Indian	1.5	1.8
Pakistani	0.9	1.3
Bangladeshi	0.3	0.5
Other Asian	0.4	0.4
Black or black British		
Black Caribbean	0.9	1.0
Black African	0.4	0.8
Black Other	0.3	0.2
Chinese	0.3	0.4
Other	0.5	0.4
Total non-white	5.5	7.9

* New category for 2001 census for people considering themselves to belong to more than one group

Figure 2.6
Ethnic minorities in the UK at the 1991 and 2001 censuses

- Groups of Asians, including Hindus, Muslims and Sikhs, found themselves as religious or political refugees following the division of India. They came to Britain in the 1950s.

Successive British governments have, since then, tried to restrict and control the number of non-white immigrants unless they:

- were dependants of relatives already living in the UK
- had specific jobs, especially those involving skills that were in short supply in Britain, e.g. doctors
- were British passport-holders evicted from their home country, e.g. Ugandan Asians
- could prove themselves to be genuine refugees, e.g. Bosnian Muslims.

By 2008, ethnic minorities made up over 8 per cent of the UK population, with an increasing number – almost half – having been born in Britain. This proportion is growing, not due to immigration but rather to ethnic groups having a higher birth rate and larger families.

Uneven concentrations of ethnic groups

Immigrants avoided areas that had high unemployment (Scotland and Northern Ireland) and went to large cities and conurbations (not small towns) where there were better job opportunities. The greatest concentrations of ethnic groups are in London, the West and East Midlands and West Yorkshire (Figure 2.7). There has also been a tendency for one ethnic group to concentrate in a particular area, e.g. Pakistanis in West Yorkshire and West Indians in Birmingham.

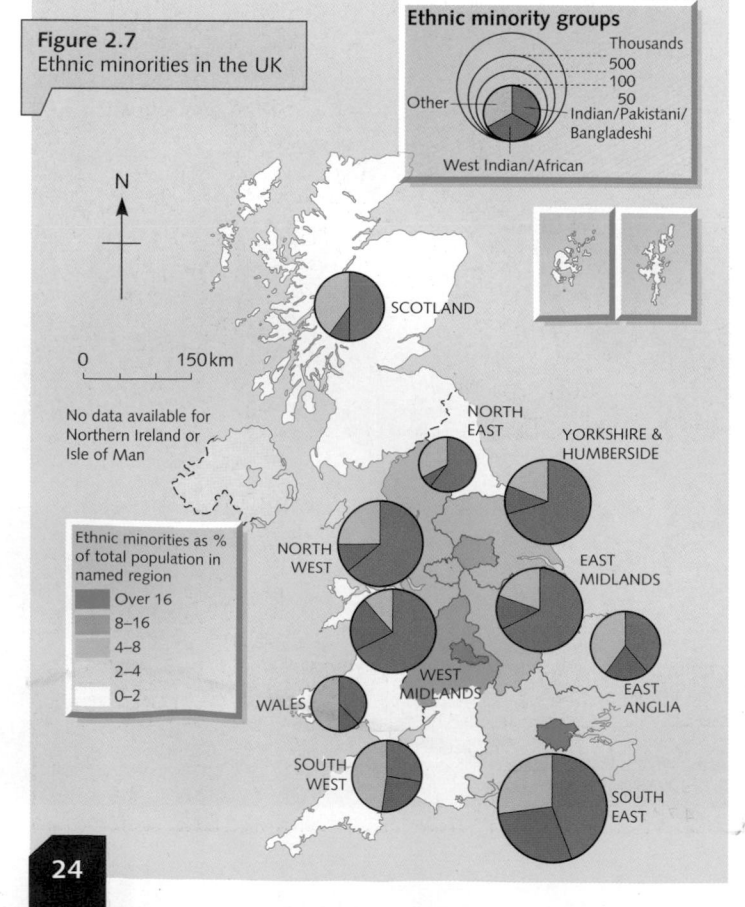

Figure 2.7
Ethnic minorities in the UK

Ethnic minority groups

Other
Indian/Pakistani/Bangladeshi
West Indian/African
Thousands
500
100
50

No data available for Northern Ireland or Isle of Man

0   150km

SCOTLAND

NORTH EAST

YORKSHIRE & HUMBERSIDE

NORTH WEST

EAST MIDLANDS

WEST MIDLANDS

WALES

EAST ANGLIA

SOUTH WEST

SOUTH EAST

Ethnic minorities as % of total population in named region
Over 16
8–16
4–8
2–4
0–2

Positive and negative effects

Two of the major present-day concerns regarding migration in Britain have been the rapid increase in the number of immigrants into the country since 2000, and the living conditions and work opportunities for several groups within the existing ethnic minorities. Migration can become, at times, a highly emotive topic provoking strong negative and positive reactions.

The majority of residents who have either arrived in Britain themselves or, increasingly, are descended from immigrants have tended to locate not only within large conurbations, but with other members of their ethnic group in inner-city areas within large urban areas (Figure 2.8). Within these areas, ethnic groups that share a similar religion, language, diet, social organisation and culture tend to congregate together as, for example, in London: Jamaicans in Brixton, Sikhs in Southall, Anguillans in Slough and Bengalis in East London.

Although this never led to the ghettos that once characterised parts of New York, it has meant that many, though not all, of these ethnic minorities have had to live in overcrowded, poor-quality housing. Such communities tend to have high birth rates and large families, increasing the degree of overcrowding; high levels of unemployment; and fewer education opportunities, leading to lower aspirations and the development of skills (Figure 2.10). A lack of personal wealth and investment by different levels of government means that public services are often inadequate. While, on the one hand, authorities speak of higher than average crime rates, especially violence and drugs, on the other residents complain of police harassment which, together, increases the lack of trust and can lead to frustration and further tension.

Yet there have been many positive sides to immigration. It has already been mentioned that West Indians were encouraged to come to work in Britain after the Second World War as, without them, London's transport system, as one example, would not have been able to operate. Since then, organisations such as the National Health Service have come to rely on highly skilled immigrants, such as doctors, and caring groups, including nurses and residential care home workers. Recently, migrants from several former Eastern European countries, especially Poland, have been welcomed to fill gaps in the UK's job market (page 28). The nation's cultural and sporting life has been enhanced by the talents of, amongst others, musicians and sportsmen and women. Again, using London as an example, people from all ethnic groups shop at markets such as in Brixton and attend events that include the Notting Hill Carnival (Figure 2.9).

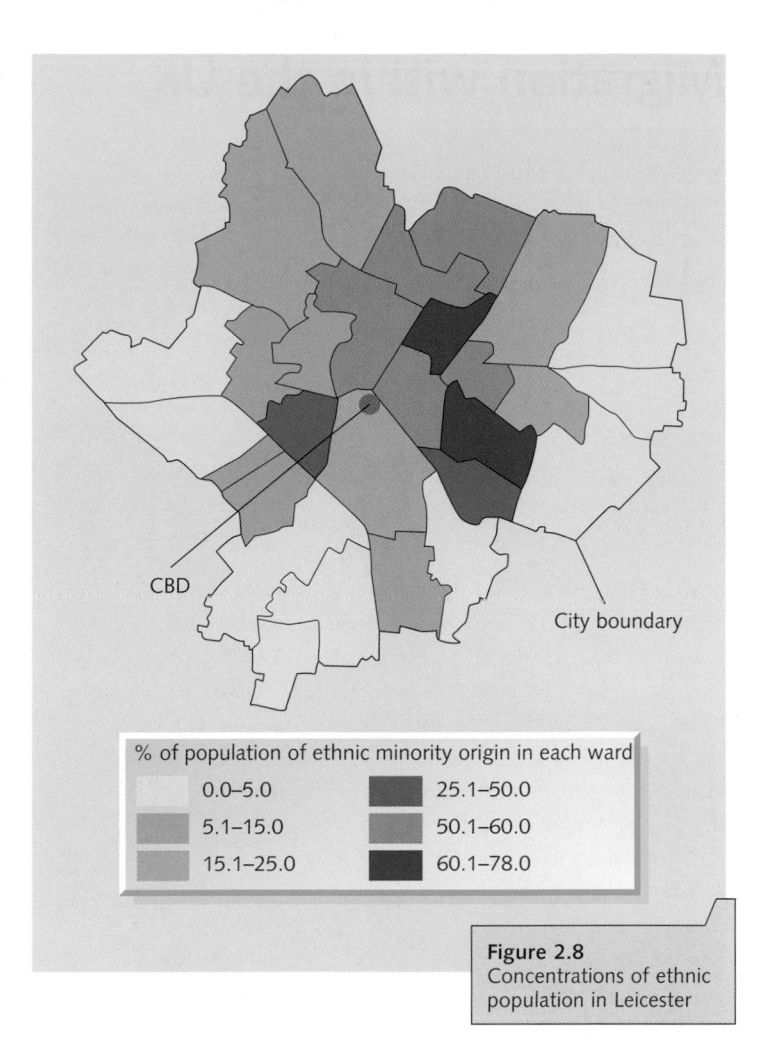

% of population of ethnic minority origin in each ward

0.0–5.0	25.1–50.0
5.1–15.0	50.1–60.0
15.1–25.0	60.1–78.0

CBD

City boundary

Figure 2.8
Concentrations of ethnic population in Leicester

Figure 2.9
The Notting Hill Carnival

Indeed it was London's image as a multicultural, global city that helped it to be selected as the venue for the 2012 Olympics.

Figure 2.10
Inequality in England and Wales, 2001 census

Group	% total population	% under 16 years	% over 65 years	Average family size	% professional/ managerial	% semi-/ unskilled	% living in conurbations
White	92.1	18	24	1.9	40	18	27
Black and Asian	7.9	38	9	3.2 Black 4.7 Asian	14	47	81

Migration within the UK

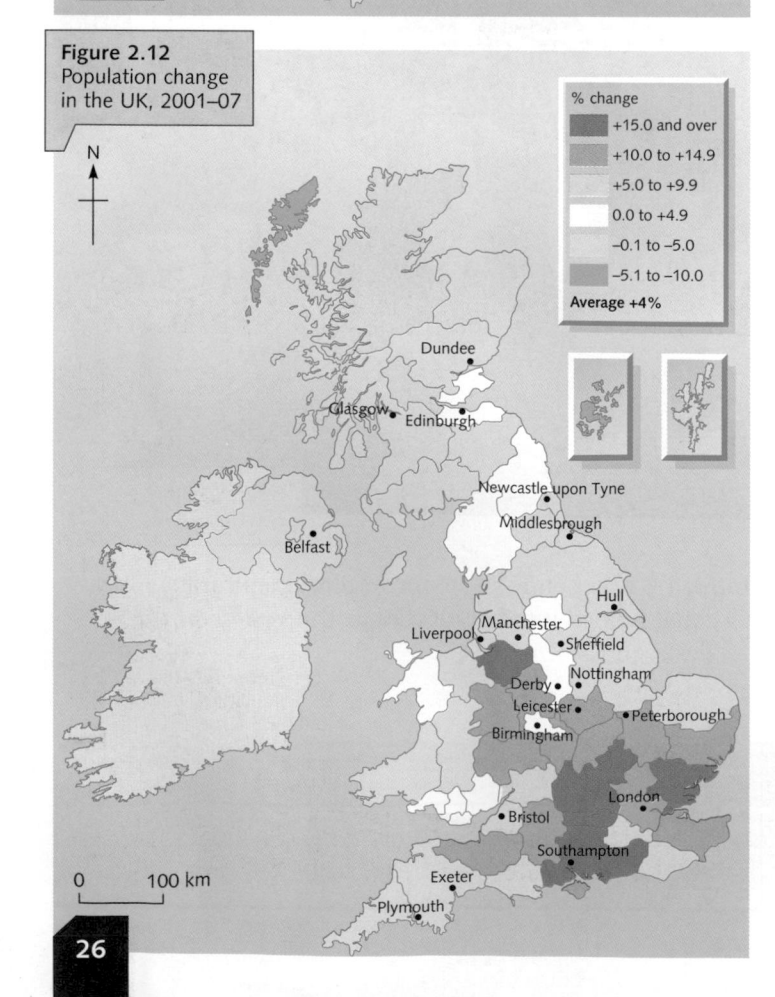

Figure 2.11
Population change in the UK, 1961–71
(boundaries adjusted to the 1974 changes)

% change

	+15.0 and over
	+10.0 to +14.9
	+5.0 to +9.9
	0.0 to +4.9
	−0.1 to −5.0
	−5.1 to −10.0

Average +5.1%

0   100 km

Figure 2.12
Population change
in the UK, 2001–07

% change

	+15.0 and over
	+10.0 to +14.9
	+5.0 to +9.9
	0.0 to +4.9
	−0.1 to −5.0
	−5.1 to −10.0

Average +4%

0   100 km

Rural-to-urban migration

During the Industrial Revolution of the nineteenth century, many people in Britain were either:

- forced to leave the countryside ('**push**' factors causing **rural depopulation**), or
- attracted to the growing towns and cities ('**pull**' factors leading to **urbanisation**).

Workers were needed in the rapidly growing:

- textile towns of Yorkshire and Lancashire
- coalfield towns in South Wales, north-east England and central Scotland where, initially, iron and, later, steel was produced
- ports of London, Liverpool and Glasgow.

Regional migration

Between about 1930 and 1980 there was a steady drift of people from the north and west of Britain to the south-east of England. Figure 2.11 shows that former industrial areas in north-east England, Lancashire, Yorkshire and central Scotland together with the more rural parts of Northern Ireland, Scotland and central Wales, all lost population. People left these areas because of **push factors**, e.g.

- older, poorer-quality housing
- exhaustion of minerals (coal/iron ore)
- decline of older industries (ships/steel/textiles)
- poorly paid, mainly manual, jobs
- poorer transport links/accessibility
- many polluted former industrial environments
- decline of older ports (Glasgow/Liverpool)
- fewer cultural amenities and social/sporting events
- colder, wetter climate.

At the same time places in the south-east increased their population. People moved here because of **pull factors**, e.g.

- newer, better-quality housing
- growth of newer/lighter/footloose industries
- better-paid, more skilled jobs
- better transport links (motorways/rail/airports)
- fewer polluted industrial environments
- growing links with Europe
- better services (shops/schools/hospitals)
- many cultural amenities, more social/sporting events
- warmer, drier climate (holidays/retirement).

Counterurbanisation

Counterurbanisation is the process by which people and employment move away from major cities and conurbations to smaller settlements (Figure 2.12). It is, therefore, movement in the reverse direction to that of rural–urban migration. The process first occurred in south-east England when families and businesses began moving out of London to live and locate in new towns and dormitory/commuter settlements. More recently, this movement has been from most urban areas to smaller, often nearby, towns and villages (suburbanised villages, page 64).

Those people who are leaving the cities (Figure 2.13) do so for economic, social and environmental reasons.

Employment As industry has declined in inner city areas, it has relocated on edge-of-city sites or in smaller rural towns. People move for promotion, for better-paid jobs or simply to find a job. The increase in footloose and high-tech industries means that many firms now have a much freer choice in where they locate (Chapter 9).

Housing When people become more affluent they are likely to move from the high-density small terraced houses and high-rise flats of the inner city to larger, modern houses with better indoor amenities, garages and gardens.

Changing family status People move as a result of an increase in family wealth or family size.

Environmental factors These include moving away from the noise, air and visual pollution created by increasing traffic and declining industry in large urban areas to quieter, less polluted environments with more open space.

Social factors People may move out of cities due to prejudice against neighbours and ethnic groups, an above-average local crime and vandalism rate, or poorer educational facilities.

The 2001 census showed that fewer people, when changing home, are making long-distance moves. Most moves are short, with over two-thirds being under 10 km and within the same county (Figure 2.14).

Cycle of change in London

Figure 2.15 shows that the population of London, the world's largest city in 1801, continued to grow rapidly throughout the nineteenth century.

After 1900, the population of the inner boroughs began to decrease as more people, aided mainly by improvements in public transport (e.g. the Underground), moved to the suburbs (page 46).

By the 1960s, even the population of the outer boroughs had begun to fall as this was the time of maximum counterurbanisation.

More recently there has been a reversal in movement. This has partly been due to wealthy people and property developers buying older, derelict property and, by refurbishing it, transforming parts of the inner areas into expensive housing and flats, e.g in Islington and the former Docklands, and partly due, in the last decade, to a rapid rise in immigration.

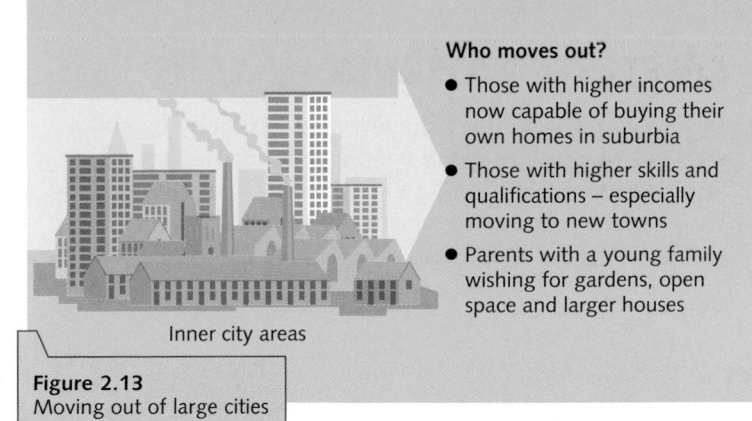

Who moves out?
- Those with higher incomes now capable of buying their own homes in suburbia
- Those with higher skills and qualifications – especially moving to new towns
- Parents with a young family wishing for gardens, open space and larger houses

Inner city areas

Figure 2.13
Moving out of large cities

13.4% 11.5% 13.8% 61.3%

- Within districts
- Between districts/within county
- Between counties/within region
- Between regions

16.8% 6.2% 8.4% 12.7% 55.9%

- 0–4 km
- 5–9 km
- 10–19 km
- 20–49 km
- 50 km and over

Figure 2.14
A classification of migration in the UK

Figure 2.15
Population change in Greater London, 1801–2008

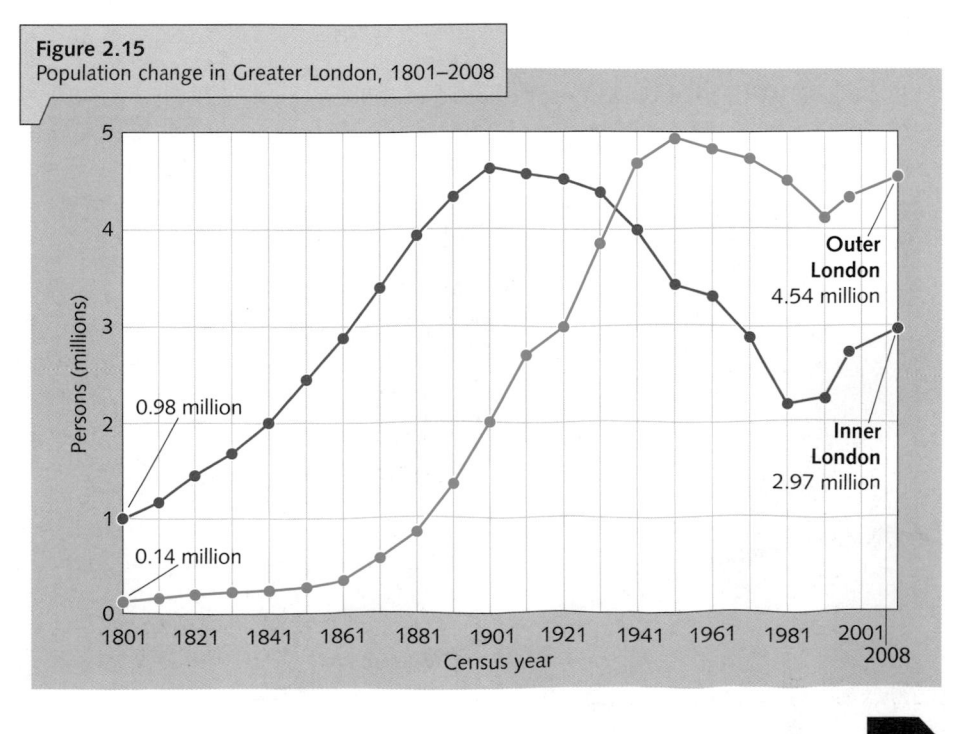

Persons (millions)

0.98 million

0.14 million

Outer London 4.54 million

Inner London 2.97 million

1801 1821 1841 1861 1881 1901 1921 1941 1961 1981 2001 2008
Census year

Migrant workers

In countries where there is a low standard of living and a shortage of jobs, groups of people will migrate to nearby, wealthier countries hoping to find work. Many migrant workers, the majority of whom are likely to be men, may move with the intention of:

- returning home after several years, having 'made their fortune' (e.g. Turks into Germany, North Africans into France, Egyptians and Jordanians into Saudi Arabia)
- bringing their family to join them a later date (e.g. Indians into the UK)
- working seasonally at harvest times (e.g. Mexicans into California, page 30).

Polish workers into the UK

In 2004, Poland and several other former Eastern European countries joined the EU. Of the existing members, only the UK and the Republic of Ireland allowed unlimited immigration from the new members. This led, in the UK, to the largest influx for centuries with, by early 2008, the arrival of over 800 000 migrant workers – an average of over 200 000 a year. Of these, an estimated 500 000 had come from Poland (Figure 2.16).

Migrants from Poland were largely welcomed as they came with a wide range of skills, many of which were currently lacking in the British workforce. At one end were people who obtained senior jobs in administration and management, such as computing, IT support and the National Health Service. At the other were those prepared to work long hours either as health care workers, as shop assistants or as manual workers in factories and on farms (Figure 2.17). In between came plumbers, electricians, bricklayers and decorators – four professions in which Britain had a severe skills shortage.

Whereas earlier migrants into Britain had tended to concentrate into specific areas within larger urban areas (page 25), it was claimed that Polish workers were living in every local authority in Britain. They set up their own radio stations, printed their own newspapers (Figure 2.18), celebrated church Mass in their language, produced Polish bread and processed other food products which they sold in their own shops (Figure 2.19).

Figure 2.16
Nationality of foreign workers, May 2004 – February 2008

Thousands

Nationality	Thousands
Poland	508 385
Slovakia	78 830
Lithuania	73 315
Latvia	37 300
Czech Republic	34 555
Hungary	25 755
Estonia	6845
Slovenia	695

Figure 2.17
Polish workers picking vegetables from the fields of East Anglia

DZIENNIK POLSKI
Prezydenci w

Figure 2.18
A Polish newspaper

Most came to Britain to find better-paid jobs as, at the time of their arrival, the average wage in the UK was several times higher than that in Poland. To some the idea was to work hard so that they could return home with sufficient money to set up their own business. To others it was a case of earning enough to live on and to send the remainder home to support their families back in Poland. The majority of immigrants, as is usually the case with migrant workers, were men, of whom over 80 per cent were aged between 18 and 34 years (Figure 2.20).

Most workers were generally well accepted by local communities in Britain, perhaps because the migrants were European themselves, perhaps because their hard work and valuable skills were appreciated. But their presence did create problems. In large numbers they could 'swamp' schools, hospitals and other services; by buying property in the lower price range they competed with local first-time buyers; those with fewer skills competed with local job-seekers as they were prepared to worker longer hours and for less pay; and money they earned is sent out of the country and so is lost to the British economy.

Figure 2.21 lists some of the advantages and disadvantages to both the home and the host country with respect to migrant workers. The same can be applied to Mexican workers into California (Case Study 2), Turks into Germany and North Africans into France.

Figure 2.19
A Polish shop

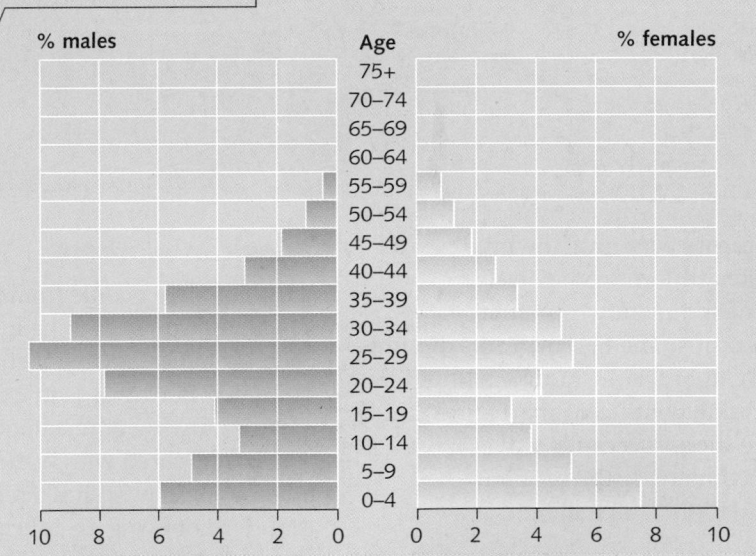

Figure 2.20
Polish immigrants in the UK

Advantages	Disadvantages
Losing country	
• Reduces pressure on jobs and resources, e.g. food	• Loses people in working-age group
• Loses people of child-bearing age, causing decline in birth rate	• Loses people most likely to have some education and skills
• Migrants develop new skills which they may take back home	• Mainly males leave, causing a division of families
• Money earned abroad is sent back to their home country	• Left with an elderly population and so a high death rate
Receiving country	
• Overcomes labour shortage	• Pressure on jobs but most likely to be the first unemployed in a recession
• Prepared to do dirty, unskilled jobs	• Low-quality, overcrowded housing lacking in basic amenities (inner-city slums – *bidonvilles* in France, *favelas* in South America)
• Prepared to work long hours for a low salary	• Ethnic groups tend not to integrate
• Cultural advantages and links	• Racial tension
• Some highly skilled migrants	• Limited skilled/educated group
• In a developing country these migrants could increase the number of skilled workers	• Lack of opportunities to practise their own religion, culture, etc.
	• Language difficulties
	• Often less healthy

Figure 2.21
Advantages and disadvantages of migrant workers

Immigrants into California

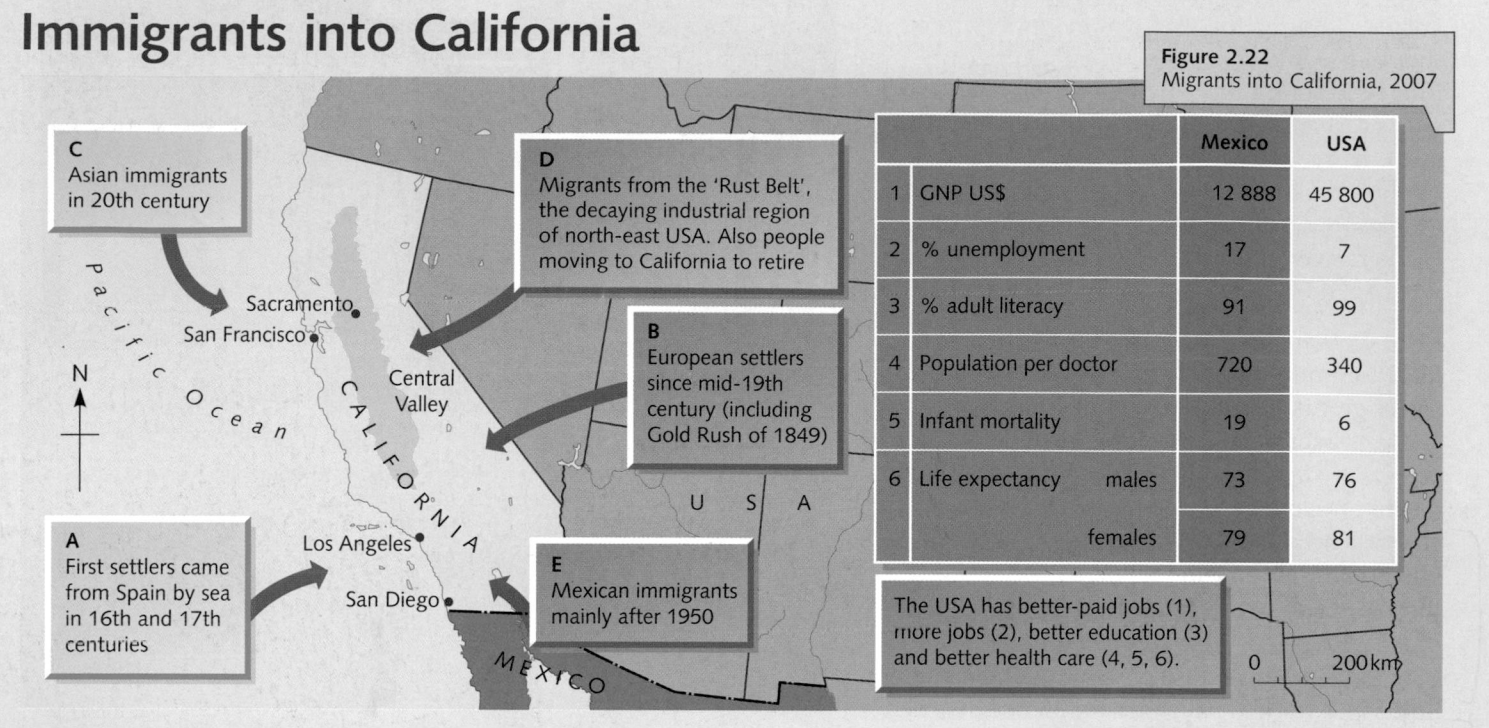

Figure 2.22
Migrants into California, 2007

C Asian immigrants in 20th century

D Migrants from the 'Rust Belt', the decaying industrial region of north-east USA. Also people moving to California to retire

B European settlers since mid-19th century (including Gold Rush of 1849)

A First settlers came from Spain by sea in 16th and 17th centuries

E Mexican immigrants mainly after 1950

		Mexico	USA
1	GNP US$	12 888	45 800
2	% unemployment	17	7
3	% adult literacy	91	99
4	Population per doctor	720	340
5	Infant mortality	19	6
6	Life expectancy males	73	76
	females	79	81

The USA has better-paid jobs (1), more jobs (2), better education (3) and better health care (4, 5, 6).

0   200 km

California was, until the mid-nineteenth century, sparsely populated mainly by Native Americans. The first migrants did not arrive in any large numbers until the mid-nineteenth century (Figure 2.22). Since then, successive waves have helped increase the resident population to 32 million:

- **Nineteenth century:** from Western Europe – external, voluntary, permanent (Figure 2.1).
- **Early twentieth century:**
  – eastern and southern Europe
  – eastern Asia
  both external, mainly voluntary/some forced, permanent.
- **1930s:** American Mid-west (the 'Dust Bowl') – internal, partly forced, permanent.
- **Since 1950:**
  – north and east of the USA – internal, voluntary, permanent
  – eastern Asia – external, voluntary and forced, permanent
  – Mexicans (Hispanics) – external, voluntary, initially seasonal/increasingly permanent.

Mexican workers

Hispanics are people from Spanish-speaking countries in Latin America. In recent years, many Hispanics have migrated northwards to form the fastest-growing ethnic minority in the USA. Compared with its rich northern neighbour, Mexico has a relatively low standard of living, insufficient jobs and poorer education and health provision. As a result many Mexicans, especially those living in villages and working on the land (Figure 2.23), migrate into the USA, although often on a temporary basis. At one time only the men migrated, returning when they had earned enough money. Now they often stay permanently, sometimes taking their families with them (some villages in Mexico have lost half their population), sometimes deserting their family.

Figure 2.23
A Mexican village

Migrants – illegal and legal

Estimates suggest that between 1 and 2 million Mexicans try each year to cross into the USA, many illegally. Although the USA has set up elaborate border security controls using horses, helicopters and other advanced detective equipment, it is possible that nearly a million migrants manage to slip through each year. As one border guard claimed, 'We catch somebody one night, return him to Mexico the next day, for him to try again that night and the next, until he is finally successful'. The number of migrants who arrive legally with their 'Green Card' issued by the Department of Immigration, rises and falls with the seasonal availability of jobs.

Jobs taken by migrant workers

Although Mexican workers are viewed as a drain on America's social security and welfare system, they are essential to the nation's economy. The migrants take the harder, dirtier, seasonal, more monotonous, more dangerous, less skilled and less well-paid jobs. For example:

- Large numbers find seasonal work on large agricultural estates, mainly in California's Central Valley, at harvest times (Figure 2.24). Despite the low pay by American standards, some migrants can earn more during three or four months in the USA than in a full year back in Mexico.
- Many others obtain employment in large urban areas, mainly Los Angeles, either in the construction industry or in hotels and restaurants.

Mexicans in Los Angeles

Many Mexicans perceive Los Angeles as the 'City of Opportunity'. The reality is very different (Figure 2.25).

Most of the immigrants are young, have little money and few qualifications. Unless they have a Green Card, they cannot work legally and are forced to take very low-paid jobs, often in the informal sector (page 148). Some have to move around to avoid detection while others find language a problem when seeking a permanent job. Many of those who do get jobs find their income is insufficient to obtain decent accommodation and are forced to live in ghettos in the poorest districts. In one such district, that of Compton, Hispanics make up nearly three-quarters of the total population (Figure 2.26).

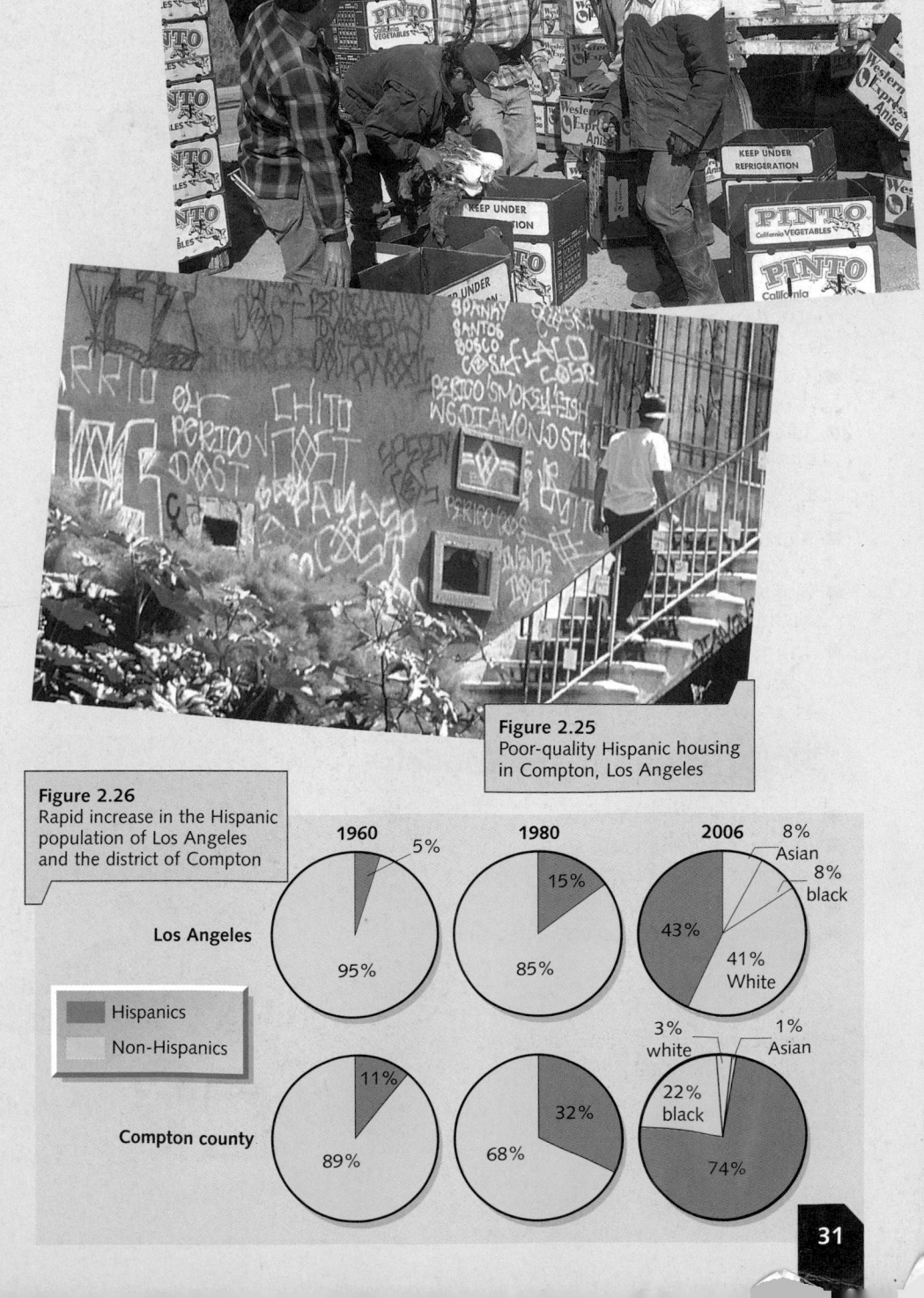

Figure 2.24
Mexican migrant workers packing vegetables in California

Figure 2.25
Poor-quality Hispanic housing in Compton, Los Angeles

Figure 2.26
Rapid increase in the Hispanic population of Los Angeles and the district of Compton

Hispanics
Non-Hispanics

Los Angeles
1960: 95%, 5%
1980: 85%, 15%
2006: 43%, 41% White, 8% Asian, 8% black

Compton county
1960: 89%, 11%
1980: 68%, 32%
2006: 74%, 22% black, 3% white, 1% Asian

Migration

Key Words and Terms

a You should know the meaning of the following terms:
- migration • migrant • immigrant • emigrant • refugee • asylum seeker
- economic migrant • ethnic group • New Commonwealth migrant
- Old Commonwealth • migrant worker • illegal immigrant.

b You should know the difference between:
- voluntary and forced migration
- internal and international migration
- refugees and economic migrants
- rural–urban migration and counterurbanisation
- cultural mix and ethnic segregation.

Key Ideas

You should know and understand the following:

- There are several types of migration.
- Migration takes place for a variety of reasons, both voluntary and forced.
- Migration takes place both within and between countries.
- Migration can create multicultural societies that may or may not live in harmony.
- Migration can affect both the population growth and the population structure of a country.
- The difference between refugees and economic migrants.
- Traditionally most migration has been a rural–urban movement but recently, especially in some MEDCs, the direction of movement has been reversed (counterurbanisation).
- There are both advantages and disadvantages of migration.
- In international migration there are gains and losses for both the losing and the receiving country.
- The causes and consequences of a major recent international migration.

Skills, Theories and Models

- Interpret choropleth maps
- Interpret and use graphs
- Interpret and use population pyramids
- Understand migration issues
- Appreciate other people's values and attitudes

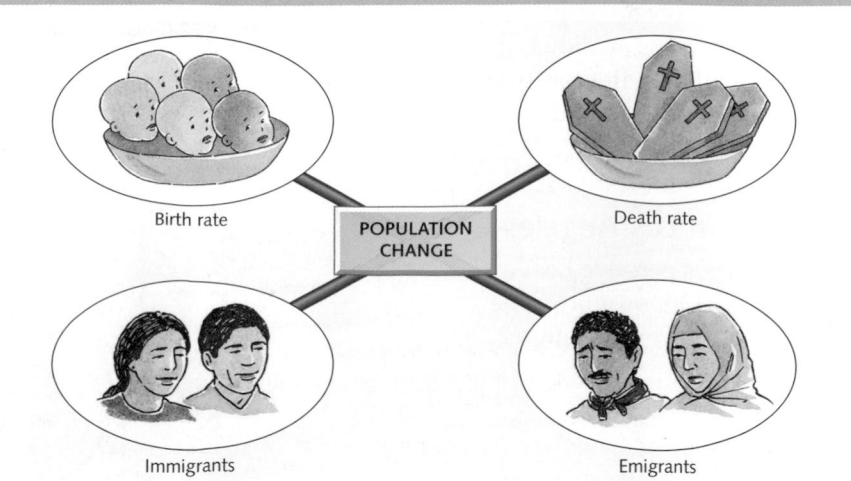

Birth rate

Death rate

POPULATION CHANGE

Immigrants

Emigrants

1 (Pages 22 and 23)

a i) Give four examples of voluntary migration. *(4)*

ii) Give four examples of forced migration. *(4)*

b What is the difference between a *refugee*, an *economic migrant* and an *illegal immigrant*? *(3)*

2 (Pages 24 and 25)

a i) What is meant by the term *New Commonwealth immigrant*? *(2)*

ii) Name two New Commonwealth countries. *(2)*

iii) Why did New Commonwealth immigrants come to Britain? *(3)*

iv) Why did most immigrants settle in large cities? *(3)*

v) Why did most immigrants settle in inner city areas? *(3)*

b i) What are the problems any immigrant might face when he or she arrives in a new country, as far as the following are concerned:
- accommodation
- employment
- language
- culture
- prejudice? *(5)*

ii) Give three possible benefits to the country that receives this immigrant. *(3)*

3 (Pages 26 and 27)

This graph shows changes in the population of inner and outer London since 1850.

a For inner London

i) Why did most people live here in 1850? *(1)*

ii) Why did it continue to grow rapidly until 1900? *(1)*

iii) Why did its population decrease after 1900? *(1)*

iv) Why has its population increased again in recent years? *(1)*

b For outer London

i) Why did so few people live here before 1900? *(1)*

ii) Why did its population increase rapidly after 1900? *(1)*

iii) Why did its population decrease between 1950 and 2000? *(1)*

4 (Pages 28 and 29)

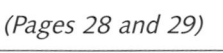

a What is meant by the term *migrant worker*? *(1)*

b i) Why did so many migrant workers leave Poland after 2004? *(2)*

ii) Why did Britain encourage workers from Poland? *(2)*

iii) Describe and give reasons for the types of jobs these migrant workers found in Britain. *(2)*

c i) How did Britain benefit from these workers? *(2)*

ii) What problems did their arrival cause in Britain? *(2)*

iii) How did Poland benefit from sending workers to Britain? *(2)*

iv) What problems were created in Poland by the loss of these workers? *(2)*

d Why did some Polish workers start to return home in 2008? *(2)*

5 (Pages 30 and 31)

a Name a state in the USA that attracts Mexican workers. *(1)*

b Why do many Mexicans try to enter the USA illegally? *(1)*

c Give two reasons why the USA:
- tries to restrict Mexican migrants
- needs to attract seasonal Mexican labour. *(4)*

d Many Mexicans move to Los Angeles. Describe the types of accommodation and jobs they are likely to find there. *(4)*

3

Settlement and urban growth

A settlement is a place where people live. It can be large or small, permanent or temporary. Although each settlement is unique, it is likely to share similar characteristics with other settlements. Settlements can be grouped together (i.e. classified) using a variety of criteria. These criteria include location (**site and situation**), shape (**patterns**), major use (**function**), and position, or **rank**, as a service centre (**hierarchy**).

Site and situation

The location of a settlement is related to its site and situation.

Site describes the point at which the town (or hamlet, village, city) is located. Factors such as local relief, soil, water supply and resources were important in choosing the initial site of a settlement.

Situation describes where the settlement is located in relation to surrounding features such as other settlements, mountains, rivers and communications. It is the situation of a settlement that determines whether or not it will continue to grow to become a large town or city or whether it remains as a small hamlet or village. Paris, for example, had the site advantage of being located on an island in the River Seine which could be defended and which made bridging easier (Figure 3.1). However, the continued growth of Paris into Europe's largest city was due to its situation in the centre of a major farming area where several routes (rivers) converged.

Figure 3.1
Ile de la Cité, Paris

Early settlements developed within a rural economy which aimed to be self-sufficient. Their sites were determined by a series of mainly physical factors. An ideal site was likely to have the benefit of several of these factors.

A **wet-point site**, especially in relatively dry areas, was essential as water is needed virtually every day throughout the year and is heavy to carry any distance. In early times, rivers were still sufficiently clean to give a safe, permanent supply. In lowland Britain, many early settlements were located at springs at the foot of chalk (Figure 3.2) or limestone escarpments.

A **dry-point** site, especially in relatively wet areas, was needed to avoid flooding or to be above unhealthy marshland, e.g. Ely in eastern England, which was built on a mound that acted as a natural island in the Fens.

Building materials, which ideally included stone, wood and clay, had to be obtained locally as these were heavy and bulky to move at a time when transport was poorly developed.

Defence against surrounding tribes was sometimes necessary. Good defensive sites may have been within a river meander, with the river giving protection on three sides, as at Durham and Shrewsbury (Figure 3.3), or on a hill with steep sides and commanding views as with Edinburgh and many Mediterranean settlements (Figure 3.4).

Figure 3.2
Spring-line settlements, South Downs

Figure 3.3
Shrewsbury

Figure 3.4
A Mediterranean hilltop village

A **fuel supply** was needed for heating and cooking, which was, in earlier times in Britain and still today in many developing countries, usually wood.

Food supplies were needed from land nearby, some of which was suitable for the rearing of animals and some for the growing of crops, e.g. sheep on the South Downs and crops in the clay vales (Figure 15.5).

Nodal points were where several valleys (natural routes) met to create a route centre, as at York and Paris, or where two rivers joined, as at Khartoum (Blue Nile and White Nile) and St Louis (Mississippi and Missouri), or which controlled routes between hills, as at Guildford and Toulouse.

Bridging points may originally have been at a ford in the river, e.g. Oxford, or where the river was shallow and narrow enough to enable a bridge to be built, e.g. Paris.

Shelter and aspect In Britain it is an advantage to be sheltered from the strong prevailing south-westerly and cold northerly winds, and to have a south-facing aspect, as this gives most sunshine, heat and light (e.g. Torquay).

Figure 3.5 shows an area of land available for early settlement. What were the advantages and disadvantages of each of the five sites labelled A to E? Were any of the sites ideal? Finally, remember that many of these early site factors may no longer be important, e.g. in Britain most settlements now have a piped water supply, have one or more shops providing food, and do not need to be defended.

Figure 3.5
Which would be the best site for an early settlement?

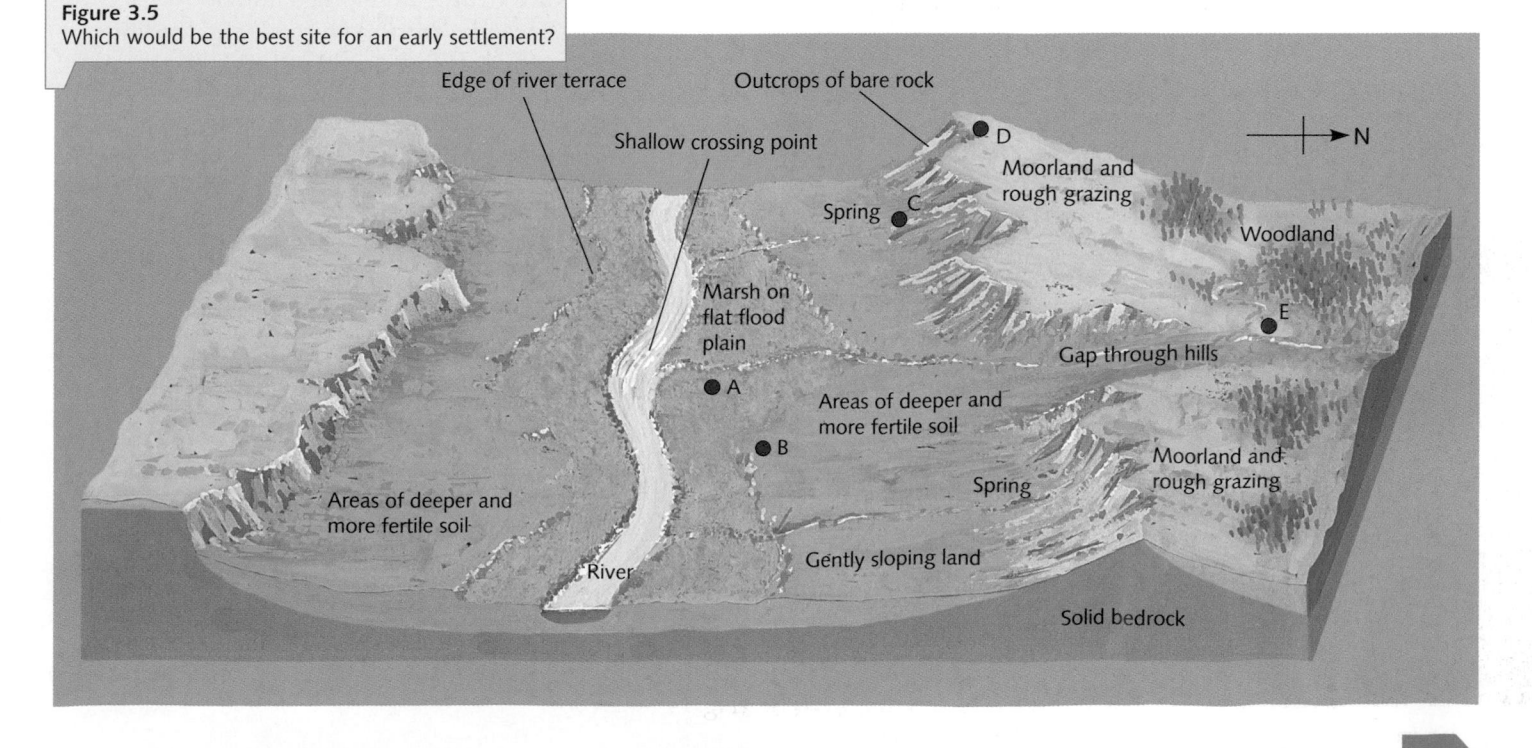

Settlement types

Settlements are, for convenience, divided into rural and urban. However, it is often difficult to see and tell the difference between the two basic types. This difficulty is partly due to problems in defining the difference between the terms **rural** and **urban** and between what constitutes a village and a town. Figure 3.6 lists the usually accepted types of rural and urban settlement.

Patterns

Geographers are interested in the patterns and shapes of villages and towns as well as in their main functions. Although villages have characteristic shapes, these vary from place to place both within Britain and across the world. Although it is unusual to find all the characteristic shapes within a small area, most can be seen in south Holderness to the east of Kingston upon Hull (see OS map, page 41).

Dispersed (Figure 3.7) This can either be:
- an **isolated**, individual building, or
- a group of two or three buildings, perhaps forming a hamlet, and separated from the next group by two or three kilometres.

Dispersed settlement occurs in an area of adverse physical difficulty where natural resources are insufficient to support more than a few people. Traditionally, most buildings are farms although increasingly some are being used as second homes or for holidays. In Britain, dispersed settlement occurs in mountainous parts of Scotland, Wales and northern England and in previously marshy areas such as the Fens.

Figure 3.6
Types of settlement

TYPES OF SETTLEMENT

- RURAL (countryside)
  - Isolated building
  - Hamlet
  - Village
  - Small market town
- URBAN (towns)
  - Larger industrial town
  - City
  - Conurbation (and/or capital city)

← increasingly rural   increasingly urban →

Nucleated (Figure 3.8) This is when several buildings were grouped together, initially often for defensive purposes and later for social and economic reasons. The nucleation of buildings into villages occurred where there was enough farmland for the inhabitants to be self-sufficient, as in the English Midlands and East Anglia, and where the water supply was reliable (Figure 3.2). Nucleated settlements often occur every 5 to 10 kilometres.

Linear or street (Figure 3.9) Linear settlement occurs where buildings are strung out along a line of communication. This may be a main road (e.g. leading out of a British city), a river valley (e.g. South Wales) or a canal or dyke (e.g. English Fens and the Netherlands).

Figure 3.7
Dispersed settlement in Northumberland

Figure 3.8
A nucleated settlement: Hooton Pagnell, in Yorkshire

Functions

The function of a settlement relates to its economic and social development and refers to its main activities (Figure 3.10). Normally, larger settlements will have more functions than smaller settlements. Large settlements tend to be multi-functional (i.e. they have several functions) although one, or possibly two or three, is likely to be predominant, e.g. London is a capital city, Cambridge is a university town and Dover is a port.

In some cases, the original function may no longer be applicable, e.g. British towns no longer have a defensive function (Figure 3.11). In other cases, functions have changed over a period of time, e.g. a Cornish fishing village may now be a tourist centre, and a former South Wales mining town a location for high-tech industry.

Figure 3.9
A linear settlement: Parson's Drove, Cambridgeshire

Figure 3.10
Types of function

	Description	UK example	World example
Market towns	Originally collecting and distribution centres for surrounding farming area. Today they may service and process agricultural machinery and produce	York	Winnipeg
Mining towns	Developed to exploit local mineral or fuel	Corby	Prudhoe Bay
Industrial–manufacturing	Where raw materials are processed into manufactured goods	Birmingham	Pittsburgh
Ports	Located on coasts, rivers and lakes for the movement of goods and people from land to sea, or vice versa	Southampton	Thunder Bay
Route centres	At the convergence of several natural routes or at nodal points resulting from economic development	Carlisle	Paris
Commercial	Providing the needs of industry and business	London	Hong Kong
Cultural/religious	Attracting people, perhaps for a short period, for educational and religious purposes	Cambridge	Rome
Administrative	Developed to control areas which may vary from a small region (county town) to a country (capital city)	Exeter	Brasilia
Residential	Where the majority of residents live but do not work	Telford	Marne-la-Valleé
Tourist resorts	Include spa towns, coastal and mountain resorts	Bath	Orlando

Early function, no longer applicable

– castle and meander = defensive

Other early functions – still applicable

– cathedral = religious
– weir = small industries
– houses = residential

More recent and present-day functions

– town hall = administrative
– railway and station; main roads and bus station; motorway; park and ride; ferry = route centre/transport
– museums = cultural
– information centre = tourism
– hospital = health
– college, schools = educational

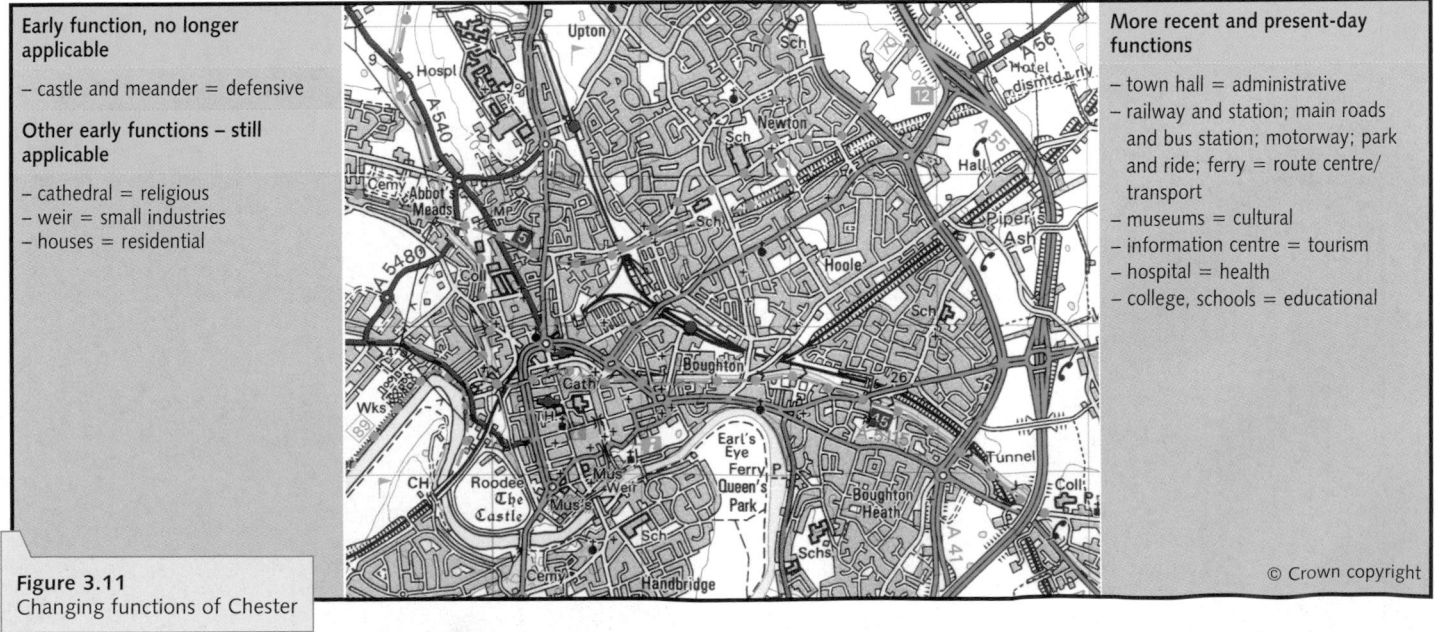

© Crown copyright

Figure 3.11
Changing functions of Chester

Hierarchies

The term **hierarchy** refers to the arrangement of settlements within a given area (e.g. a country or county) in an 'order of importance'. Isolated farms and small hamlets form the base of the hierarchy pyramid, with the largest and/or capital city at the top (Figure 3.12). Three different methods to determine the 'order of importance' in the hierarchy have been based on:

1 the population size of a settlement
2 the range and number of services provided by a settlement
3 the sphere of influence, or market area, of a settlement.

1 Population size

Early attempts to determine a settlement hierarchy were based on size. However, no one has been able to produce a widely accepted division between, for example, a hamlet and a village or a village and a town. Indeed so-called villages in places like India and China are often as large as many British towns. Figure 3.12 lists the conventional hierarchy, as applied to Britain, in terms of types of settlement. However, the cut-off points for divisions based on population size and the distance between settlements, use generalised, arbitrary figures. Notice that the larger the settlement the fewer there are in number and the greater the distance between them.

2 Range and number of services

Villages provide a limited range and number of services. Services that do exist are those likely to be used daily (the village shop) or which reduce the need to travel to other places (a primary school). In Figure 3.13, where the hierarchy is based on services, each place in the hierarchy is likely to have all the services of settlements below them.

Capital	Cathedrals, government buildings, banking HQ, railway termini, museums and art galleries, large theatre, shopping centre, several universities, international airport
City	Large railway station, large shopping complex, cathedral, opticians and jewellers, large hospital, large football team, university, theatre, county hall, airport
Large town	Several shopping areas/arcades, hypermarket, railway station, bus station, hotels, banks, small hospital, small football team
Small town	Town hall, doctor, several churches/chapels, cafés and restaurants, small secondary school, railway station, several shops
Village	Church, post office, public house, shop for daily goods, small junior school, village hall
Hamlet	Perhaps none, or public telephone

Figure 3.13
Hierarchy of settlements according to services

3 Sphere of influence

The sphere of influence, or **market area**, may be defined as the area served by a particular settlement. The area of the sphere of influence depends on the size and services of a town and its surrounding settlements, the transport facilities available and the level of competition from rival settlements.

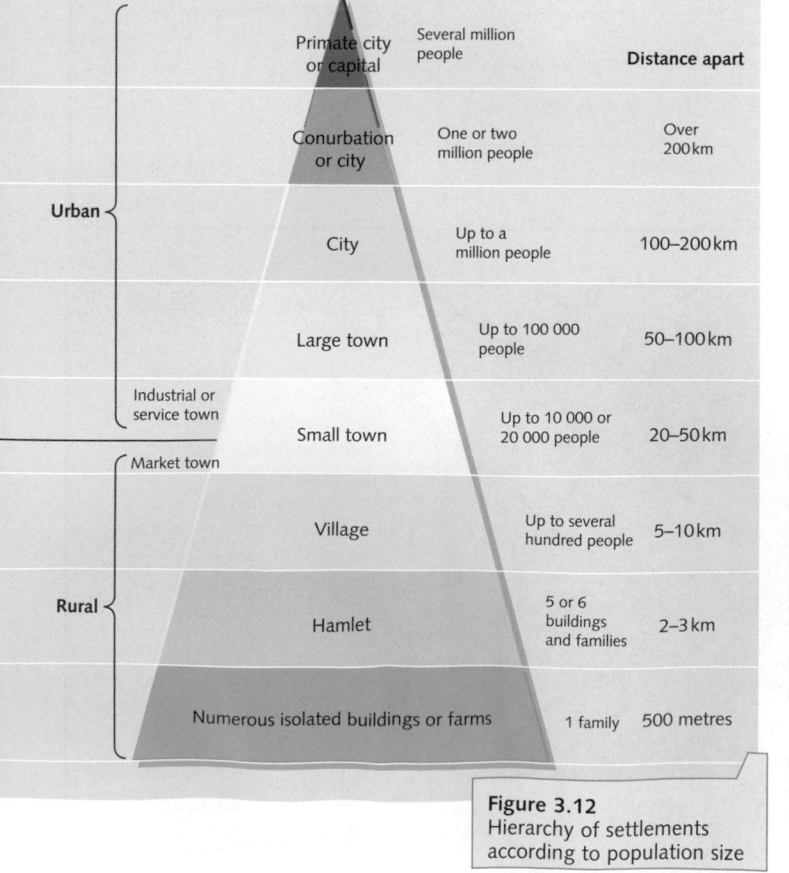

Figure 3.12
Hierarchy of settlements according to population size

Figure 3.14
Market areas of Exeter

Two main ideas should be noted:

- A **threshold population** is the minimum number of people needed to ensure that demand is great enough for a special service to be offered to the people living in that area. For example, estimates suggest that 350 people are needed to make a village shop successful, 2500 for a single doctor to be available, and 10 000 people for a secondary school. Boots the Chemist prefers a threshold of 10 000 people in its market area, Marks & Spencer 50 000 people and Sainsbury's 60 000.

- **Range** is the maximum distance that people are prepared to travel to obtain a service. Figure 3.14 shows that people are not prepared to travel far to a corner shop or even to a chemist, but are prepared to travel much greater distances to shop at a hypermarket or, as these are visited less frequently, a furniture store or theatre.

Each settlement which provides a service is known as a **central place**. A central place provides goods and services for its own inhabitants and to people living in the surrounding area. The larger the settlement the more services it will provide and the more people it will serve. Large towns and cities will therefore have larger spheres of influence than smaller villages (Figure 3.15).

Changes in time

Few settlements remain constant in size. Villages near to large cities tend to increase in size as they become more suburbanised (page 64). Villages in more isolated areas, in contrast, tend to lose population. This means that villages increasing in size are likely to gain additional or improved services (a larger school, more shops) while villages declining in size will lose services (village shop, bus service).

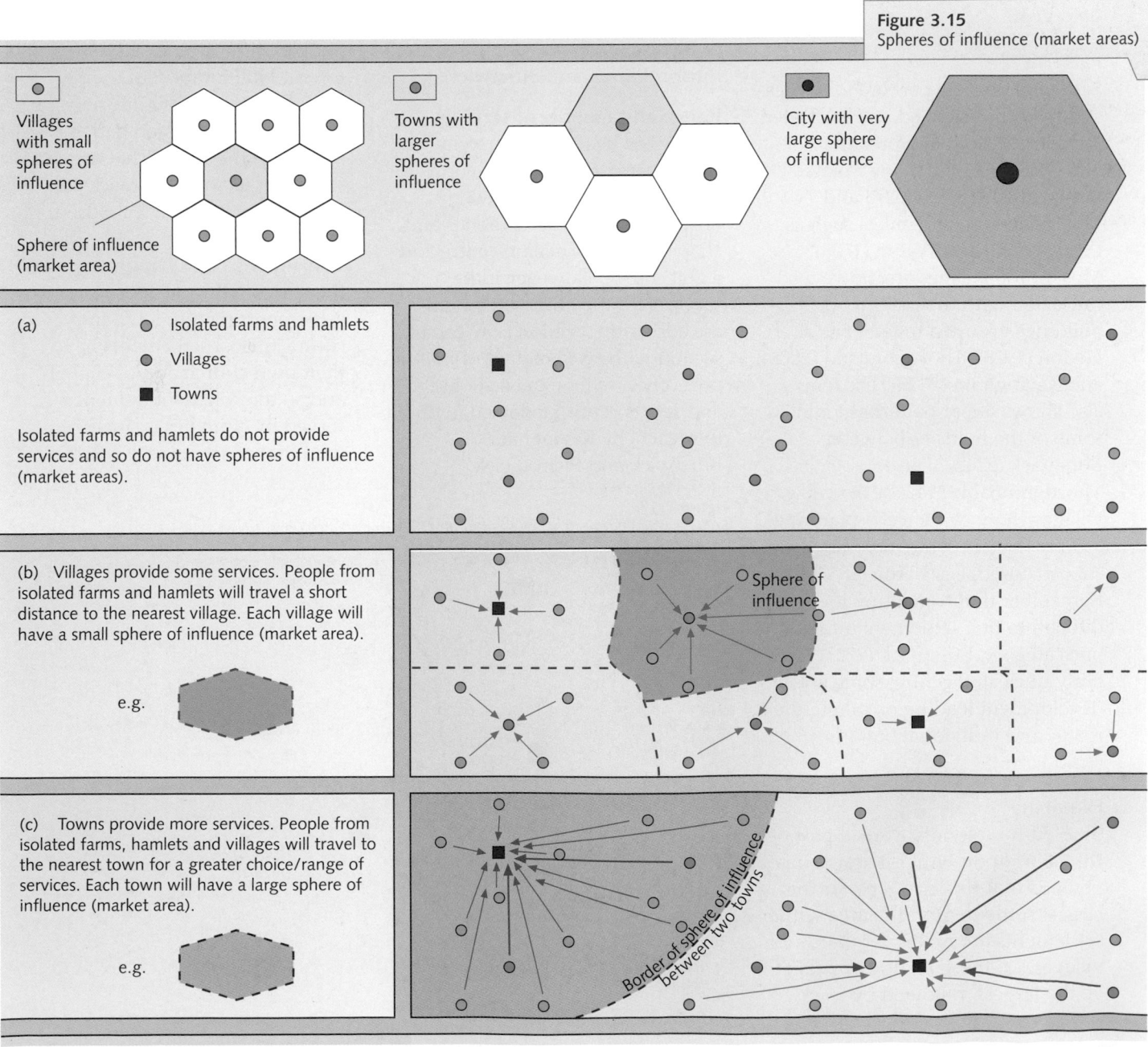

Figure 3.15
Spheres of influence (market areas)

Villages with small spheres of influence

Towns with larger spheres of influence

City with very large sphere of influence

Sphere of influence (market area)

(a)
- Isolated farms and hamlets
- Villages
- Towns

Isolated farms and hamlets do not provide services and so do not have spheres of influence (market areas).

(b) Villages provide some services. People from isolated farms and hamlets will travel a short distance to the nearest village. Each village will have a small sphere of influence (market area).

e.g.

Sphere of influence

(c) Towns provide more services. People from isolated farms, hamlets and villages will travel to the nearest town for a greater choice/range of services. Each town will have a large sphere of influence (market area).

e.g.

Border of sphere of influence between two towns

Settlement and the OS map

The map on page 41 is part of a 1 : 50 000 OS map (sheet 107). It shows that part of Holderness to the east of Kingston upon Hull.

Site

Although the map shows the area in 2002, some of the early site factors for the various settlements are still visible. Figure 3.16 shows, using map evidence only, some of the physical and human site factors for Hedon (GR1928). Other factors, either for Hedon or any other settlement under study, might be found by using older maps and documents or by making a special field visit.

Patterns

Settlement in this part of Holderness shows both dispersed and nucleated patterns. Despite the proximity of Hull, there are still many isolated farms, e.g. 193268, 250278 and 192326, and small hamlets such as Lelley (2032) and Wyton (1733). Most of the villages are, however, nucleated with the majority of buildings grouped together as at Hedon (1928), Thorngumbald (2026) and Keyingham (2425). The map also shows linear patterns along some of the roads as between Burstwick (224273) and Thorngumbald (214262). Several villages show evidence of planning as they have become increasingly suburbanised, e.g. Bilton (1533). Remember that sometimes it is difficult to fit a settlement into just one category. Burstwick (2227) has a fairly nucleated centre, some linear development leading out along the roads, and individual buildings (Stud Farm).

Hierarchy

Size Although only a small part of Kingston upon Hull is shown, it is obvious that this city is by far the largest settlement in the area. Other settlements are no bigger than villages. Of these Hedon appears to be the largest, followed by Thorngumbald and Keyingham (which seem about equal), Preston, Burton Pidsea and Burstwick.

Range and number of services

Hull again heads the hierarchy with – using map evidence only – seven schools, a college, two hospitals, numerous churches and chapels, an information centre and a golf course. However, if the services for the other settlements are added up, Preston now comes second with six (a church, chapel, post office, school, public house and sports centre) followed, with four each, by Keyingham, Burstwick and Elstronwick.

Sphere of influence Most people in the area are likely to visit Hull for most of their services. Outside of Hull, only Burstwick and Preston had schools in the 1990s. That means, for example, that Burstwick will not only serve its own children, but also those from Thorngumbald, Keyingham and the surrounding smaller settlements (Figure 3.17). In contrast, most settlements have their own church and/or chapel and so the sphere of influence served by churches is smaller than that of schools.

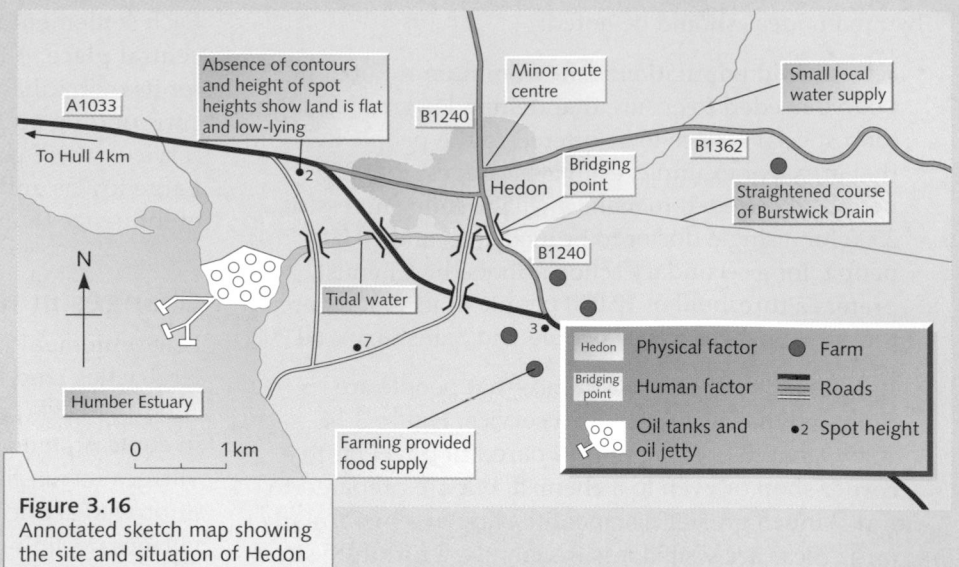

Figure 3.16
Annotated sketch map showing the site and situation of Hedon

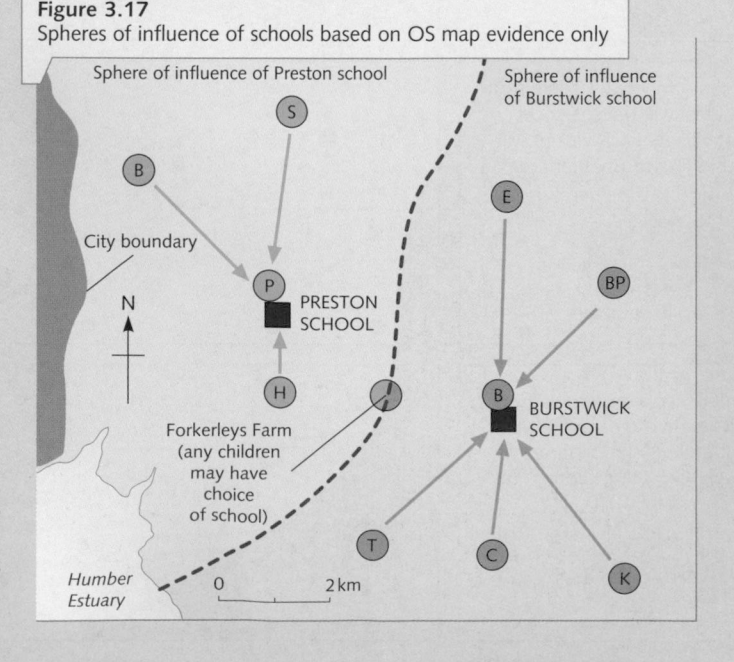

Figure 3.17
Spheres of influence of schools based on OS map evidence only

© Crown copyright

Figure 3.18
Part of the OS map sheet 107 Kingston upon Hull

VEHICLE FERRY FROM
KINGSTON UPON HULL TO
Rotterdam (Europoort) 12 hrs
Zeebrugge   14 hrs

Urban land use models

Urbanisation means an increase in the proportion of people living in towns and cities. Although towns were important even in the early civilisations of Mesopotamia and in the valleys of the Nile, Indus and Huang-He (China), most people tended to live and work in rural areas. It was not until the rapid growth of industry in the nineteenth century that large-scale urbanisation began in parts of western Europe and north-eastern USA. During the twentieth century, people continued to move to urban areas mainly for:

• more and better-paid jobs
• nearness to places of work and entertainment
• better housing, services (schools and hospitals) and shopping facilities.

Urban land use models

A model is a theoretical framework which may not actually exist, but which helps to explain the reality. It has been suggested that towns do not grow in a haphazard way, but rather they tend to develop with recognisable shapes and patterns. Although each urban area is unique, with its own distinctive pattern, it is likely to share certain generalised characteristics with other settlements. Two of the earliest land use models to be put forward, and which are still the easiest to apply, are shown in Figure 3.19.

• **Burgess** claimed that in the centre of all towns and cities there was a **central business district (CBD)**. He suggested, initially using Chicago as his example, that towns grew outwards from this CBD in a concentric pattern. The resultant circles were based on the age of houses and the wealth of their occupants, with building becoming newer and the occupants more wealthy with increasing distance from the CBD.

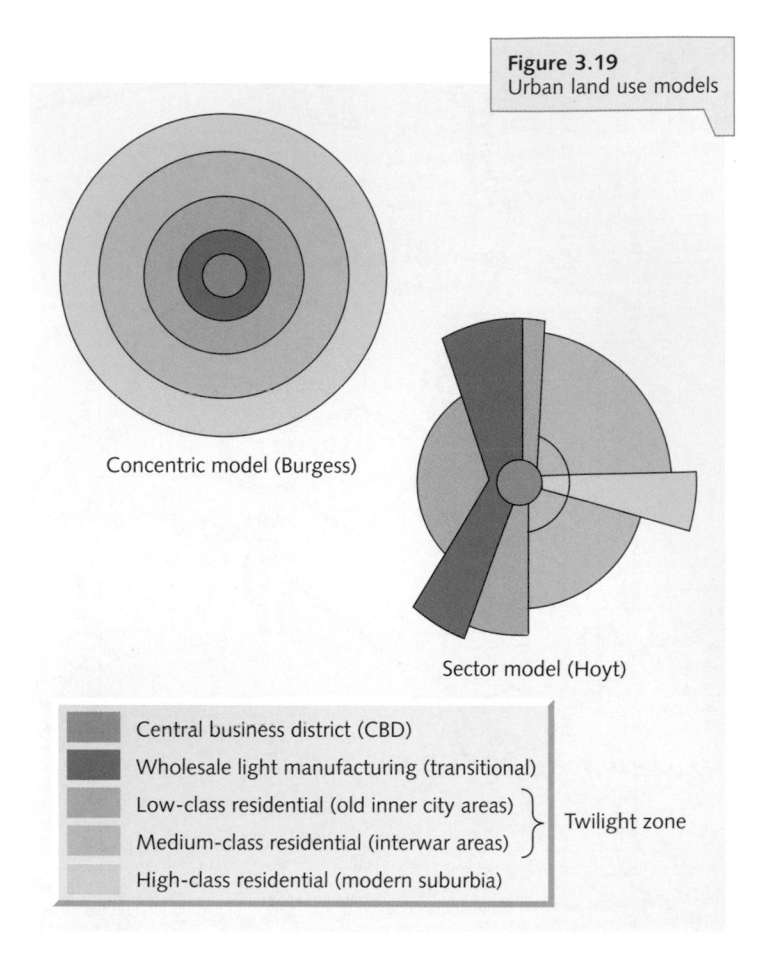

Figure 3.19
Urban land use models

Concentric model (Burgess)

Sector model (Hoyt)

■	Central business district (CBD)
■	Wholesale light manufacturing (transitional)
■	Low-class residential (old inner city areas)
■	Medium-class residential (interwar areas)
■	High-class residential (modern suburbia)

Twilight zone (brackets Low-class and Medium-class residential)

• **Hoyt** proposed his model after the development of public transport. He suggested that urban areas developed in sectors, or wedges, alongside main transport routes into and out of a city. He also claimed that if, for example, industry and low-cost housing developed in one part of a town in the nineteenth century, then newer industry and modern low-cost housing would also locate in the same sector.

Urban land use and functional zones

Each of the zones shown in Figure 3.19 has a function. The four main types of function are shops and offices, industry, housing, and open space. The location of each zone and the distribution of each functional zone are related to several factors.

Land values and space Land values are highest and available sites more limited in the CBD where competition for land is greatest. As land values decrease rapidly towards the urban boundary then both the amount of space and the number of available sites increase (Figure 3.20).

Age As towns developed outwards, the oldest buildings were near to the city centre (although many of these have now been replaced) and the newest ones in the outskirts.

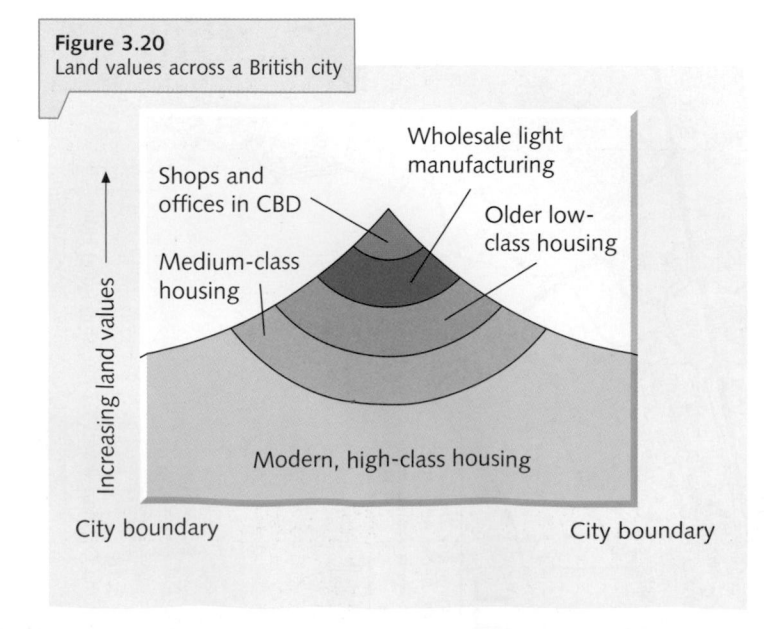

Figure 3.20
Land values across a British city

Shops and offices in CBD

Wholesale light manufacturing

Medium-class housing

Older low-class housing

Modern, high-class housing

Increasing land values

City boundary   City boundary

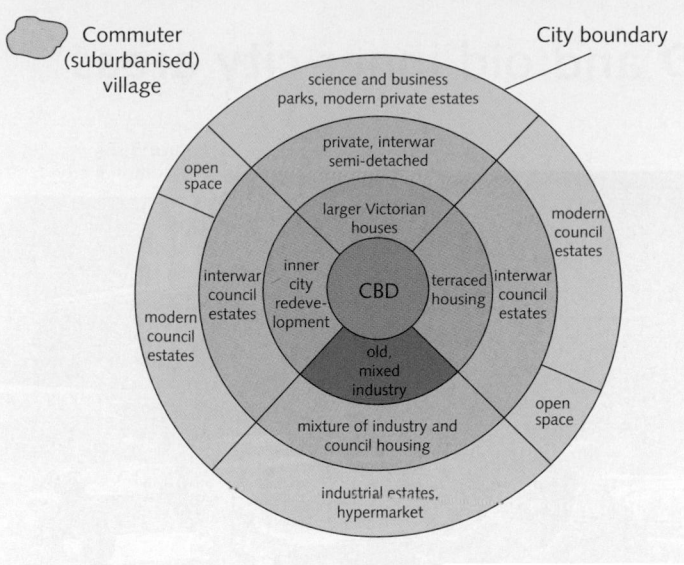

Figure 3.21
Land use in a modern British city showing a combination of the concentric and sector models

Changes in demand Land use and function change with time. For example:

- Nineteenth-century industry was located next to the CBD whereas modern industry prefers edge-of-city sites.
- The main land use demand in the nineteenth century was for industry and low-cost housing. Today it is for industry, shops and better-quality housing, all in a more pleasant environment, and open space.

Figure 3.21 is a more realistic model showing land use patterns and functional zones in a British city. Figure 3.22 is a transect, or cross-section, across a city. Figure 3.23 is a simplified land use map of Burnley in Lancashire.

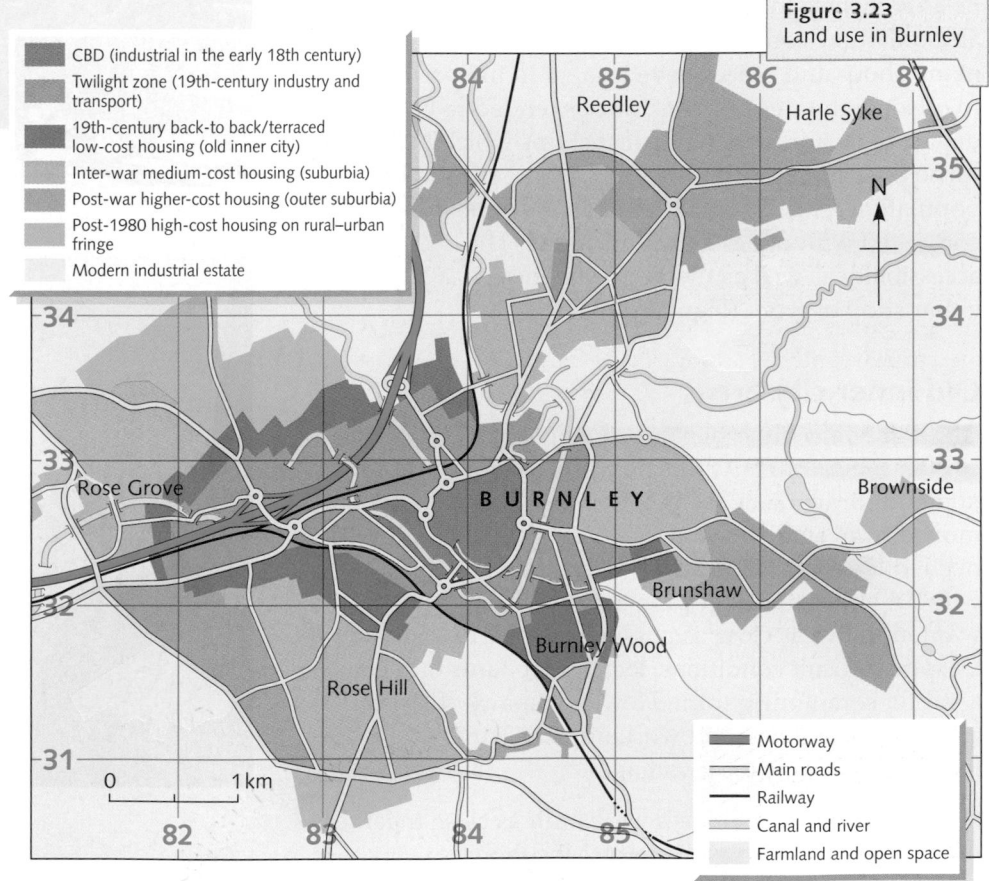

Figure 3.23
Land use in Burnley

Legend:
- CBD (industrial in the early 18th century)
- Twilight zone (19th-century industry and transport)
- 19th-century back-to back/terraced low-cost housing (old inner city)
- Inter-war medium-cost housing (suburbia)
- Post-war higher-cost housing (outer suburbia)
- Post-1980 high-cost housing on rural–urban fringe
- Modern industrial estate

- Motorway
- Main roads
- Railway
- Canal and river
- Farmland and open space

Accessibility The CBD, where the main routes from the suburbs and surrounding towns meet, has been the easiest place to reach from all parts of the city although this ease is now often reduced due to increased congestion.

Wealth of the inhabitants The poorer members of the community tend to live in cheaper housing near to the CBD (with its shops) and the inner city (where most jobs used to be found). These people are less likely to be able to afford the higher transport (private or public) and housing costs of places nearer the city boundary.

Figure 3.22
Transect across a typical British city

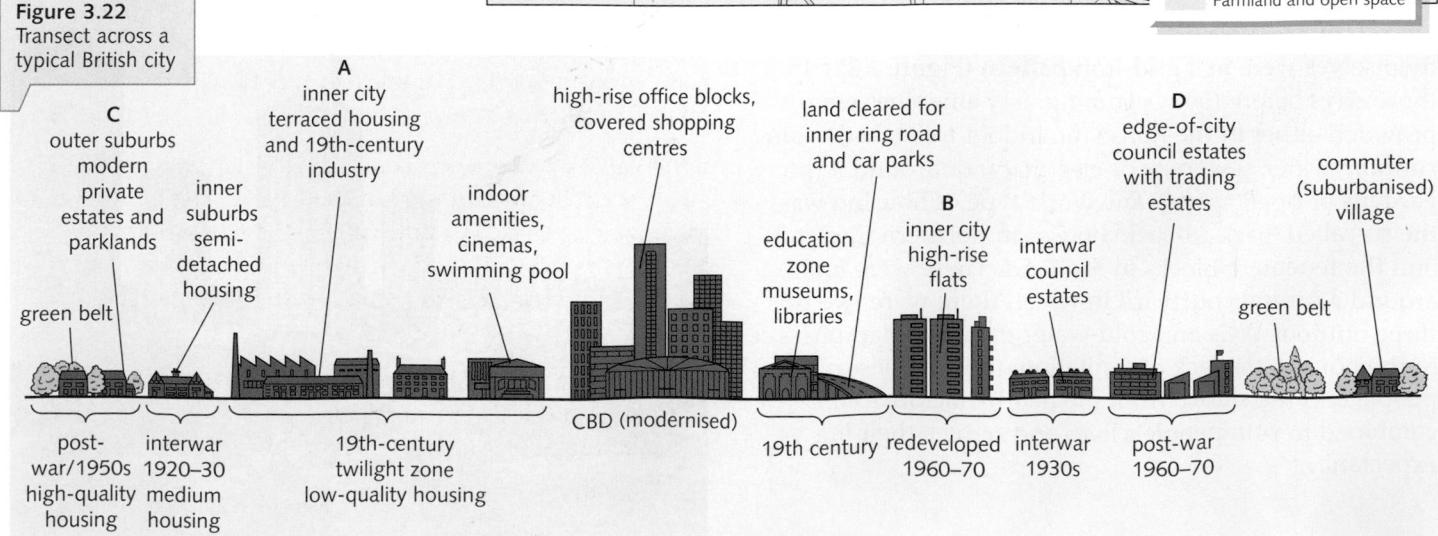

Traditional land use in the CBD and old inner city areas

The CBD

The CBD became the most accessible part of a town or city. Because most road and rail routes met there, it meant that it could easily be reached by people living throughout the urban area and beyond. Its accessibility made it a prime site for several types of land use which, over time, increasingly competed with each other to locate here. This competition, together with the limited amount of space that was available, pushed up land values (page 42).

The major land users in the CBD became shops, banks and offices. They located here mainly because they needed to be as accessible to the largest number of people possible and only they could afford the high cost of land. Even so, many shops and offices have located in high-rise buildings to offset high rates and rent. Shops tended to be large department stores and specialist shops which had a high turnover, a high profit margin or a large threshold population (page 39). In time, CBDs became increasingly congested with pedestrians and traffic. This reduced their accessibility as well as their attractiveness as a place to either shop or work (Figure 3.24).

Old inner city areas

Most inner city areas developed along with industry in the nineteenth century. As industry grew, so too did the demand for workers. As an increasing number of people moved from rural areas to the towns for work, they needed low-cost houses in which to live. At that time, without either public or private transport, people also wanted to live as close as possible to their place of work. Despite difficult conditions, local inhabitants often created a strong community spirit. However, many of the early advantages of living and working in an inner city have long since become disadvantages.

Nineteenth-century houses were built as close together as possible creating a **high density**. Most houses were built in long straight rows and in terraces (Figure 3.25). The streets themselves were in a grid-iron pattern (Figure 3.35). In those days before town planning, few amenities were provided either in the house (no indoor toilet, bathroom, running water, sewerage or electricity) or around it (no gardens or open space). The worst type of housing was the so-called 'back-to-back' houses in northern England and the tenement blocks in Scotland. These were built around a central courtyard in which there were two or three outdoor WCs and cold-water taps. The dampness of the houses, the lack of sanitation, the closeness of the people and the smoke from the nearby factories all combined to ruin people's lives and reduce their life expectancy.

Figure 3.24
A traditional city centre

Figure 3.25
Nineteenth-century terraces

Figure 3.26
High-rise residential blocks

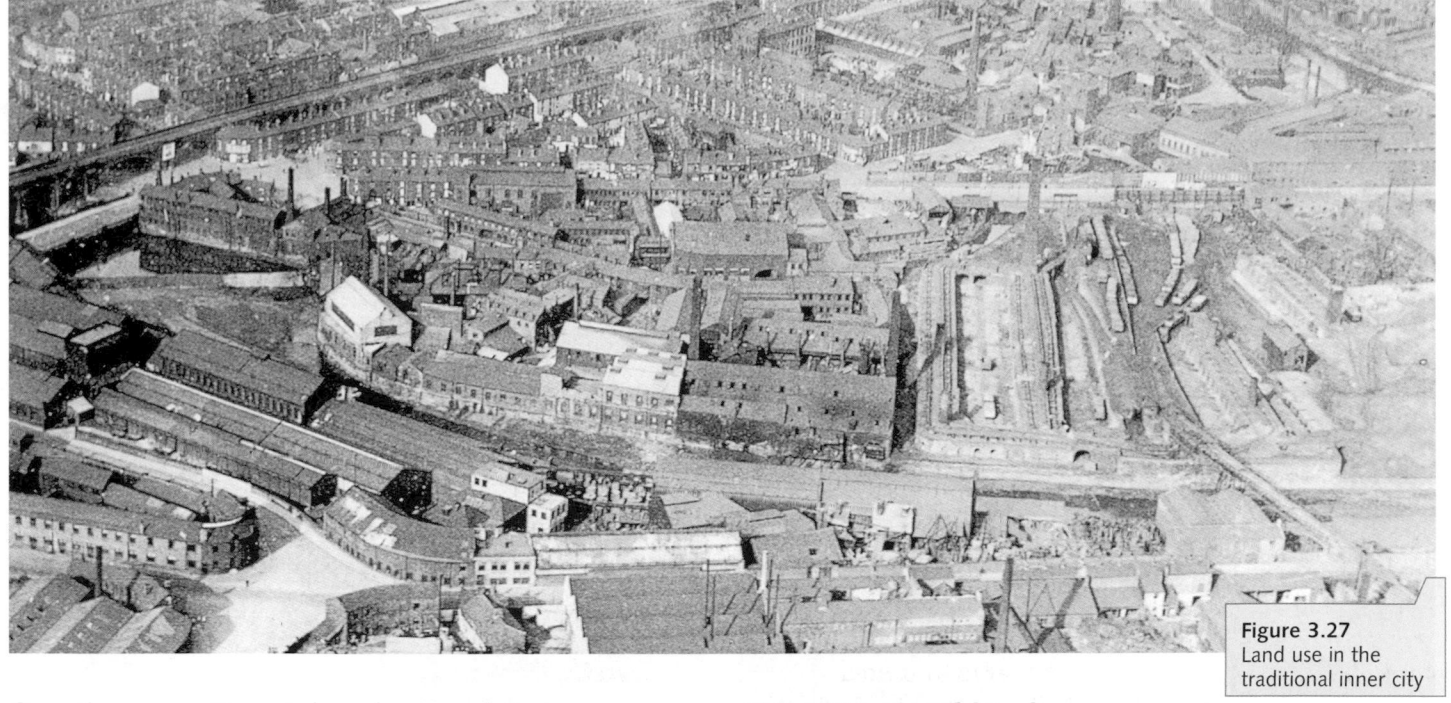

Figure 3.27
Land use in the traditional inner city

Over the years, attempts have been made to improve living conditions, initially – as in the 1960s – by bulldozing large areas (**slum clearance**) and building high-rise flats (**urban redevelopment** – Figure 3.26) and later by improving existing properties (**urban renewal**). High-rise flats, despite having modern amenities (a bathroom, running water and a WC), created as many social problems as they solved. Many inner city areas still experience cramped housing conditions and higher than average poverty, crime and unemployment. In some places, older terraced houses may have been abandoned, boarded up and even vandalised and some of the high-rise flats pulled down.

Industry

Inner cities were characterised by large factories built during the Industrial Revolution. They were located (Figure 3.27):

* on the nearest available land to the town centre and where there was enough space for large buildings
* next to canals and, after the 1840s, railways which were needed to transport the heavy and bulky raw materials and the finished manufactured goods
* beside rivers which, initially used as a source of power, provided water for washing and cooling and a means of disposing of waste
* next to land that could be used to house the large number of workers who were needed.

Since then many factories have been forced to close (Chapter 9) either due to a lack of space for expansion and modernisation, or due to narrow, congested roads. Some factories, like the terraced housing, have been left empty, while others have been pulled down to leave large areas of derelict land (**brownfield sites**, page 65).

Other types of land use

Inner cities also contained canals, railways and, later and leading to the CBD, main roads (Figure 3.27). They also included numerous corner shops (Figure 3.28). Corner shops served a small area, no more than a few streets, and were visited frequently by people needing everyday low-order, convenience goods. The shops opened for long, irregular hours and provided a social meeting place for local residents. Open space was extremely limited, partly because land near to the CBD was too expensive to leave unused and partly because few people had, due to long hours of work, spare time for recreation.

Figure 3.28
The traditional corner shop

Traditional land use in suburbia and the rural–urban fringe

Figure 3.29
Inter-war housing

Figure 3.30
Suburban shopping parade

Suburbia

Suburbia was the name given to areas that developed during the inter-war period (1920s and 1930s). Urban areas grew rapidly at this time due mainly to the introduction of public transport, the increase in popularity of the private car and, in London, the extension of the Underground. This outward growth, known as **urban sprawl**, led to the growth of numerous private, car-dependent suburbs. Many inter-war houses were semi-detached and now correspond to Burgess's zone of medium-cost housing (Figure 3.19). The houses were characterised by their bay windows, garages and front and back gardens (Figure 3.29).

As each new housing estate was built, its distance from the CBD and its shops increased. This led to the growth of small shopping parades within the estate (Figure 3.30). These parades, which local residents visited several times a week, saved people time in travelling and reduced transport costs. Each parade was likely to include a sub-post office and newsagent, a small chain store (providing mainly low-order convenience goods), several specialist shops (butcher and chemist) and one or two supplying non-essential services (hairdresser). Most shopping parades had space for limited car parking.

Apart from individual houses having their own gardens, many inter-war estates had their own small park or play area. This was due to the decrease in land values away from the CBD (Figure 3.20), the increase in land available towards the edge of the urban area, and the gradual introduction of town planning. The inter-war estates rarely, if ever, had industry nearby and so residents, especially in London, had to travel long distances to their place of work – a process known as **commuting**.

The rural–urban fringe

After the 1960s, urban sprawl continued with land on the rural–urban fringe being used mainly for:
- one of two types of housing – private estates or outer-city council housing
- new industrial and/or trading estates.

Most of the new private estates had low-density, high-quality housing (Figure 3.31). The houses, many of which were large and detached, incorporated modern amenities both indoors (central heating, double glazing, deluxe bathroom and kitchen) and outside (large gardens back and front, and even a double garage). They were built either in a cul-de-sac or along winding roads that were usually tree-lined and relatively free of traffic. The estates also contained small areas left as open space.

Figure 3.31
A modern estate on the rural–urban fringe

Figure 3.32
An outer city council estate

Outer city council estates were created mainly during the 1950s and 1960s when local councils cleared the worst of the slums and poor-quality housing from the inner city areas. Many of the evicted residents were re-housed on what were then **greenfield** edge-of-city sites. Accommodation usually consisted of a mixture of high-rise tower blocks, low-rise flats and single-storey terraces (Figure 3.32). Although the density was higher than on the private estates, it was lower than in the inner-city areas. While many homes were small and most lacked gardens and garages, they did have indoor amenities including modern kitchens and bathrooms.

By the end of the twentieth century, an increasing number of land users saw the rural–urban fringe as the ideal location for future development. Compared with other places nearer the city centre, this location has:
- less congestion and easier access, especially with other urban areas
- a more attractive and less polluted environment.

There is, however, an increase in conflict between those who wish to see the economic development and extension of the urban area and those who wish to protect the rural environment that surrounds it (Figure 3.33).

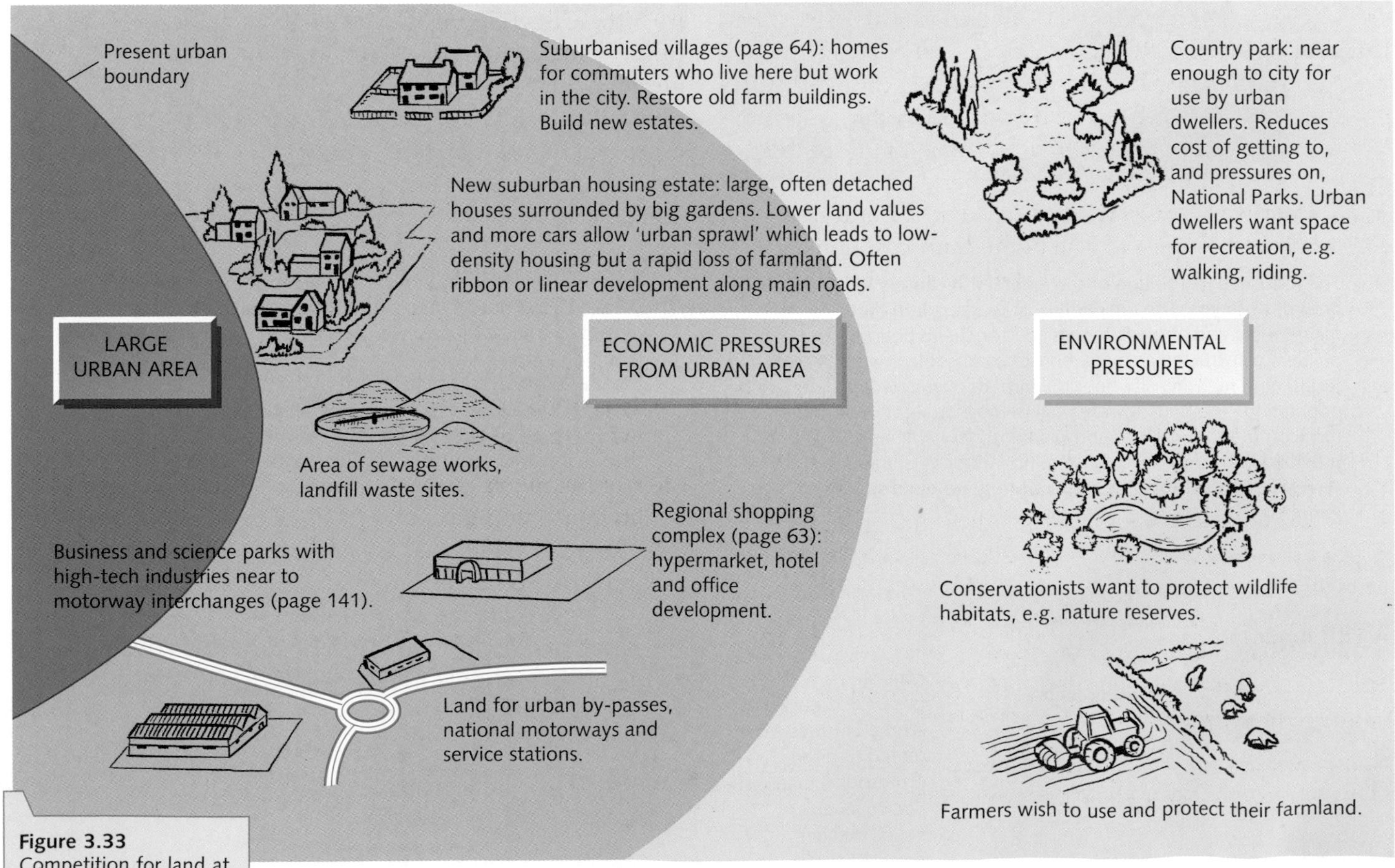

Present urban boundary

LARGE URBAN AREA

ECONOMIC PRESSURES FROM URBAN AREA

ENVIRONMENTAL PRESSURES

Suburbanised villages (page 64): homes for commuters who live here but work in the city. Restore old farm buildings. Build new estates.

New suburban housing estate: large, often detached houses surrounded by big gardens. Lower land values and more cars allow 'urban sprawl' which leads to low-density housing but a rapid loss of farmland. Often ribbon or linear development along main roads.

Area of sewage works, landfill waste sites.

Business and science parks with high-tech industries near to motorway interchanges (page 141).

Regional shopping complex (page 63): hypermarket, hotel and office development.

Land for urban by-passes, national motorways and service stations.

Country park: near enough to city for use by urban dwellers. Reduces cost of getting to, and pressures on, National Parks. Urban dwellers want space for recreation, e.g. walking, riding.

Conservationists want to protect wildlife habitats, e.g. nature reserves.

Farmers wish to use and protect their farmland.

Figure 3.33
Competition for land at the urban–rural fringe

Residential environments in British cities

Cities are constantly changing. The descriptions of the four residential types below were typical for the early 1980s. Since then, many old inner city areas have either been totally redeveloped or have undergone piecemeal improvements, while suburban areas have much more

traffic and houses have been extended. You can use these two pages to see just how much change has taken place since then in similar types of residential area in your own local town or city. Similarly, the census data for 1981 has long since been out of date. It has been left here, however,

A Old inner city area

Figure 3.34
Physical appearance

Figure 3.35
Land use

Housing
Open space and gardens
Industry
Transport
Wasteland
Services (schools, shops)

Figure 3.36
Description

The Industrial Revolution of the nineteenth century led to the growth of towns. The rapid influx of workers into these towns meant a big and immediate demand for cheap housing, and so builders constructed as many houses as possible in a small area, resulting in high-density housing with an overcrowded population. The houses were built in long, straight rows and in terraces. In those days of non-planning, few amenities were provided either in the house (e.g. no indoor WC, bathroom, sewerage or electricity) or around it (e.g. no open space and no gardens).

B Inner city redevelopment

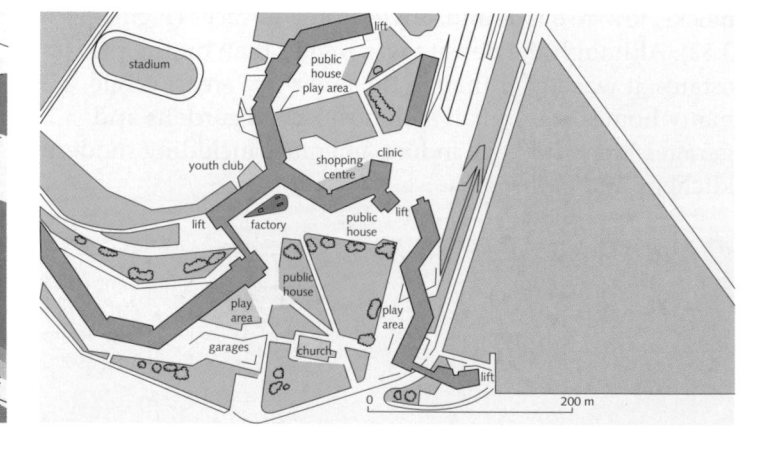

When in the 1950s and 1960s vast areas of inner cities were cleared by bulldozers, many of the displaced inhabitants either moved to council estates near the city boundary, or were rehoused in huge high-rise tower blocks which were created on the sites of the old terraced houses. Although these high-rise buildings contained most modern amenities, the apartments had to be reached by lifts which led to narrow, dark corridors. Also, despite the areas of greenery between the flats, there was still a very high housing density.

Figure 3.37
Census data

A Old inner city area

Housing tenure

Socio-economic groups

B Inner city redevelopment

council
owner-occupied
rented unfurnished
rented furnished
professional managerial
non-manual
skilled manual
semi-skilled
unskilled

0   20   40   60   80   100
% of households

0   20   40   60   80   100
% of households

so that by using the internet (which was not available in the 1980s), you can compare data for the areas shown and your present-day local areas – again to determine these changes. The next chapter helps to explain why some of these changes have occurred and what these areas now look like. What do you think will be likely changes in the future?

C Suburbia

The rapid outward growth of cities began with the introduction of public transport, and accelerated with the popularity of the private car. This outward growth (also known as urban sprawl) led to the construction of numerous private, 'car-based' suburbs.

The houses built in the outer suburbs before the Second World War are characterised by their front and back gardens. Usually they have garages and are semi-detached with bay windows. The more recent estates have housing which differs in both style and type, but they remain well planned and spacious.

D Outer city council estate

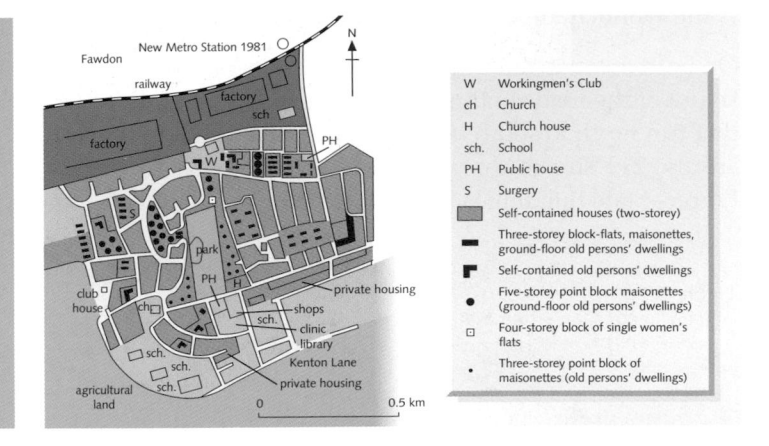

As local councils cleared the worst of the slums from their inner city areas in the 1950s and 1960s, many residents were rehoused on large council estates on the fringes of the city. Attempts were made to vary the type and size of accommodation:
• High-rise tower blocks, often 10–12 storeys high.
• Low-rise tower blocks, usually 3–5 storeys high. These were built nearer the city boundaries, where there was more open space.
• Single-storey terraces with some gardens and car parking space.

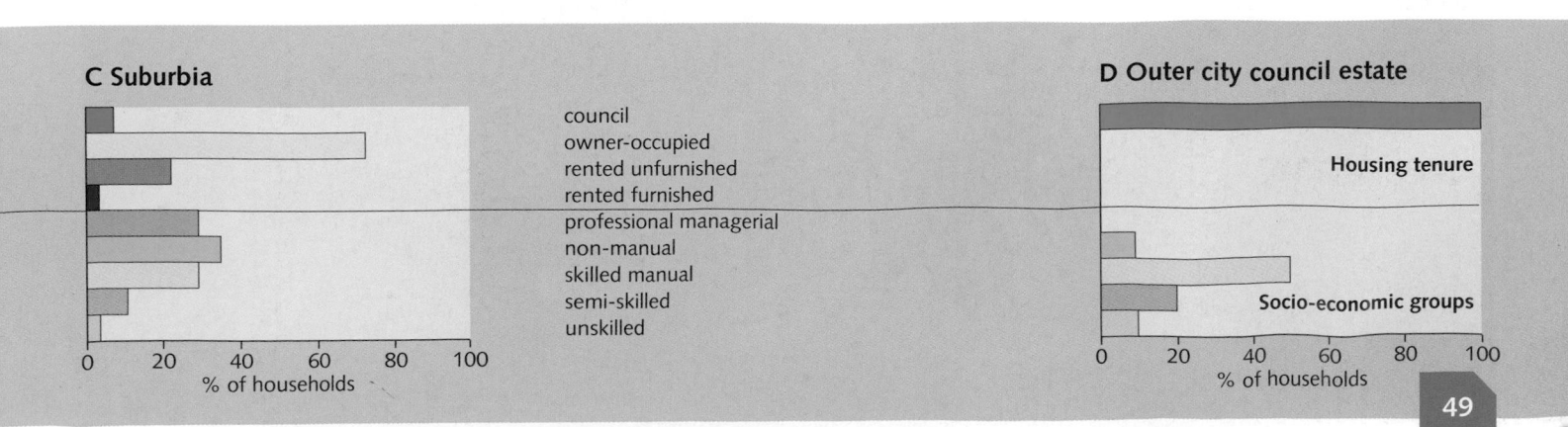

C Suburbia

council
owner-occupied
rented unfurnished
rented furnished
professional managerial
non-manual
skilled manual
semi-skilled
unskilled

% of households

D Outer city council estate

Housing tenure

Socio-economic groups

% of households

New York

Manhattan Island

Manhattan Island was the original site of New York. To most present-day tourists and overseas visitors, Manhattan *is* New York, although in reality it is only one of five boroughs (the others are Brooklyn, Queens, the Bronx and Staten Island – Figure 3.38). Figure 3.39 is a photo of the island (looking north) with Long Island, which provides the major residential and recreational areas for New Yorkers, to the east (to the right of the photo). Skyscraper development has taken place in two distinct zones:

- Midtown Manhattan (MM on Figure 3.38) which includes the main hotels (the Plaza and Waldorf Astoria), shops (Fifth Avenue and Park Avenue), theatres (Broadway) and prestige buildings (Empire State, Chrysler and United Nations)

- Downtown Manhattan (DM on Figure 3.38) which includes the financial and banking district (Wall Street and the Stock Exchange) – and, until 11 September 2001, it also included the Twin Towers of the World Trade Center.

On both the map and photo of Manhattan, notice the grid-iron layout of the roads to the north of Chinatown. There are 12 'avenues' that run north to south between the East and Hudson Rivers. There are also 219 'streets' that run across the island from east to west. The Empire State Building is 'on Fifth Avenue at 34th Street'.

Figure 3.38
Manhattan Island

Legend:
- **MM** Midtown Manhattan skyscrapers
- **DM** Downtown Manhattan skyscrapers
- Piers/docks
- Tunnels (road/rail)
- Bridges

Buildings:
- ☐1 Empire State
- ☐2 Chrysler
- ☐3 United Nations
- ☐4 Wall Street

Map labels: Henry Hudson Bridge, 219th street, George Washington Bridge, Bronx, Harlem, MANHATTAN ISLAND, New Jersey, 110th street, Triborough Bridge, 0 1km, Central Park, Lincoln Tunnel, 34th street, MM 1 2 3, Queensboro Bridge, Queens, 1st street, Queens-Midtown Tunnel, Holland Tunnel, Chinatown, Williamsburg Bridge, DM 4, Manhattan Bridge, Brooklyn-Battery Tunnel, Brooklyn Bridge, Brooklyn, Hudson River, East River, N

Figure 3.39
Aerial view of Manhattan Island

Photo labels: N, NEW JERSEY, CENTRAL PARK, MM, 1, 2, 3, HUDSON RIVER, DM, 4, EAST RIVER

Problems resulting from the growth of New York

New York had become, by the middle of the twentieth century, the world's largest city (estimates in 2007 suggested it had fallen to third). Its size exaggerated problems that were shared by most other large cities in developed countries. This case study gives a negative view of New York because it concentrates on **problems** rather than on seeing what **solutions** have been attempted and with what success.

High cost of land In the central areas, especially in the CBD districts of Middle and Downtown Manhattan, this has led to skyscraper development (Figure 3.40). High land values do mean, however, that only highly successful firms can locate here (e.g. large banks and giant oil-corporations). Many smaller companies have been forced out and even some of the larger ones are seeking less expensive and environmentally more attractive sites.

Urban decay Many areas of inner city housing, such as Harlem, were built, as in Britain, in the late nineteenth century. The tenement blocks, flats and terraced housing suffered through years of neglect (Figure 3.41). Some of the worst areas became 'ghettos' with few properties having modern amenities such as hot water, bathroom and WC. Most people rented their homes and if they could not afford the high rent, their houses became empty and vandalised. The ghettos became home for the very poorest families – ethnic minorities, the unskilled and the unemployed. They experienced overcrowding both in terms of the number of houses per square kilometre and the number of people per house.

Immigrants America has always welcomed large numbers of immigrants but has not always been able to offer them equal opportunities. Over a period of time, many immigrants find jobs and, as their wealth increases, are able to move away from the poorest areas, leaving them vacant for the next wave of immigrants. This process, known as **centrifugal movement**, is illustrated in Figure 3.42. Immigrants and poor families who cannot find work or only earn low wages may find themselves trapped in the so-called **vicious circle of poverty** (Figure 3.43). In 1990, estimates suggested that there were over 50 000 homeless people in New York, most of them in Manhattan.

Figure 3.41
Poor-quality housing in Manhattan

Figure 3.40
The Empire State Building

Figure 3.42
Centrifugal movement of immigrants

Early 19th century	Late 19th century	Early 20th century	End of 20th century

CBD

Inner city or twilight/blight zone – very low income or on relief

Inner suburbs – below average income

Outer suburbs – above average income

Suburbanised villages – high income

Applied to New York:

A Established inhabitants of north-west European descent (British, German, French)

B Immigrants and their descendants from poorer European countries (Irish, Italian, Eastern European)

C Blacks from south-east of the USA and immigrants from the Caribbean (Puerto Rico)

D Recent immigrants, including refugees

1 Harlem (tenement blocks and flats)

2 Bronx and Brooklyn (apartment blocks and flats)

3 Queens, Yonkers, western Long Island (detached)

4 Eastern Long Island and the states of New York, New Jersey and Connecticut (detached)

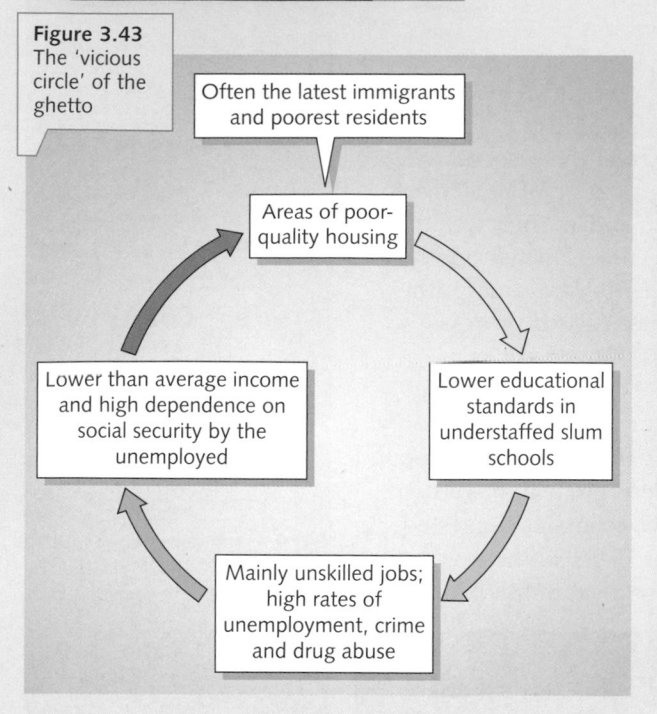

Figure 3.43
The 'vicious circle' of the ghetto

- Often the latest immigrants and poorest residents
- Areas of poor-quality housing
- Lower educational standards in understaffed slum schools
- Mainly unskilled jobs; high rates of unemployment, crime and drug abuse
- Lower than average income and high dependence on social security by the unemployed

Figure 3.44
Graffiti on the New York subway

Figure 3.45
Canyon under the Chrysler building

Traffic congestion Each morning up to 2 million commuters (people living outside Manhattan but working there) travel into the area. At night the direction of movement is reversed. About 75 per cent of commuters travel by public transport, mainly the underground, or 'subway' (Figure 3.44). Roads become blocked with cars, buses and yellow taxis. Although the rivers that surround Manhattan made it an ideal site for the first settlers, they now add to the congestion, especially where they are crossed by bridges, tunnels and ferries (Figure 3.38).

Unemployment During the 1980s up to 1.5 million New Yorkers (one in every seven of the workforce) were unemployed. Unemployment was mainly due to a decline in the port, in associated port industries and, especially, in the clothing industry. New industries tend to be high-tech and in finance – jobs that require skills not possessed by many of the unemployed.

Crime By the late 1980s, crime had become, arguably, the greatest single problem. Street violence, subway muggings, drug-related crimes and murder (on average, one every five hours) had turned certain parts of

Manhattan into 'no-go' areas. Ethnic groups tended to congregate into their own small communities, and racial tension was high.

Pollution New York has a severe refuse collection and disposal problem. Vehicles cause air and noise pollution, while run-down houses and graffiti cause visual pollution.

Water supply Water has to be pumped a distance of almost 200 km.

Climate New York becomes so hot in summer that many of the more wealthy residents have a second home in order to escape from the high temperatures and humidity. Winters, in contrast, can be extremely cold with blizzards (which add to traffic congestion). The tall skyscrapers create 'canyons' down which strong winds may funnel and where the sun may never shine (Figure 3.45).

In the next chapter we look at attempted solutions to urban problems. It should be noted here that during his eight years in office (his term finished in early 2002), Mayor

Guiliano of New York brought major changes to the streets of New York. He vastly increased the number of policemen on the streets (successfully reducing crime), created jobs by improving houses and tidying up the streets, and reduced the number of homeless.

Settlement and urban growth

Key Words and Terms

a You should know the meaning of the following terms:
- function of a settlement • settlement hierarchy • sphere of influence
- threshold population • range of goods/services • land use model
- land values • functional zones of a city • central business district
- twilight zone • old inner city • suburbia • rural–urban fringe
- urban sprawl • urban decay • centrifugal movement.

b You should know the difference between:
- site and situation of a settlement
- dispersed, nucleated and linear settlements
- Burgess and Hoyt urban land use models
- high-density and low-density housing
- urban redevelopment and urban renewal
- brownfield and greenfield sites.

Key Ideas

You should know and understand the following:
- Physical and economic factors affect the location, shape and growth of settlements.
- Many site factors that were important in the past are now no longer important.
- Settlement patterns can be dispersed, nucleated or linear.
- Settlements usually develop more than one main function.
- The function(s) of a settlement may change over a period of time.
- The hierarchy of a settlement is linked to population size, the range and number of services, and the sphere of influence.
- The provision of goods and services differs in settlements of different sizes.
- There is a recognisable pattern of land use in urban areas in MEDCs.
- Settlements in MEDCs usually show a common pattern of distinct zones.
- Urban growth in MEDCs can create numerous social, economic and environmental problems.

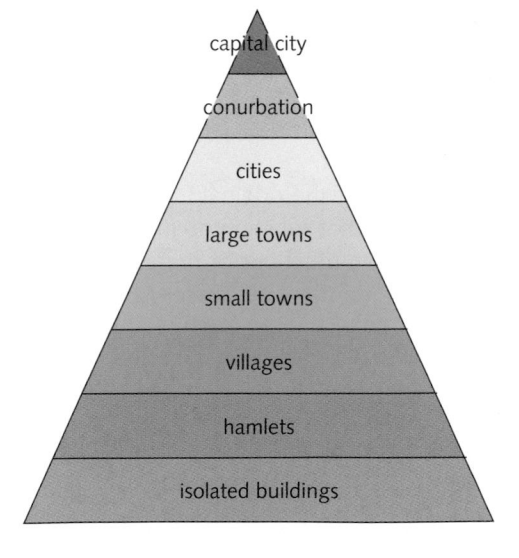

Skills, Theories and Models

- Interpret and use OS maps
- Interpret and use maps at a range of scales
- Analyse and use aerial photos
- Interpret and use graphs
- Interpret and use diagrams
- Use census data
- Understand settlement hierarchies
- Recognise the distinctive character of places
- Interpret and use land use models
- The Burgess and Hoyt models

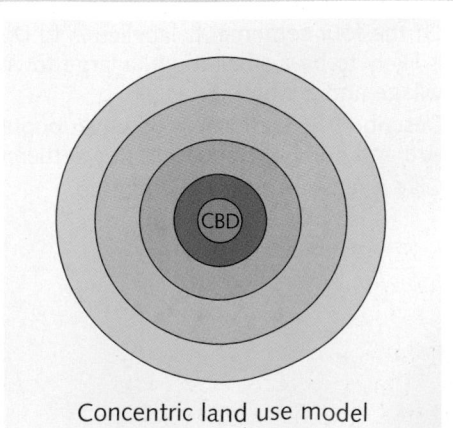

Concentric land use model

1 (Pages 34 and 35)

Imagine that a group of settlers has sailed up the river shown in Figure 3.5. As their leader you have to choose the best site for a village. Your scouts have reported good possibilities at A, B, C, D and E.

a Copy the table. It lists several important factors that you will have to consider before choosing the best site.

i) Complete the table for each possible site. Give a score of 1 to 5 for each factor for each site.

1 = Very good; 2 = Good; 3 = Average; 4 = Poor; 5 = Very poor (5)

ii) State which you consider to be the best site (the one with the lowest score). Give reasons for your answer. (5)

b i) With help from the map, write a paragraph to describe the location of the chosen site. (4)

ii) What might be the main problems of living at this site? (2)

Resource	Site A	Site B	Site C	Site D	Site E
Water					
Crop land					
Grazing land					
Fuel					
Building materials					
Defence					
Flat land that does not flood					
Total					

2 (Pages 36 and 37)

a Which of A, B and C is an example of:
- a dispersed settlement
- a nucleated settlement
- a linear settlement? (2)

b Give a definition for each of the settlements named in part (**a**). (3)

c i) What was the original function of Chester? (1)

ii) Why is that function no longer important? (1)

iii) Give five functions of present-day Chester. (5)

d i) What are the main present-day functions of your nearest large town or city? (3)

ii) How have these functions changed over a period of time? (3)

3 (Page 38)

Refer to the graph.

a i) What is the population size of settlement A? (1)

ii) How many services does settlement A provide? (1)

iii) Of the four settlements labelled A to D, which is likely to be a small town, a large town, a village and a city? (3)

b i) Describe the relationship between population size and number of services in a settlement. (1)

ii) Give a reason for this relationship. (1)

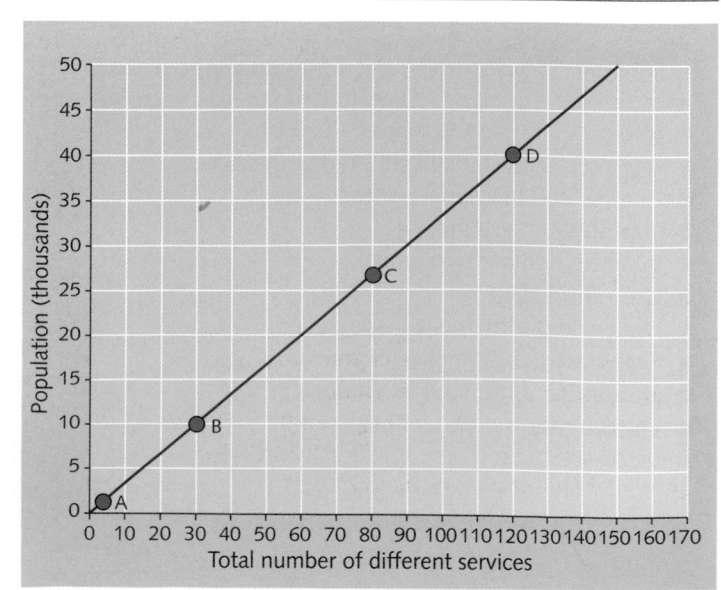

4 *(Pages 38 and 39)*

a i) Write out this list of settlements in order of size from largest to smallest:

town – village – city – conurbation – hamlet *(2)*

ii) Name two services that you might find in a village. *(1)*

iii) Name two services that you might find in a town, but **not** in a village. *(1)*

b Why does a village have fewer services than a town? *(4)*

c Complete the following sentences by using **one** of the terms in brackets.

i) Where something is organised in order of importance it is called a (service centre/hierarchy).

ii) The area from which people travel for a service is called a (market area/hierarchy).

iii) The minimum number of people in a market area needed to support a service is called the (range/threshold population).

iv) The market area is also known as the (range/sphere of influence). *(4)*

d The map below shows where people come from to use shops in towns A and B.

i) Copy the map and complete the market area for each town. *(2)*

ii) Why do market areas overlap? *(2)*

e i) What is the greatest range, in kilometres, of shoppers to:

• town A
• town B? *(2)*

ii) If a hypermarket was to be built at Y, how would this affect the market areas of towns A and B? *(2)*

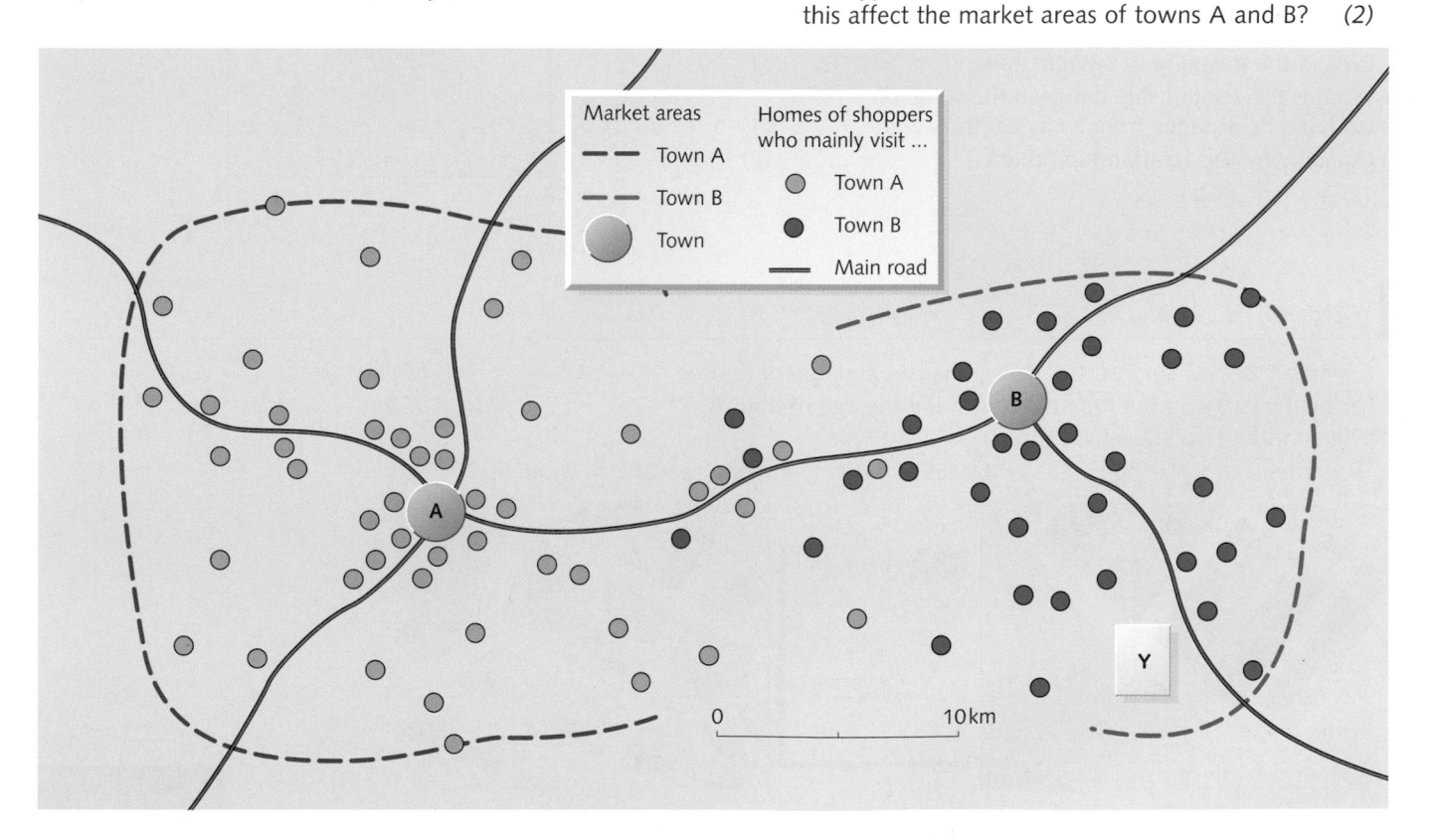

5 *(Page 41)*

An Ordnance Survey map shows some of the services found in settlements. Locate on Figure 3.18 the ten settlements of Bilton, Wyton, Hedon, Preston, Burstwick, Thorngumbald, Elstronwick, Keyingham, Lelley and Burton Pidsea.

a Draw a table and give each settlement one point for every church, chapel, school, public house, post office, hotel, public telephone and sports centre shown on the OS map. *(10)*

b i) Rank the settlements in order, putting the one with most services (points) first and the one with fewest services (points) last. *(5)*

ii) Do you think there is a very close, close or not very close relationship between the number of services and the size of the ten settlements? *(1)*

iii) What other services may these settlements have that are not shown on the Ordnance Survey map? *(2)*

6 *(Pages 42 and 43)*

a Complete the key to diagram A by adding each of the following statements to the correct letter:
  - oldest, poorest-quality housing intermixed with industry
  - large shops and office blocks
  - newest and most expensive housing
  - traditionally the centre of light manufacturing
  - mainly inter-war, medium-cost housing. *(5)*

b Diagram B shows land use zones and land values in a city in the developed world.
  i) Which of the five zones is the CBD? *(1)*
  ii) What do the letters CBD stand for? *(1)*
  iii) Name two kinds of land use found in the CBD. *(1)*
  iv) Give two reasons why there is little open space in zone B. *(2)*
  v) Which zone is most likely to have large houses with *gardens*? *(1)*

c i) Explain the meaning of *twilight zone*. *(1)*
  ii) What is the relationship between the value of land and its distance from a city centre? *(1)*
  iii) Explain why this relationship occurs. *(2)*

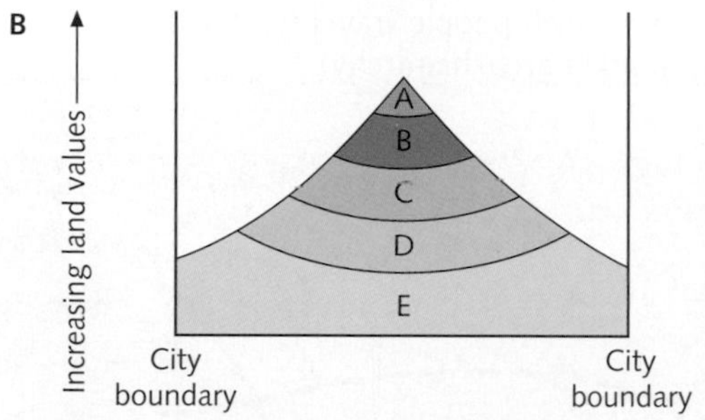

7 *(Pages 44, 46, 48 and 49)*

a i) For map 1 below, suggest the type of houses and age of houses. *(2)*
  ii) For map 2, suggest the type of houses and the age of houses. *(2)*
  iii) Which of the two areas has more:
  - gardens
  - car parking
  - places of work
  - open space
  - residents? *(5)*

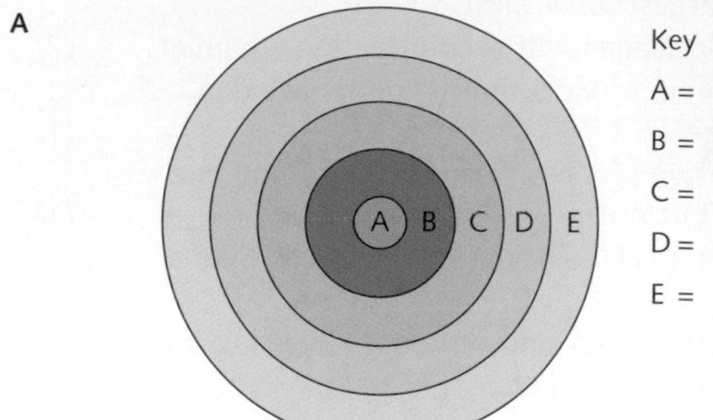

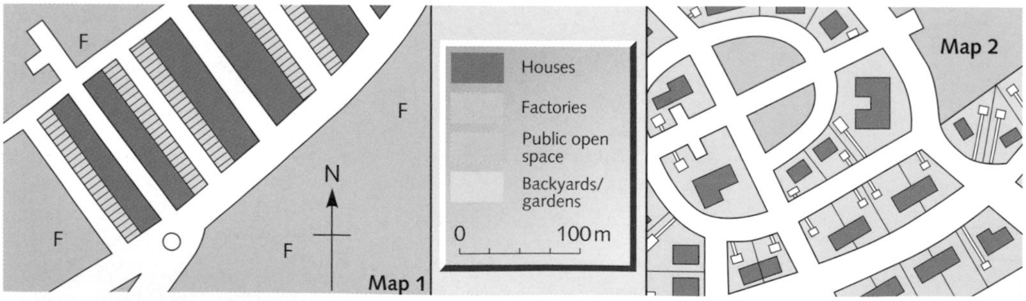

Key:
- Houses
- Factories
- Public open space
- Backyards/gardens

0   100 m

Map 1   Map 2

b Make a copy of, and then complete, the table for residential areas A and C (or any two of the four) described on pages 48 and 49. *(8)*

c For the two areas you described in part (**b**), give one reason for the differences in:
  - type, design and age of housing
  - road pattern
  - land use
  - socio-economic groups
  - household amenities
  - quality of the environment. *(6)*

	First area	Second area
Named example		
Location in city		
Type, appearance, age of housing		
2 main types of land use		
Main type of tenure (ownership)		
Road pattern		
2 main types of socio-economic group		

8 *(Pages 44, 46, 48 and 49)*

a i) Which set of statistics below, A, B or C, is most likely to represent the area located at 1 on the map? Give three reasons for your answer. *(1 + 3)*

ii) Which set of statistics, A, B or C, is most likely to represent the area located at 3 on the map? Give three reasons for your answer. *(1 + 3)*

b i) Give two advantages of living in area 1. *(2)*

ii) Give two disadvantages of living in area 1. *(2)*

iii) Give two advantages of living in area 3. *(2)*

iv) Give two disadvantages of living in area 3. *(2)*

c How may changes in family wealth, family size or the age of the family affect the type of area where people might live? *(3)*

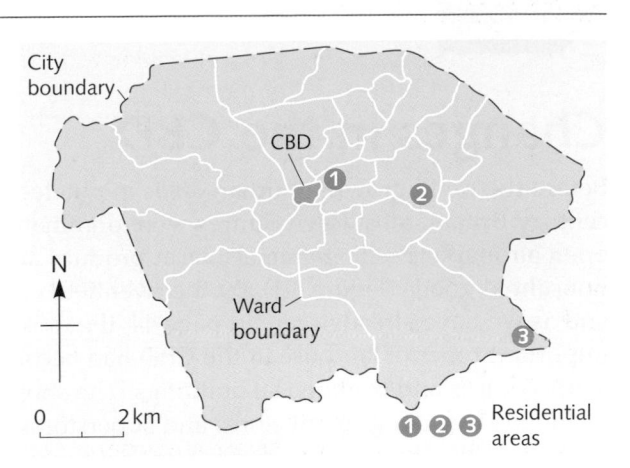

Residential areas	Housing tenure			Housing density		Housing quality
	% owner occupied	% owned by the council	% rented	Households wih over 1.5 persons per room	Households with fewer than 0.5 persons per room	% households which share or lack a WC
Set of statistics A	98	0	2	1	84	0
Set of statistics B	24	34	42	19	35	19
Set of statistics C	62	14	24	8	14	10

9 *(Pages 44 to 49)*

a Make a large copy of the table. Complete it by describing the traditional types of land use found in the CBD, in old inner cities, in suburbia and at the rural–urban fringe. *(5 × 4 = 20)*

b Give reasons for changes in land use from the centre of a city to the edge of a city. *(5)*

Land use / Zone	CBD	Old inner city	Suburbia	Rural–urban fringe
Housing				
Shops				
Industry				
Open space				
Transport				

10 *(Pages 50 to 52)*

a i) What type of buildings are found in both Downtown Manhattan and Midtown Manhattan? *(1)*

ii) Describe the road pattern that covers most of Manhattan Island. *(2)*

iii) Why was Manhattan Island an advantage to early settlers but a disadvantage to present-day New Yorkers? *(2)*

b The star diagram lists ten problems faced by people living in a large city in a developed country. Describe six of these problems with reference to either New York or any other developed city that you have studied. *(6 × 2 = 12)*

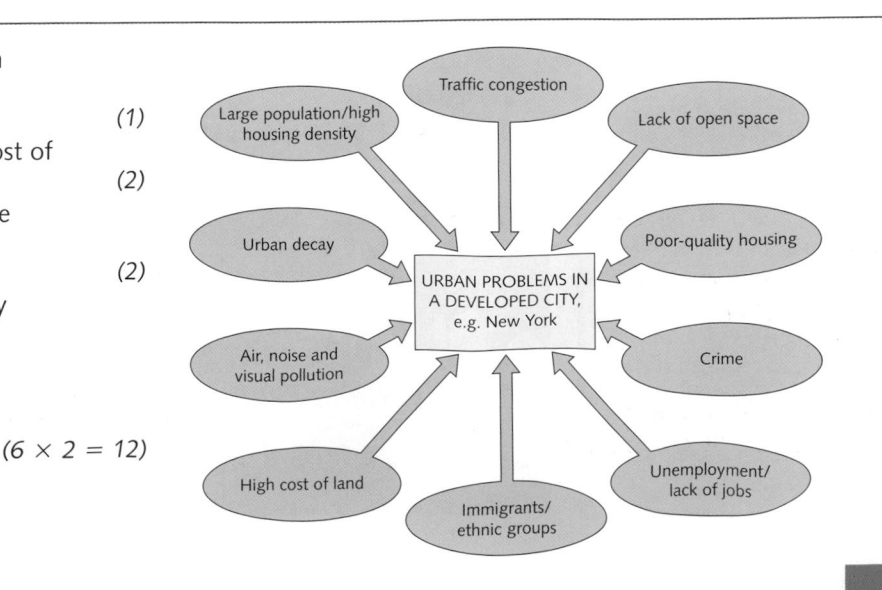

Urban change

Changes in the CBD

Before the rapid growth of urban areas in nineteenth-century Britain, most town centres were dominated by open air markets selling animals, farm produce and household goods (Figure 4.1). By the twentieth century, and as we have already seen on page 44, the most important types of land use in the CBD had become shops, offices and commercial buildings. The shops included large department stores and superstores as well as those selling comparison goods (e.g. clothes and shoes), specialist items (e.g. jewellery and electrical goods) and food on a relatively small scale (bakers, grocers, butchers and fishmongers). Commercial buildings included banks and building societies. Offices located here for ease of access, proximity to banks and, sometimes, for prestige.

City centres also came to include places of entertainment (theatres and cinemas), culture (museums and art galleries) and public buildings (town or city hall). The CBD, because it was most accessible for everyone living in and around the town or city, became the most frequently visited and, consequently, the busiest (see Figure 3.24). As a result:

* many of the older properties were in need of improvement and modernisation
* the narrow streets, built before the invention of the car, bus and delivery lorry, were often congested
* the large volume of traffic caused air pollution and was a danger to the health and safety of shoppers and other pedestrians

Figure 4.1
The farmers' market in Newark, Nottinghamshire

* there was insufficient space for car parks, and buses found it difficult to keep to time
* increasingly taller buildings were constructed to try to offset the high rates and rents that resulted from the high land values
* larger shops moved out and re-located on less congested sites in either old inner city areas (e.g. DIY, furniture and carpet shops) or on the rural–urban fringe (e.g. electrical shops and hypermarkets, pages 62 and 63)
* smaller shops (butchers and grocers) were forced to close as the area became increasingly expensive
* there was a need for modern, carefully planned city centres that considered social, economic and environmental needs.

During the later part of the twentieth century and continuing into the twenty-first, major changes have occurred in most city centres. Some of the reasons for these changes are given in Figure 4.2.

* To reduce traffic congestion and pollution

* To create a safer and a more pleasant environment

* To satisfy the demand for more leisure amenities, especially during the evening

* To segregate shoppers and traffic

Improved technology (fax, email, internet) has encouraged home shopping (internet, mail order) and reduced the need for offices to be centrally located (online banking, easy movement of information)

* To revitalise CBDs, many of which had suffered economically from developments on the rural–urban fringe

Figure 4.2
The need for change in the city centre

Recent changes

Many city centres have had to undergo several changes in an attempt to reduce the loss of shops, hotels and offices to out-of-town locations.

The first major change to occur in most city centres was the creation of **pedestrianised** zones (Figure 4.3). Pedestrianised zones are either traffic-free or allow only limited access to delivery vehicles and public transport. The advantage to shoppers and city centre workers has been a reduction in accidents and traffic pollution (air, noise and visual) and an improvement in safety and in the environment.

Later came the development of **shopping malls** (Figure 4.4). Shopping malls, which are also pedestrianised, are built under cover so that shoppers are protected from the weather and have shorter distances to travel between shops. Shopping malls allow shoppers to compare styles and prices and to stay warm and dry.

More recently has come an increased demand for:
- a more relaxing atmosphere within the city centre
- an increase in leisure amenities, especially in the evenings.

This is being achieved by:
- providing sitting areas, erecting hanging baskets, planting flowerbeds and shrubs and adding small areas of grass
- providing cafés with outside tables, small restaurants and theme bars
- employing more staff to ensure that the place is clean and undamaged.

There has also been an increase in travel agents and estate agents as well as in specialist shops selling clothes and mobile phones, and in opticians.

Figure 4.3
Pedestrianisation of the city centre: Bath

Figure 4.4
A shopping mall: Queensgate Shopping Centre, Peterborough

Figure 4.5
A modern city centre: Cardiff

Changes in old inner city areas

We have already seen on pages 44 and 45 what the traditional land use in inner cities used to be and how, during the 1960s, large areas were bulldozed and rebuilt (**urban redevelopment**) or, later, existing areas were improved (**urban regeneration**). During the 1980s, the government set up **Urban Development Corporations** (**UDCs**) in an attempt to regenerate those inner city areas that had large amounts of derelict and unused land and buildings. UDCs had the power to acquire, reclaim and use land, to convert buildings and to promote industrial, housing and community developments.

The first of 13 eventual UDCs in England and Wales were in the London Docklands (LDDC), which was to include Canary Wharf, and in Merseyside (MDC) with the redevelopment of Liverpool's river frontage (Figure 9.16). Although the last UDC was wound up in 2000, problems remained in various inner city areas.

Regeneration in east London and the 2012 Olympics

The prestigious Canary Wharf complex was part of the London Docklands redevelopment that also saw major improvements in housing, job opportunities, transport links, services and what had been, until then, a rundown environment in the borough of Tower Hamlets.

Yet, as the 2001 census showed, not all of London's East End benefited from the achievements of the LDDC. Data from that census showed that Canning Town, in the adjacent borough of Newham (Figure 4.6), was, at that time, the poorest and most deprived area in the country not only for its poor quality of housing but also for the number of people unemployed, lacking in educational qualifications and job skills or having a limiting illness or disability. It was this urgent need to regenerate places such as Canning Town and Stratford that became a centrepiece in London's bid for the 2012 Olympic Games.

Figure 4.6
Derelict land in the east of London

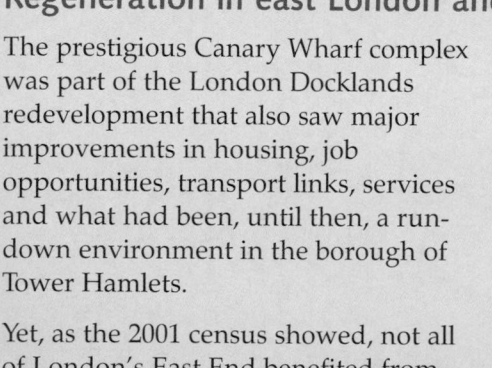

Figure 4.8
London 2012

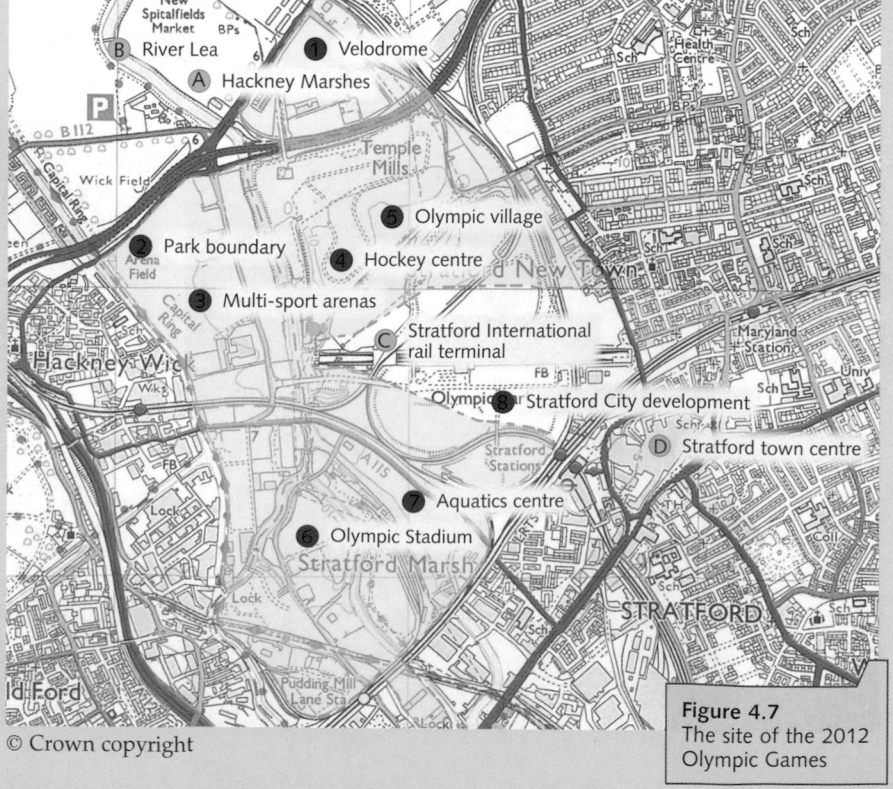

© Crown copyright

Figure 4.7
The site of the 2012 Olympic Games

Why was London chosen?

London's bid for the 2012 Olympic Games was not based solely on its plans to use the event as a way of regenerating a deprived inner city area but also because:

- the plans would satisfy the Olympic criteria that regeneration schemes must have a sustainable impact that will last long beyond the period of the Games themselves
- the long-term benefits would outweigh the total costs
- London is already a global city with one of the world's most culturally and ethnically mixed populations (page 25)
- children would be likely to benefit the most, helping, through sport, to link the nations of the world.

What was the site like in 2005?

The chosen site was along the River Lea. The land was a mixture of industrial estates with many of the firms in the service sector (page 92); university halls of residence; low-cost housing; and large tracts of waste land and derelict buildings (Figure 4.6). However, it had the advantage of being near to Stratford with its nine surface and underground rail links and, after 2009, an international station on the high-speed Channel Tunnel rail link, bringing it within two hours of Brussels and Paris.

What will it be like in, and beyond, 2012?

Figure 4.9 shows, as a computer image, the location of the Olympic Village, the main Olympic Stadium and various specialised sporting venues. The view is looking south, unlike the map in Figure 4.7 which, as is customary, has north at the top. The Olympic Village will host 17 000 athletes and officials. After the Olympics, the Village will be converted into 3500 mainly affordable homes while, on the overall site, a further 9000 new homes, half of which will be also be affordable, will be built. The area will be left with its improved transport system together with a new primary health care centre and an academy school. Some of the sporting facilities, such as the multi-sports arena at Hackney Wick, will either be dismantled or relocated to leave a large urban park alongside the River Lea. The river itself will be cleaned and its resultant wildlife protected.

Figure 4.9
The perceived Olympic Village and stadiums

Changes at the rural–urban fringe

We have also already seen how towns and cities grew rapidly during the 1920s and 1930s, a process referred to as **urban sprawl** (page 46). In an attempt to control urban growth, **green belts** were created by Act of Parliament in 1947. A green belt was defined as an area of land around an urban area where the development of houses and other buildings was to be severely restricted and where the open character of the countryside was to be preserved for farming and recreation. Planning permission was normally not meant to be granted for development proposals but in reality planners often came under considerable pressure to release land for more housing, jobs and roads. Most large British cities are surrounded by a green belt.

By the end of the twentieth century, there was increasing competition for land at the rural–urban fringe (Figure 3.33). The main reasons are that at the fringe there is:
- cheaper land (Figure 3.20)
- less traffic congestion and pollution
- a better road infrastructure that gives easier access to surrounding settlements and other urban areas
- a pleasanter environment with more open space.

Places at the rural–urban fringe that have not yet been built upon are known as **greenfield sites** (Figure 4.10). These sites, still used for farming or recreation, are under constant threat for:
- housing development as urban sprawl continues and nearby villages become suburbanised (pages 64 and 65)
- science and business parks with their mainly high-tech firms (page 141)
- retail parks and regional shopping centres with hypermarkets, superstores, DIY stores and discount warehouses (Figure 4.11) – the MetroCentre section (opposite) gives location factors explaining why large shopping centres are attracted to edge-of-city locations
- office development now that modern technology and easier access to, and the transfer of, information allows firms a freer (footloose) choice of location (page 137)
- hotels and conference centres which, like science parks, often stand in several hectares of landscaped grounds
- road development schemes, including motorways and urban by-passes
- sewage works and landfill sites for urban waste
- recreational areas such as country parks, playing fields and new sports stadiums.

Figure 4.10
Settlement on the rural–urban fringe of a UK city

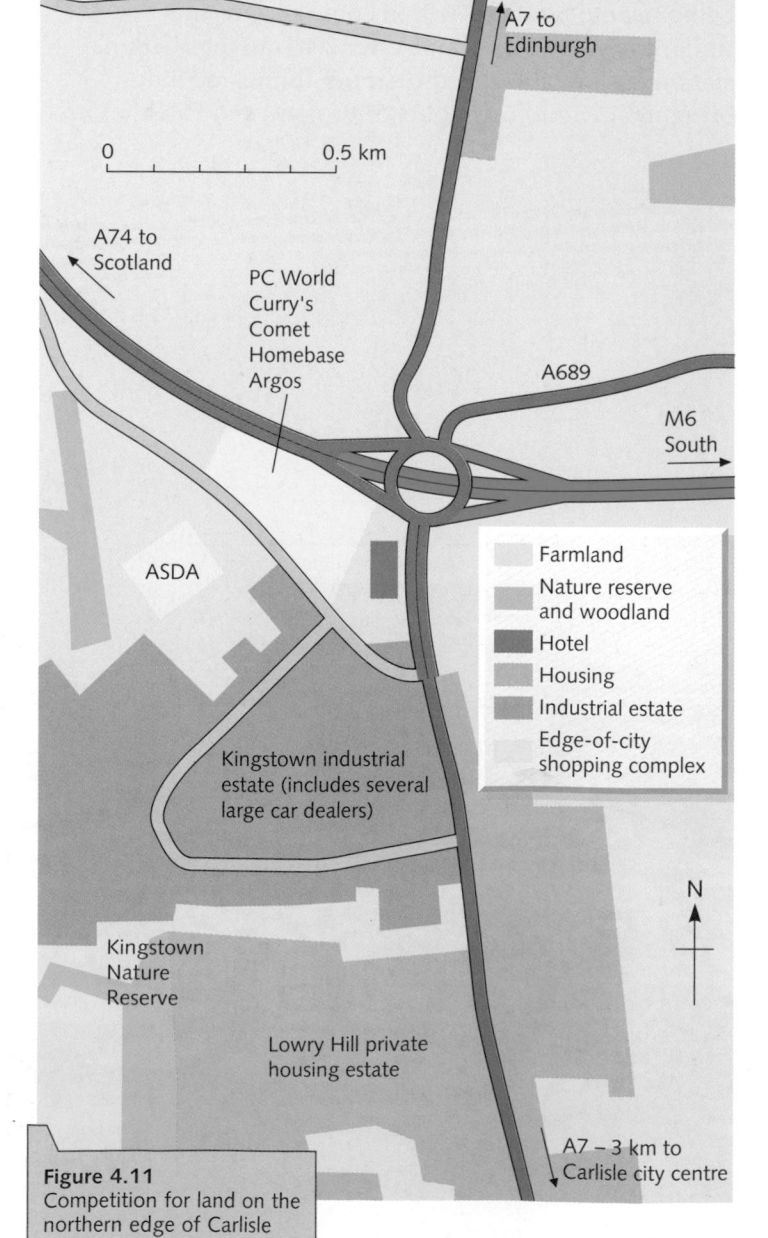

Figure 4.11
Competition for land on the northern edge of Carlisle

Out-of-town shopping centres

MetroCentre in Gateshead

Family shopping has evolved from the corner shop to the supermarket, and from the hypermarket to the 'out-of-town' shopping centre. Sir John Hall, whose brainchild is the MetroCentre, claims that since the 1960s shopping has evolved in three stages around central malls: the Arndale Centres of the 1960s, the Brent Cross and Eldon Square complexes of the late 1970s, and the MetroCentres of the 1980s. The main concept of the MetroCentre is to create a day out for the family, with the emphasis on family shopping and associated leisure activities.

The site for Gateshead's MetroCentre (Figure 4.12) was surveyed in 1980, and at that time received little interest from the city centre 'magnet' (or 'anchor') shops. However, by the time plans were published in 1983, retailing had changed and the success of out-of-town DIY shops led such retailing outlets as Marks & Spencer to reconsider their future policy. Indeed the MetroCentre was Marks & Spencer's first out-of-city location.

The scheme

There is free parking for 10 000 cars, with special facilities for the disabled driver, and new bus and rail stations for the non-motorist. Inside there are over 340 shops and 50 eating places.

Much attention has been paid to creating a pleasant shopping environment (Figure 4.14) – wide, tree-lined malls, air conditioning, one kilometre of glazed roof to let in natural light (supplemented by modern lighting in 'old world' lamps), numerous seats for relaxing, window-boxes, and escalators and lifts for the disabled. A market effect has been created by traders selling goods from decorative street barrows, and there is a wide variety of places to eat in. Leisure is a vital part of the scheme. There is a ten-screen cinema, a crèche for children, a space city for computer and space enthusiasts, a covered fantasy-land with all the attractions of the fair without the worries of the British climate, and a children's village with children's shops. A 150-room hotel, IKEA, DFS and ASDA have also located on the complex.

Figure 4.12
The site of Gateshead's MetroCentre beside a dual-carriageway (foreground), mainline railway and the River Tyne (behind)

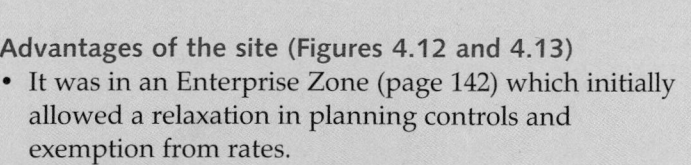

Figure 4.13
Location of the MetroCentre

Figure 4.14
The emphasis inside the MetroCentre is on a pleasant, bright layout based on two-tiered malls

Advantages of the site (Figures 4.12 and 4.13)

- It was in an Enterprise Zone (page 142) which initially allowed a relaxation in planning controls and exemption from rates.
- The area was previously marshland and relatively cheap to buy. The 47 hectare site provided the space for later expansion.
- It is adjacent to the western by-pass (2 km of frontage) which links with the North East's modern road network – essential for an out-of-town location (Figure 4.13).
- 2.6 million people live within one hour's drive.
- It is adjacent to a main railway line, with its own railway station (69 trains a day).

Suburbanised villages

In parts of the developed world there has been a reversal of the movement to large urban areas, with groups of people moving out into surrounding villages. This movement, a process known as **counterurbanisation**, has led to changes in the characteristics of these settlements and to them being called **suburbanised** (because they increasingly adopt some of the features of urban areas).

Some are also known as **commuter** or **dormitory** settlements as many of their residents have to travel to nearby towns and cities for work. Figure 4.15 shows who is likely to move into these villages and how the villages are likely to change. It also lists some of the benefits and problems resulting from suburbanisation.

Figure 4.15 Changes in a suburbanised village

Characteristic	Original village	Suburbanised village
Population structure	An ageing population; mostly born in village; labouring/manual groups	Young/middle-aged married couples with children; very few born in village; wealthy professional/ executive groups; some wealthy retired people
Community/social	Close-knit community (many people were related) have lived here for several generations	Local community swamped by newcomers; division between local people and newcomers; may be deserted during the day (commuters absent)
Housing	Detached, stone-built houses/cottages with slate/thatch roofs; some farms, many over 200 years old; old barns	Many new detached houses, semi-detached houses and bungalows; renovated barns and cottages; expensive estates
Services	Village shop; small junior school; public house; village hall; church	More shops; enlarged school; modern public houses; restaurants; garage
Transport	Bus service (limited); some cars; narrow, winding roads	Good bus service (unless reduced by private car); most families have one or two cars; improved roads
Employment	Farming and other primary activities (forestry, mining); low-paid local jobs	New light industry (high-tech and food processing); good salaries; local shops and transport
Environment	Quiet, relatively pollution-free	Increase in noise and pollution, especially from traffic; loss of farmland/open space

Thurston, Suffolk

Thurston is located 5 km east of Bury St Edmunds. First mentioned in the Domesday Book of 1086, most of its residents worked on farms or in quarries until the coming of the railways in 1846. However, it was not until the 1950s, when people began to move out of London, that the village began to grow rapidly. The Heath (1960s) and Malting Farm (1980s) were sold for housing (Figure 4.16a) and the nearby A14 became a dual-carriageway. By 1994 the village had grown considerably (from 600 residents in 1884 to over 3000 in 2000) and it had become increasingly suburbanised. Figure 4.16b shows that although the smithy and several of the named farms of 1884 had disappeared, the village still retained some of its former services as well as gaining several new ones. The two maps also show changes in both the function of the settlement and in the types of job available.

Figure 4.16 Thurston in 1884 and 1994

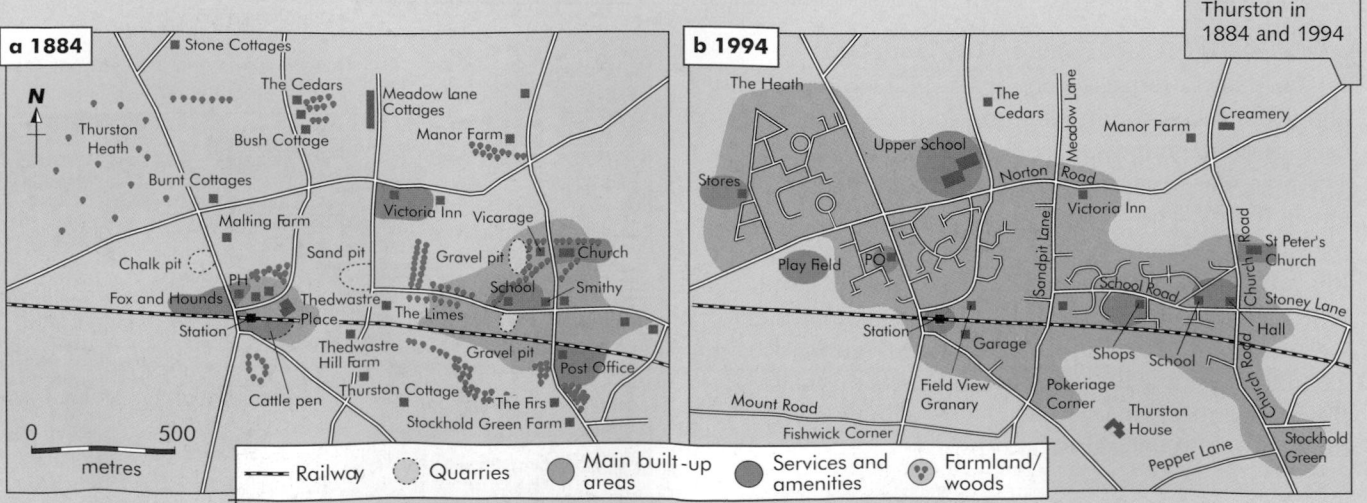

Brownfield and greenfield sites

Britain is short of houses. Estimates suggest that 3 million new homes could be needed by 2028. Assuming that this prediction is correct, the question then is, 'Where will all these new houses be built?' (Figure 4.17). Initially the government announced that:

- 50 per cent would be on **brownfield sites**, i.e. disused and derelict land in existing urban areas
- 50 per cent would be on **greenfield sites**, i.e. in the countryside, including places in green belts (page 62).

After a major public outcry to 'save the countryside', the government revised the figures to 60 per cent brownfield and 40 per cent greenfield. However, the National Data Base shows a mismatch between:

- the south-east of England, where most houses are needed but where brownfield sites are limited
- the Midlands and the North, where most brownfield sites are available but where demand for new housing is less.

Figure 4.18 gives some of the arguments as to why some groups of people would prefer most of the new developments to be on brownfield sites and why other groups favour greenfield sites.

Figure 4.17
Extract from The *Daliy Telegraph*, 17 July 2008

Three million new homes

THE government plans to force local councils to allow development on previously protected land in order to achieve their aim of providing 3 million new homes by 2028. Ministers want 33 000 new houses to be built each year, almost 100 every day, in the South East alone for the next 20 years. Of these a high proportion should either be affordable or social homes. Released documents reveal that the green belts around Oxford and Guildford are to be reviewed and that expansion into London's green belt may also be required.

Despite growing opposition to their recently announced 10 'eco-towns' (low-energy, carbon neutral settlements each with between 5000 and 15 000 homes, one-third of which will be affordable housing), the government is set to push ahead with its large house-building programme regardless of the global credit crises and the fact that several of the country's leading house-builders are in difficulty.

Figure 4.18
Brownfield and greenfield sites

Why build on brownfield sites?
Groups such as the Council for the Protection of Rural England (CPRE) and Friends of the Earth argue that:

- there are already three-quarters of a million unoccupied houses in cities that could be upgraded
- a further 1.3 million could be created by either subdividing large houses or using empty space above shops and offices
- according to the database, 1.3 million homes could be built on vacant and derelict land and another 0.3 million by re-using old industrial and commercial premises
- urban living reduces the need to use the car and maintains services, especially retailing, in city centres.

The government assumes that 80 per cent of the demand for new houses will come from single-parent families who prefer, or need, to live in cities.

Why build on greenfield sites?
Developers claim that:

- most British people want to own their own home, complete with garden, set in a rural, or semi-rural location (unlike most people on the continent who prefer to rent an apartment)
- people are healthier and generally have a better quality of life in rural areas
- at present, for every three people moving into cities (almost 160 000 a year), five move out into the countryside (250 000 a year) – a net loss to urban areas of 90 000 a year
- greenfield sites are cheaper to build on than brownfield sites as they are likely to have lower land values and are less likely to be in need of clearing-up operations (especially if the proposed development is on a former industrial site).

Transport in urban areas

Traffic in general and the car in particular create numerous problems in urban areas. Most families in the more developed countries of Western Europe, North America and Japan own their own car (Figure 4.19). The car gives individual families greater mobility and, because it allows them to travel 'from door to door', gives them greater freedom of movement and choice in where they live, work, shop, are educated and find their recreation. However, it is the widespread use of the car that is bringing traffic in many large cities to a standstill (Figure 4.20) and which is a major cause of urban environmental, economic and social problems (Figure 4.23).

Commuting

Urban traffic problems are often at their worst during the early morning and late afternoons when commuters are travelling to and from work. A **commuter** is a person who lives in a smaller town or village in the area that surrounds a larger town or city, and who travels to that town or city for work. The term can also be used for residents who live in the suburbs of a large town or city, especially London. The increase in car ownership and the improvements in the road network means that commuters can live further from their place of work. This has led to large **commuter hinterlands** (the areas around large cities) where commuters live.

A more recent trend is a 'reversed' flow of commuters. This group includes:
- the less skilled and poorly paid, together with members of ethnic minorities, who live in low-cost inner city housing and who travel long distances to their place of work on edge-of-city industrial and retail parks
- some well-paid people who have moved into regenerated areas near to the city centre but who work on science and business parks on the urban–rural fringe.

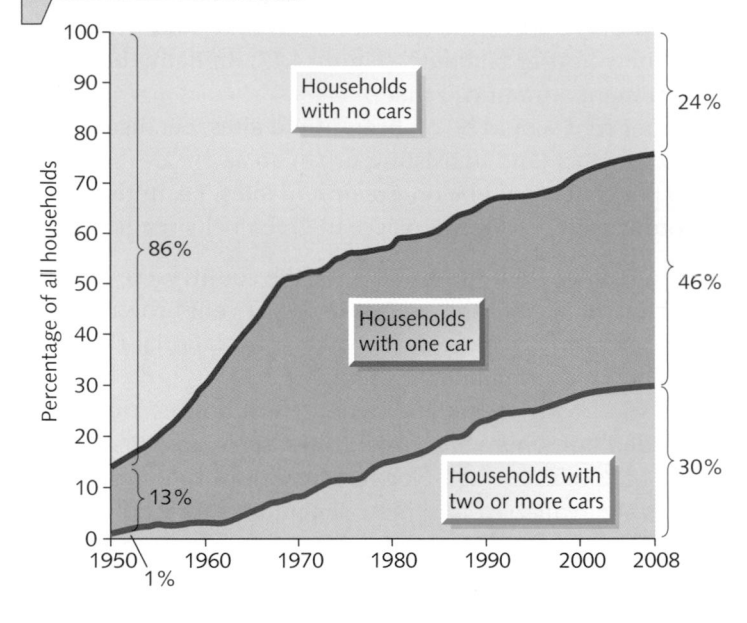

Figure 4.19
Increase in car ownership, 1950–2008

Figure 4.20
Traffic in Los Angeles, California

Figure 4.21
Smog over Los Angeles

	Yesterday's reading	Forecast
Nitrogen dioxide	40 ppb	Very good
Less than 50 = Very good, 50–99 = Good, 100–299 = Poor 300+ = Very poor		
Ozone	28 ppb	Very good
Less than 50 = Very good, 51–89 = Good, 90–179 = Poor 180+ = Very poor		
Sulphur	5 ppb	Very good
Less than 60 = Very good, 60–124 = Good, 125–399 = Poor 400+ = Very poor		

Air quality is calculated in ppb (parts per billion)

London Evening Standard, 9 June 1997

Figure 4.22
Air quality in London, 9 June 1997

Why has traffic in urban areas increased?		• Greater affluence and increased car ownership. Many families own two cars (Figure 4.19). • People commuting to work and travelling to city centres or edge of cities for shopping and entertainment. • Reduction in public transport at expense of private cars. • Increased volume of road freight, e.g. delivery lorries.
What are the damaging effects of increased traffic in urban areas?	• Environmental	• Air pollution from vehicle exhausts – especially in large cities like Los Angeles (Figure 4.21), Athens and Tokyo. Some cities, e.g. London, issue daily air quality forecasts/reports (Figure 4.22). • Noise pollution from cars, lorries and buses. • Visual pollution of motorways (Figure 4.20). • Loss of land for road widening and car parks.
	• Economic	• Congestion, especially at peak times (rush hour). In central London, even though many people travel by public transport, the average speed of traffic in 2000 was the same as it was in 1900, i.e. 18 km/hr. • Time wasted sitting in traffic jams and in gridlock conditions, looking for parking spaces and avoiding unloading lorries. • Cost of building and maintaining roads; delays due to repairs of roads and underground utilities (electricity, gas, water). • Cost of petrol/diesel and the use of a non-renewable resource (oil).
	• Social	• Danger of accidents. • Health problems: increase in stress both to drivers and pedestrians; respiratory illnesses (e.g. asthma) caused by fumes and low-level ozone. • Destruction of property for new or wider roads and car parks. • Damage to foundations caused by traffic vibration. • Overcrowding on commuter trains and buses.
How can transport systems in urban areas be managed to reduce the damaging effects of increased traffic?		• Exclude, reduce or accommodate traffic by schemes such as traffic-free zones, Park and Ride schemes and urban motorways. • Try to reduce pollution, especially from vehicle exhausts. • Improve public transport, e.g. supertrams in Sheffield, Metrolink in Manchester, metros in Tyne and Wear, Singapore and Hong Kong. These rapid transport systems have resulted from improved technology. • Smaller, more affordable and sustainable schemes, e.g. Park and Ride, cycle tracks, traffic calming, speed cameras. • Future schemes?: road tolls, charges to drive into large cities.

Figure 4.23
The impact of traffic on urban areas

Everybody agrees that traffic problems in urban areas need drastic solutions, but people are less likely to agree on what those solutions might be. People have become increasingly concerned about the effects of transport on human health, the quality of the environment, global warming, time wasted sitting in near gridlock conditions, and the use of a non-renewable source of energy (Figure 4.23). At the same time, people wish to travel as safely, quickly and cheaply as possible. Future plans must be sustainable (page 184), i.e. they should improve people's mobility without damaging the environment in which they live.

In the UK, the DoE (Department of the Environment) has produced what it calls a 'New Deal for Transport' (Figure 4.24). Various towns and cities in Britain have tried to implement this ideal by introducing a range of traffic schemes. These schemes include 'Park and Ride' where car owners park their cars on the edges of urban areas and travel by public transport to the CBD, supertrams (as in Manchester and Sheffield), by imposing a congestion charge (London), introducing cycle tracks, bus and no-car lanes, and traffic calming measures (speed ramps, rumble strips and road narrowing) in residential areas.

Figure 4.24
The New Deal for Transport in the UK

The New Deal for Transport means:		The ten-year transport plan
• Cleaner air to breathe by tackling traffic fumes	£180 billion	One-third from private sector, two-thirds from public sector
• Thriving town centres by cutting the stranglehold of traffic	£59 billion	On roads, including widening 580 km of roads, e.g. A1, M6, and 100 by-passes
• Quality places to live where people are the priority	£60 billion	For railways, including 25 inner city light railways and a London orbital railway
• Increasing prosperity backed by a modern transport system	£26 billion	For local services
• Easier and safer to walk and cycle	£35 billion	For other schemes
• Revitalised towns and cities through better town planning.		

Lille – an integrated traffic system

Lille, with a population of just under 1 million, is the fourth largest city in France. It is located near to the Belgian border some 300 km north-east of Paris. Based on large local reserves of coal and smaller deposits of iron ore, Lille became the centre of a major industrial area in the nineteenth century. These resources led to the growth of the steel, engineering and textile industries. However, by the 1950s, as in other old industrial areas of Western Europe, the region around Lille experienced rapid industrial decline. It was left with high levels of unemployment and large areas of derelict land, together with poor housing and inadequate modern services.

During the early 1980s the development of the French TGV high-speed rail network and the proposed Channel Tunnel rail link gave Lille an unrivalled opportunity to exploit its position as a focus for land-based transport. The city's mayor at that time fought against the original plan which recommended that the new rail link should by-pass the city. Instead he persuaded the French railways to build the new Lille-Europe station for the Eurostar trains in the centre of the city, rather than on the outskirts, on a site adjacent to the Lille-Flandres station which was already being used by regional and national trains (Figure 4.25). At the Gare Lille-Europe, a 400 m long window reveals both the TGV and the city centre (compare this with south-east England where local residents did all they could to divert Eurostar trains from urban areas).

The narrow-gauge VAL metro system, centred on the Gare Lille-Flandres, has two lines (Figure 4.25). The first line was opened in 1983. The second line, which has been in operation since 1989, was extended, and in 2000 Lille had the longest automatic metro system in the world. The tiny two-carriage computer-controlled trains run underground through the city centre (85 per cent of the total track) and on elevated tracks in the suburbs. They run at one-minute intervals during rush hours and at up to six-minute intervals at other times of the day. Energy-saving escalators only switch on when passengers step onto them. All metro tickets are single-fare and are issued from machines. The metro has no conductors and no ticket staff – just security cameras! The system safely transported 750 million passengers in its first 15 years.

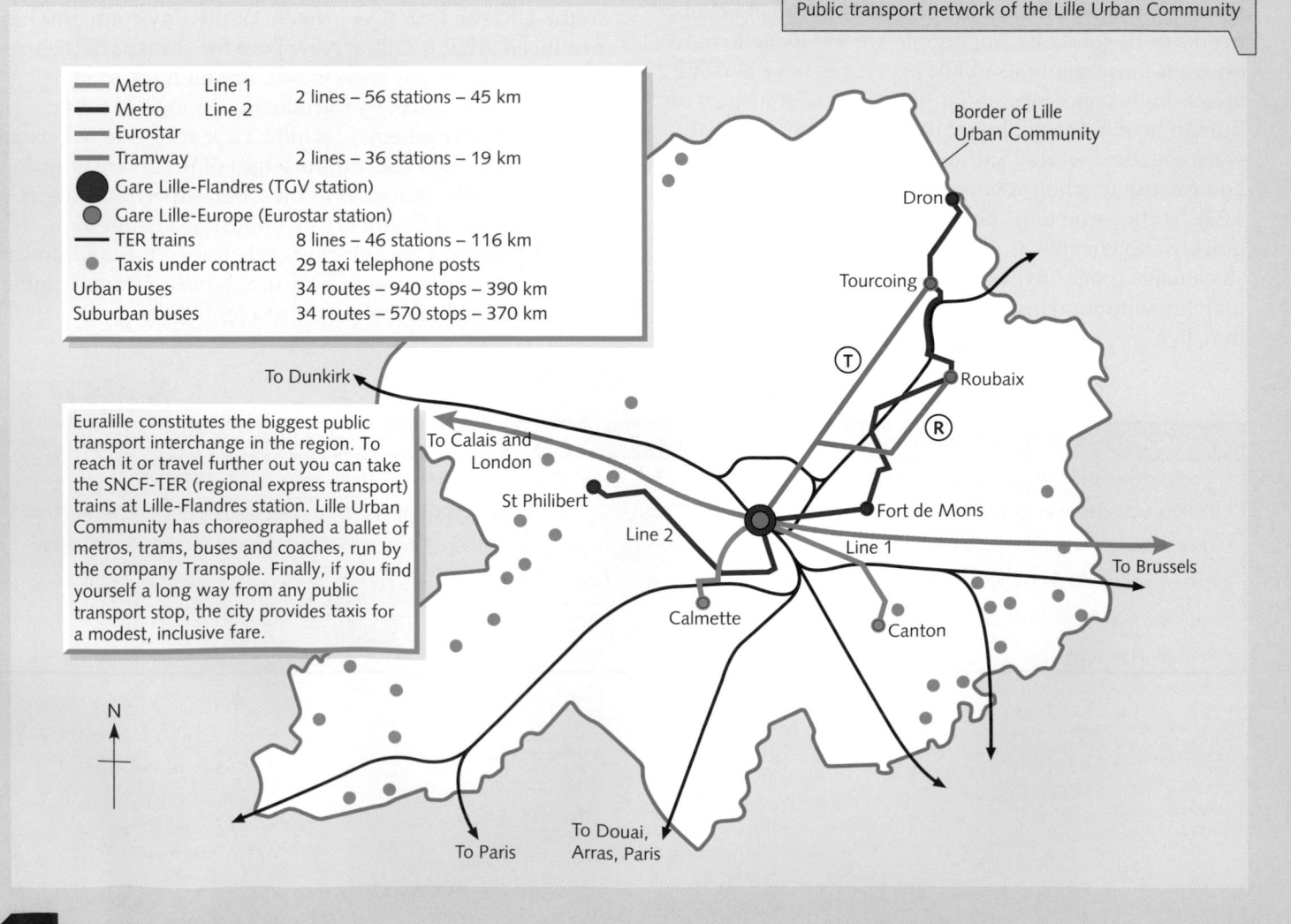

Figure 4.25
Public transport network of the Lille Urban Community

Above the TGV station are the twin business centre towers of the Crédit Lyonnais bank and the Lilleurope office block (Figure 4.26). Next to these, and bordered by an express highway that serves as an inner ring road, is the Euralille complex. Euralille has hotels, a large shopping centre (which includes Carrefour), an exhibition hall, a conference centre and a 5500 seat venue for rock concerts and other spectaculars. Around this complex are six car parks with spaces for over 6100 vehicles (Figure 4.27). Two tram routes, using newly designated coaches, run north-eastwards from the two main railway stations to the towns of Tourcoing and Roubaix (Figure 4.25). Residents also have the option of 34 urban and 34 suburban bus routes and, for those in the suburbs, access to a linking taxi service.

Figure 4.26
The two business centre towers: Crédit Lyonnais (right) and Lilleurope (left)

Figure 4.27
Euralille – an integrated transport system

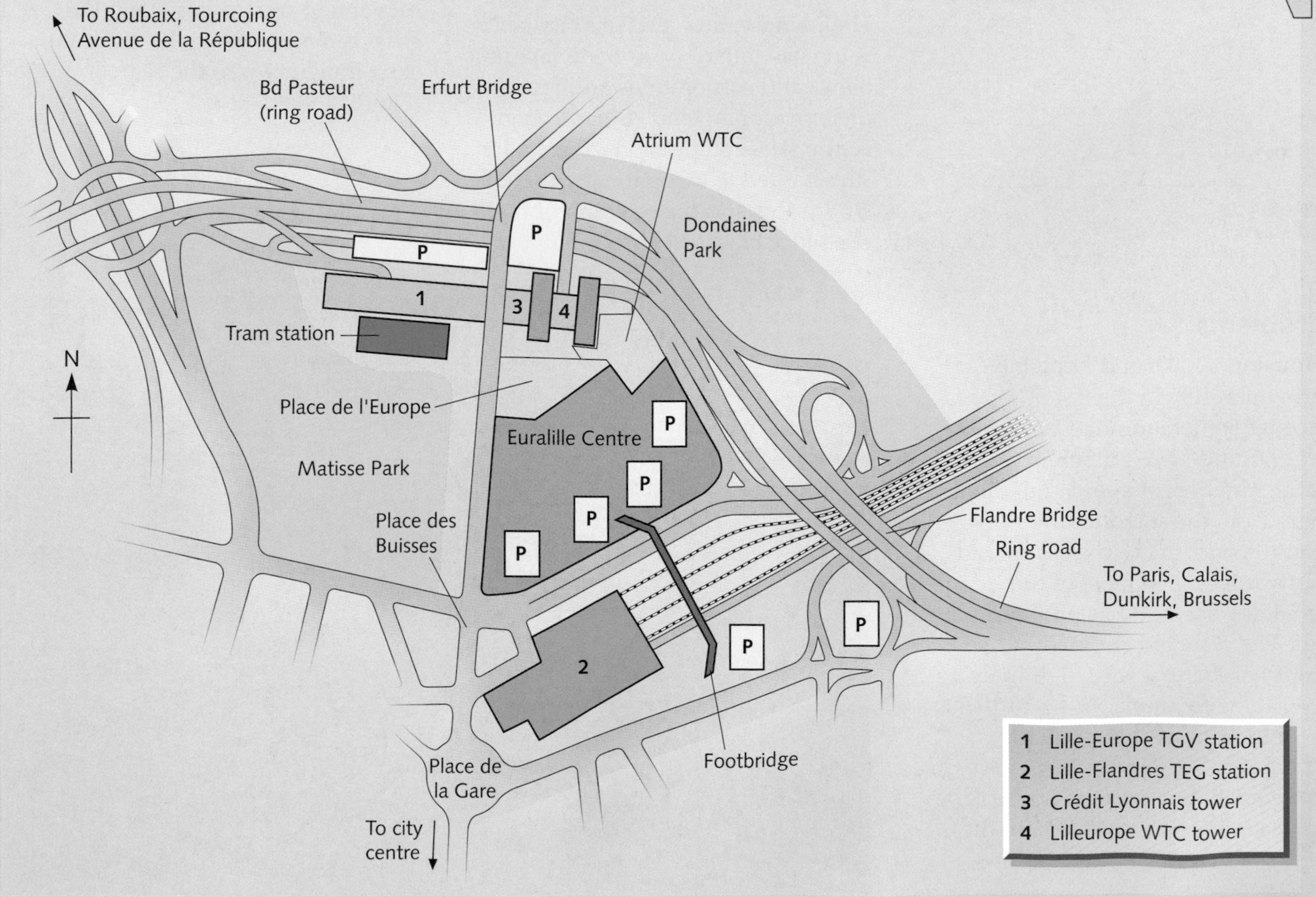

To Roubaix, Tourcoing
Avenue de la République

Bd Pasteur
(ring road)

Erfurt Bridge

Atrium WTC

Dondaines
Park

P

1

Tram station

3 4

Place de l'Europe

Euralille Centre

P

Matisse Park

P

Place des
Buisses

P

P

Flandre Bridge

Ring road

To Paris, Calais,
Dunkirk, Brussels

N

P

2

P

P

Footbridge

Place de
la Gare

To city
centre

1 Lille-Europe TGV station
2 Lille-Flandres TEG station
3 Crédit Lyonnais tower
4 Lilleurope WTC tower

Osaka–Kobe

The twin cities of Osaka–Kobe are located in the Kansai region of the Japanese island of Honshu (Figure 4.28). This region, as elsewhere in Japan, has limited flat land. Osaka and Kobe lie between the steep-sided mountains of central Honshu and Osaka Bay. Kansai, with a population of over 9 million, has become a world leader in education, science, business, technology and industry ... but this success has not been achieved without creating problems.

Figure 4.29
Osaka central commercial district

Figure 4.28
Location of Osaka–Kobe

Problems

Housing Osaka and Kobe have grown rapidly in the last 50 years (the agglomeration had 5.12 million in 1995 and 12.26 million in 2005). As large numbers of people migrated here from the surrounding countryside, the two cities spread outwards so that most of the available flat land has now been used up. The resultant competition for land has led to a very high housing density with, in places, up to 10 000 people living within a square kilometre. The houses themselves are very small. The average house of 4.5 m² is eight times smaller than the average UK house. Many houses consist of only one main room, where the occupants have to cook, eat, work and sleep, and a small bathroom. There is no space for gardens and very little for parks or recreation (Figure 4.29).

High land values Osaka's city centre is the headquarters of many Japanese banks and corporations and is a major shopping and entertainment centre. The competition for prime sites has led to exceptionally high land prices and the building of high-rise office blocks (Figure 4.29).

Transport The unplanned growth of Osaka–Kobe and a rapid increase in road traffic has caused major transport problems. Roads are congested and often in a state of near gridlock. Emissions from vehicle exhausts cause severe air pollution. Movement westwards along the coast is made difficult by mountains extending down to the sea (Figure 4.31).

Figure 4.30
Osaka port

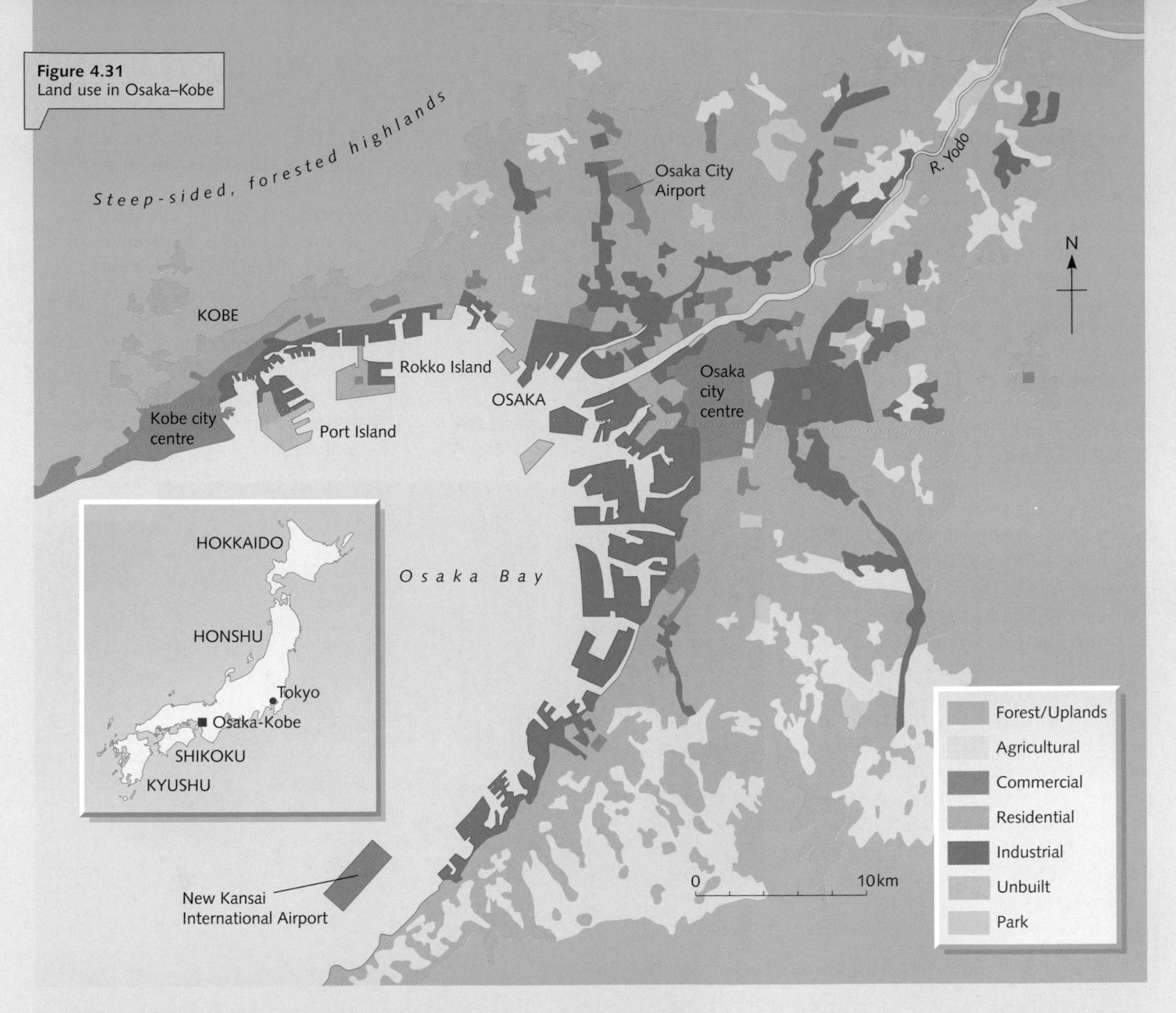

Figure 4.31
Land use in Osaka–Kobe

Steep-sided, forested highlands

Osaka City Airport

R. Yodo

N

KOBE

Rokko Island

OSAKA

Osaka city centre

Kobe city centre

Port Island

HOKKAIDO

Osaka Bay

HONSHU

Tokyo

Osaka-Kobe

SHIKOKU

KYUSHU

New Kansai International Airport

Forest/Uplands

Agricultural

Commercial

Residential

Industrial

Unbuilt

Park

0   10km

Port Up to 1400 ships a day enter the docks at Osaka (Figure 4.30). Many bring in oil and other raw materials. The port also exports large amounts of Japanese manufactured goods (page 194).

Industry Kansai is a major industrial region. Much of the industry is concentrated along Osaka Bay (Figure 4.31) where large companies manufacture incoming raw materials. As Japan has to import over 99 per cent of its oil, coal and iron ore, the port area has large oil refineries, steelworks and other industries processing imported raw materials. Once again, a major problem is a lack of space, for new docks, for enlarging the container port, for storing goods being imported or exported, and land

needed by processing industries. Other industries, many of them small, are located alongside either the River Yodo or the outer city motorways. Even today 90 per cent of Japanese companies are small, often family concerns making parts for larger transnationals (page 144). Most of the components are used by car assembly or electronic and other high-tech companies.

Pollution The large volume of road traffic, smoke from some of the older heavy industries and rubbish incinerated at sea all cause serious air pollution. Rubbish dumped at sea and oil spilled from ships, especially oiltankers, pollutes the water of Osaka Bay.

Physical environment Japan is located at a destructive plate margin (page 264). As the Pacific Plate is pushed under the Eurasian Plate and destroyed, it forms volcanoes and fold mountains. Osaka and Kobe are hemmed in by these mountains which, although not very high, are steep and forested. Destructive margins also experience severe earthquakes, the most recent in Japan destroying much of Kobe in January 1995 (page 270). Japan is also in the path of tropical storms known as typhoons. Typhoons, with heavy rain and hurricane-force winds, can affect Japan 30 times in an average year (page 216).

71

Some solutions

Housing Port Island (Figure 4.32) and Rokko Island have both been recently reclaimed from the sea in Osaka Bay (Figure 4.31). Much of Rokko Island is a new town with housing in the form of flats. Even before these flats were built there was a huge waiting list and the first occupants were those whose names were drawn from a hat. Compared with those in Osaka and Kobe, the houses are spacious (Figure 4.33), the majority having two large and one smaller room. The residential areas are near to schools, shopping centres, hospitals, parks and places of entertainment. Traffic is, where possible, segregated from the housing districts, while a new railway has been built to take commuters to Kobe. The main problem with living on Rokko Island is the break with Japanese tradition: there is an absence of temples and shrines, and women now have to go to work as two salaries are usually needed to pay for the expensive housing.

Port development Land has also been reclaimed from the sea for port extensions. Kobe's new container port, on Port Island, extends 7 km out to sea (Figure 4.32).

Industry Industrial development has taken place at four locations:

1 On land adjacent to Osaka Bay, previously used for the Nippon steelworks.
2 On land reclaimed from Osaka Bay by levelling inland areas behind Osaka, transporting the material and depositing it on the sea-bed.
3 By creating up to 11 'science cities' (some on the newly levelled sites) each with its high-tech industries and research centres. The centres are being financed by leading Japanese corporations in an attempt to maintain their country's global supremacy in this field.
4 Where space permits, alongside major motorways.

Figure 4.33
Inside a modern home

Figure 4.32
Port Island, looking north to Kobe and the forested highlands

Figure 4.34
Kansai International Airport

Transport

- The Shinkansen (wrongly translated into English as 'the bullet train'), provides the most reliable rail service in the world (Figure 4.35). One of three Shinkansen lines passes through Osaka and Kobe. The latest models can travel at speeds of up to 300 km/hr and can reach Tokyo, 553 km to the east, in about $2\frac{1}{2}$ hours. The trains carry 275 million people per year, run at 7 minute intervals, are computer controlled, arrive prompt to the second and have never been involved in a serious accident. The route to the west passes through a series of tunnels cut through the coastal mountains.

- In an attempt to improve road traffic along the northern shore of the Inland Sea, a new road has been built linking Osaka and Kobe with the islands of Awaji and Shikoku (Figure 4.28). The Akashi Bridge, 4 km in length, was the world's largest when it was built. A proposed extension to the road to the southern island of Kyushu would create a 400 km by-pass of the urban-industrial centres on the mainland.

- Kansai International Airport, built on an artificial island in Osaka Bay, opened in 1994 (Figure 4.34). The airport has a main runway of 3500 metres, operates 24 hours a day and is connected to Osaka by a 13 km road and rail causeway. The terminal, the world's largest, can handle over 30 million passengers a year.

Pollution Japan, with money available, has made serious attempts to clean up some of the pollution caused during its rapid industrialisation. The Inland Sea, once a 'dead' sea, now has fish and oyster farms. The new high-tech industries are less polluting than the older, heavier industries. Nuclear energy, despite its risks, causes less air pollution than fossil fuels.

The environment Earthquakes and typhoons cannot be controlled but attempts can be made to reduce their effects. New buildings are designed to withstand earthquakes – as indeed many did in the 1995 Kobe 'quake (page 270). The Akashi Bridge is a fine example of how the Japanese use their technology. The foundations of the bridge have been sunk to a great depth under the sea-bed to resist earthquakes; the foundations and supports have been strengthened to combat the fast sea-currents; and the superstructure has been designed to withstand 130 km/hr typhoon winds.

Figure 4.35
The Shinkansen

Urban change

Key Words and Terms

a You should know the meaning of the following terms:
- central business district (CBD) • retailing
- Urban Development Corporations (UDCs) • rural–urban fringe
- urban sprawl • green belts • suburbanised villages
- counterurbanisation • commuting • integrated traffic system.

b You should know the difference between:
- comparison and convenience goods
- pedestrian precincts and shopping malls
- urban redevelopment and urban regeneration
- brownfield and greenfield sites
- environmental, economic and social factors.

Key Ideas

You should know and understand the following:
- All parts (zones) of urban areas undergo constant change.
- Why there is competition for land in the CBD and at the rural–urban fringe.
- About contemporary solutions to problems in the CBD.
- About contemporary measures/strategies to regenerate inner city areas.
- Why large shopping centres locate on edge-of-city sites.
- The process of counterurbanisation can lead to the suburbanisation of villages.
- About the conflict between brownfield and greenfield sites for future housing development.
- Urban traffic causes adverse environmental, economic and social effects which can only be overcome by careful planning and management.
- About one integrated transport system.

Skills, Theories and Models

- Interpret and use a variety of maps
- Analyse and use photos
- Understand how and why changes happen in places
- Understand the issues resulting from change
- Understand how development can cause conflict
- Appreciate other people's values and attitudes
- Recognise the distinctive character of places

1 *(Pages 58 and 59)*

a i) Using this diagram only, give two reasons why shops locate in the CBD. (2)

ii) What types of shop are likely to be found in the CBD? (2)

iii) What types of land use, other than shops, are likely to be found in a CBD? (2)

b i) Describe the appearance of a shopping mall. (2)

ii) Give three reasons why shopping malls were built. (3)

iii) Describe three changes, other than shopping malls, that have taken place recently in many CBDs. (3)

c i) What are the main problems in the CBD of your nearest town or city? (3)

ii) What has been done to try to reduce these problems? (3)

iii) Which groups of people have benefited from these changes? Explain why. (4)

iv) Which groups of people feel disadvantaged by these changes? Explain why. (4)

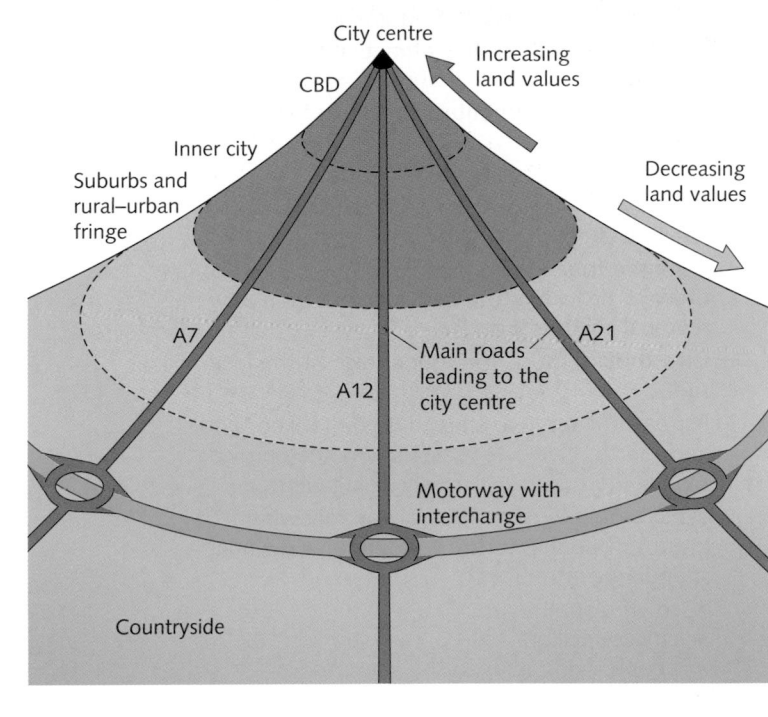

2 *(Pages 60 and 61)*

a Give the map reference for i) the Olympic Village ii) the Olympic Stadium iii) Stratford International rail terminal. (3)

b Why, in 2001, were parts of east London among the poorest and most deprived in the country? (3)

c i) Why was the Lea Valley chosen as London's proposed site for its 2012 Olympic Games bid? (3)

ii) Using the words in the Olympic logo below, explain why London's bid for the 2012 Olympic Games was successful. (5)

d i) Using Figures 4.7 and 4.9, describe the changes that will take place in the Lea Valley by the time the Olympics open. (3)

ii) What further changes will take place in the area after the Games close? (3)

Regeneration

Sustainable development

Costs

Global city

Children

3 *(Pages 62 and 63)*

a i) Where are you likely to find greenfield sites? *(1)*

ii) Why are greenfield sites attractive for urban development? *(3)*

iii) List five likely competitors for land use on greenfield sites. *(5)*

b i) What is a regional shopping centre? *(1)*

ii) What types of shop are likely to be found in a regional shopping centre? *(3)*

c i) Rank the four sites labelled J to M on the map in order of providing the best location for a new regional shopping centre. *(4)*

ii) Give three reasons for your choice of the ideal site. *(3)*

iii) For each of the two sites that you ranked third and fourth, give two reasons why they provide a poor site for a regional shopping centre. *(4)*

iv) What would be the views of the following groups of people to the building of a regional shopping centre at N? *(6)*
- town planners
- city councillors
- people living near to the site
- shopkeepers in the city centre
- long-distance lorry drivers
- the elderly and disabled.

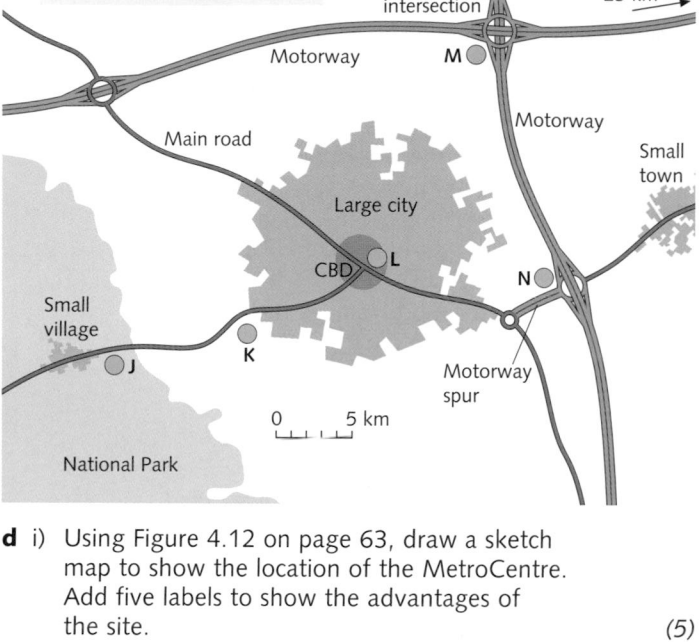

d i) Using Figure 4.12 on page 63, draw a sketch map to show the location of the MetroCentre. Add five labels to show the advantages of the site. *(5)*

ii) List the factors that make the MetroCentre attractive for:
- shoppers
- a family day out. *(4)*

4 *(Page 64)*

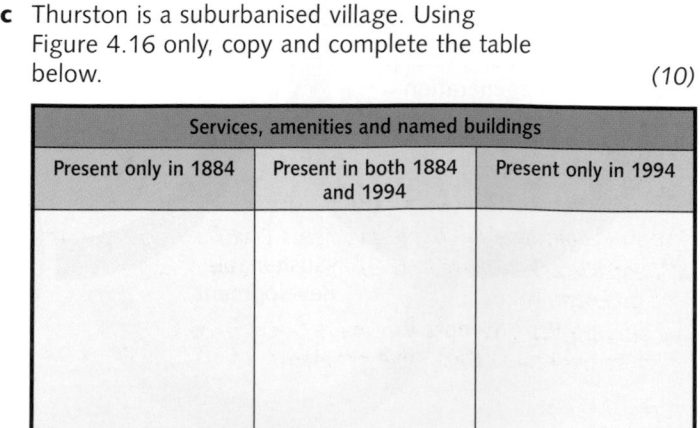

a The diagram above shows the occupations of people living in a village (not the same one as shown in Figure 4.16) in 1900 and 2000.

i) Copy and complete the table below using information from the diagram. *(8)*

ii) What was the main occupation in:
- 1900
- 2000?

Give one reason for the change in the main occupation. *(3)*

	Agri-culture	Manu-facturing	Trans-port	Distri-bution	Professional workers	Other services
1900						
2000						

b Using Figure 4.15 and your own knowledge, describe the likely changes in that village between 1900 and 2000 in terms of:
- type and design of houses
- public and private transport
- age and occupations of the inhabitants
- services and amenities available
- land use within and around the original village
- quality of life of the original inhabitants. *(2 × 6)*

c Thurston is a suburbanised village. Using Figure 4.16 only, copy and complete the table below. *(10)*

Services, amenities and named buildings		
Present only in 1884	Present in both 1884 and 1994	Present only in 1994

5 (Page 65)

a Estimates suggest that Britain will need 3 million new houses by 2028.
  i) What proportion of this total does the government suggest is built on brownfield sites? (1)
  ii) Which part of Britain has most need for these houses? (1)
  iii) Which part of Britain has least available brownfield sites? (1)

b It is proposed to build 1000 new homes in, or around, your local town or city.
  i) Explain why developers would prefer to build most, or all, of the new houses on greenfield sites. (5)
  ii) Explain why conservation groups and people living in nearby villages want the new houses to be built on brownfield sites. (5)

6 (Pages 66 to 69)

a With reference to the map:
  i) What is a commuter? (1)
  ii) Why do more people commute into Amsterdam than out of Amsterdam? (1)
  iii) From which district do most of Amsterdam's commuters come? (1)
  iv) Why do you think most of Amsterdam's commuters come from this district? (1)
  v) From which district do the least number of commuters to Amsterdam come? (1)
  vi) Why do you think so few commute from this district? (1)
  vii) How many people commute from Alkmaar? (1)

b i) Give four reasons why city centres often experience traffic congestion. (4)
  ii) Look at Figure 4.23 on page 67 and then copy and complete the table below to show the damaging effects of traffic in urban areas. (9)

c i) Describe two attempts that have been made in your local town or city to reduce the problems created by traffic. (4)
  ii) With reference to Lille, or any other developed city outside the UK, describe the attempts made to overcome its traffic problems. (6)

	Environmental effects	Economic effects	Social effects
1			
2			
3			

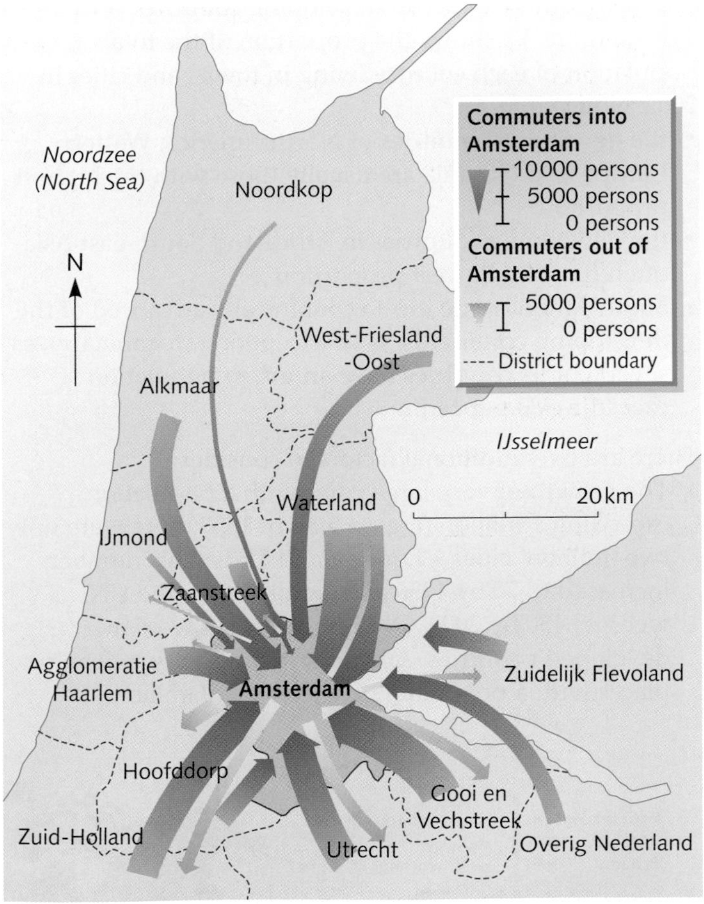

Map labels: Noordzee (North Sea); Noordkop; N; West-Friesland -Oost; Alkmaar; IJsselmeer; Waterland; 0 20km; IJmond; Zaanstreek; Agglomeratie Haarlem; Amsterdam; Zuidelijk Flevoland; Hoofddorp; Gooi en Vechstreek; Zuid-Holland; Utrecht; Overig Nederland

Map key: Commuters into Amsterdam — 10000 persons / 5000 persons / 0 persons; Commuters out of Amsterdam — 5000 persons / 0 persons; ---- District boundary

7 (Pages 70 to 73)

The star diagram shows five problems that face Osaka–Kobe.

a Describe briefly why each of the five factors poses a problem. (10)

b Describe the attempts that have been made to try to overcome these five problems. (10)

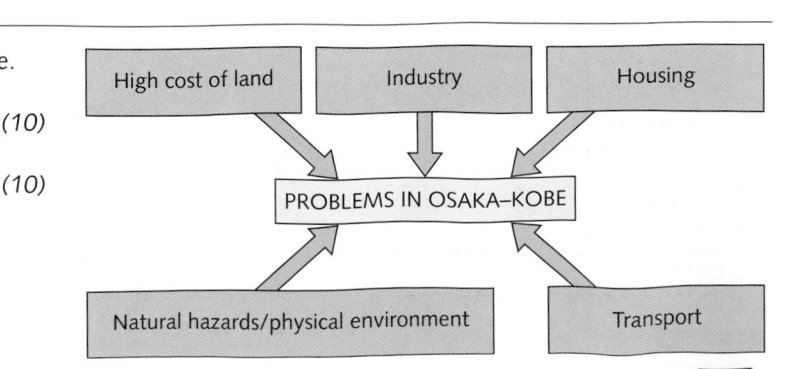

High cost of land | Industry | Housing

PROBLEMS IN OSAKA–KOBE

Natural hazards/physical environment | Transport

Growth of world cities

Urbanisation is the proportion of people living in towns and cities. Whereas in 1800 only 3 per cent of the world's population lived in urban areas, by 1950 this proportion had risen to 29 per cent and by 2008 it had reached 50 per cent (Figure 5.1). Although most developed countries have experienced urbanisation since the early 1800s, the process has only gathered pace in developing countries since the 1950s. Figure 5.2 shows the proportion of the total population of each country living in towns and cities in 2002. Notice that:

- the developed countries of North America, Western Europe and Oceania are usually those with the highest proportion
- the developing countries in Africa and South-east Asia often have the lowest proportion
- South America, the most economically advanced of the developing continents, is an exception (an **anomaly**), as several of its countries have an urban population exceeding 80 per cent.

There are two additional factors to consider:

1 The growth of very large cities with a population exceeding 1 million (Figure 5.3). In 1850 there were only two 'million' cities – London and Paris. This number increased to 70 by 1950 and, according to the UN website, 450 by 2009. Of these, only 114 were in developed countries. A new term, **megacity**, refers to places with a population in excess of 10 million.

2 The change in the distribution and location of large cities. Prior to 1950, most of the 'million' cities were in developed countries and in temperate latitudes north of the Equator (i.e. 40° to 50°N in Europe and North America). Since then there has been a dramatic increase in 'million' cities in developing countries, the majority of which lie within the tropics (Figure 5.4).

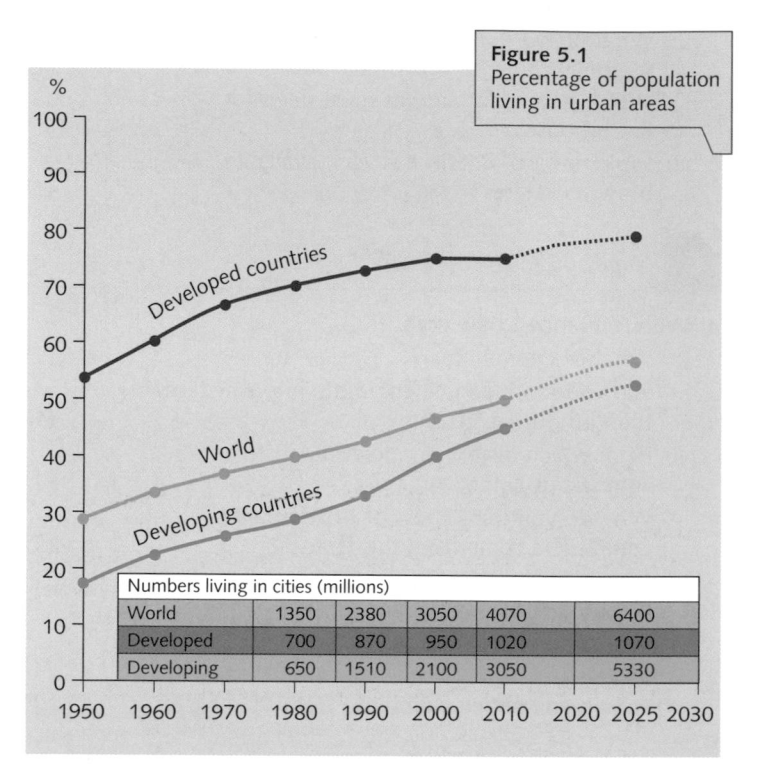

Figure 5.1 Percentage of population living in urban areas

Numbers living in cities (millions)					
World	1350	2380	3050	4070	6400
Developed	700	870	950	1020	1070
Developing	650	1510	2100	3050	5330

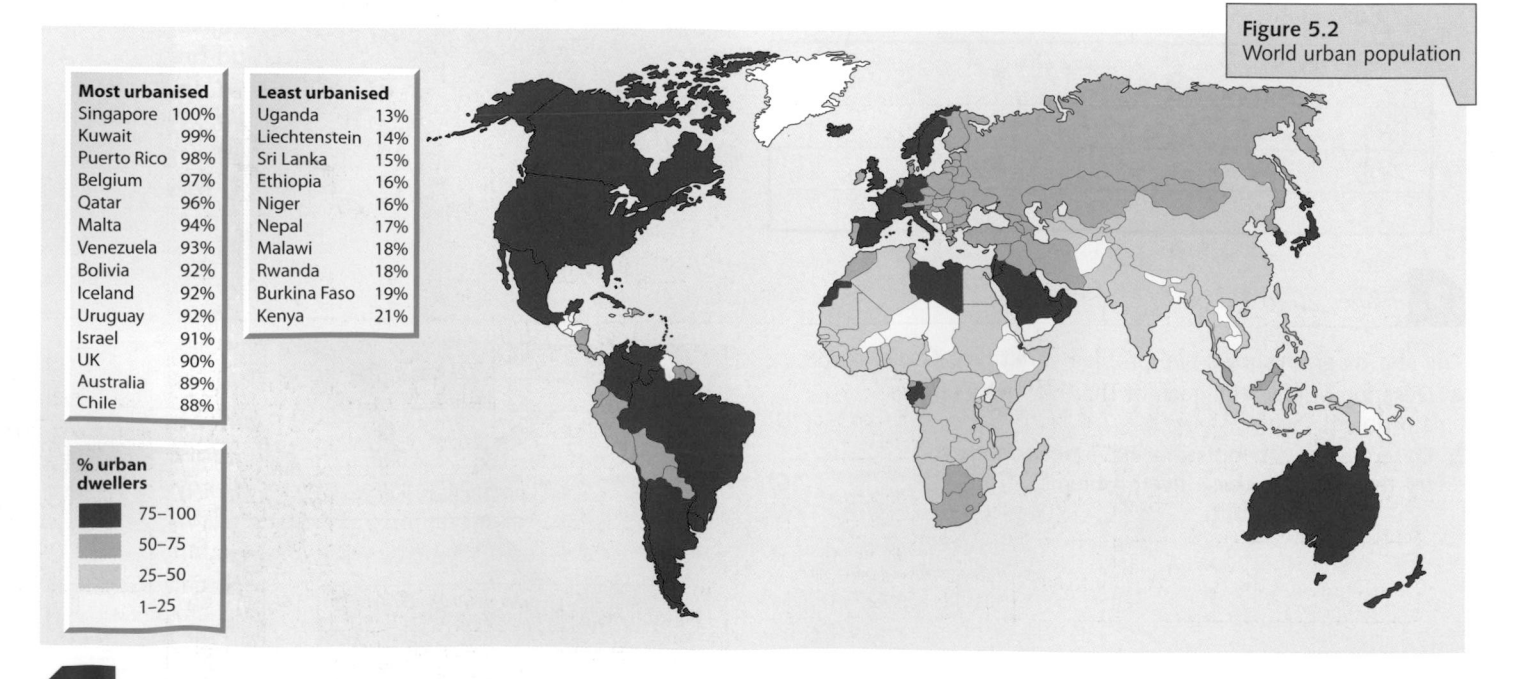

Figure 5.2 World urban population

Most urbanised		Least urbanised	
Singapore	100%	Uganda	13%
Kuwait	99%	Liechtenstein	14%
Puerto Rico	98%	Sri Lanka	15%
Belgium	97%	Ethiopia	16%
Qatar	96%	Niger	16%
Malta	94%	Nepal	17%
Venezuela	93%	Malawi	18%
Bolivia	92%	Rwanda	18%
Iceland	92%	Burkina Faso	19%
Uruguay	92%	Kenya	21%
Israel	91%		
UK	90%		
Australia	89%		
Chile	88%		

% urban dwellers
- 75–100
- 50–75
- 25–50
- 1–25

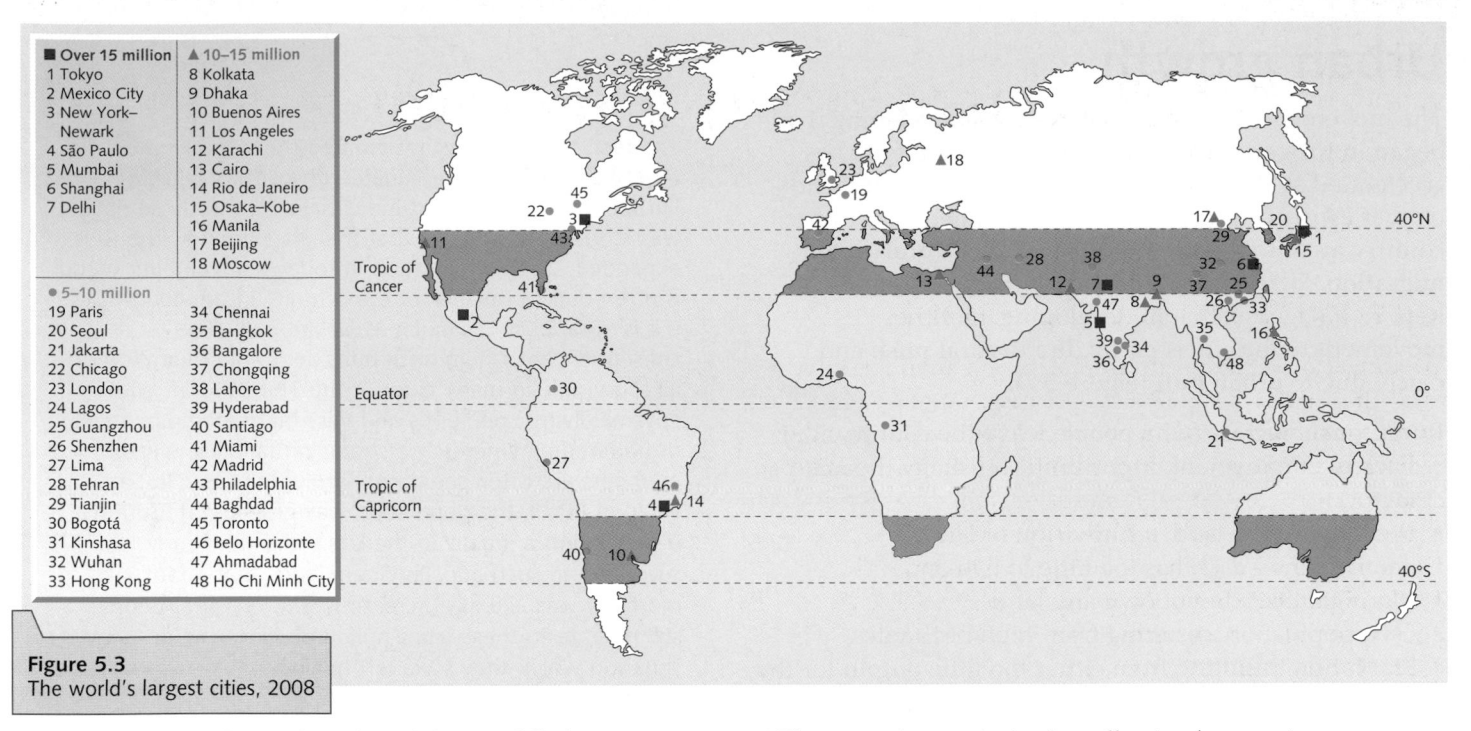

Figure 5.3
The world's largest cities, 2008

■ Over 15 million	▲ 10–15 million
1 Tokyo	8 Kolkata
2 Mexico City	9 Dhaka
3 New York–Newark	10 Buenos Aires
4 São Paulo	11 Los Angeles
5 Mumbai	12 Karachi
6 Shanghai	13 Cairo
7 Delhi	14 Rio de Janeiro
	15 Osaka–Kobe
	16 Manila
	17 Beijing
	18 Moscow

● 5–10 million	
19 Paris	34 Chennai
20 Seoul	35 Bangkok
21 Jakarta	36 Bangalore
22 Chicago	37 Chongqing
23 London	38 Lahore
24 Lagos	39 Hyderabad
25 Guangzhou	40 Santiago
26 Shenzhen	41 Miami
27 Lima	42 Madrid
28 Tehran	43 Philadelphia
29 Tianjin	44 Baghdad
30 Bogotá	45 Toronto
31 Kinshasa	46 Belo Horizonte
32 Wuhan	47 Ahmadabad
33 Hong Kong	48 Ho Chi Minh City

Figure 5.5 gives the rank order of the world's largest agglomerations over a period of years. In 1970 half of these cities were located in the industrialised, developed countries in North America and Europe. By 2008, however, 14 of the 18 largest cities were in developing countries in Asia, Latin America and Africa (Figure 5.3). The figures for world cities can, at the best, only be estimates. This is because:

- There are variations between countries as to how they define and delimit an urban area, e.g. is it Liverpool and Manchester or Merseyside and Greater Manchester?

- There are inaccuracies in collecting/processing population data, e.g. birth/death rate projections, or a failure to obtain everyone's census details, e.g. street dwellers, illegal immigrants (page 10).

Since the 1970s the growth rate of most cities in developed countries has slowed down, with some having even declined. During this time the two fastest-growing urban areas were São Paulo and Mexico City, both in Latin America, although recently even they have grown at a less fast rate. At present the fastest-growing cities are in southern Asia. The population of Dhaka (Bangladesh) doubled between 1895 and 1995 and that of Karachi, Delhi and Bangkok between 1985 and 2000. Can you imagine all the inhabitants of a place such as Liverpool or Glasgow suddenly arriving in Dhaka each year? Think of the problems this must pose for the newcomers, the existing inhabitants and the city authorities.

Figure 5.4
Growth and location of 'million' cities

Number of million cities (y-axis): 0–325

Lines: Total number (est. 320); Between tropics and 40° N or 40° S; North of 40° N; Within tropics

51.4	91.8	270.0	482.2	1112.1	Population in million cities (millions)
1800	2300	3100	4432	6010	World population (millions)
2.86	4.01	8.71	10.88	18.51	% population in million cities

(Year: 1920, 1940, 1960, 1980, 2000)

Figure 5.5
The world's largest agglomerations (population in millions)

Rank order	1970	1985	2007
1	New York 16.5	Tokyo 23.0	Tokyo 35.2
2	Tokyo 13.4	Mexico City 18.7	Mexico City 18.7
3	London 10.5	New York 18.2	New York 18.6
4	Shanghai 10.0	São Paulo 16.8	São Paulo 18.3
5	Mexico City 8.6	Shanghai 13.3	Mumbai 18.2
6	Los Angeles 8.4	Los Angeles 12.8	Delhi 15.0
7	Buenos Aires 8.4	Buenos Aires 11.6	Shanghai 14.5
8	Paris 8.4	Rio de Janeiro 11.1	Kolkata 14.3
9	São Paulo 7.1	Kolkata 9.2	Dhaka 12.6
10	Moscow 7.1	Mumbai 8.2	Buenos Aires 12.5

North America

Asia

Latin America

Europe and former USSR

Urban growth

The movement of people to cities in the developing world began in the early twentieth century. Since then it has accelerated so that many places are expanding at a rate of over 25 per cent every decade. The movement from country areas to towns and cities is called **rural–urban migration**. Figure 5.6 is a quotation from the Brandt Report. It suggests that in developing countries, movement to the city is partly due to **rural push** and partly due to **urban pull** (Figure 5.7).

Rural 'push' factors (why people leave the countryside):

- Lack of employment opportunities (usually the main factor).
- Pressure on the land, e.g. division of land among sons – each has too little to live on.
- Many families do not own any land.
- Overpopulation, resulting from high birth rates.
- Starvation, resulting from either too little output for the people of the area, or crop failure. Often, it may also be caused by a change in agriculture – from producing crops for local/family consumption to a system that produces cash/plantation crops for consumption in the developed world (page 111).
- Limited food production due to overgrazing, or to misuse of the land resulting in soil erosion or exhaustion (page 254).
- Mechanisation has caused a reduction in jobs available on the land together with, in many areas, reduced yields.
- Extreme physical conditions such as aridity, rugged mountains, cold, heat and dense vegetation.

The rush to the towns has created the same kind of misery as existed in the nineteenth-century cities of Europe and America. But industrialisation in those days was labour-intensive, so that the cities grew as the jobs expanded. .... The migration in today's developing world is often due to the lack of opportunity in the countryside – it is 'rural push' as much as 'urban pull'. The consequences of high birth rates and rapid migration are all too visible in many cities of the Third World, with abysmal living conditions and very high unemployment or underemployment. The strains on families, whose members are often separated, are very heavy. In São Paulo in Brazil, the population was growing at around 6–7 per cent annually in the late sixties and early seventies, in such appalling conditions that infant mortality was actually increasing. The fact that people still migrate to these cities only underlines the desperate situation which they have left behind.

Figure 5.6
Extract from the Brandt Report

- Farming is hard work with long hours and little pay. In developing countries a lack of money means a lack of machinery, pesticides and fertiliser.
- Local communities (Amazon Amerindians) forced to move.
- Lack of services (schools, hospitals).
- Lack of investment as money available to the government will be spent on urban areas.

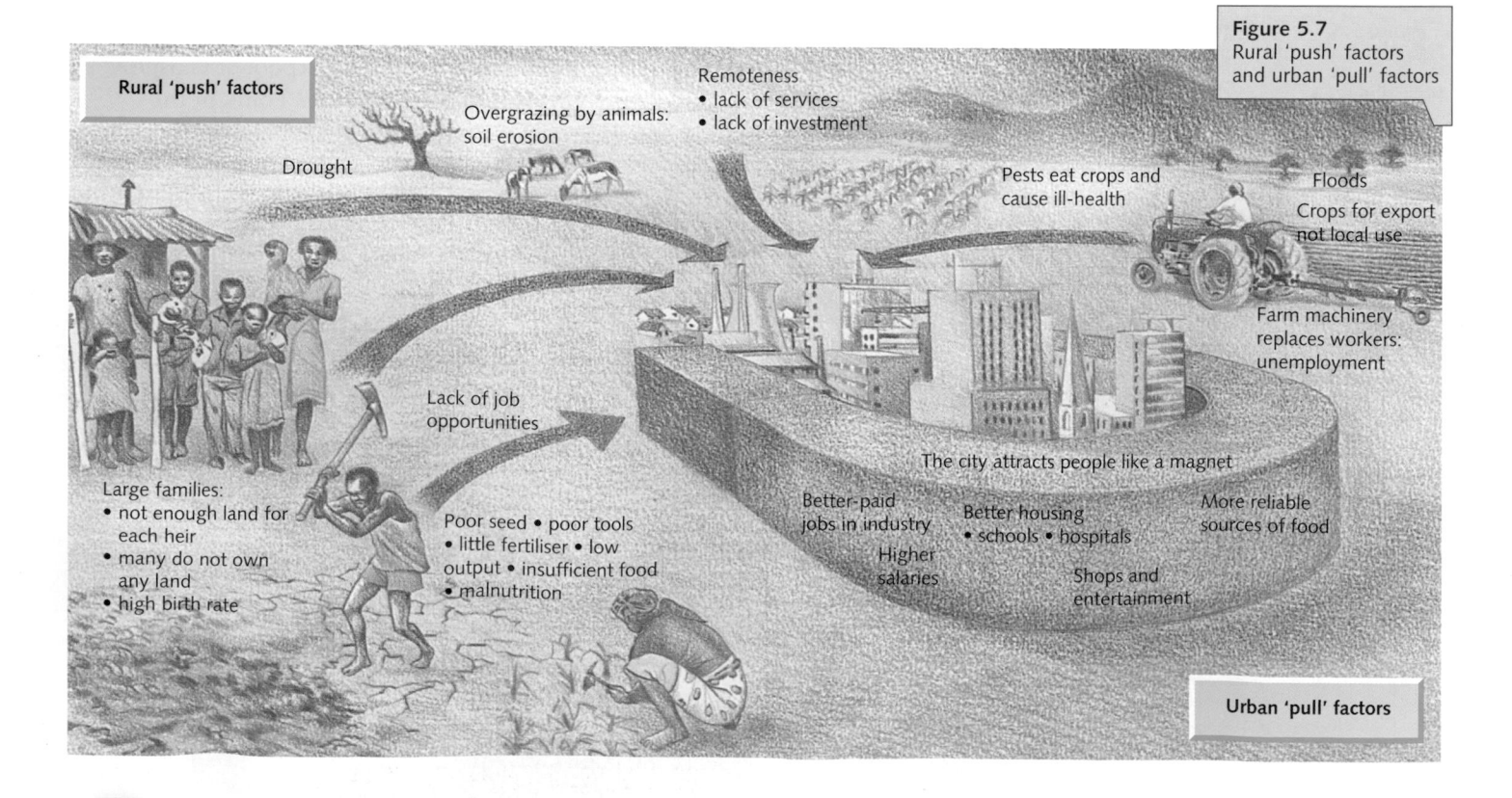

Figure 5.7
Rural 'push' factors and urban 'pull' factors

Rural 'push' factors

Overgrazing by animals: soil erosion

Drought

Remoteness
• lack of services
• lack of investment

Pests eat crops and cause ill-health

Floods

Crops for export not local use

Farm machinery replaces workers: unemployment

Lack of job opportunities

Large families:
• not enough land for each heir
• many do not own any land
• high birth rate

Poor seed • poor tools
• little fertiliser • low output • insufficient food
• malnutrition

The city attracts people like a magnet

Better-paid jobs in industry

Better housing
• schools • hospitals

More reliable sources of food

Higher salaries

Shops and entertainment

Urban 'pull' factors

Urban 'pull' factors (why people move to the city):

- They are looking for better-paid jobs. Factory workers get about three times the wages of farm workers.
- They expect to be housed more comfortably and to have a higher quality of life.
- They have a better chance of services such as schools, medical treatment and entertainment.
- They are attracted to the 'bright lights'.
- There are more reliable sources of food.
- Religious and political activities can be carried on more safely in larger cities.

In most developing countries the 'pull' factors are often greater than the 'push' factors. Figure 5.7 presents the perceptions of the city held by many rural families, i.e. what they think, expect or were led to believe the city is like. The reality is often very different (Figure 5.8). As many of the new arrivals are unlikely to have much, if any, money they will be unable to buy or rent a house – even if one was available – and will probably have to make a temporary shelter using cheap or waste materials.

The gulf between the rich (who are relatively few in number) and the poor (many of whom were migrants from surrounding rural areas) is much greater than in developed countries. This means that the contrast between the well-off and the poorest areas is much greater in terms of the quality and density of housing, the quality of both life and the environment, and the provision of amenities.

Urban land use and functional zones

We have already seen (page 42) that cities in developed countries do not grow in a haphazard way, but tend to develop recognisable patterns. Although each city will have its own distinctive pattern, it is likely to share characteristics with other urban areas.

Similarly, cities in developing countries also develop their own distinctive pattern (Figure 5.9) although this differs from the model for developed cities in several ways:

- Most of the better-off housing is located near to the city centre.
- The quality of housing decreases rapidly towards the city boundary where many people are forced to live as squatters in **shanty settlements** or, using the UN term, **informal settlements**.
- Industry tends to locate along main roads leading into the city centre.

The land use pattern in developing cities is, like that of developed cities, more complex than that shown in the model. Figure 5.10, showing land use and functional zones, is more realistic.

Figure 5.8
New arrivals to developing cities often encounter conditions like these in Ho Chi Minh City, Vietnam

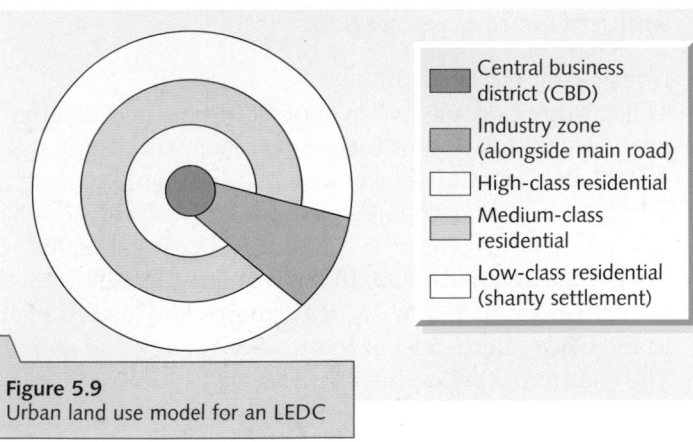

Central business district (CBD)

Industry zone (alongside main road)

High-class residential

Medium-class residential

Low-class residential (shanty settlement)

Figure 5.9
Urban land use model for an LEDC

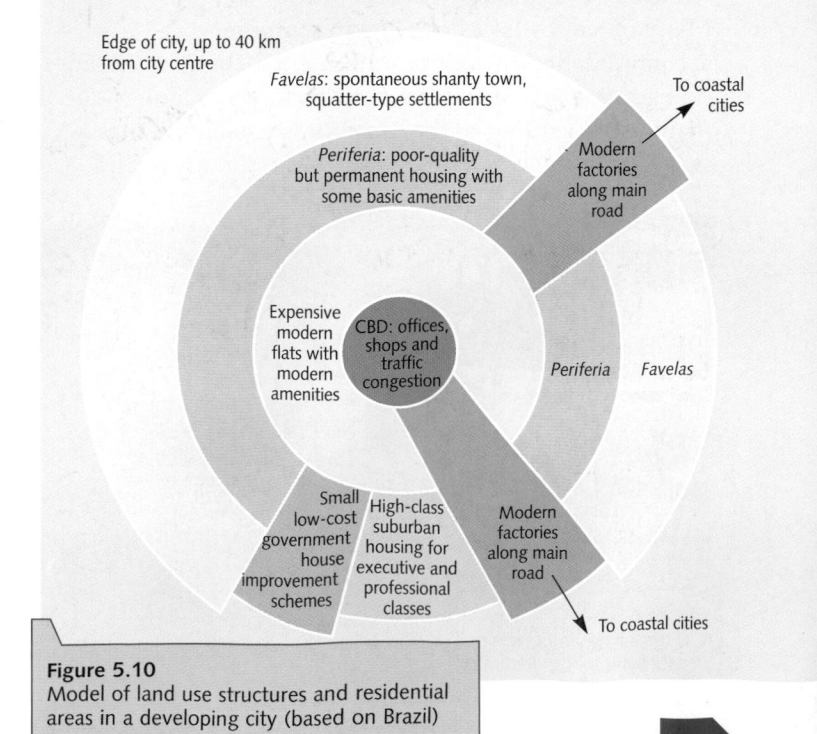

Edge of city, up to 40 km from city centre

Favelas: spontaneous shanty town, squatter-type settlements

Periferia: poor-quality but permanent housing with some basic amenities

To coastal cities

Modern factories along main road

Expensive modern flats with modern amenities

CBD: offices, shops and traffic congestion

Periferia   *Favelas*

Small low-cost government house improvement schemes

High-class suburban housing for executive and professional classes

Modern factories along main road

To coastal cities

Figure 5.10
Model of land use structures and residential areas in a developing city (based on Brazil)

Differences in residential areas

In most cities in developing countries there is an extreme contrast between the living conditions of the well-off, who usually live near to the city centre, and those of the poor, who are forced to live towards the city boundary.

This contrast can be seen in places such as São Paulo (below) and Rio de Janeiro (page 86) in Brazil, and Kolkata (pages 84 and 85) in India.

São Paulo

São Paulo is the largest, richest and most industrialised city in Brazil. Its growth has resulted from a massive influx of immigrants, initially from Europe (after 1850) and more recently from rural Brazil, and a high birth rate. This rapid growth has been accompanied by increased social and economic segregation, with the poor being relegated to the city's periphery with its shanty settlements (*favelas*) and tenement slums (*corticos*). With the city's population having grown from 7 million in 1970 to 18.3 million by 2008, the authorities have had an almost impossible task in providing sufficient housing and services.

Housing for the well-off

This group of people live in expensive housing ranging from elegant apartment complexes, each with its own social and recreational facilities, to Californian-style detached houses with large gardens and individual swimming pools (Figure 5.11). Family size is generally limited to two children, with housemaids. These properties, protected by security guards, are located near to the CBD where most of the residents work and shop. The children, who are healthy and well educated, are likely to find well-paid jobs in the future.

Housing for the poor (*favelas*)

Many of São Paulo's poor live in temporary accommodation in shanty settlements. This is sometimes on vacant space next to modern factories or alongside main roads leading to the city, but is usually on the outskirts of the city. These people are 'squatters' and

Figure 5.12
A *favela* in São Paulo – a 'slum of despair'

have no legal right to the land they occupy. The rapid growth of these spontaneous settlements, or *favelas* as they are known in Brazil, is common to cities in the developing world (Figures 5.8, 5.12, 5.17 and 5.23). *Favelas* are found on land that has little economic value and which the well-off find unsuitable for development. *Favelas* often develop on steep hillsides which are liable to landslides, or on badly drained, unhealthy valley floors.

The housing is often a collection of primitive shacks made from any material available – wood, corrugated iron, cardboard or sacking. Some may only have one room in which the family has to live, eat and sleep. Others may have two rooms, one being used to sleep the family which is likely to consist of at least six children. Most houses lack such basic amenities as electricity, clean running water, toilets and mains sewerage. Any empty space between houses will soon be occupied either by later migrants or, as there is no refuse collection, with rubbish. *Favelas* are overcrowded and have a high housing density.

Figure 5.11
Californian-style housing for the rich in São Paulo

Self-help schemes

Over 100 million people living in developing cities have no shelter of any kind, while over one-third live in squatter settlements – sites only vacant because they are subject to flooding, landslips or industrial pollution. Although most local authorities would probably prefer to remove shanty settlements from their cities, few can find the necessary resources that would be needed to provide alternative accommodation. As a result, shanty settlements become permanent. Self-help and site and services schemes seem, therefore, the only hope for the squatters to improve their homes (Figures 5.13 and 5.14). The most immediate needs of the poor are often simple: a plot on which to build, a small loan to improve or extend the house, cheap building materials, and basic services (see Places below). There is no need for advanced technology or expensive accommodation (page 151).

Unfortunately, despite the introduction of self-help schemes, local authorities in developing cities like São Paulo (below) and Rio de Janeiro (pages 87 and 88) rarely can keep pace with the continuous and large numbers of newcomers.

Housing improvements

There are two local government-assisted schemes in São Paulo aimed at improving the quality of life in a *favela*.

Low-cost improvements (Figure 5.13)
Existing homes may be improved by rebuilding the houses with cheap and quick and easy-to-use breeze-blocks. A water tank on the roof collects rainwater and is connected to the water supply and, in turn, to an outside wash basin and an indoor bathroom/toilet. Electricity and mains sewerage are added. Most inhabitants of this type of housing, which is found in the peripheral parts of São Paulo, will have some type of employment enabling them to pay a low rent.

Figure 5.13
Low-cost improvements to housing in São Paulo

Breeze-blocks · Roofing tiles · Water tank · Electricity wires · Sink · Streets improved · Bathroom with toilet · Underground sewer · Living and sleeping quarters with concrete floor · Improved road

Self-help schemes (Figure 5.14)
Groups of people are encouraged to help build their own homes. Each group will do basic work such as digging ditches to take the water and sewage pipes. The local authority will then provide breeze-blocks and roofing tiles, and the group provide the labour. The money which this saves the authorities can be used to provide amenities such as electricity, a clean water supply, tarred roads and a community centre. The advantages of self-help schemes are that they can be done in stages, they can create a community spirit and, as the cost of building is relatively cheap, more houses can be provided.

Figure 5.14
A community self-help housing project in São Paulo – a 'slum of hope'

Problems in developing cities

Kolkata

Kolkata is one of the most notorious examples of the way in which problems are created when cities grow too quickly. In describing these problems, which in Kolkata are severe, it is easy to provide a negative stereotype of life in a developing city. However, despite the poverty and squalor, and even with the scale and range of the problems facing the city, and its lack of resources, the authorities have made real improvements in the urban environment.

Location and growth Kolkata is situated on flat, swampy land alongside the River Hooghly, one of numerous rivers that form the Ganges delta (Figure 5.15). It developed during colonial times as a major port and is now India's largest city. Its population of almost 7 million in 1970 had increased to over 14 million by 2008.

Physical problems With the highest point of the city being only 10 metres above sea-level, there is a constant threat of flooding. Flooding can result from heavy rainfall associated with the annual monsoon (page 214) or be brought by tropical storms (typhoons – page 216). At these times, surplus storm water (often contaminated by sewage) has to be pumped away.

Housing Many families have no home other than the pavement. Reports suggest that over a quarter of a million people are forced to sleep in the open, covered only by bamboo, sacking, polythene or cardboard (Figure 5.16). Another 2 million residents live in *bustees* (Figure 5.17). *Bustees* are collections of houses built of non-permanent materials, such as wattle, with tiled roofs and

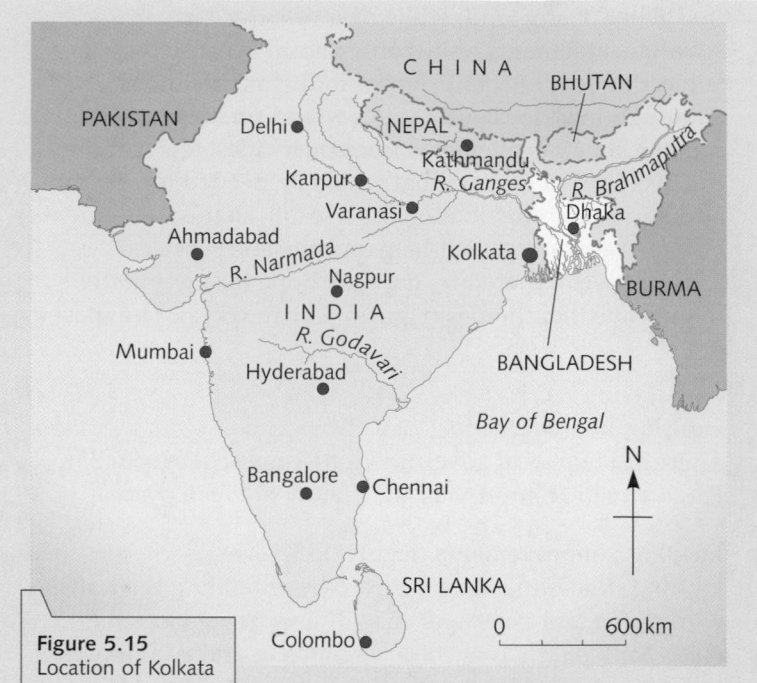

Figure 5.15
Location of Kolkata

mud floors – not the best materials to withstand the heavy monsoon rains. The houses, packed closely together, are separated by narrow alleys. Inside there is often only one room, and that is usually no larger than an average British bathroom. In this room the family, which can number up to eight, live, eat and sleep. Despite this overcrowding, the insides of the dwellings are clean and tidy. The houses belong to landlords who rent them out to *bustee* dwellers – the tenants are evicted if they cannot pay the rent. Estimates suggest a population density in some *bustees* of over 150 000 people per km².

Figure 5.16
Street dwellers in Kolkata

Figure 5.17
In the *bustees*

Figure 5.18
Problems with clean water

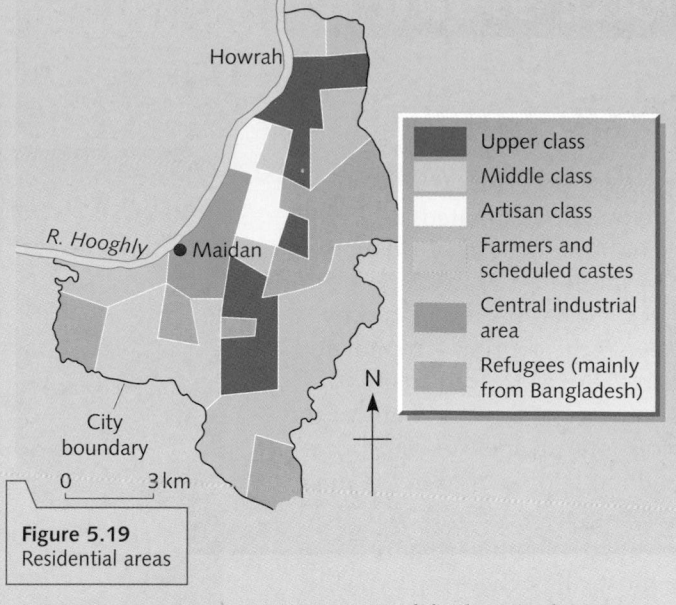

Howrah

- Upper class
- Middle class
- Artisan class
- Farmers and scheduled castes
- Central industrial area
- Refugees (mainly from Bangladesh)

R. Hooghly ● Maidan

N

City boundary

0   3 km

Figure 5.19
Residential areas

Water supply, sanitation and health Although three-quarters of Kolkata's population has access to piped water, it is not uncommon, especially in the *bustees*, for a single street tap to be used by 35 to 45 families (Figure 5.18). Sanitation is almost non-existent in the *bustees* and an estimated one-third of the city's population is without a toilet of any kind. Here human effluent runs down the narrow lanes. Elsewhere, the 100-year-old and more drains and sewage pipes tend to crack, spilling their contents onto the streets. Sewage often contaminates drinking water, especially after heavy rain. Rubbish, dumped in the streets, is rarely collected. A survey in 1991 recorded an infant mortality rate of 123 per 1000 live births. The major causes of death were cholera, typhoid, dysentery, tetanus and measles – diseases linked to poor sanitary conditions and overcrowding.

Figure 5.20
A stall in a Kolkata *bustee*

Segregation Figure 5.19 is a simplified map showing segregation between peoples of different caste, language, religion and occupations. The original Bengali-speaking Hindus live apart both from non-Bengalis and from the later Hindu refugees from Bangladesh.

Provision of services The provision of electricity, clean water, schools and hospitals, together with the collection of rubbish, all require considerable amounts of money – a resource in short supply to the authorities.

Transport Most people have either to walk or to use the overcrowded public transport system. The buses, many of which are old, are insufficient to carry everyone, and passengers can be seen hanging on to the outsides. Rickshaws are cheap but their numbers and slowness add to the congestion. Traffic noise, congestion and pollution continue despite the addition of a second bridge over the Hooghly and an underground system.

Employment Those with jobs tend to work in the informal sector (page 148) and use their home as a place of work. Often the fronts of houses are 'opened up' to allow the occupants to sell wood, food and clothes (Figure 5.20). Although few people are totally unemployed, many jobs only occupy a few hours a week and provide a very low income.

The Kolkata Metropolitan Development Authority This was set up in 1970. Since then it has attempted to make the *bustees* more habitable by paving some of the alleys, digging extra drains and providing more water taps and public toilets. Despite financial support from many voluntary agencies, a lack of money and high birth and immigration rates have combined to slow down progress in providing better housing, services, transport and more jobs.

Rio de Janeiro

Location and growth

Rio de Janeiro is situated around the huge natural harbour of Guanabara Bay in south-east Brazil (Figure 5.21). Although it has been replaced by Brasilia as the country's capital and by São Paulo as the centre of industry and commerce, Rio is still one of the world's largest cities. In 2008, some 7.2 million people lived in the main urban area and 11.8 million in the whole agglomeration (Figure 5.22). Like most cities, there are two sides to Rio:

- the beaches of Copacabana (page 149) and Ipanema backed by luxury housing
- the problems of rapid urban growth including housing, traffic, pollution and crime.

Problems

Housing Apart from an estimated half-million homeless street dwellers, over 1 million people live in *favelas* and another 1 million in poor-quality local authority housing (*periferia*). *Favelas* are informal, shanty-type settlements (Figure 5.23). The largest of over 600 *favelas* are Morro de Alemao (Figure 5.21, bottom left) and Roçinha, each with a population of over 100 000. The official definition of a *favela* is 'a residential area of 60 or more families living in accommodation that lacks basic services (no running water, sewerage or electricity) and who have no legal right to the land on which they live'. The houses are constructed from any materials available – wood, corrugated iron, broken bricks and tiles (Figure 5.23). Most *favelas* are built on hillsides considered too steep for normal houses. The most favoured sites are at the foot of the slope near to the main roads and water supply – although these places may receive sewage running

downhill in open drains from houses built higher up. People living near the hilltops have to carry everything they need, including water. When it rains on the steep slopes, flash floods and mudslides can carry away the flimsy houses – storms in 1988 caused over 200 deaths (Figure 5.24). Several attempts have been made to clear *favelas* (one before the visit of the Pope) but, with nowhere else to go, the evicted residents soon return. In some cases the authorities built new homes (*conjuntos habitacionais* – Figure 5.22) to replace those they had cleared, but these were little

better than the original shacks. The local authority now accepts the existence of *favelas* and is working with residents' associations to improve conditions there (pages 87 and 88). Despite the problems, living in a *favela* has its positive side. There is often a strong community spirit, a rich street life and plenty of football and samba music.

Crime Many residents and visitors see crime as one of Rio's main problems. The *favelas* are perceived, certainly by non-residents, to be areas associated with organised

Figure 5.21
Rio de Janeiro, Sugar Loaf Mountain and Guanabara Bay

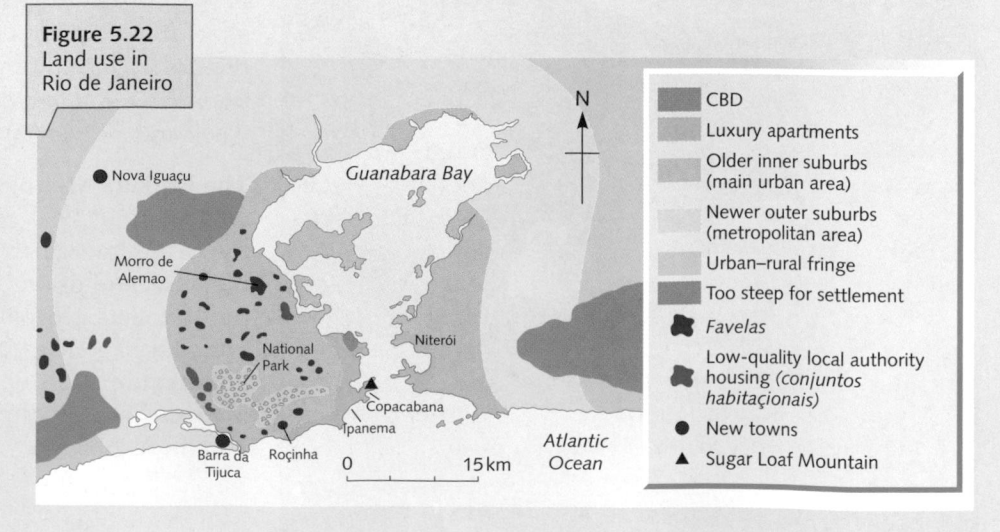

Figure 5.22
Land use in Rio de Janeiro

Nova Iguaçu

Guanabara Bay

Morro de Alemão

National Park

Niterói

Copacabana

Ipanema

Barra da Tijuca

Roçinha

Atlantic Ocean

0   15 km

N

| CBD |
| Luxury apartments |
| Older inner suburbs (main urban area) |
| Newer outer suburbs (metropolitan area) |
| Urban–rural fringe |
| Too steep for settlement |
| *Favelas* |
| Low-quality local authority housing (*conjuntos habitaçionais*) |
| New towns |
| Sugar Loaf Mountain |

crime, violence and drug trafficking (residents, on the other hand, claim that crime has decreased and the community spirit has increased). Many well-off Rio residents are now moving out of the city to places such as Barra da Tijuca (page 88) which they see as a cleaner, safer environment for their family. Tourists to Rio's famous beaches are warned not to take valuables with them nor to wear jewellery or watches.

Traffic Although the mountains add to Rio's attractiveness, they also hem in the city and force traffic along a limited number of routes. For much of the day there is severe congestion, pollution and, even through the night, noise.

Pollution An industrial haze, intensified by traffic fumes, often hangs over much of Guanabara Bay. Along the coast, the beaches and sea are also polluted. A city the size of Rio produces huge amounts of waste. In *favelas* the rubbish is unlikely to be collected and its presence, together with possible polluted water supplies and the sewage in open drains, causes health hazards (e.g. an outbreak of cholera in 1992).

Figure 5.23
A *favela* on a steep hillside

Figure 5.24
Landslide on a Rio hillside

Attempts to solve some of Rio's problems

Self-help housing schemes – Roçinha

The residents of Roçinha (Figure 5.22) have slowly transformed their *favela* into a small city. Most of the original temporary wooden buildings have been upgraded to brick and tile and many have been extended to utilise every square centimetre (Figure 5.25). Many residents have lived here since the *favela* developed in the 1950s and have set up their own shops and small industries – the so-called 'informal sector' (page 148) –

and created their own places of entertainment. The authorities, having accepted the existence and permanence of *favelas* such as Roçinha, now work with local residents associations. They have added electricity (satellite TV has reached here too!), paved and lit some of the steeper streets and added water pipes. Improvements are, however, restricted by the high density of housing and the steepness of the hillsides.

Figure 5.25
A self-help scheme in Roçinha

The city authority's Favela Bairro project

In the 1990s the city authorities set aside £200 million to improve living conditions in 60 of the 600 *favelas* within their boundary. Initially 16 mid-size *favelas* were chosen because:

- being smaller, the problems were perceived to be easier and cheaper to tackle than in larger *favelas* such as Morro de Alemao and Roçinha
- they had fewer drug, crime and unemployment problems
- some were more visible from the well-off areas and to tourists to Rio.

The authorities claimed that they wished to transform the *favelas* socially and culturally and to integrate them as part of the city. The plan included:

- replacing buildings that were either made of wood or built on dangerous slopes with brick-built housing – the new houses are much larger (5 x 4 metres) and have a yard of equal size
- widening selected streets so that emergency services and waste collection vehicles can gain access
- laying pavements, concrete paths, water pipes and electricity cables
- improving sanitation, adding health facilities and providing sports areas
- using labour from within each *favela* so that residents can develop and use new skills (Figure 5.25) – in return the residents have to pay taxes to the authorities.

The new town of Barra da Tijuca

In an attempt to find more space and a safer place to live, many of Rio's more wealthy residents have begun to move out of the city. This has led to a population loss in central Rio, a situation normally only associated with cities in developed countries (counterurbanisation, page 64).

The nearest available land to central Rio was 20 km along the coast to the south. However, due to the mountains between here and Rio extending down to the sea, the area had remained largely inaccessible and uninhabited until the 1970s. Then a four-lane motorway was built through tunnels under the mountains and raised on stilts over the sea. By 2000, the new town of Barra da Tijuca had a population of almost 140 000.

Barra da Tijuca is the 'new' Rio. It is not a suburb but a self-contained town built alongside the coastal motorway (Figure 5.26). It has 5 km of shops (the largest complex in South America), schools, hospitals, offices and places of entertainment, including a concert hall and cinemas. It has a culture based around the beach (20 km of sand and surf), shopping malls, expensive restaurants and leisure centres. The new residents live in spacious and luxury accommodation. Three-quarters of this accommodation is in 10- to 30-storey high-rise apartments. These apartments are protected by security guards and contain every possible modern amenity and gadget. The remainder of the accommodation is single- and double-storey detached houses. Although both adult members of the family usually work through choice, they have often to seek high-paid jobs to pay for their expensive accommodation and lifestyle. Despite each family having its own car, there is an efficient local bus service linking the apartments, shops and leisure amenities.

However, Barra already has its own new *favelas*. This is because the low-paid housekeepers, cooks, cleaners and gardeners, employed by the wealthy residents, also need somewhere to live.

Figure 5.26
View of Barra da Tijuca

Urbanisation in developing countries

Key Words and Terms

a You should know the meaning of the following terms:
  - urbanisation • million cities • rural–urban migration • functional zones
  - shanty settlements • informal settlement • squatter settlements
  - *favelas/bustees* • high-density housing • self-help schemes.

b You should know the difference between:
  - rural push factors and urban pull factors
  - *favelas* and *periferia* housing
  - self-help and local authority housing schemes.

Key Ideas

You should know and understand the following:
- The meaning and causes of urbanisation.
- The world pattern/distribution of million cities.
- There is a recognisable pattern of land use in urban areas in LEDCs.
- There are inequalities in residential areas in urban areas in LEDCs.
- The pattern of land use differs between urban areas in MEDCs and LEDCs.
- Rapid urban growth in cities in LEDCs causes major problems, including the growth of shanty settlements.
- There are major social, economic and environmental problems in shanty/squatter settlements (*favelas/bustees*) in cities in LEDCs.
- The attempts to improve living conditions/quality of life in shanty/squatter settlements through self-help and local authority schemes.
- Contemporary solutions to problems in a named city in an LEDC.

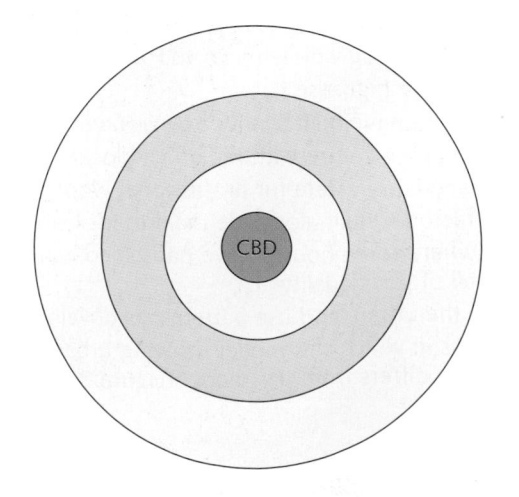

Skills, Theories and Models

- Interpret and use graphs
- Interpret choropleth maps
- Use photos to describe places
- Interpret and use the LEDC land use model
- Understand the concept of push–pull factors
- Understand the process and effects of change
- Recognise the distinctive character of places

1 *(Pages 78 and 79)*

a The map below shows predicted urban population changes 1980–2008.

i) Estimate the urban population of Europe in 1980 and in 2008. *(2)*

ii) Which two regions are expected to show the greatest urban population growth between 1980 and 2008? *(2)*

iii) State two differences in the urban population growth between North America and Latin America. *(2)*

b What evidence does the map show that urbanisation is faster in poorer parts of the world? *(3)*

c i) What is meant by the term 'million city'? *(1)*

ii) How many 'million cities' were there in:
  • 1920 • 1960 • 2008? *(3)*

iii) Name the two fastest-growing cities in the 1970s. *(2)*

iv) Name three of the fastest-growing cities in the 1990s. *(3)*

v) Using Figure 5.3, complete the top table to show the location of world cities with a population of over 5 million. *(5)*

vi) Using Figure 5.5, complete the bottom table to show the changing location of the world's largest cities. *(3)*

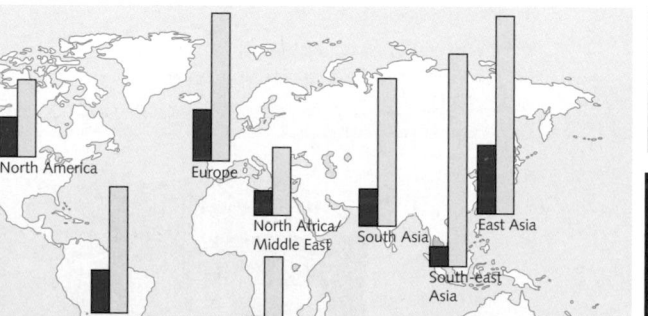

Europe	North America	Asia	Latin America	Africa

		1970	1985	2007
Developed continents	Europe			
	North America			
Developing continents	Latin America			
	Asia			

2 *(Pages 80 and 81)*

A feature of cities in the developing world is that they are growing very rapidly, partly due to people moving to them from surrounding rural areas.

a i) What is this movement from the countryside called? *(1)*

ii) Give four reasons why people may wish to move into a big city from the surrounding countryside (urban 'pull' factors). *(3)*

iii) Give four reasons why people may have to move away from the countryside (rural 'push' factors). *(3)*

b Name four problems likely to occur in urban areas when large numbers of people move into them. *(4)*

c How does the reality of living in a city in a developing country differ from the perception of life in that city held by people living in rural areas? *(2)*

3 *(Page 81)*

The diagram on the right is an incomplete model of a city in the developing world.

a Match up the following with letters A to F on the model:
  • Modern, luxury high-rise flats
  • Squatters who have built shanty towns (*favelas*)
  • A large shopping centre with tall office blocks
  • A suburban luxury estate for professional workers
  • Modern factories built alongside main roads leading out of the city
  • An area where some houses have had piped water and electricity added, yet are still of poor quality *(6)*

b Explain how the urban land use pattern has developed. *(3)*

c Give two ways in which this model showing urban land use in a developing city differs from the model (Figure 3.19) for a developed city. *(2)*

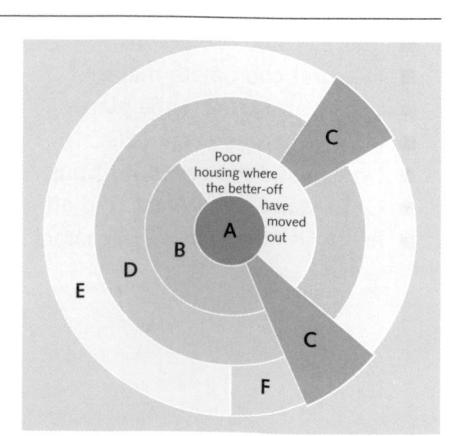

4 (Pages 82 and 84 to 87)

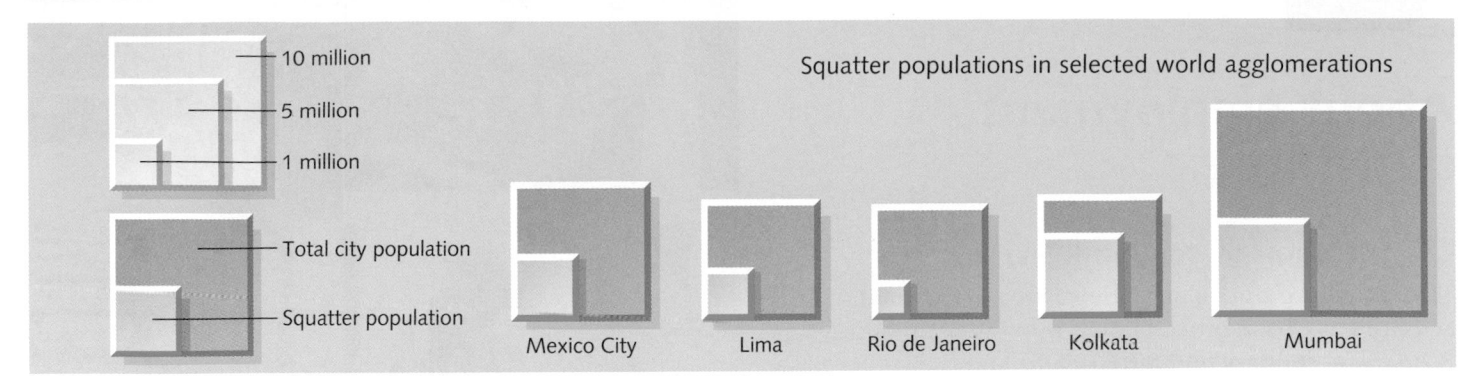

Squatter populations in selected world agglomerations

10 million
5 million
1 million

Total city population

Squatter population

Mexico City   Lima   Rio de Janeiro   Kolkata   Mumbai

a i) Name the city with the largest population. (1)
   ii) Estimate the squatter population of:
   • Rio de Janeiro • Lima • Kolkata. (3)
b What is a shanty settlement? (2)
c What are shanty settlements called in:
   • Brazil • Kolkata? (1)
d Name one city in a less economically developed country where many people live in shanty settlements. (1)

e Give two reasons why many people in your chosen country live in shanty settlements. (2)
f In what part of the city are shanty settlements most likely to be found? (2)
g Describe the housing conditions likely to be found in shanty settlements in your chosen city. (4)
h Describe three other problems likely to be found in the shanty settlement. (3)

5 (Pages 83, 85, 87 and 88)

Many shanty/squatter settlements have been improved through self-help schemes. These schemes are simple, low-cost housing projects where, with help from the authorities, local people are encouraged to improve their living conditions.

a Name a city where the authorities have encouraged self-help schemes. (1)
b Using Figure 5.13, describe the improvements which have been made in:
   • the building • the provision of services. (2 + 2)

c Describe what is being done in a named city to improve the living conditions of people who live there. Refer to efforts of:
   • the local inhabitants
   • the local authorities. (2 + 2)
d Give two advantages and two disadvantages of self-help schemes. (2 + 2)
e How successful do you think the self-help schemes have been? Give reasons for your answer. (3)

6 (Page 88)

a The sketch below has been simplified from Figure 5.26. Make a simple copy of the sketch and on it label the following:
   • the sea
   • inland lake/lagoon
   • the motorway
   • coastal mountains
   • main shopping area
   • high-rise luxury apartments
   • open space (two areas)
   • a large car parking area. (8)

b Where do you think you might find:
   • the most expensive housing
   • the least expensive housing (*favelas*)?
   Give a reason for each answer. (2 + 2)

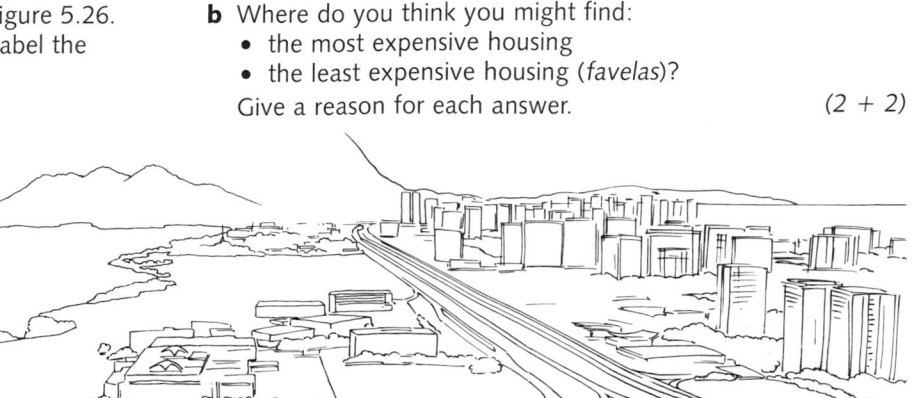

6 Employment structures

About employment structures

Classification of economic activities

Traditionally, industry and other types of **economic activity** have been broken down into the main groups of **primary**, **secondary** and **tertiary** although, since the 1980s, **quaternary** has been added to make a fourth group (Figure 6.1).

- **Primary** industries extract raw materials directly from the earth or sea. Examples include farming, fishing, forestry and mining.
- **Secondary** industries process and manufacture the primary products, e.g. steelmaking and furniture manufacture. They also include the construction industry and the assembly of component parts made by other secondary industries, e.g. car assembly.
- **Tertiary** industries provide a service. These include education, health, office work, retailing, transport and entertainment.
- **Quaternary** industries provide information and expertise. They include the relatively new micro-electronics industries.

Employment structure

The proportion of people working in each of the primary, secondary and tertiary sectors is called the **employment structure**. The figure for each group is given as a percentage of the total (Figures 6.2, 6.4 and 6.6). Employment structures change over a period of time (Figure 6.2) and vary from place to place (Figures 6.3 and 6.5).

Figure 6.1
Employment sectors:
(a) Primary
(b) Secondary
(c) Tertiary
(d) Quaternary

Changes over time

Figure 6.2 gives employment structures for the UK and shows how they have changed over a period of time. Two hundred years ago, before there was accurate data, most working people were employed in the primary sector and made their living from the land. The majority were farmers while others either made things for use in farming, such as ploughs, or from items produced by farmers, such as bread.

One hundred years ago, and as a result of the Industrial Revolution, fewer people worked in the primary sector (fewer farmers) but many more in the secondary sector (e.g. in steelworks, shipyards and textile mills). Further changes took place in the twentieth century. Farming and industry became more mechanised and needed fewer workers. Coal became less abundant and industry faced increasing competition from overseas. At the same time, more people were needed to work in the tertiary sector, especially in schools, hospitals, shops, offices and in transport.

In 2008, 2 per cent were engaged in the primary sector, 27 per cent in the secondary sector and 71 per cent in the tertiary sector.

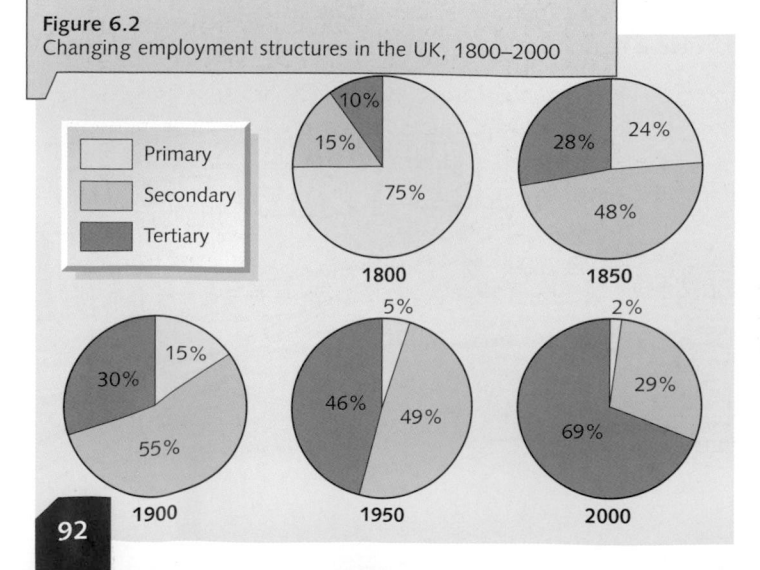

Figure 6.2
Changing employment structures in the UK, 1800–2000

- Primary
- Secondary
- Tertiary

1800: 10%, 15%, 75%
1850: 24%, 28%, 48%
1900: 15%, 30%, 55%
1950: 5%, 46%, 49%
2000: 2%, 29%, 69%

Differences between places

Employment structures can also be used to show differences between places. Figure 6.3 shows:
- changes in employment structure between the economic regions of the UK
- the number of people actually employed in each region.

The map shows that:
- the south-east region employs most people
- all regions have their highest percentage of workers in the tertiary sector and their lowest percentage in the primary sector
- the south-east has the lowest proportion in both the primary and secondary sectors and the highest proportion in the tertiary sector
- the two Midland regions appear to have the highest percentage in secondary employment.

You should realise, however, that within each region there are likely to be large differences between places. For example, as shown in Figure 6.4, one place may be a market town, one an industrial centre and one a holiday resort. Each place will, therefore, have a different population structure.

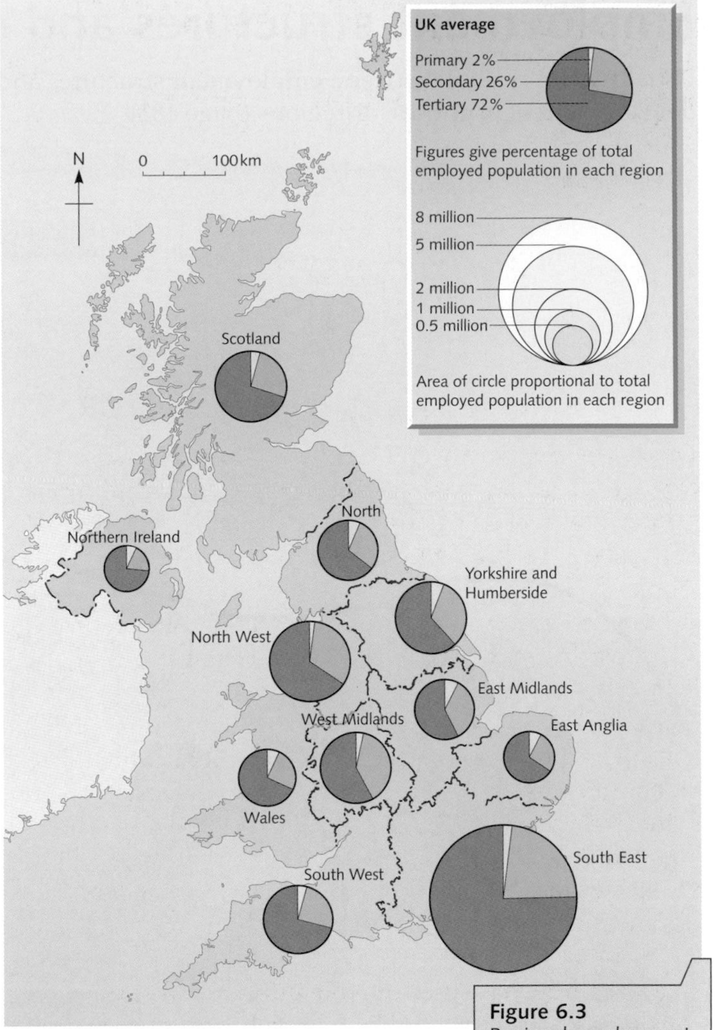

Figure 6.3
Regional employment structures in the UK

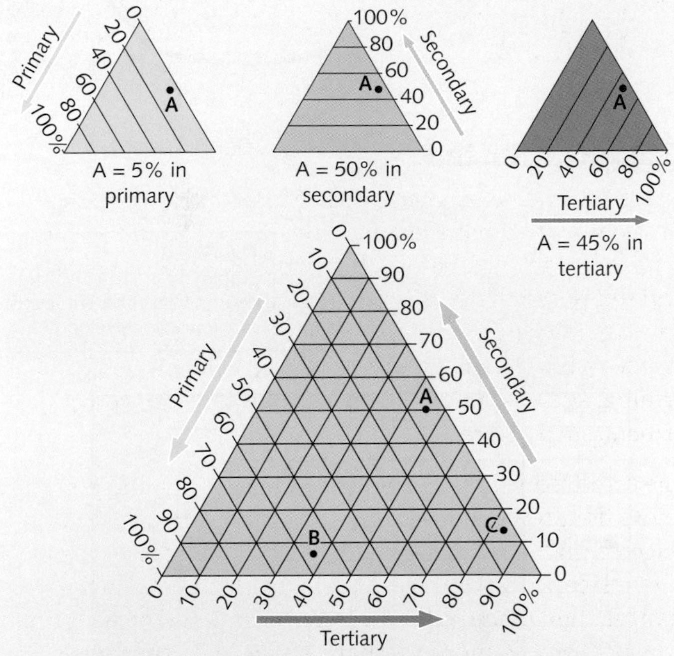

Triangular graphs

A triangular graph is an equilateral triangle with each of its three 'bases' divided into percentage scales. Each base represents one of the three variables of primary, secondary and tertiary activities. It is convenient, though not essential, to make the sides of the triangle 10 cm long (so that 1 cm = 10%).

Figure 6.4 shows how the three variables are plotted to show the employment structure of Town A. The figure for the primary sector is found by using the left-hand scale (see the yellow graph), for secondary activities by using the right-hand scale (blue graph) and for tertiary activities the base (orange graph). The answer for Town A is given underneath the graph. Try to complete the table for Towns B and C. These three towns represent (but not necessarily in this order) a small market town, a holiday resort and an industrial town. Match the figures with the appropriate letter (and be able to justify your answer).

Town	Primary	Secondary	Tertiary
A	5	50	45
B			
C			

*Note that the numbers in each row add up to 100.

Figure 6.4
Employment structures: the triangular graph

Employment structures and development

Usually there is a link between employment structures and levels of economic development of countries, or regions (page 181).

(page 181)

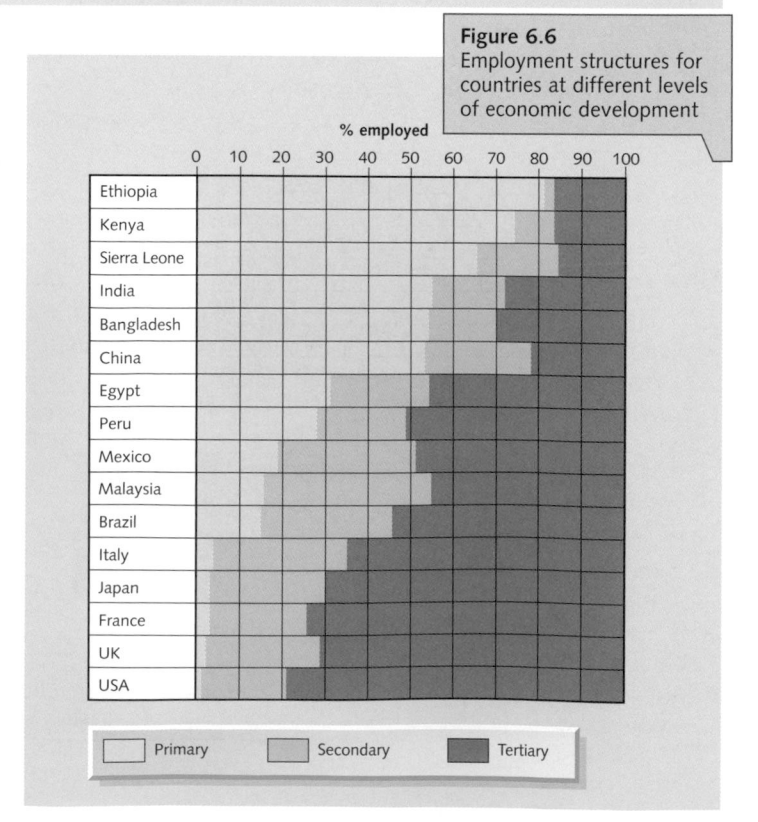

Figure 6.5
Employment structures and economic development

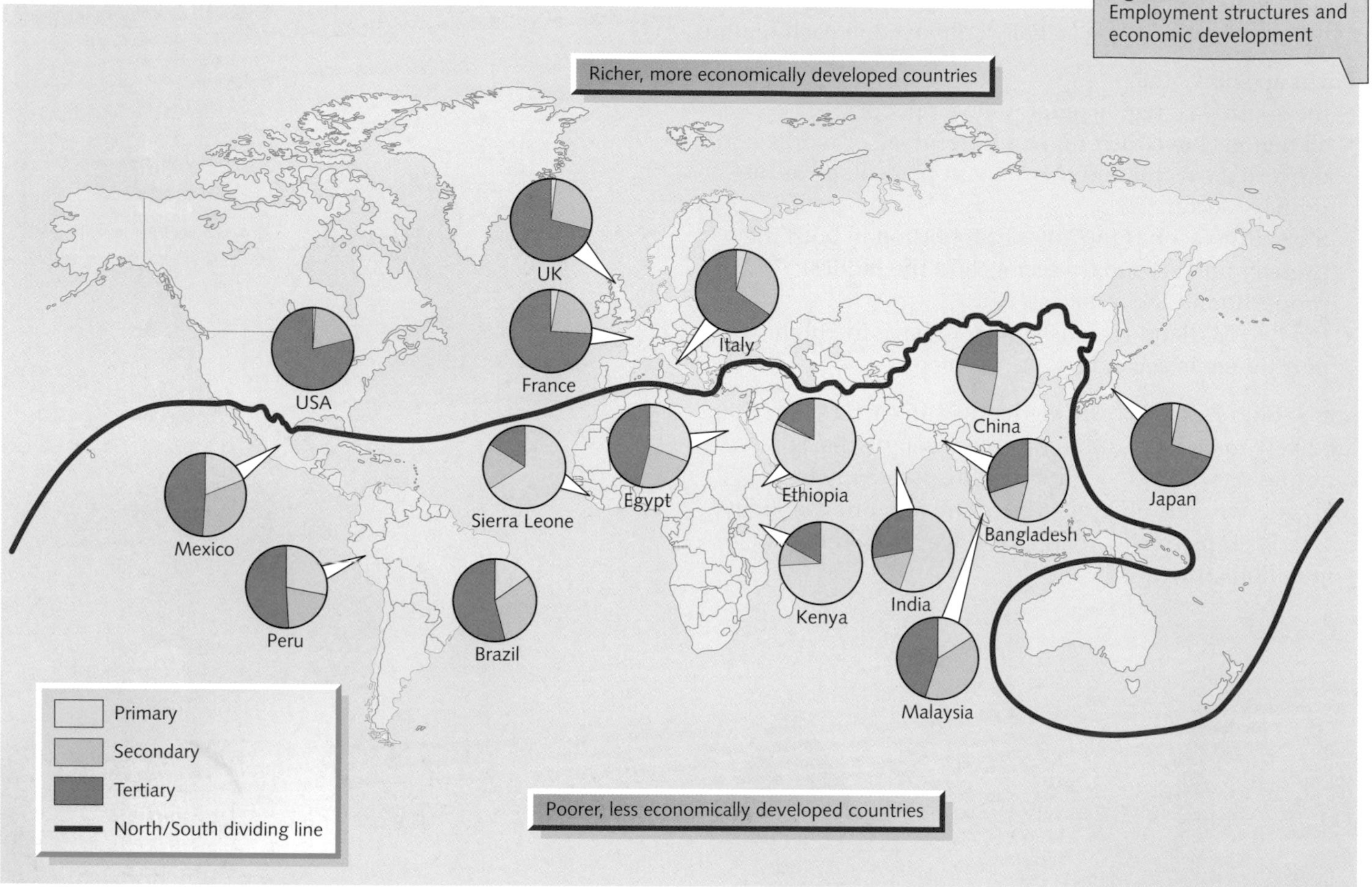

Richer, more economically developed countries

UK
France
USA
Italy
Mexico
Sierra Leone
Egypt
Ethiopia
China
Japan
Peru
Brazil
Kenya
India
Bangladesh
Malaysia

Primary
Secondary
Tertiary
North/South dividing line

Poorer, less economically developed countries

As shown on Figure 6.5:

- The richer, industrialised more economically developed countries (MEDCs) have a very high percentage in the tertiary sector, a high percentage in the secondary sector and a very low percentage in the primary sector.
- The poorer, least industrialised less economically developed countries (LEDCs) have, by contrast, a very high percentage in the primary sector (most are farmers), and a low percentage in both the secondary and service sectors.

Figure 6.6 has ranked a selection of countries according to the proportion of people engaged in the primary sector. You should refer to this graph when you look at the GDP of these countries on page 180.

Figure 6.6 is a percentage bar graph. This is a horizontal bar, again ideally drawn 10 cm long so that it is easy to divide into 10 per cent segments (1 cm = 10%). The bar is then divided into three to indicate the percentage of the working population in each of the primary, secondary and tertiary sectors.

Figure 6.6
Employment structures for countries at different levels of economic development

% employed

0 10 20 30 40 50 60 70 80 90 100

Ethiopia
Kenya
Sierra Leone
India
Bangladesh
China
Egypt
Peru
Mexico
Malaysia
Brazil
Italy
Japan
France
UK
USA

Primary
Secondary
Tertiary

Employment structures

Key Words and Terms

You should know the meaning of the following terms:
- employment structures
- economic activities
- primary activities/sector
- secondary activities/sector
- tertiary activities/sector
- quaternary.

Key Ideas

You should know and understand the following:
- Economic activity can be classified into (traditionally) three and (more recently) four groups.
- Employment structures change over time and vary between places.
- Employment structures can be used as an indicator of levels of economic development.

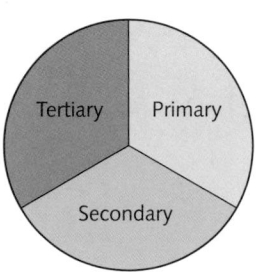

Skills, Theories and Models

- Interpret and use pie, triangular and divided bar graphs
- Understand the process and effects of change
- Classification of economic activities

QUESTIONS

1 (Pages 92 and 94)

a i) Describe the changes in the UK's employment structure between 1800 and 2000. (5)

ii) Give reasons for these changes. (6)

b i) Describe the differences in employment structure between the four most economically developed and the four least economically developed countries shown on Figures 6.5 and 6.6. (4)

ii) Give reasons for these differences. (4)

2 (Page 93)

The triangular graph shows the percentage of the working population in each of the three main sectors of activity for selected countries. Using the graph:

a Complete the table above the graph. (5)

b Which country has 42 per cent in primary, 21 per cent in secondary and 37 per cent in tertiary? (1)

c Name the four most economically developed countries. (4)

d Name the four least economically developed countries. (4)

e Give three differences between the figures for the two sets of countries. (3)

f Give three reasons for the differences between the two sets of figures. (3)

	Japan	Brazil	Kenya
% primary sector	3		81
% secondary sector		43	
% tertiary sector	56		

❶ Bangladesh
❷ Brazil
❸ Egypt
❹ Nepal
❺ India
❻ Italy
❼ Japan
❽ Kenya
❾ UK
❿ USA

Farming systems and types

Farming is an industry and operates like other industries. It is a **system** with **inputs** into the farm, **processes** which take place on the farm and **outputs** from the farm (Figure 7.1).

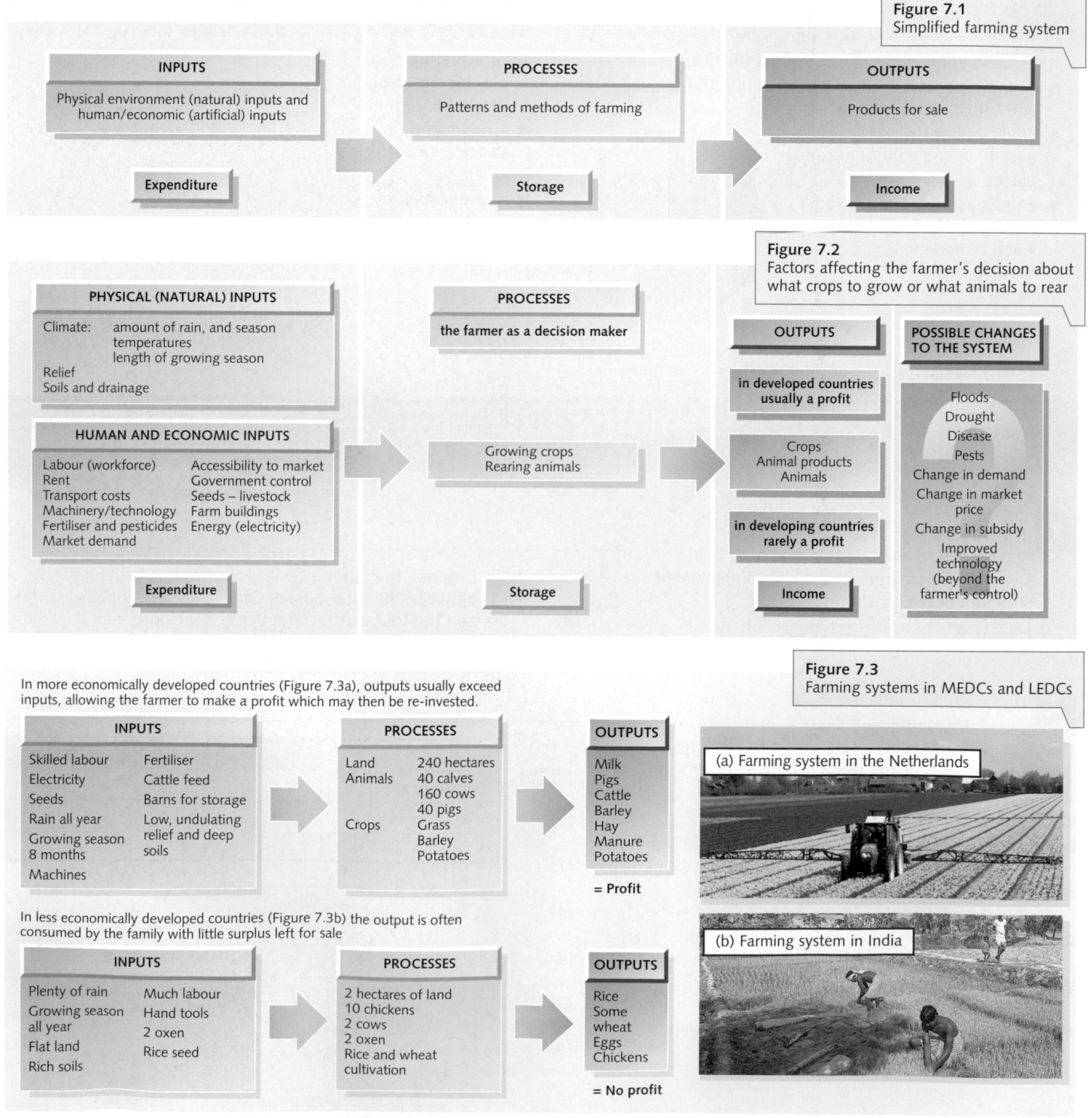

Figure 7.1
Simplified farming system

INPUTS	PROCESSES	OUTPUTS
Physical environment (natural) inputs and human/economic (artificial) inputs	Patterns and methods of farming	Products for sale
Expenditure	Storage	Income

Figure 7.2
Factors affecting the farmer's decision about what crops to grow or what animals to rear

PHYSICAL (NATURAL) INPUTS
Climate: amount of rain, and season
              temperatures
              length of growing season
Relief
Soils and drainage

HUMAN AND ECONOMIC INPUTS
Labour (workforce)   Accessibility to market
Rent   Government control
Transport costs   Seeds – livestock
Machinery/technology   Farm buildings
Fertiliser and pesticides   Energy (electricity)
Market demand

Expenditure

PROCESSES
the farmer as a decision maker

Growing crops
Rearing animals

Storage

OUTPUTS

in developed countries usually a profit

Crops
Animal products
Animals

in developing countries rarely a profit

Income

POSSIBLE CHANGES TO THE SYSTEM
Floods
Drought
Disease
Pests
Change in demand
Change in market price
Change in subsidy
Improved technology
(beyond the farmer's control)

In more economically developed countries (Figure 7.3a), outputs usually exceed inputs, allowing the farmer to make a profit which may then be re-invested.

Figure 7.3
Farming systems in MEDCs and LEDCs

INPUTS		PROCESSES		OUTPUTS
Skilled labour	Fertiliser	Land	240 hectares	Milk
Electricity	Cattle feed	Animals	40 calves	Pigs
Seeds	Barns for storage		160 cows	Cattle
Rain all year	Low, undulating		40 pigs	Barley
Growing season 8 months	relief and deep soils	Crops	Grass	Hay
Machines			Barley	Manure
			Potatoes	Potatoes

= Profit

(a) Farming system in the Netherlands

In less economically developed countries (Figure 7.3b) the output is often consumed by the family with little surplus left for sale

INPUTS		PROCESSES	OUTPUTS
Plenty of rain	Much labour	2 hectares of land	Rice
Growing season all year	Hand tools	10 chickens	Some wheat
Flat land	2 oxen	2 cows	Eggs
Rich soils	Rice seed	2 oxen	Chickens
		Rice and wheat cultivation	

= No profit

(b) Farming system in India

The farmer as a decision maker

Each individual farmer's decision on what crops to grow or animals to rear, and which methods to use to maximise outputs, depends on an understanding of the most favourable physical and economic conditions for the farm (Figure 7.2). Sometimes the farmer may have several choices and so the decision may depend on individual likes and expertise. On other occasions the choice may be limited by extreme physical conditions or economic and political pressures.

Classification of types of farming

The classification shown in Figure 7.4 is based on the following criteria:

Specialisation This includes **arable** (the growing of crops), **pastoral** (the rearing of animals) and **mixed** (both crops and animals) farming.

Economic status **Commercial** farming is the growing of crops or rearing of animals for sale (i.e. outputs exceed inputs). **Subsistence** farming is when just sufficient food is provided for the farmer's own family (i.e. outputs may be the same or less than the inputs and so the family may struggle for survival).

Intensity of land use This depends on the ratio between land, labour and capital (money). **Extensive** farming is where the farm size is very large in comparison with

either the amount of money spent on it (Amazon Basin) or the numbers working there (American Prairies). **Intensive** farming is when the farm size is small in comparison with either the numbers working there (Ganges Delta) or the amount of money spent on it (Denmark).

Land tenure **Shifting** (and **nomadic**) cultivation is where farmers move from one area to another. **Sedentary** is where farming and settlement is permanent.

Remember that the map in Figure 7.4 is simplified. It only shows the generalised world location of the main types of farming. It does not show local variations, transitions between the main farming types nor if several types occur within the same area.

	Type of farming	Named example
1	Nomadic hunting and collecting	Australian aborigines
2	Nomadic herding	Maasai in Kenya, Sahel countries
3	Shifting cultivation	Amerindians of Amazon Basin
4	Intensive subsistence agriculture	Rice in the Ganges Delta
5	Plantation agriculture	Oil palm in Malaysia
6	Livestock ranching (commercial pastoral)	Beef on the Pampas
7	Cereal cultivation (commercial grain)	Canadian Prairies, Russian Steppes
8	Mixed farming	Netherlands, Denmark
9	'Mediterranean' agriculture	Southern Italy, southern Spain
10	Irrigation	Nile Valley, California
11	Unsuitable for agriculture	Sahara Desert

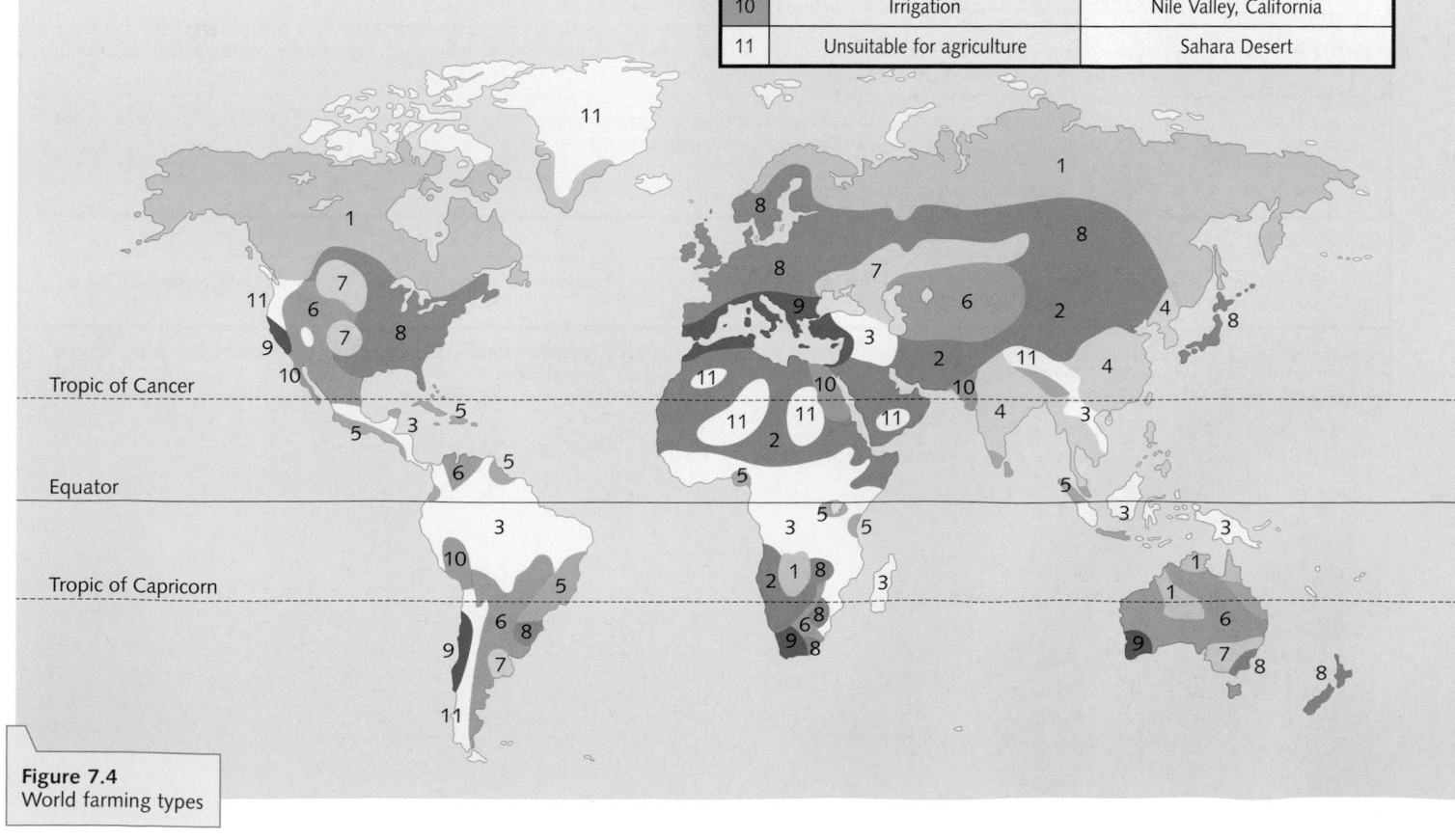

Figure 7.4
World farming types

Farming in the UK

Factors affecting farming

The location of different types of farming at all scales depends on the interaction of three factors: physical (environmental); human (social) and economic; and political (Figure 7.5).

Ideally, individual farmers should have a knowledge and understanding of each of these factors in their local area.

They can then make decisions as to which crops to grow and/or which animals to rear in order to give them the greatest profit. Sometimes the farmer may have several choices and so the decision may depend on individual preferences, traditions and expertise – the so-called **behavioural factor**.

Figure 7.5
Factors affecting the location and distribution of farming types in the UK

Physical (environmental) factors	
Relief and altitude	Usually the flatter and the more low-lying the land, the more efficient and commercial is the farm (arable in East Anglia). Output tends to decrease as the land gets steeper and higher and farming becomes less commercial and more extensive (pastoral in the Lake District).
Soils	The deeper and richer the soil, the more intensive and commercial the farming (alluvium of the Fens). Ideally soils should be well drained yet capable of retaining water.
Temperature/sunshine (pages 201 and 207)	In Scotland, summers are cool and the growing season is too short for most cereals. Moving south, temperatures, the amount of sunshine and the length of the growing season all increase. Aspect is an important local factor (maximum sunlight, protection against frost and wind).
Rainfall/water supply (pages 202 and 207)	Areas with adequate and reliable rainfall throughout the year tend to produce good grass for rearing animals (western Britain). Drier areas to the east grow cereals (East Anglia) and fruit (Kent).
Human (social) and economic inputs	
Land ownership/tenure	Many British farmers own their own farms but some are tenant farmers (the former are usually more commercial). A small, but increasing, number are run by processing companies (frozen foods in East Anglia).
Size of farms/fields (page 106)	Farms increase in size as larger, more efficient farmers buy up smaller, less successful farms. Field size has also increased, especially where hedgerows have been removed (eastern England). The EU encourages larger farms and larger fields.
Competition for land (page 62)	Many traditional farming areas, especially near the rural–urban fringe, are under threat from urban sprawl and the demand for new roads, industry, housing and recreation.
Transport and markets	Perishable goods need to be produced near to markets for freshness and bulky goods near to markets due to their weight. Fruit and vegetables (market gardening) are grown near most large urban areas.
Capital (money)	By world standards the UK is well-off, so farmers can find money to improve their farm buildings and machinery and to buy fertiliser and good-quality seed and animals.
Mechanisation/technology	The increased use of labour-saving machinery and computers increases outputs but reduces the need for farm workers. Farms in the south and east tend to be more mechanised than those in the north and west.
Processors/retailers	Demands on standards and prices set by food processors and supermarket chains.
Political factors	
Government/EU policies (pages 106 and 107)	Governments have provided grants for new stock and machinery and subsidies to guarantee a fixed income. This now comes from the EU's Common Agricultural Policy (CAP).
Variable inputs	Farmers are vulnerable to changes in government/EU policies, market prices and market demand. They are also affected by changes in the weather (floods, drought and frost) and by disease (foot and mouth, potato blight).

Distribution of the main farming types in the UK

It is essential, if British farmers are to make a profit and earn a living, that they are as efficient as possible. When looking at the different physical conditions of the British Isles (i.e. climate, soils and relief) and then taking into consideration various human and economic factors (mainly transport, markets and capital), it can be seen that different parts of the UK favour different types of farming.

Figures 7.6 and 7.7 show that there are five main types of farming in the UK. However, the map has been simplified to make interpretation easier. It suggests that all farmers within a certain location specialise in just one type of farming, e.g. sheep, cattle or cereals. The map hides local variations and fails to show that most farmers will keep some animals and grow some crops.

Arable farms are found in the east where summers are sunny and warm, there is less rain, the land is low-lying and flatter, soils are deep and fertile, transport systems are good, and where there are large urban markets nearby.

Cattle farms are important where summers are cool and winters are mild, there is plenty of rain throughout the year, the land is low-lying and relatively flat, transport is good and urban markets are within easy reach.

Hill sheep farms predominate in those parts of Scotland, Wales and northern England where the land is high and steep, soils are poor, temperatures are lower, rainfall is heavy, transport systems are poorly developed and urban markets are not within easy reach.

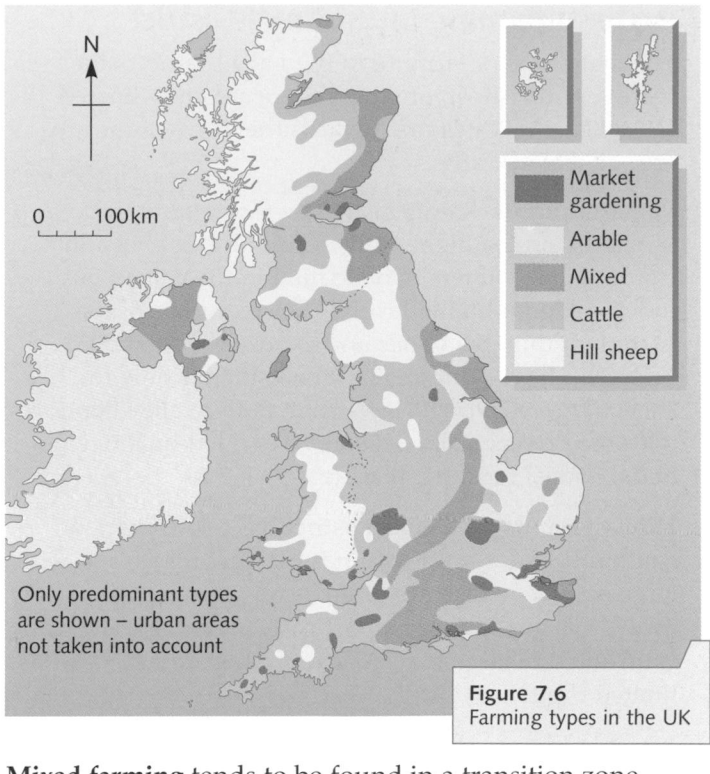

Figure 7.6
Farming types in the UK

Only predominant types are shown – urban areas not taken into account

Legend: Market gardening, Arable, Mixed, Cattle, Hill sheep

Mixed farming tends to be found in a transition zone between the crop-growing areas in the east and the animal-rearing areas in the west.

Market gardening is important near large urban areas and where transport links are good. As some market garden produce can be grown under cover in artificial conditions (flowers, tomatoes), human factors tend to be more important than physical factors.

Figure 7.7
The variety of UK farms:
(a) Arable
(b) Cattle
(c) Hill sheep
(d) Mixed
(e) Market gardening

99

Pastoral farming in the Lake District

The main type of farming in the Lake District is the rearing of hill sheep of which there are an estimated 1.5 million. Valley floors towards the edges of the area are suitable for cattle.

Physical inputs Rainfall is heavy throughout the year, especially on the higher fells. Summers are cool and cloudy while winters may be mild in the valleys but much colder at higher levels. The steep valley sides, with their thin, poor soils and exposed rock, and the high fells, have poor-quality grass suitable only for sheep (Figure 7.9). The low-lying, flatter valley floors, with their deeper, alluvial soil, have good-quality grass and are ideal for cattle rearing.

Human inputs The Lake District has relatively poor communications both within it (narrow roads) and with large markets which are often a considerable distance away (this increases transport costs and time). The land is unsuited to large-scale mechanisation and the area has limited capital.

Variable inputs These include occasional heavy snowfalls at lambing time; changes in government (mainly EU) policies; and outbreaks of disease (sheep affected by radioactive fallout after Chernobyl in 1986); and restrictions on the movement of animals (cattle due to BSE in 1996, sheep due to foot and mouth in 2001).

Processes On the hill farm, these include lambing, dipping, shearing, collecting hay and maintaining dry-stone walls.

Outputs Young lambs, lamb (meat), wool and hay.

Figure 7.8
A Lake District farm

Recent changes These include:
- a fall in the price of lambs, lamb and wool
- marginal land being taken out of production
- a decrease in the number of farms
- more farms turning to other sources of income, mainly tourism, e.g. B&B, camping and caravan sites, craft shops (page 106).

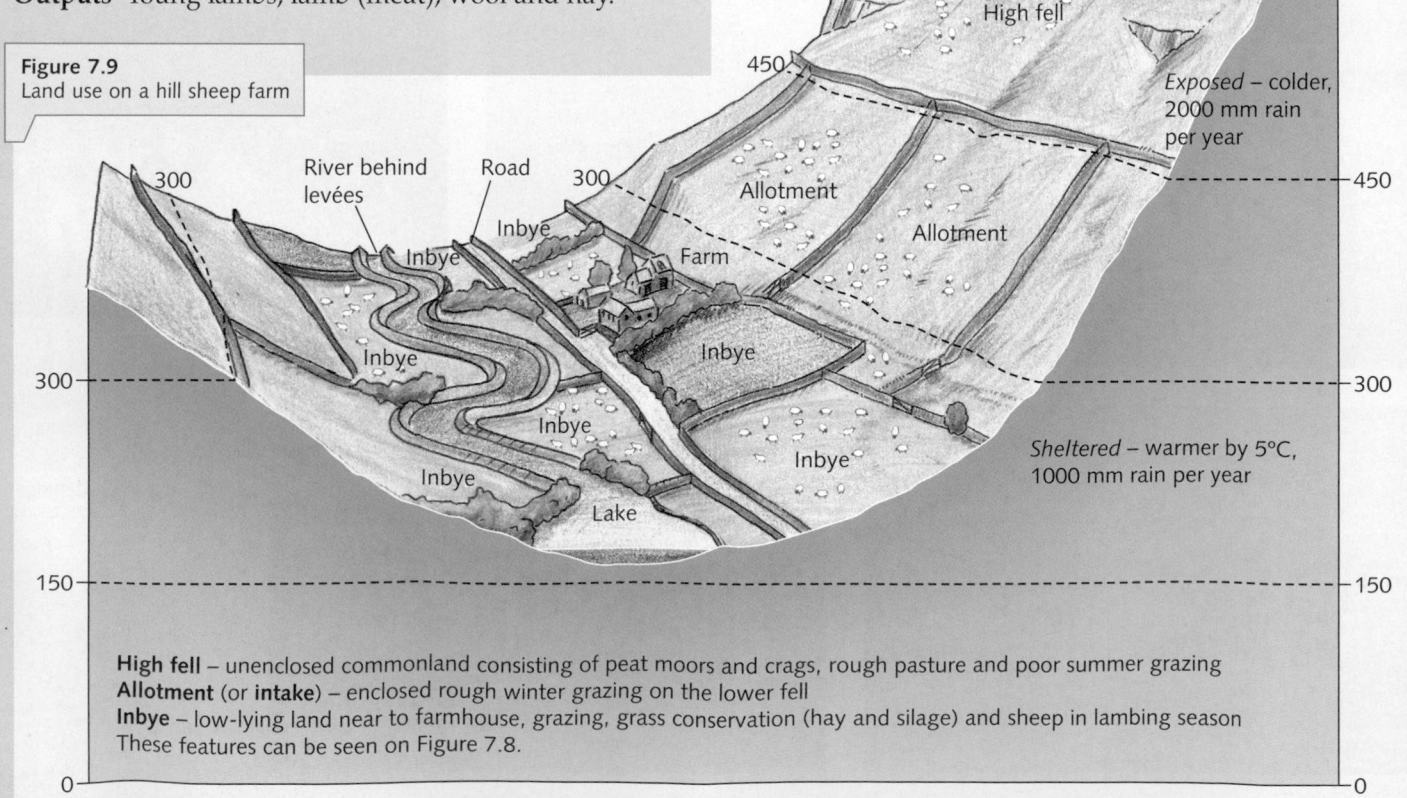

Figure 7.9
Land use on a hill sheep farm

Height (metres)

600

High fell

Exposed – colder, 2000 mm rain per year

450

Allotment

Allotment

River behind levées

Road

300

300

Inbye

Farm

Inbye

Inbye

Inbye

Inbye

Inbye

Sheltered – warmer by 5°C, 1000 mm rain per year

Inbye

Lake

300

150

600

450

300

150

High fell – unenclosed commonland consisting of peat moors and crags, rough pasture and poor summer grazing
Allotment (or **intake**) – enclosed rough winter grazing on the lower fell
Inbye – low-lying land near to farmhouse, grazing, grass conservation (hay and silage) and sheep in lambing season
These features can be seen on Figure 7.8.

0

0

Arable farming in East Anglia

The main type of farming in East Anglia is the commercial growing of cereals, mainly wheat and barley, together with root crops (potatoes and sugar beet) and vegetables (peas and beans).

Physical inputs Rainfall is the lowest in the UK (under 650 mm a year) but fortunately most comes during the summer growing season. The warm, sunny summers are ideal for ripening crops, while frosts during the cold winters help break up the soil. The land is gently undulating and low-lying. The soils, which are deep, fertile and well drained, are mainly either alluvium deposited by rivers, or boulder clay deposited on chalk during the Ice Age.

Human inputs The flatness of the land has allowed a good transport system to develop (road and rail) linking the region with large, nearby markets in south-east England (saves time and costs, easier to transport both bulky and perishable goods – Figure 7.11). The land is ideal for large-scale machinery (combine harvesters, sprayers) and the region has access to considerable capital.

Variable inputs These include possible droughts during summer; changes in government (mainly EU) policies; and disease affecting cereals or other crops.

Processes These include ploughing, harrowing, weeding, applying fertiliser and pesticides, weeding, harvesting and maintaining field boundaries and machinery.

Outputs Wheat, barley, sugar beet, potatoes, peas and beans.

Figure 7.10
Extensive cereal growing in East Anglia

Recent changes These include:
- larger farms – encouraged by the EU
- larger fields (for larger machinery) due to removal of hedgerows (page 108)
- increased use of fertiliser (page 108)
- reduced subsidies (to reduce overproduction in the EU) (page 106)
- the ending of subsidies for set-aside land (page 106).

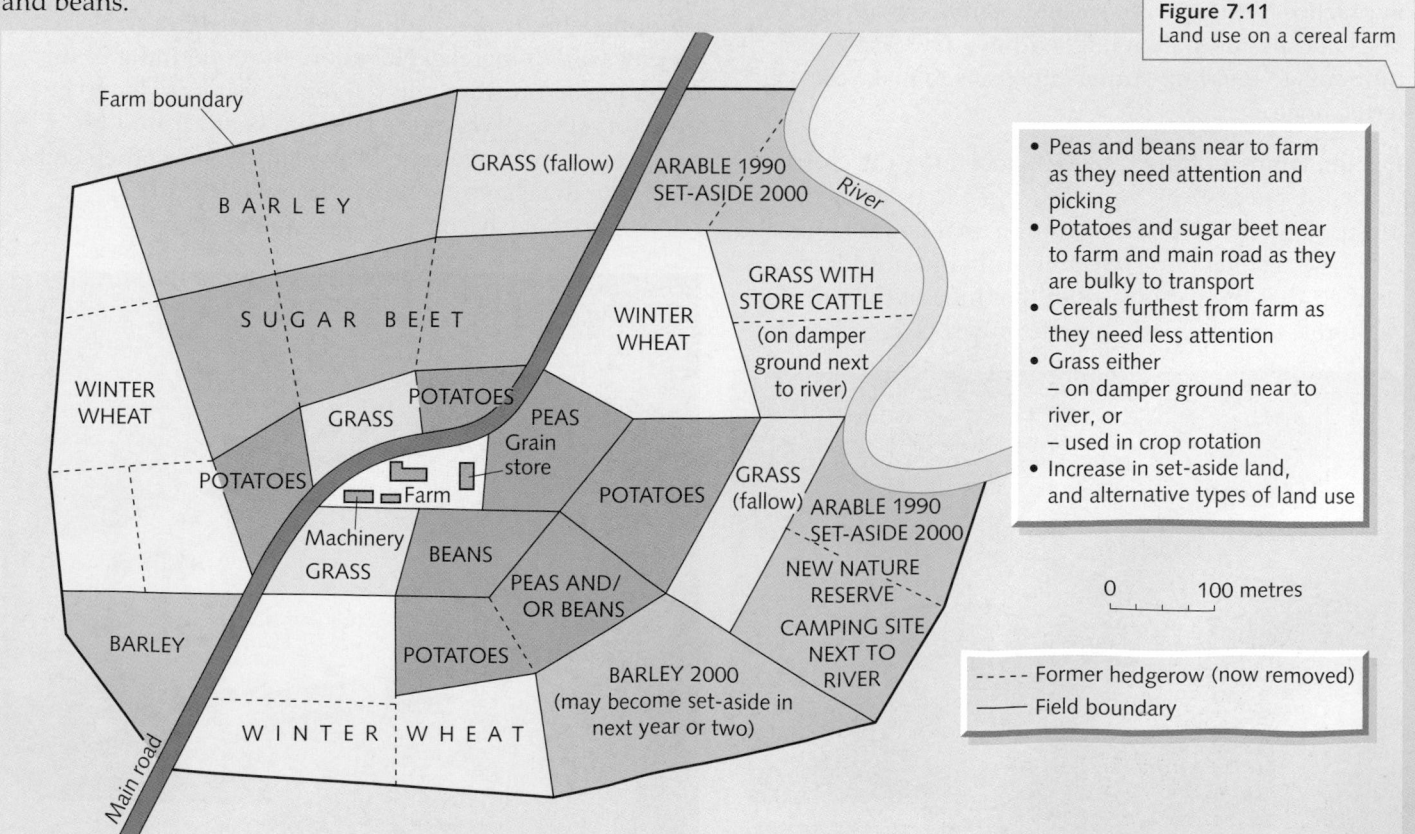

Figure 7.11
Land use on a cereal farm

Farm boundary

BARLEY

GRASS (fallow)

ARABLE 1990
SET-ASIDE 2000

River

SUGAR BEET

WINTER WHEAT

GRASS WITH STORE CATTLE
(on damper ground next to river)

WINTER WHEAT

POTATOES

GRASS

PEAS

Grain store

Farm

Machinery

GRASS

BEANS

POTATOES

PEAS AND/OR BEANS

POTATOES

GRASS (fallow)

ARABLE 1990
SET-ASIDE 2000

NEW NATURE RESERVE

CAMPING SITE NEXT TO RIVER

BARLEY 2000
(may become set-aside in next year or two)

BARLEY

WINTER WHEAT

Main road

- Peas and beans near to farm as they need attention and picking
- Potatoes and sugar beet near to farm and main road as they are bulky to transport
- Cereals furthest from farm as they need less attention
- Grass either
  – on damper ground near to river, or
  – used in crop rotation
- Increase in set-aside land, and alternative types of land use

0   100 metres

----- Former hedgerow (now removed)
——— Field boundary

Tropical agriculture

Subsistence farming is the traditional type of agriculture in many of the less economically developed countries and isolated regions that lie within the tropics. It is a farming system with relatively few inputs – such as fertiliser, machinery or capital (Figure 7.1) – and simple processes and low outputs, perhaps only sufficient to satisfy the immediate needs of the family or local community. The most extreme form of subsistence farming is shifting cultivation which is still practised in parts of the tropical rainforests (Figure 7.4).

Shifting cultivation in the Amazon rainforest

Amerindian tribes use machetes to clear small areas of rainforest. The felled trees and undergrowth are then burnt – hence the alternative name of 'slash and burn' (Figure 7.12). Houses, built from tree trunks and thatched with palm leaves, are built in a circle with the meeting house in the centre (Figure 7.13). Crops are grown in small plots, known as *chagras*. Manioc, the main crop, together with yams, peppers, pumpkins and beans, are planted into what is initially a very rich soil (Figure 7.14). The fertility of the rainforest soil depends on the rapid and unbroken recycling of nutrients (Figure 14.10). However, without the protective tree cover and the addition of fallen leaves, the main source of nutrients, the heavy afternoon rains (page 212) soon wash the nutrients out of the soil leaving it infertile. Within four or five years the land has to be abandoned, forcing the tribe to shift to a different area of forest where they have to clear new areas for their crops and build new houses for themselves. Although this method of farming may appear to be a wasteful use of the land, in practice it is the reverse as the rainforest quickly re-establishes itself allowing, within a few years, nutrients to build up sufficiently so as to make the soil fertile once again.

It is the women of the tribe who look after the crops while the men, after clearing the area to be farmed, seek to supplement the diet by hunting, mainly for tapirs and monkeys, fishing and collecting fruit and nuts. It is claimed that the various tribes that inhabit the Amazon rainforest are now probably the only society across the world that still lives in harmony with its environment and which provides a true example of sustainable development.

Recent changes continue to see the traditional Amerindian way of life threatened by the destruction of the rainforest (page 236). As land is cleared by giant logging companies or for new highways, beef cattle ranches, hydro-electric schemes and mineral exploitation, the Amerindians are either pushed further into the rainforest or forced to live on reservations. Where the land has been cleared for ranching, farms have soon been abandoned as their soils, not given the chance to recover as with shifting cultivation, have become infertile and eroded.

Figure 7.12
Slash and burn – a shifting cultivator clearing the rainforest

Figure 7.14
Hoeing the *chagra*

Figure 7.13
A traditional village within the cleared rainforest

Commercial farming is the growing of crops or the rearing of animals for sale at a profit. This means that in this farming system the income earned from outputs must exceed the money spent on inputs. Commercial farming was introduced into developing countries located within the tropics when they were colonies of a richer, more developed country (Figure 7.4). Crops such as sugar cane, rubber, bananas, tea, coffee, cotton and tobacco were grown on large estates, known as plantations, which relied on the use of cheap local labour.

Plantation agriculture in Malaysia

Plantations in what was then Malaya were first developed by British settlers in the early nineteenth century. As in many other tropical areas, large areas of forest were cleared and a single crop, in this case rubber, was planted in rows (Figure 7.15). Rubber, a typical plantation crop in that it is grown for export, thrives in a hot, wet climate, growing best on foothills of mountains where there is good drainage. As on all plantations, rubber needs a high capital input to clear the land, provide the trees and construct roads. It also requires much manual and cheap labour as the mature trees have to be 'tapped' daily in order to obtain the sap, or latex, needed to process rubber. The labourers, who are also needed to clear the forest, work in the nurseries and plant new trees, were provided by poorer Malays and by immigrants, mainly from India.

The Malaysian government took control of all the country's rubber plantations in the early 1970s, a time that happened to coincide with the rapidly falling world demand for the product, mainly caused by increased competition from synthetic rubber. Today, there are relatively few rubber plantations still operating in Malaysia and many of the remaining workers are forced to live below the poverty line (page 326).

Fortunately for Malaysia, this was also a time when the global demand for oil palm was growing. Oil palm, which now covers over 80 per cent of the country's plantations (Figure 7.16), has many advantages over rubber. These include higher yields and prices as well as lower production costs as labour is less intensive (Figure 7.17). This is partly due to being able to spray pesticides and apply fertiliser mechanically. Oil palm is also more versatile than rubber because, apart from providing an edible oil and being used in a wide range of foodstuffs, it can be used in the manufacture of soap, cosmetics and paint. It has also, in the last few years, become an important source for biofuel, mainly by countries in the EU (page 125). However, this has meant that more land, previously used for growing food for Malaysia's own domestic consumption, is increasingly being converted into oil palm plantations.

Figure 7.15
A rubber plantation

Figure 7.16
An oil palm plantation

		Rubber	Oil palm
Tonnes per hectare		2	25
Years for trees to mature		6 to 7	4 to 5
Labour intensive		High	Lower
Prices (income)		Low	High and rising
% world trade	1950	42	5
	2007	12	43
Uses		Limited	Numerous

Figure 7.17
Difference between rubber and oil palm

103

The effects of the global market on subsistence farming

Subsistence farming has always been the mainstay of agriculture in developing countries, especially those that lie within the tropics (Figure 7.4). It is a system that allows farmers to grow sufficient food for their family and local community in places that often lack capital and technology and where transport may be underdeveloped. But subsistence farming is under threat, not from within these countries themselves, but from a growing global market, the demands of supermarkets in richer countries, the introduction of new forms of technology that might not be appropriate, and improvements in world transport. While these changes may benefit people in developed countries, they can lead to food shortages, in places like Kenya, and to environmental problems, as in Egypt.

Kenya

Where rainfall is more reliable, crops are grown in small plots of land known as *shambas* (Figure 7.18). The most important crop is maize which may be grown under the shade of small banana trees. Beans, yams and millet are also grown, as are vegetables and tropical fruits. The land is farmed intensively and none is wasted.

But this part of Kenya can produce fruit, vegetables and flowers at a time when it is winter in northern Europe and where demand for fresh produce now remains high throughout the year. This has resulted in large transnational corporations and supermarkets from the EU replacing the former *shambas* with large estates that grow specialised crops. One example is an estate on the shores of Lake Naivasha that grows over 25 million carnations in a six-month period (Figures 7.19 and 7.20).

- The estate, which is well managed, employs 15 000 people but it is debatable whether the change from subsistence farming to the growing of flowers for export has improved the standard of living of local people.
- This estate, like many others, provides free accommodation, education and medical care for its employees but the work is extremely hard and poorly paid and it is uncertain whether the former small-scale farmer now has a better way of life or not.
- The increase in exports brings valuable currency to Kenya but, as more farmland is turned over to cash (export) crops, less land is available to provide food for the country's own people and so shortages are becoming more frequent.

Figure 7.18
Kenyan *shambas*

- Supermarkets, including those in the UK, are desperate to obtain reliable sources of fresh produce but they aim to pay the lowest possible price (not always a case of fair trade – page 330).
- New technology means that the flowers can be dried, graded and cooled to 4°C, enabling them to arrive fresh at their destination such as Covent Garden in London. Water is also extracted for irrigation from Lake Naivasha but some is lost through evapotranspiration, which has caused the lake to shrink in size, while the water that is returned is polluted with fertiliser and pesticides.
- Improvements in transport mean that flowers picked one afternoon can be sent along improved roads and in cooled lorries to Nairobi airport ready for an overnight flight to the EU where they can be sold the next day, but there has been little improvement in transport for local people.

Figure 7.19
Estate near Lake Naivasha growing carnations

Figure 7.20
Extent of the carnation estate

Egypt

Egypt became one of the world's earliest civilisations, mainly due to the annual flooding of the Nile and the silt that the river then deposited across its floodplain. Present-day satellite images of Egypt show a narrow, continuous strip of land extending on either side of the Nile until it widens to form the triangular shape of the river's delta (Figure 7.21). To either side of this strip of land is desert.

Throughout the 2000 years of the Pharaohs, Egyptians were more than self-supporting. Painted reliefs and hieroglyphics found in tombs along the valley give an accurate account of farming and the way of life in Ancient Egypt. From these it is known that the Ancient Egyptians grew wheat, millet, maize, flax, cotton and sugar cane as well as a wide range of fruit and vegetables. Peasants used farm tools and irrigated their fields using a technology appropriate for that time and which is still used along parts of the valley.

Without the Nile, Egypt would not exist but while its water brought life to the country, the river also posed problems. If the annual flood was too high, lives and property were lost, and if it was too low then not enough crops could be grown to feed the people. Once the flood subsided, farmers built irrigation channels and planted their crops. As the river level continued to fall, water had to be raised onto the fields – a task done by either the saquia wheel (Figure 7.22a), the shaduf (Figure 7.22b) or the Archimedes screw. This way of life changed with the building of the Aswan Dam and now floodwater is held back to be released more evenly. This means that the flooding is controlled and water for crops is available throughout the year (Figure 7.23).

However, the dam has also created problems. Silt is no longer spread naturally over the fields and so, with the added burden of an extra crop now grown each year, the soil becomes less fertile. To maintain its fertility, farmers have to buy fertiliser which they can ill afford and which is later washed back to pollute the river. Irrigation means that water is on the earth's surface for longer than during a flood. As the water evaporates under the hot sun, it leaves behind a concentration of salt, a process known as **salinisation**, which is harmful to the roots of crops. Also, as in Kenya, more crops are being grown for export, leaving fewer for domestic use.

Figure 7.21
The Nile delta: satellite view

a

b

Figure 7.22
(a) Saquia wheel
(b) Shaduf

Figure 7.23
Intensive farming in the Nile delta

Changes in farming in the EU

Farming, like other industries, is dynamic and always changing. These changes may result from political decisions made by the EU, economic changes in regional or global economies, social pressures in farming communities, and environmental factors resulting from climatic change. Sometimes these changes may have advantages while at others they may be detrimental to either the farmer and the farming community, the people who need feeding, or the local environment.

All the present 27 members of the EU are meant to implement the Common Agricultural Policy (CAP) which was set up in 1962. The basic aims of the CAP were to:
- create a single market in which agricultural goods could easily be transferred
- make the EU self-sufficient by giving preference to produce grown within member countries
- give financial support to EU farmers, mainly through guaranteed prices (subsidies) and markets
- help to maintain jobs and improve the standard of living in farming areas, especially in marginal areas such as hill farms in the UK
- keep consumer prices stable.

Although most of these aims were fulfilled by the 1990s, there was increasing concern over both the running and the effects of the CAP. These included:
- the 70 per cent of the total EU's budget being spent on agriculture
- overproduction of certain products due to the payment of subsidies and improvements in technology
- a decrease in the number of farms, as those that were larger and more profitable took over those that were smaller and less efficient
- a decrease in the number of farm workers, due to improvements in technology (improved machinery and the introduction of computers)
- the protection of EU produce which restricted imports from less economically developed countries and the development of fair trade (page 330)
- insufficient care of the environment, for example the draining of wetland habitats and the removal of hedgerows when enlarging field sizes in places such as eastern England.

Figure 7.25 lists some of the benefits and some of the problems still facing EU countries over the CAP. It should also be pointed out that there is often conflict between member states, for example between those, including the UK, that believe too much money is spent on the CAP and those, such as France, that wish to maintain existing levels.

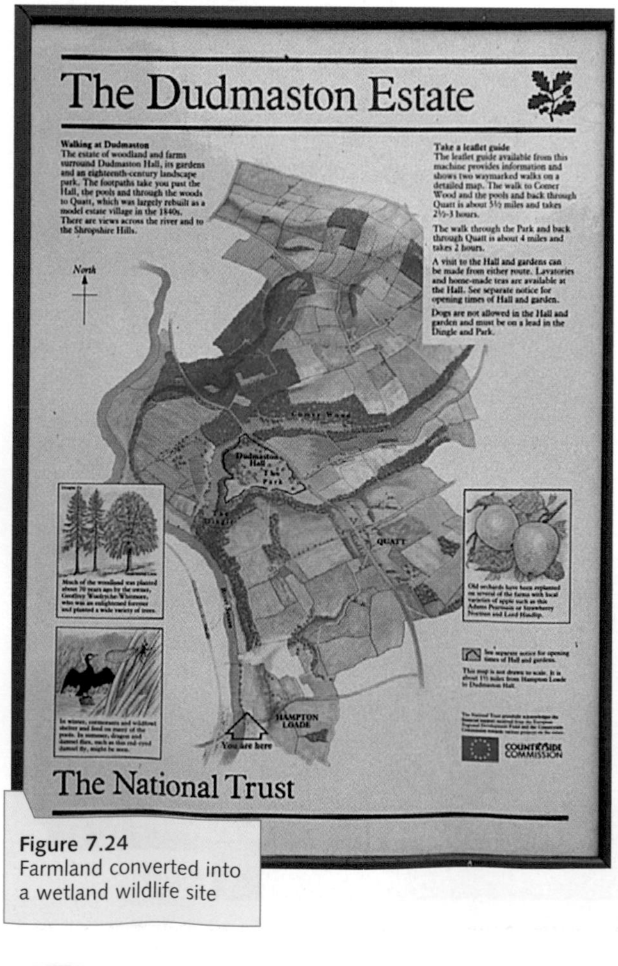

Figure 7.24
Farmland converted into a wetland wildlife site

Achievements

Achieved a larger measure of self-sufficiency. This reduces the costs and unreliability of imports.

Created higher yields due to input of capital for machinery and fertiliser.

In NW Europe the average farm size has increased almost to the recommended level.

Amalgamation of fields – in parts of France the number of fields has been reduced to one-eighth of the 1950 total.

Production has changed according to demands, e.g. less wheat and potatoes and more sugar beet and animal products.

Subsidies to hill farmers have reduced rural depopulation.

Poorer farmers gain opportunity to receive a second income by working in nearby factories ('5 o'clock farmers') or from tourism.

Higher income for farmers.

Subsidies have reduced the risk of even higher unemployment in rural areas.

Reduced reliance on crops imported from developing countries which themselves have food shortage. A surplus one year can offset a possible crop failure in another year.

Problems

An increase in food prices, especially in the net importing EU countries of Germany and the UK.

A tendency to overproduce which created food surpluses.

Dumping unwanted produce on less developed countries.

Increased gap between the favoured 'core' agriculture regions and the periphery.

Peripheral farm units still very small and often uneconomic.

High costs of subsidies. 'Industrial' countries such as the UK object to 70% of the EU budget being spent on agriculture.

'Five o'clock farmers' spend insufficient time on their farms. In France 15% and in Germany 30% of farmers have a second income.

Destruction of hedges to create larger fields destroys wildlife and increases the risk of soil erosion.

By reducing imports from developing countries the latter's main source of income is lost thus increasing the trade gap between the two areas.

Figure 7.25
A balance sheet showing some of the achievements and some of the problems still to be faced by the EU's Common Agricultural Policy

Changes since 2000

- **Subsidies** were introduced in the early days of the EU to guarantee farmers a minimum price and an assured market for their produce. Farmers tended, as a result, to overproduce and the payment of subsidies became an increasing drain on EU finances. Since 1992 there has been a progressive reduction in the level of subsidies paid until, in 2006, it was announced that future payments for farm produce would gradually be phased out. Farmers are now only receiving payments if they meet environmental and animal standards and keep their land in good condition – the so-called 'health check'. The EU is, in other words, encouraging farmers to become the 'guardians of the countryside'.

- **Quotas** were introduced in 1984 to reduce milk output and farmers who, at present, produce too much milk are fined. Quotas are also to be phased out but, to reduce the impact on dairy farmers, not until 2015 and after five years of annual quota increases.

- **Set-aside** was initially introduced on a voluntary basis, but later enforced, to try to reduce overproduction, especially of cereal crops. Farmers who took 15 per cent of their land out of cultivation were given £20 a hectare. Instead of growing crops, farmers had either to leave their land fallow (pasture), turn it into woodland (under the Farm Woodland Management Scheme) or **diversify** into other non-agricultural land uses (Figure 7.28). By the turn of this century, an increasing number of British farmers other than cereal growers were also forced into diversification due to such factors as the high value of the pound, rising costs of animal feed and fertiliser, the reduction in subsidies and a loss of overseas markets. Diversification included, for example, creating trout farms (Figure 7.26), caravan parks (Figure 7.27), wildlife habitats (Figure 7.24), golf courses and nature trails.

- When economic conditions changed in 2007–08 and the global price of food products rapidly increased, the EU reduced the set-aside rate to zero, allowing British farmers, for example, to bring up to 5 million hectares back into production.

Figure 7.26
Kilnsey Park Trout Farm, Yorkshire

Figure 7.27
Caravan park

Figure 7.28
Diversification of farm use

Recreation	• Creation of new golf courses (possible conflict as a development on greenbelt land). • Mountain-bike access and motorcycle scrambling (conflict with other countryside users). • Riding stables; trout farms (Figure 7.26).
Accommodation	• Campsites and caravan parks (conflict over visual appearance – Figure 7.27). • Holiday villages for families to stay in, surrounded by activity opportunities (conflict with local people)
Environment and wildlife	• Restoration of wetland ecosystems/re-creation of ponds (Figure 7.24). • Planting trees/creating new areas of woodland. • Creation of nature trails, complete with information boards.

Farming and the environment

The use of chemicals

- **Pesticide** is defined as all chemicals applied to crops to control pests, diseases and weeds. The United Nations claimed in the 1960s that one-third of the world's crops were lost each year due to these three factors. Scientists have estimated that without pesticides, yields of cereal crops could be reduced by up to 45 per cent within two years. Unfortunately, pesticides can also affect non-harmful wildlife such as bees.

- **Fertiliser** is a mineral compound containing one or more of the six main nutrients needed for successful plant growth. The average soil rarely contains sufficient nutrients, especially nitrogen, phosphorus and potassium, to provide either a healthy crop or a high yield. Although it is expensive to use, chemical fertiliser replaces nutrients that have been removed from the soil. There is, however, considerable concern when nitrate is washed (**leached**) through the soil into:
  – rivers where, being a fertiliser, it causes a rapid growth of algae and other plants which use up oxygen leaving insufficient for fish life – the process of **eutrophication** (the effects of eutrophication can be seen in Figure 7.29)
  – underground domestic water supplies where it can be harmful to human health.

- **Phosphate**, released from farm slurry (animal manure) and untreated human sewage, can also pollute water supplies. Slurry can be 100 times more polluting than household waste (e.g. the Norfolk Broads).

The loss of wildlife habitats

The intensification of farmland during the second half of the twentieth century led to loss of important wildlife habitats, e.g. wetlands, moorlands and hedgerows.

Figure 7.29
Algal growth caused by excess nitrate in the water

The removal of hedgerows

Although most of Britain's hedgerows were only planted by farming communities in the eighteenth century, they are now often perceived to be part of our natural environment. Modern farming, especially in arable areas, uses large machines that are easier to work if the fields are large. This meant that between 1945 and 1990 over 25 per cent of Britain's hedges were cleared (in parts of East Anglia the figure was over 60 per cent) in order to create larger fields (Figure 7.30). Hedges were cleared because:

- they are costly and time-consuming to maintain
- they take up space that could be used to grow crops, and so cost the farmer money
- they limit the size of field machinery
- trees get in the way of mechanised hedge-trimmers.

In contrast, Figure 7.31 shows some of the advantages of well-maintained hedgerows.

Figure 7.30
A hedgeless landscape in eastern England

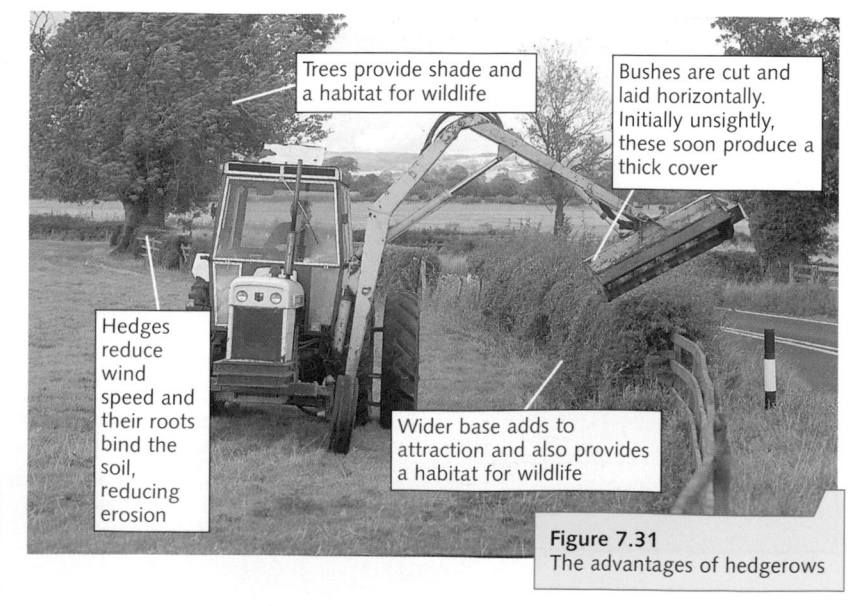

Trees provide shade and a habitat for wildlife

Bushes are cut and laid horizontally. Initially unsightly, these soon produce a thick cover

Hedges reduce wind speed and their roots bind the soil, reducing erosion

Wider base adds to attraction and also provides a habitat for wildlife

Figure 7.31
The advantages of hedgerows

Organic farming

Since the mid-1980s a small, but increasing, number of British farmers have turned to organic farming. Compared with conventional farming, organic farming is self-sustaining in that it does not misuse soil and water resources. Instead of using chemical (artificial) fertiliser and pesticides, it favours natural fertiliser such as animal and green manure, and mineral fertiliser such as fish and bonemeal. These natural fertilisers put organic matter back into the soil, enabling it to retain moisture during dry periods, allowing better drainage and aeration during wetter spells, and reducing the risk of soil erosion and exhaustion (page 254). Organic farming involves the intensive use of both land and labour and the use of fallow land. It is less likely to harm the environment as there are no nitrates to run off into rivers and no pesticides to harm wildlife.

Organic farming has its problems:
- using no chemical fertiliser means that, initially, yields are lower
- using no pesticide means more weeding has to be done
- farmers have to wait several years before they can market their goods as 'organic'
- the extra work means organic produce is more expensive for the shopper to buy.

Genetically modified (GM) crops

GM crops were first developed for commercial production in 1996 (Figure 7.32). Two-thirds of GM crops are produced in the USA, and 99 per cent by the USA, Argentina, Canada and China. GM crops include corn (maize), oilseed rape, sugar beet and soya beans. Although seen by scientists as a solution to food shortages in developing countries, especially in Africa, and a means of reducing food bills in developed countries, their production in the UK is being strongly opposed by conservation groups.

Genetic modification

Genetic modification involves taking genes from one species and inserting them into another to give it new qualities, such as improved resistance to pests, heat or cold. Advocates say the technique is sound, safe and can help to increase crop yields or improve animal breeds. Detractors argue that there is a risk that genes engineered in plants and animals will be transferred to species in the wild, irreversibly altering the world's ecosystems.

Figure 7.32
From *Geographical Magazine*, March 2002

Environmental Stewardship Scheme (ESS)

Figure 7.33 shows the location of 22 Environmentally Sensitive Areas (ESAs) chosen because of 'their high landscape, wildlife or historic value' and where it was felt that the land was under threat from changing farming practices. Farmers living in these areas were invited to join a scheme by which they received payment based on two levels of environmental commitment (replacing the payments previously paid in the form of subsidies). The lower level was paid on condition that farmers maintained the existing landscape, the higher level if they made environmental improvements such as replanting hedges or restoring ponds and traditional farm buildings. This, in 2005, became the Environmental Stewardship Scheme (ESS).

The ESS is a joint farming–environmental initiative whose main objectives are to:
- conserve wildlife
- maintain and enhance landscape quality and character
- protect the historic environment and natural resources
- promote public access and understanding of the countryside.

Within these objectives are two secondary aims: genetic conservation and flood management.

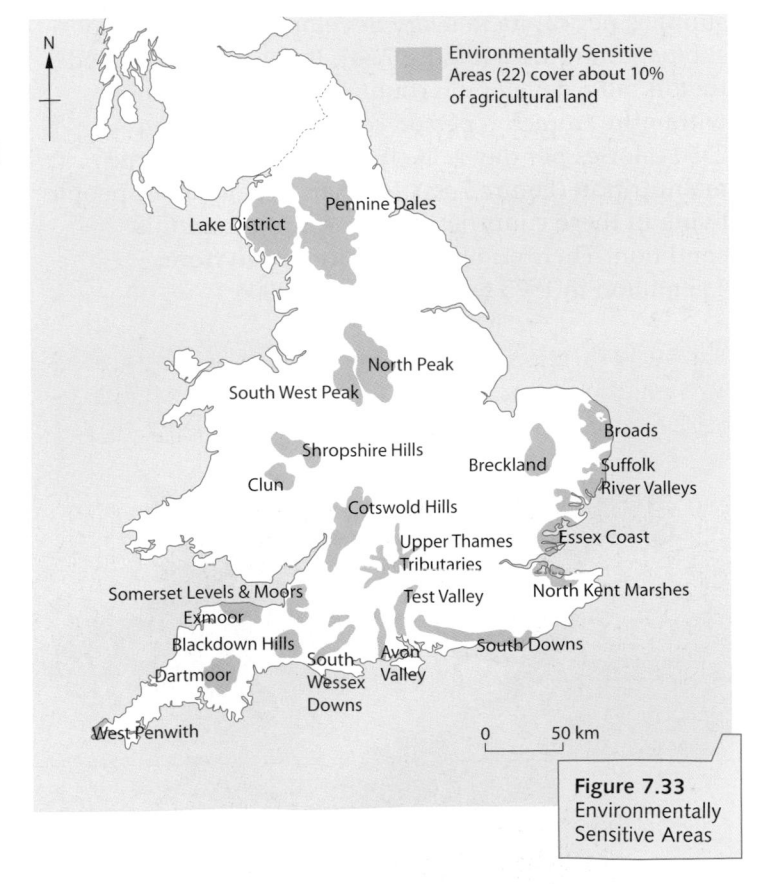

Figure 7.33
Environmentally Sensitive Areas

Food supply and malnutrition

The steady increase in global food production since the 1960s has led (with the exception of Africa where there has been a 10 per cent decrease) to a fall in the **proportion** of underfed people. At the same time, however, increases in world population and world poverty have meant a sharp increase in the **number** of people suffering from **malnutrition**. Malnutrition is caused by deficiencies in diet, either in amount (quantity) or type (quality). Until the 1970s, it was believed that malnutrition resulted from the population growing more rapidly than food supplies. Today it is attributed to poverty, as large numbers of the world's population are unable to afford to buy an adequate diet. While malnutrition results in starvation only under extreme conditions, it does reduce people's capacity to work and their resistance to disease. In children, it can retard mental and physical development, and cause illness.

Dietary energy supply (DES)

DES is the number of calories per capita (i.e. per person) available each day in a country. Like GDP (page 180) it does not take into account differences between individuals or between areas within a country. Between 1970 and 2006 there was an increase in available food supplies per capita in every developing region except sub-Saharan Africa (Figure 7.34). It has been estimated that in most developing countries, especially those within the tropics, a person consuming less than 2350 calories per day is likely to experience chronic malnutrition (Figure 7.35). In 2006, 17 per cent of people living in these countries were suffering from this condition. Their numbers have increased from 435 million in 1975 to 630 million in 2006.

This increase is mainly due to human factors (e.g. civil wars, political instability and international debt) rather than to physical causes (e.g. natural disaster such as drought).

Figure 7.35 shows that people living in developed countries need more calories per day than those in developing countries. This is partly because there is a greater proportion of adults in developed countries (children have smaller needs) and partly because most developed countries are in cooler latitudes (where more energy is needed for body heating).

Region	1970	1980	1990	2000	2006
Sub-Saharan Africa	35	36	37	42	31
Near East and North Africa	23	10	5	12	8
Central America and Caribbean	24	15	13	11	10
South America	17	12	12	9	9
South Asia	34	30	24	22	21
East Asia	35	22	17	16	14
China	46	28	16	14	12
All developing regions	36	26	20	19	17

Figure 7.34
Percentage of chronically underfed, 1970–2006

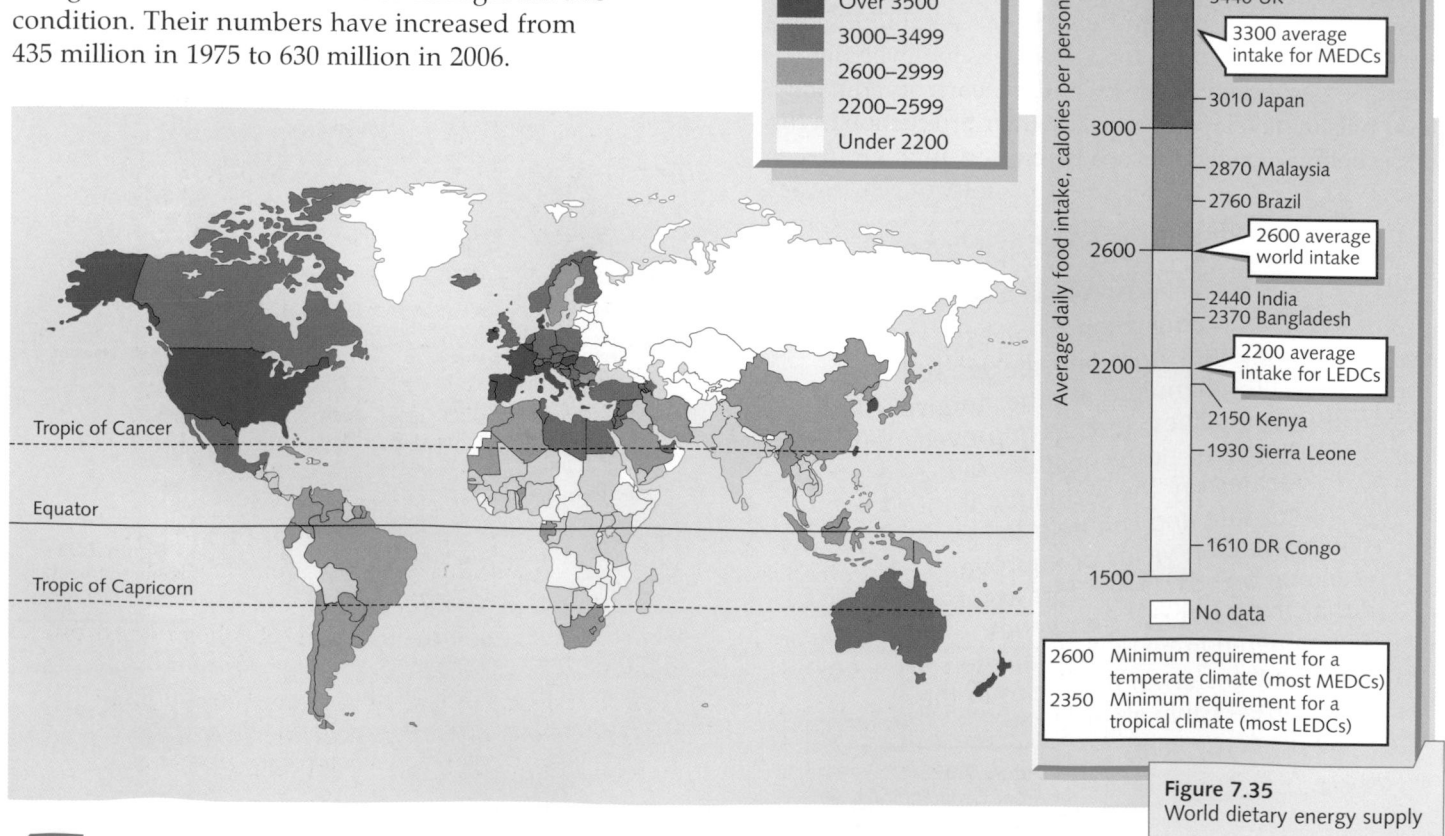

Kilocalories consumed per person per day

- Over 3500
- 3000–3499
- 2600–2999
- 2200–2599
- Under 2200

Tropic of Cancer

Equator

Tropic of Capricorn

Average daily food intake, calories per person

- 3770 USA
- 3670 Italy
- 3440 UK
- 3300 average intake for MEDCs
- 3010 Japan
- 2870 Malaysia
- 2760 Brazil
- 2600 average world intake
- 2440 India
- 2370 Bangladesh
- 2200 average intake for LEDCs
- 2150 Kenya
- 1930 Sierra Leone
- 1610 DR Congo

No data

| 2600 | Minimum requirement for a temperate climate (most MEDCs) |
| 2350 | Minimum requirement for a tropical climate (most LEDCs) |

Figure 7.35
World dietary energy supply

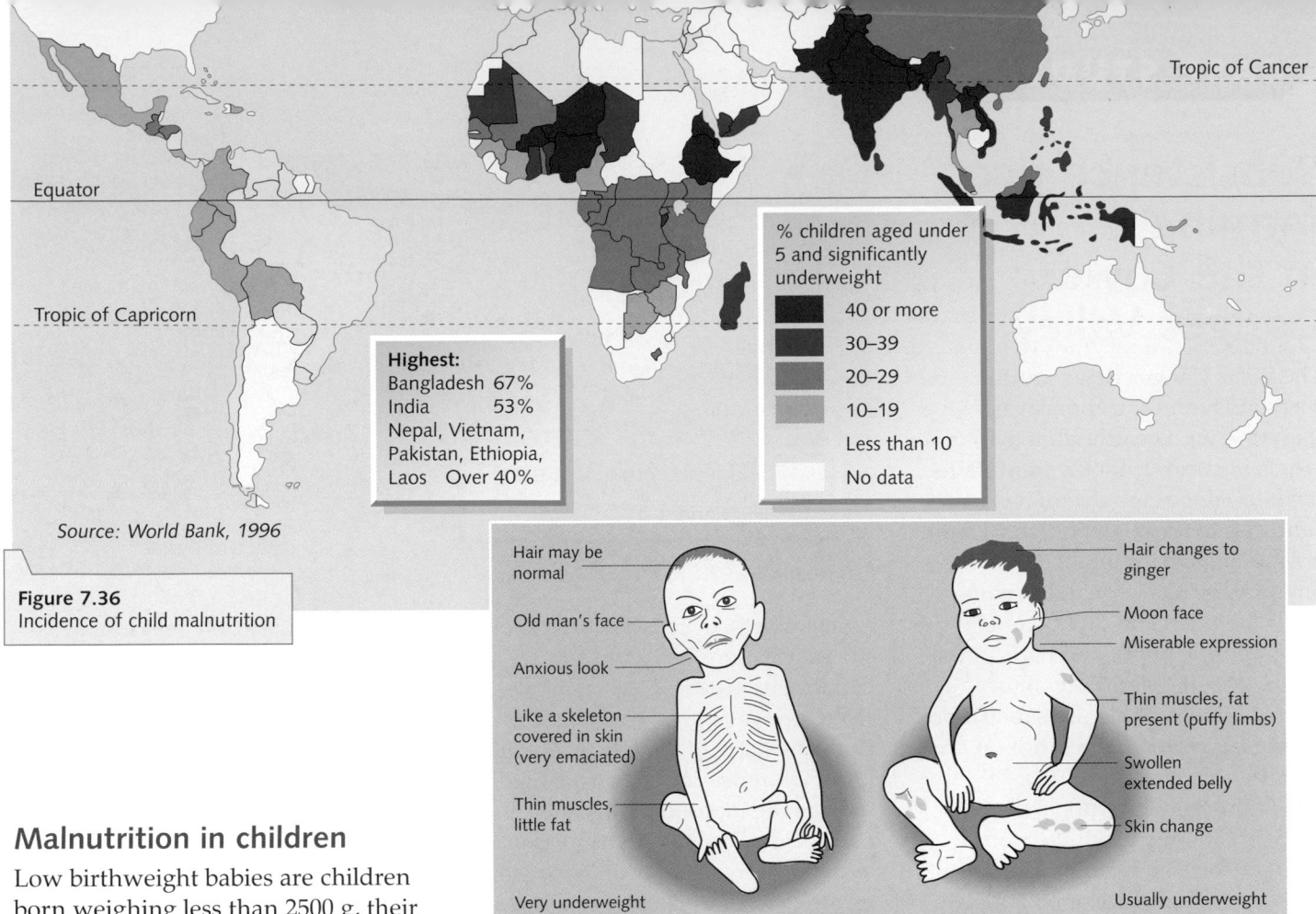

Equator

Tropic of Capricorn

Highest:
Bangladesh 67%
India   53%
Nepal, Vietnam,
Pakistan, Ethiopia,
Laos   Over 40%

% children aged under
5 and significantly
underweight

	40 or more
	30–39
	20–29
	10–19
	Less than 10
	No data

Source: World Bank, 1996

Figure 7.36
Incidence of child malnutrition

Hair may be normal

Old man's face

Anxious look

Like a skeleton covered in skin (very emaciated)

Thin muscles, little fat

Very underweight

Marasmus
(especially in first year)

Hair changes to ginger

Moon face

Miserable expression

Thin muscles, fat present (puffy limbs)

Swollen extended belly

Skin change

Usually underweight

Kwashiorkor
(especially years 1–5)

Figure 7.37
The effects of marasmus and kwashiorkor

Malnutrition in children

Low birthweight babies are children born weighing less than 2500 g, their low weight being attributed to maternal malnutrition. Low birthweight children are often prone to a shortened lifetime full of health problems, including retarded development and susceptibility to disease.

Children under the age of 5 are particularly susceptible to malnutrition. In 2006, 28 per cent of children in this age-group in the developing world were considered to be underweight (Figure 7.36). The percentage is, surprisingly, highest in southern and South-east Asia where, despite improvements in food supply per capita since 1970 (Figure 7.34), 46 per cent of children in this region are underweight in contrast with 26 per cent in sub-Saharan Africa (and under 2 per cent in developed countries).

Children fall ill either because their diet contains too few proteins, which are particularly important during early stages of growth, or too few calories. The two major protein deficiency diseases are **marasmus** and **kwashiorkor** (Figure 7.37). Marasmus is most common in children in their first year of life. Kwashiorkor results from a predominance of cereals (e.g. rice) and a deficiency of protein (e.g. milk, eggs and meat). Two of several diseases resulting from a lack of vitamins are beri-beri and rickets. Beri-beri, due to a lack of vitamin B, can lead to a wasting and paralysis of limbs. Rickets, caused by a deficiency in vitamin D, causes deformities in bones, legs and the spine.

Why do some people in sub-Saharan Africa suffer from malnutrition?

- The high birth rate and falling death rate means there are many more people to be fed.
- Few farmers have the money to buy high-yielding seeds, fertiliser, pesticides or machinery, or to implement irrigation schemes.
- When food is scarce, neither governments nor people can afford to buy high-priced surplus from overseas.
- During colonial times, European companies established commercial crops for their own profit instead of encouraging subsistence crops for local use. Although now independent, some governments still give tax concessions to overseas transnationals, allowing them to continue to grow these crops.
- The soil has been overused in the past and few nutrients remain. In places soil erosion has led to desertification (page 256).
- Many areas receive small and unreliable amounts of rainfall (pages 224 and 256).
- Pests and diseases destroy crops and stored grain.
- Often there is not enough protein in the diet.
- In many countries there is political instability.

Subsistence rice farming in the Lower Ganges Valley

The River Ganges flows south-eastwards from the Himalayas (Figure 7.38). The **alluvium** (silt) which it carries from the mountains has been deposited over many centuries to form, east of New Delhi, a flat plain and, where it enters the Bay of Bengal, a large delta (page 285). The plain and delta is one of the most densely populated parts of the world (Figure 1.1). Many of the people who live here are **subsistence** farmers, growing mainly rice on an intensive scale. Despite their exceptionally hard work, many farmers can only produce enough food for their own family and, perhaps, village. Rice, with a high nutritional value, can form over three-quarters of the total local diet and is a **sustainable** form of farming.

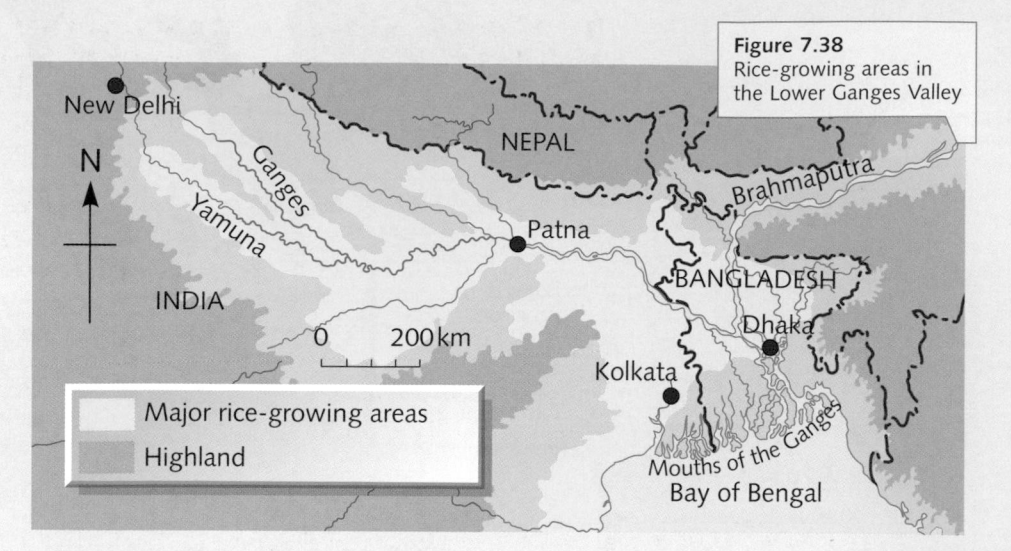

Figure 7.38
Rice-growing areas in the Lower Ganges Valley

☐ Major rice-growing areas
▨ Highland

Physical inputs

Wet padi, a variety of rice, needs a rich soil. It is grown in silt which is deposited annually by the Ganges and its tributaries during the monsoon floods. This part of India and Bangladesh has high temperatures, over 21°C, throughout the year (Figure 7.39), and the continuous growing season allows two crops to be grown annually on the same piece of land. Rice, initially grown in nurseries, is transplanted as soon as the monsoon rains flood the padi-fields. During the dry season, when there is often insufficient water for rice, either vegetables or a cereal crop is grown.

Human inputs

Rice growing is labour-intensive. Much manual effort is needed to construct the bunds (embankments) around the padi-fields, to build (where needed) irrigation canals, to prepare the fields, and to plant, weed and harvest the crop. Many farms, especially nearer the delta, are very small. They may only measure 1 hectare (the size of a football pitch) and be divided into 12 or 15 plots. The smallness of the farms and the poverty of the people means that hand-labour has to be used rather than machines (Figure 7.40). Water buffalo (oxen) provide manure and are used in preparing the padi-fields.

Processes

Planting rice in nursery; ploughing and transplanting rice into padi-field; harvesting rice, planting winter wheat, harvesting wheat; growing vegetables; looking after chickens.

Outputs

Rice and some wheat, vegetables, chickens (eggs/meat).

Temp °C / Rainfall mm

Monsoon rains begin. Rice planted in padi-fields

Second crop harvested

Still warm enough for crops to grow

Rice harvested in drier weather

Second crop planted

J F M A M J J A S O N D

Figure 7.39
Climate graph for Kolkata

Figure 7.40
Rice cultivation on the floodplain of the River Ganges

Recent changes

Land reform Many farms in the Ganges Basin are very small and broken up into tiny plots that are spread over a wide area. This makes efficient farming difficult. Many of the poorest farm workers have no land at all and live in poverty. In contrast, a few wealthy people own much of the farmland. Three aims of land reform are to increase farm size for those with little land, to give any surplus land to landless farm labourers, and to set an upper limit to the amount of land that a wealthy family can own.

Green Revolution and HYVs The Green Revolution refers to the application of modern, Western-type farming techniques to LEDCs. It began with the development of high-yielding varieties (HYVs) of cereals – new varieties of maize and wheat in Mexico and rice in the Philippines (the latter increased yields by 6 per cent in the first year). Although the new seeds were faster growing and disease resistant, they needed large amounts of fertiliser and pesticides, making farming **less sustainable**. The successes and failures of the Green Revolution are given in Figure 7.41.

Appropriate technology This is technology suited to the needs, skills, knowledge and wealth of local people (pages 127 and 150). In the Ganges Basin this includes the building of simple, easy-to-maintain water pumps, projects that use human labour rather than machines such as tractors, low-cost irrigation schemes instead of building large dams, and the use of animal manure rather than chemical fertiliser. Such schemes are said to be **sustainable** (page 184) because they are affordable and help to improve people's standard of living and quality of life without damaging the environment.

Figure 7.41
Successes and failures of the Green Revolution

Successes
- HYVs have increased food production. For example, India, which used to experience food shortages until the 1960s, became self-sufficient in cereals.
- The increase in yields led to a fall in food prices.
- Faster-growing varieties allow an extra crop to be grown each year.
- Yields are more reliable as many new varieties are more disease-resistant.
- Higher yields allow other crops, notably vegetables, to be grown, adding variety to the local diet.
- HYVs allow the production of some commercial crops.
- HYVs are not so tall as traditional varieties, enabling them to withstand wind and rain.
- Many of the more well-off farmers who could afford seed, fertiliser and tractors, have become richer.

Failures
- HYVs need large amounts of fertiliser and pesticides which increase costs, encourage the growth of weeds and can harm water supplies (page 108).
- HYVs need a more reliable and controlled supply of water. They are more vulnerable to drought and to waterlogging. Irrigation, where used, increases costs and can cause salinisation (page 105).
- HYVs are more susceptible to attacks by pests and diseases.
- Many of the poorer farmers who do not own the land they farm and cannot afford to buy seed, fertiliser and tractors, have become much poorer.
- Mechanisation has increased rural unemployment and migration to the towns.
- Farming has become less sustainable.

Farming

Key Words and Terms

a You should know the meaning of the following terms:
- market gardening • subsidies • quotas • set-aside • diversification
- Common Agricultural Policy (CAP) • eutrophication • organic farming
- genetically modified (GM) crops • irrigation • malnutrition • calories
- Green Revolution • high-yielding varieties (HYVs)
- appropriate technology • sustainable development.

b You should know the difference between:
- arable and pastoral farming
- commercial and subsistence farming
- intensive and extensive farming
- fertiliser and pesticides.

Key Ideas

You should know and understand the following:
- Farming is a system with inputs, processes and outputs.
- The farmer is a decision maker.
- The type of farming is affected by physical/environmental, economic/social and human/political factors.
- Farming varies from place to place both in the UK, in the EU and across the world.
- Farming is dynamic and constantly changing.
- The demands for cash crops and by supermarkets has an impact on subsistence agriculture.
- Farming changes due to developments in transport and technology and the effects of political decisions.
- Farming can benefit and have adverse effects on the environment.
- There can be a link between population growth and food supplies and shortages.
- Contemporary solutions to food supplies depend on sustainable development.
- Many parts of the world still suffer the effects of malnutrition.

Skills, Theories and Models

- Interpret and use diagrams
- Analyse and use statistical data
- Interpret and use a variety of maps
- Recognise the distinctive character of places
- Explain how places are interdependent (CAP)
- Understand the idea of sustainable development
- Understand the causes and effects of change
- Classification of farming types
- The farming system

INPUTS — Physical / Human → PROCESSES → OUTPUTS

1 *(Page 96)*

Study the farming systems for farms A and B. One is a subsistence farm in India, the other a commercial farm in south-west England.

a i) What is the size of farm A? *(1)*

   ii) How many animals are kept on farm A? *(1)*

   iii) What happens to the outputs from farm A? *(1)*

b The boxes labelled 'Inputs' and 'Outputs' for farm A have been left empty. From the following list, fill in the two boxes to show four likely inputs and four likely outputs: barley, fertiliser, hay, labour, machinery, fattened pigs, manure, milk. *(8)*

c Explain the differences between the two farms under the following headings:

Land, Labour, Machinery, Animals, Crops. *(5)*

d Is farm A or farm B the one located in India? *(1)*

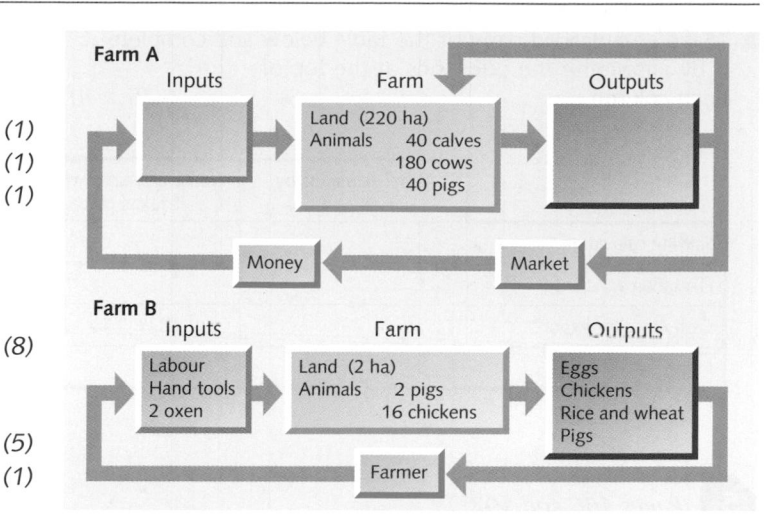

2 *(Pages 96, 100 and 101)*

The partly completed farming system is for a hill sheep farm in northern England.

a i) Match each of the following with the letters A, B, C, D, E and F on the diagram: EU subsidies, labour, lamb, shearing, soil, wool. *(6)*

   ii) Name two other likely inputs into the farm. *(2)*

   iii) Name two possible hazards that might affect this farm. *(2)*

   iv) Why is it important for the farmer to re-invest? *(1)*

b Draw a farming system for a cereal farm in East Anglia. Name two physical inputs, two human/economic inputs, two processes and two outputs. *(8)*

c For each farm, describe how it has been affected by two recent changes. *(2 + 2)*

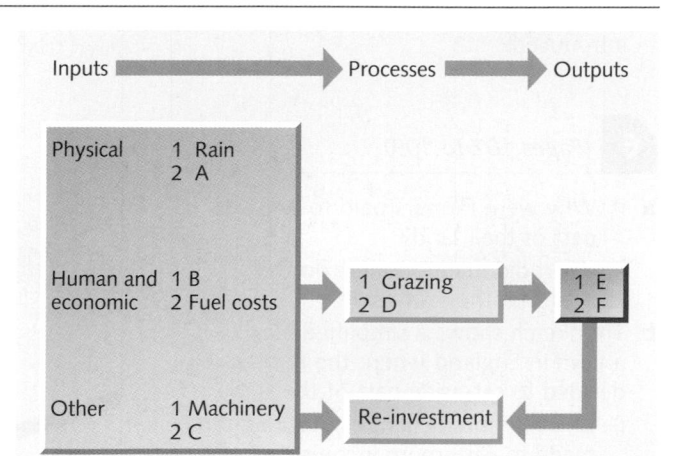

3 *(Pages 98 to 101)*

Look at the simplified map showing the location of the main farming types in the UK.

a Match up each of the following types of farming with the letters A, B, C and D on the map: arable, hill sheep, mixed, pastoral. *(4)*

b The type of farming in an area is often the result of farmers acting as decision makers. Describe three physical factors and three human/economic factors that may influence a farmer's decision. *(6)*

c On an enlarged copy of the table below, show how the factors affecting farming in the Lake District (or elsewhere in western Britain) differ from those in East Anglia (or elsewhere in eastern Britain). *(10)*

Factors	Farming in the Lake District (or western Britain)	Farming in East Anglia (or eastern Britain)
Climate		
Relief		
Soils		
Markets		
Transport		
Machinery		

4 (Pages 102 to 105)

a Make an enlarged copy of the table below and complete it by answering the questions at the top of each column. (5 × 4)

b How has the demand for cash crops, or by supermarkets, in developed countries affected each type of farming? (4)

	What is meant by	Name one area where it takes place	Name the main crop(s)	Disadvantages to local farmers	Advantages to local farmers
Shifting cultivation					
Plantation agriculture					
Cash crops for export					
Irrigation					

5 (Pages 106 and 107)

a What were the five aims of the Common Agricultural Policy (CAP) when it was set up in 1962? (5)

b Why, at a later date, were i) subsidies ii) quotas introduced? (2)

c State three ways in which you think the CAP has been successful and three problems that still remain. (3 + 3)

6 (Pages 107 to 109)

a i) Why were farmers paid to 'set-aside' part of their land? (1)

   ii) What did many farmers do with land that they set-aside? (1)

b The sketch shows a simplified plan of a farm in England where the farmer decided to set-aside part of the land.

   i) Describe three changes that might be made to earn more income. (3)

   ii) Describe three changes that might be made to improve the quality of the environment. (3)

c The farmer has just decided to join the Environmental Stewardship Scheme.

   i) What is the Environmental Stewardship Scheme? (1)

   ii) How might farmers earn a higher level of payment if they join? (1)

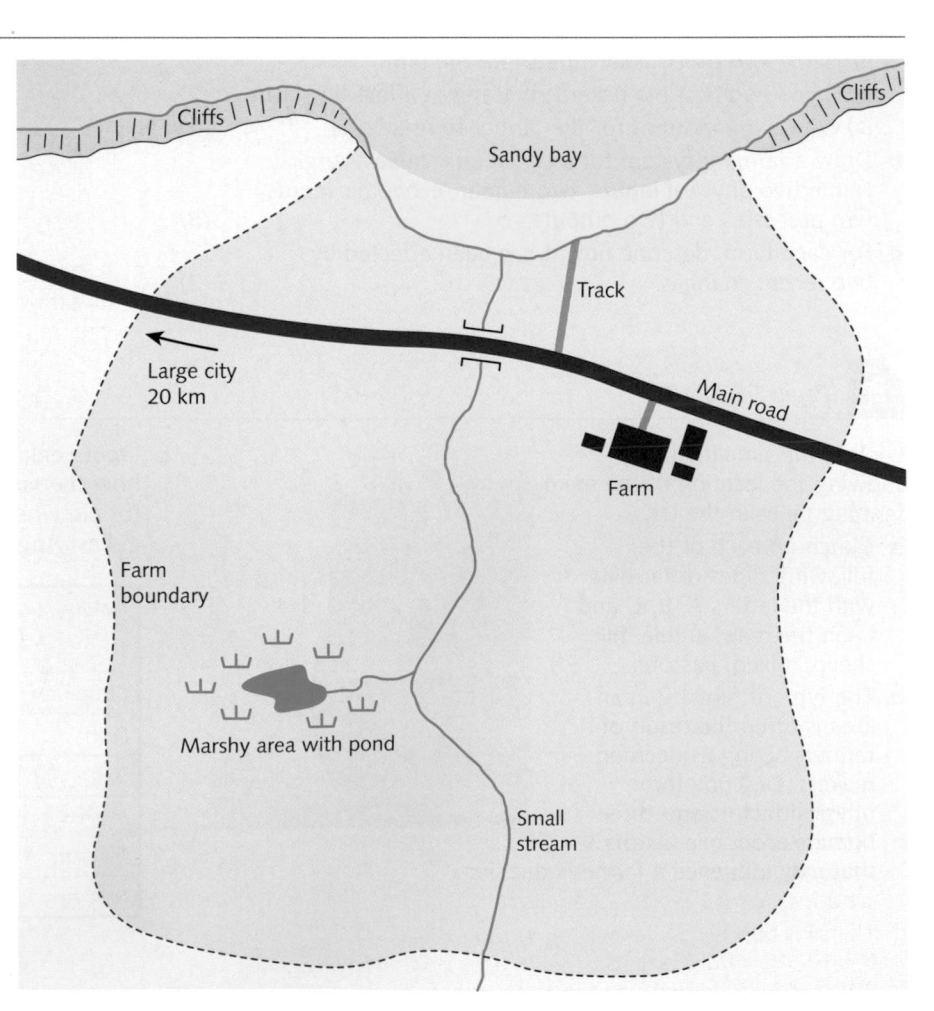

7 *(Pages 108 and 109)*

The sketch below shows a farming landscape in eastern England.

a i) What is one advantage and one disadvantage to the farmer who uses nitrate (chemical) fertiliser? *(2)*

ii) Describe how nitrate fertiliser may reach rivers. *(1)*

iii) How does an increase in nitrate in rivers affect:
  • fish • people? *(2)*

b i) Why is it an advantage to some farmers to clear hedgerows? *(3)*

ii) Give three ways by which the removal of hedgerows can harm the environment. *(3)*

c Describe two other aspects of modern farming which it is claimed might harm the environment. *(2)*

d People have different opinions about farming changes like the ones shown in the land-sketch. Describe the likely contrasting views of the following people:
  • an organic farmer
  • a Defra employee
  • a large-scale cereal farmer in eastern England
  • a member of a wildlife conservation group. *(4)*

New buildings constructed – marshland damaged and views changed

New crops introduced – sunflowers, oilseed rape

Bigger farms and larger fields

60% hedgerows cleared since 1945 – loss of wildlife habitats

Modern technology – tractors, harvesters, pesticide sprayers

Many ponds and old dykes filled in – loss of habitats for wildlife

GM crops

Increased chemical use, especially fertiliser – some pollution of water systems

Use of nitrates (Kg per hectare, 1970–1990)

8 *(Pages 110 and 111)*

a i) How many calories does an average adult need in:
  • the tropics • the UK? *(2)*

ii) Which continent suffers most from malnutrition? *(1)*

iii) Which region has most underweight children aged under 5? *(1)*

b i) What effect can malnutrition have on children? *(2)*

ii) Describe two child illnesses caused by malnutrition. *(4)*

c On a star diagram, list six reasons why people in many parts of sub-Saharan Africa suffer from malnutrition. *(6)*

9 *(Pages 112 and 113)*

a Describe farming in the lower Ganges Basin under the headings: Physical inputs, Human/economic outputs, Processes, Outputs. *(8)*

b i) Why was the Green Revolution introduced in this region? *(2)*

ii) Describe three advantages of the Green Revolution to local farmers. *(3)*

iii) Describe three ways in which the Green Revolution failed local farmers. *(3)*

c Why is appropriate technology more suited to the subsistence farmers of the Ganges Basin than the introduction of Western-type methods? *(3)*

d The two farmers A and B lived in the same village in India in the 1960s. Why did Farmer A become very wealthy by 2000 while Farmer B became even poorer? *(4)*

I'm quite wealthy. I have enough rice to feed my family and sell some as well. I've also got some savings.

I'm very poor, have no savings and have only enough food to feed my family.

What are resources?

Resources can be defined as features of the environment which are needed and used by people. The term usually refers to **natural resources** which occur in the air, in water or on the land. These resources include raw materials (minerals and fuels), climate, vegetation and soils. Sometimes the term is widened (though not in this chapter) to include **human resources** such as labour, skills, machinery and capital (Figure 8.1).

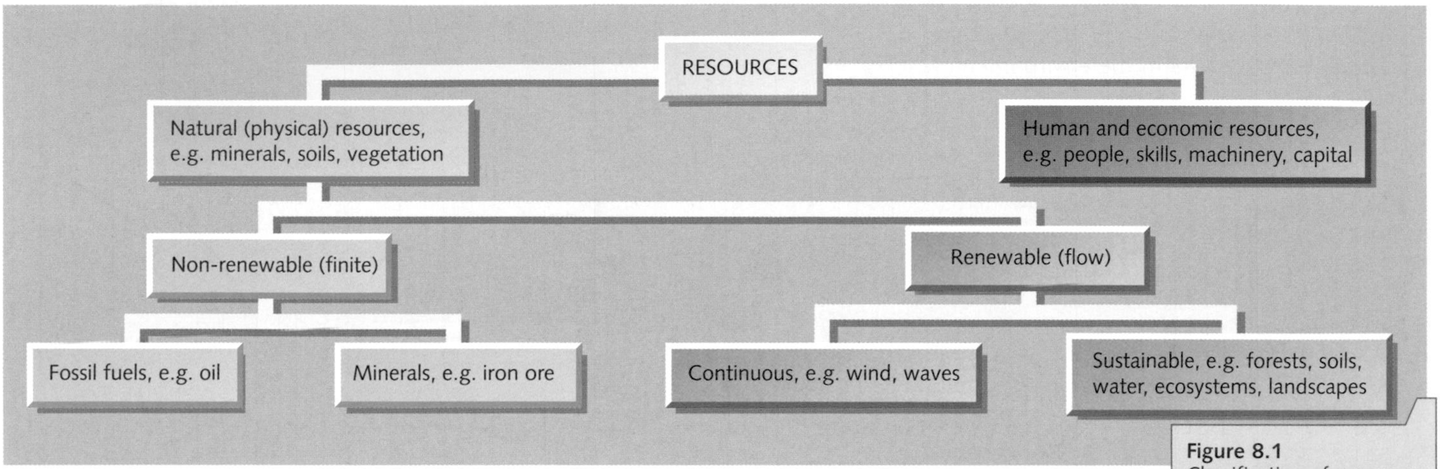

Figure 8.1
Classification of resources

Natural resources are commonly subdivided into two groups:

1 **Non-renewable resources** are said to be finite or non-sustainable as their exploitation and use will eventually lead to their exhaustion, e.g. fossil fuels and minerals.

2 **Renewable resources** can either be:
   • **a flow of nature** so that, being continuous, they can be used over and over again, e.g. solar and geothermal energy, wind and water power
   • **sustainable**, which means they are renewable and self-generating if left to nature, e.g. clean water, trees, fish, wildlife, soils, ecosystems and landscapes. However, if these sustainable resources are used carelessly or are over-used by people, then either:
     – their value may be reduced (i.e. degraded), as when soils lose their fertility and are eroded, water supplies are polluted or trees die due to acid rain, or
     – their existence is threatened, as with overfishing, deforestation, the draining of wetlands or the development of scenic areas.

Why is the demand for resources increasing?

The demand for, and the use of, the world's resources continues to grow at an increasingly faster rate. This is mainly due to:

• **population growth** as the number of people in the world continues to increase

• **economic development** as more countries try to develop industrially and economically and attempt to raise their standard of living and quality of life

• **increasing wealth**, especially in the more economically developed countries

• **technological advances**.

How can the Earth's resources be protected?

The combined effects of population growth, economic development, increasing wealth and technological advances means that there is a growing need to manage and protect the Earth's resources (Figure 8.2). This might be achieved through **sustainable development** (page 184). Sustainable development, which is improving people's standard of living and quality of life without wasting resources or spoiling the environment, includes:

• conservation, e.g. wildlife and scenery
• recycling, e.g. glass and waste paper
• greater efficiency in existing resource use, e.g. home insulation
• developing renewable resources, e.g. the wind, waves, tides and the sun
• controlling pollution, e.g. reducing emissions from vehicles, power stations and farm waste
• using appropriate technology (Figure 11.7), e.g. low-cost energy schemes and local building materials.

This ...

... or this?

Figure 8.2
Making the choice

Energy resources

The sun is the primary source of the Earth's energy. Without energy, nothing can live and no work can be done. Green plants convert energy, through the process of photosynthesis, into a form that can be used by people.

Non-renewable resources

In 2008, coal, oil and natural gas accounted for 88 per cent of the world's commercially consumed energy (Figure 8.3a). These are forms of stored solar energy produced by photosynthesis in plants over thousands of years. As these three types of energy, referred to as **fossil fuels**, take so long to form and be replaced, they are regarded as non-renewable. Each year the world (in reality mainly the developed countries) consumes an amount of fossil fuel that took nature some 1 million years to provide – a rate far in excess of their replacement. Fossil fuels have, in the past, been relatively easy to obtain and cheap to use, but they have become major polluters of the environment.

Two other non-renewable sources of energy should be noted. Nuclear energy uses uranium and so is not a fossil fuel. Fuelwood is a non-commercial source of energy. If included with the other forms of energy, it provides an estimated 14 per cent of the world's and 35 per cent of developing countries' energy requirements (Figure 8.3b).

Renewable (alternative) resources

Renewable resources of energy, which are mainly forces of nature which can be used over and over again, are considered to be sustainable. At present only running water (hydro-electricity) is a significant source of renewable energy on a global scale (Figure 8.3a). Other sources, often more important on a local scale, are the sun (solar), the wind, vegetation waste (biomass) and heat from the Earth (geothermal). As yet, economic and technical problems tend to restrict the conversion of these sources of energy on a large scale. However, in time the world is likely to have to look to these and other sources of renewable energy (e.g. waves and tides) as the supply of fossil fuels becomes exhausted.

Figure 8.3
Global energy consumption

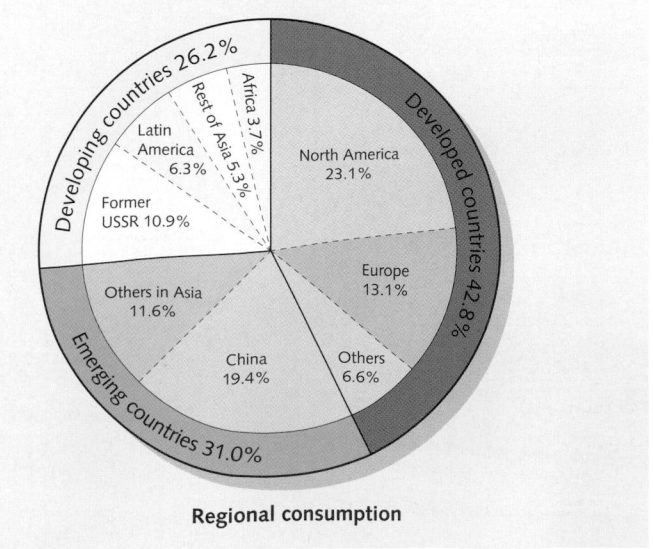

Nuclear 3.6%

Others 2%

HEP 6.4%

Oil 35.6%

Natural gas 23.8%

Coal 28.6%

Commercial consumption

Developing countries 26.2%

Rest of Asia 5.3%

Africa 3.7%

Latin America 6.3%

North America 23.1%

Former USSR 10.9%

Developed countries 42.8%

Others in Asia 11.6%

Europe 13.1%

China 19.4%

Others 6.6%

Emerging countries 31.0%

Regional consumption

Non-renewable energy resources

Coal

The fortunes of coal in Western Europe have declined since the 1960s when the cheaper price of oil and the exhaustion of the most easily obtainable coal meant a decline in both production and the workforce. Rising oil prices, after the 1974 Middle East War, led to the exploitation of lower-cost coal resources in the USA, Australia and South Africa. The recessions of the early 1980s and 1990s sealed the fate of most mining communities in the UK and Belgium. Globally, coal production is increasing although little is used by developing countries, apart from China which produces and consumes 40 per cent of the world's total.

Advantages Reserves are likely to last for over 300 years. Improved technology has increased the output per worker, allowed deeper mining with fewer workers, and made conversion to electricity more efficient. Coal is used for electricity, heating and making coke.

Disadvantages The most easily accessible deposits have been used up and production costs have risen. With increased competition from other types of energy, it means that whereas in 1980, 70 per cent of the UK's electricity was generated using coal, by 2008 it had fallen to 18.1 per cent. The burning of coal causes air pollution and, by releasing carbon dioxide, contributes to global warming (page 218). Deep mining can be dangerous, while opencast mining temporarily harms the environment (Figure 8.4). Coal is heavy and bulky to transport.

Oil and natural gas

Many industrialised countries have come to rely upon either oil or natural gas as their main source of energy. Very few of them, the UK being an exception, have sufficient reserves of their own.

Advantages Oil and gas are more efficient to burn, easier to transport and distribute (by pipeline and tanker), and less harmful to the environment than coal, with gas being even cheaper and cleaner than oil. They are safer than nuclear energy. Both are used for electricity, with gas increasingly the most favoured of all the non-renewable sources. During the 1990s most new electricity power stations were gas-fired, while many former coal-fired stations were converted to gas – the so-called 'dash-for-gas' policy. Oil is the basis of the huge petrochemical industry.

Disadvantages Reserves may only last another 50–70 years. New fields are increasingly difficult to discover and exploit (Figure 8.5). Terminals and refineries take up much space and there is the danger of spillage (oil), leaks (gas), explosions and fire. The burning of gas and oil releases, respectively, nitrogen oxide and sulphur dioxide which contribute to acid rain (page 221). Both oil and gas are subject to sudden international price changes and are vulnerable to political, economic and military pressures.

Figure 8.4
Opencast coal mining

Figure 8.5
A marine oil platform

- Only very limited raw materials are needed, e.g. 50 tonnes of uranium per year compared with 540 tonnes of coal per hour needed for coal-fired stations.
- Oil and natural gas could be exhausted by the mid twenty-first century. Coal is difficult to obtain and dirty to use.
- Numerous safeguards make the risks of any accident minimal.
- Nuclear waste is limited and can be stored underground.
- Nearly all the money spent in Britain on energy research has been on nuclear power.
- Nuclear energy schemes have the support of large firms and government departments.
- Nuclear power is believed to contribute less than conventional fuels to the greenhouse effect and acid rain.
- Renewable forms of energy will not be sufficient to meet the UK's demands in the near future.

- It is not clear how safe it is. So far there have been no serious accidents in Britain, as, for example, at Chernobyl in the Ukraine, but there have been several leaks at Sellafield.
- There is a large conservationist lobby which claims that one accident may kill many, and ruin an area of ground for hundreds of years. The Irish Sea is increasingly contaminated.
- Many people think that Britain should concentrate on using renewable forms of energy rather than those that are non-renewable.
- Nuclear power cannot be used for two of industry's major demands, heating and transport, as costs are too high.
- There is less demand for energy by industry as declining industries (such as steel) used more energy than those that are replacing them (such as micro-electronics).
- Potential health risks. The high incidence of leukaemia around Sellafield and Dounreay has been linked to the proximity of the power station.
- Nuclear waste can remain radioactive for many years. There are problems with reprocessing and then storing nuclear waste.
- The cost of decommissioning old power stations, the first of which closed in 1989 at Berkeley, is extremely high.

Figure 8.6
The nuclear energy debate

Nuclear energy

Few debates have generated such strong emotions as that on the merits of nuclear power (Figure 8.6). Several older industrialised countries (e.g. France and Belgium) and many newly industrialised countries (e.g. Japan, South Korea and Taiwan) which lack sufficient fossil fuels of their own have turned increasingly to nuclear power, despite grave fears over its safety (Figure 8.7).

Fuelwood

In Africa, trees have been called the 'staff of life' due to their vital role in preserving the environment and in providing rural communities with their basic needs (shelter, food, fuel and shade). Yet here, as in many rural areas in other developing regions, trees are being removed at an ever faster rate. Collecting fuelwood is a time-consuming job for women and children (Figure 8.8). Each day they may have to walk many kilometres to find enough wood upon which to cook their meals (women also have to do the farming and look after large families).

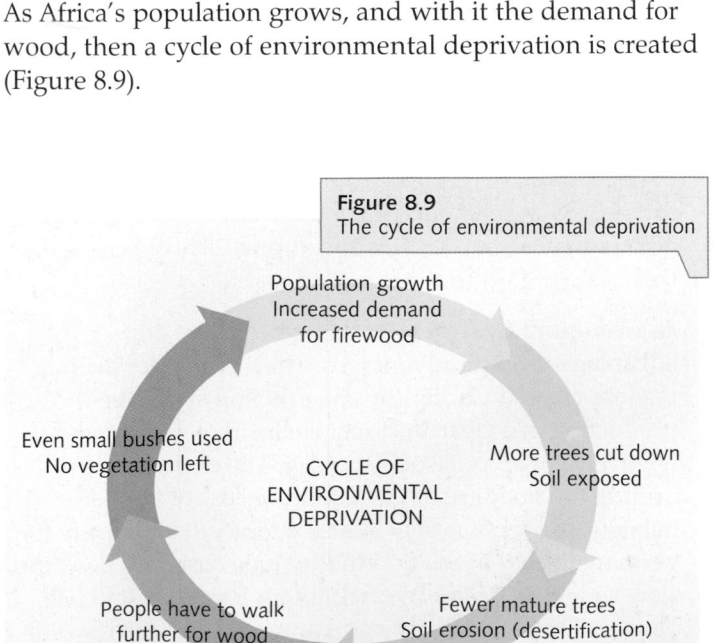

Figure 8.7
Nuclear reactors at Wakasa Bay, Japan

As Africa's population grows, and with it the demand for wood, then a cycle of environmental deprivation is created (Figure 8.9).

Figure 8.8
Collecting wood for fuel, Kenya

Figure 8.9
The cycle of environmental deprivation

Population growth
Increased demand for firewood

More trees cut down
Soil exposed

CYCLE OF ENVIRONMENTAL DEPRIVATION

Fewer mature trees
Soil erosion (desertification)
Farmland becomes desert

People have to walk further for wood

Even small bushes used
No vegetation left

Renewable energy resources

Hydro-electric power (HEP)

Hydro-electricity generates the highest proportion of the renewable types of energy. Although accounting for only 6.4 per cent of the world's total commercial energy (Figure 8.3a), in many countries it accounts for over 80 per cent, e.g. Paraguay 100 per cent, Norway 96 per cent and Brazil 86 per cent. It is, as the examples suggest, important to both developed and developing countries – providing they have a constant supply of fast-flowing water. Figure 8.10 shows some of the factors ideally needed for the location of a hydro-electric power station. Hydro-electricity can be generated at a natural waterfall (e.g. Niagara Falls), by building a dam across a valley (e.g. the Three Gorges Dam on the Yangtze – pages 293–5, and at Itaipù – Figure 8.12) or where water flows rapidly down a hillside (e.g. Norway).

Advantages Hydro-electric power is renewable, is often produced in highland areas where the population is sparse, is a relatively cheap form of electricity and creates only limited pollution. Dams, where built to store water, reduce the risks of flooding and water shortages.

Disadvantages Dams are very expensive to build, large areas of farmland and wildlife habitats may have to be flooded forcing people and animals to move, unsightly pylons can cause visual pollution, and there is always the possibility of the dam collapsing. Silt, previously spread

Figure 8.10
Factors involved in the location of an HEP station

- Heavy precipitation (relief rainfall) over high mountains
- Snow and glaciers provide spring meltwater
- Large drainage basin traps more water
- Natural glacial lake or reservoir provides constant supply of water
- Impervious rock prevents water soaking through, and gives solid foundations
- Site of former waterfall provides a head of water
- Steep-sided glaciated valley helps dam construction
- Nearby industrial and domestic demand

over farmland, will be deposited in the lake. Quite recently (1966) it has been shown that if an area is flooded, the decaying vegetation can release methane and carbon dioxide – two greenhouse gases.

PLACES

Itaipù, Brazil

In 1982 a dam was completed across the River Parana at Itaipù, on the border of Brazil and Paraguay (Figure 8.11). The lake behind the dam is 180 km long and 5 km wide (Figure 8.12). The 18 turbines fitted at Itaipù make it the world's largest HEP scheme. Although the venture was a joint one between Paraguay and Brazil, Paraguay receives its annual needs from just one turbine. By agreement, Brazil purchases the remainder of Paraguay's share at a reduced price. The electricity is then transmitted to the São Paulo area.

As with other hydro-electric schemes, there are advantages and disadvantages. The advantages include the jobs created during the construction stage, the production of a clean and reliable form of energy, and the relatively cheap price of electricity. The disadvantages include the flooding of farmland, the loss of wildlife habitats, the relocation of 42 000 people, the relatively few permanent jobs in relation to the huge cost and, now, the slow silting-up of the lake. Although Paraguay and São Paulo get their electricity and Paraguay gets much-needed income, there has been little benefit to the local residents of Itaipù.

Figure 8.11
Location of Itaipù

- Itaipù Reservoir
- BRAZIL
- Itaipù Dam
- Friendship Bridge
- Foz do Iguaçu
- River Parana
- PARAGUAY
- Iguaçu National Park
- River Iguaçu
- Iguaçu Falls
- ARGENTINA
- 0 — 10km
- N

Figure 8.12
The Itaipù Dam in Brazil is 190 metres high

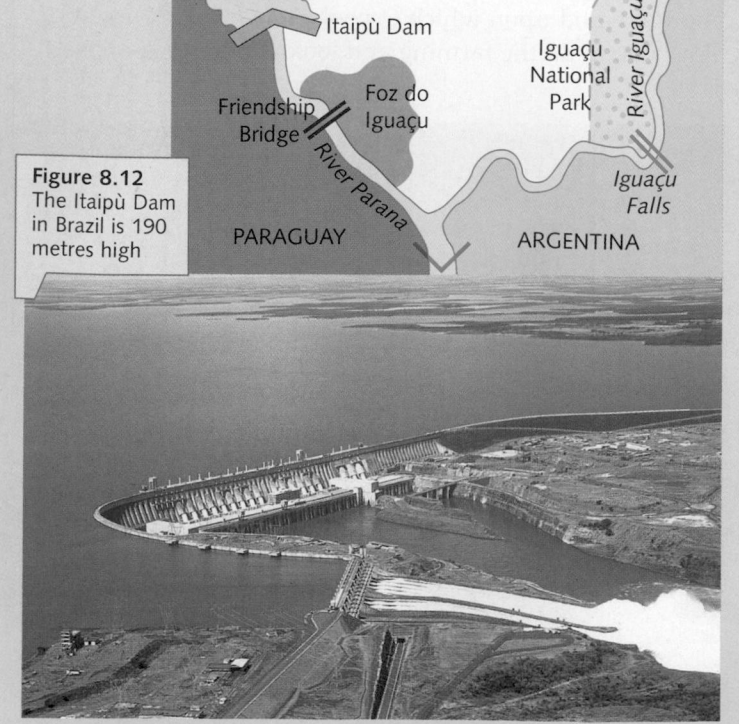

Geothermal energy

Geothermal means 'heat from the Earth'. Heat is stored in magma or in rocks beneath the Earth's surface. In volcanic areas, such as New Zealand, Iceland, Japan and central America, the heated rocks are near to the surface. In a natural geothermal system, water, usually originating as rain, seeps downwards through cracks and cavities. Where it comes into contact with heated rock, it is warmed and rises to the surface where it is ejected as geysers, hot springs or steam (Figure 8.13). In some areas of known heated rock, cold water is artificially pumped downwards through boreholes before returning to the surface, usually as steam (Figure 8.14). The advantages of geothermal energy include being renewable, providing a constant supply and being relatively pollution free. The disadvantages include the high cost of construction and maintenance to produce useful energy, its limitation to volcanic areas, the threat to power stations from eruptions and earthquakes, and the emission of sulphuric gases.

Figure 8.13
Geothermal activity in New Zealand

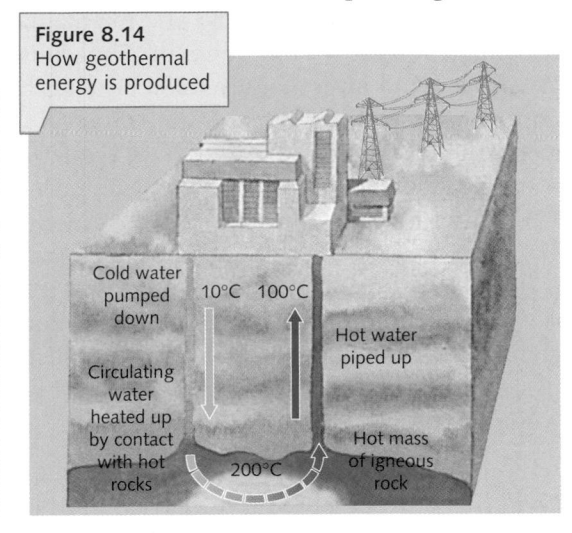

Figure 8.14
How geothermal energy is produced

Cold water pumped down

Circulating water heated up by contact with hot rocks

10°C   100°C

Hot water piped up

Hot mass of igneous rock

200°C

New Zealand

The Taupo volcanic area in New Zealand's North Island contains several geothermal fields (Figure 8.15). Some of these fields are used to generate electricity (e.g. Wairakei and Ohaaki), some are tourist attractions (e.g. Rotorua and Wairakei), some are used for domestic heating (e.g. Rotorua and Taupo) and some are used for industry (e.g. the pulp and paper mill at Kawerau).

Wairakei, commissioned in 1958, is the second oldest geothermal power station in the world. A power station was proposed following two dry years when there was insufficient water to generate hydro-electricity. Wairakei was chosen because it was in the centre of a geothermal field and was next to the Waikato river which could provide water for cooling. Approximately 50 production wells have been drilled in the area (Figure 8.16). On reaching the surface the steam is piped 2 km to the power station. Cold water from the river is then sprayed into condensers to help condense the steam (it takes 40 kg of water to condense 1 kg of steam). At present, Wairakei produces 4 per cent of New Zealand's electricity.

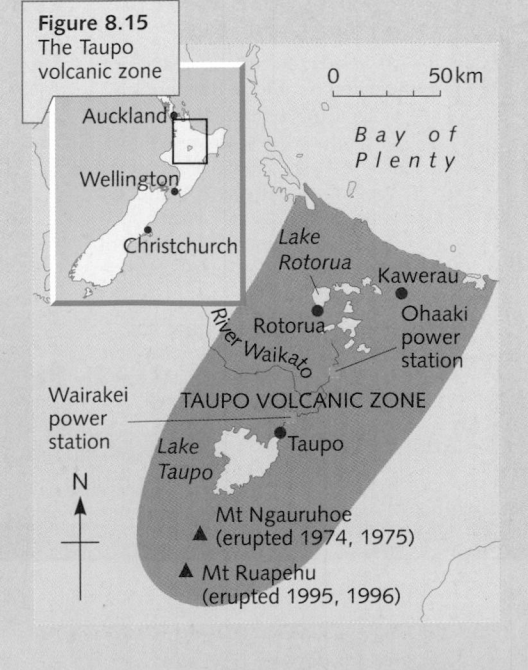

Figure 8.15
The Taupo volcanic zone

Auckland
Wellington
Christchurch

0   50km

Bay of Plenty

Lake Rotorua
Kawerau
Rotorua
Ohaaki power station
River Waikato
Wairakei power station
TAUPO VOLCANIC ZONE
Lake Taupo
Taupo

N

Mt Ngauruhoe (erupted 1974, 1975)

Mt Ruapehu (erupted 1995, 1996)

Figure 8.16
Wairakei

123

Wind

Although windmills were used in Britain during the Middle Ages to grind corn, few attempts were made until recently to generate this form of energy. Now, apart from hydro-electricity, it is the only renewable option being developed commercially in the UK. Wind turbines, to be at their most efficient, need to be in areas with high and regular wind speeds. Such sites are usually found on exposed coasts or in upland areas in the more remote parts of western Britain (Figure 8.17). As 50-metre high turbines are expensive to build and maintain, it is an advantage to group a minimum of 25 machines together to form a 'wind farm' (Figure 8.18). Britain's first wind farm was opened in 1991 near Camelford in Cornwall. The farm, on moorland 250 metres above sea-level and where average wind speeds are 27 km/hr, generates enough electricity for 3000 homes. By early 2009, when there were 188 wind farms (7 of which were offshore) and over 2400 turbines in operation, wind still produced less than 2 per cent of the UK's generated energy. With another 173 wind farms (16 will be offshore) either under construction or projected, Britain still hopes that 6 per cent of its energy will come from the wind by 2010.

Although the first large-scale wind farms were built in California, Europe now produces one-third more energy from wind power than does the USA. Denmark relies on wind power more than any other country, although in 2008 it was overtaken by the UK in terms of total production.

- Terrestrial wind farm
- Offshore wind farm

0   200 km

Figure 8.17
Wind farms in the UK, 2008

As with other forms of energy, wind power has its advantages and disadvantages (Figure 8.18).

Figure 8.18
Wind farm in California, USA

Advantages

- It is safe (does not give off radioactive emissions) and clean (does not give off chemical emissions). Unlike fossil fuels, it does not contribute to global warming or acid rain.
- It has a minimal effect on local ecosystems.
- Winds are much stronger in winter, which coincides with the peak demand for electricity.
- After the initial expense of building a wind farm, the production of electricity from this renewable source is relatively cheap.
- Wind farms provide a source of income for farmers and may attract small industries to rural parts of Britain where job opportunities are at present limited.
- Future wind farms are likely to be built off-shore.

Disadvantages

- Wind does not blow all the time. At present electricity generated during storms cannot be stored for use during calm periods.
- Groups of 30-metre tall turbines spoil the scenic attraction of the countryside and can affect wildlife, especially birds.
- As yet it is an expensive and not very efficient form of energy. Around 7000 turbines are needed to produce the same amount of electricity as one nuclear power station. As many as 100 000 wind farms may be needed if Britain is to generate 20 per cent of its total energy supply from the wind.
- Wind farms are noisy and can interrupt radio and TV reception for people living nearby. They can also affect property values.

Solar energy

Estimates suggest that the annual amount of energy received from the sun is 7000 times greater than the world's current demand for energy. Solar energy can be used to generate electricity through solar panels and photo-voltaic cells (Figure 8.19). It has the advantages of being safe, pollution free, efficient and limitless in supply. Unfortunately technology has yet to find a cheap and efficient way of constructing solar 'stations', although several large transnational companies (page 144) are trying to do just that. In Britain, the solar option is further hindered by the weather. Winter, when demand is highest, is when the UK gets most cloud, shorter hours of daylight and, when it does shine, the sun is at a low angle in the sky (Figure 12.2). Although solar energy is only likely to provide 1 per cent of the UK's energy requirements, its potential is far greater in developing countries, many of which lie in the warmer, sunnier tropics and where, as in India, many rural areas are too isolated for mains electricity.

Power from hydrogen, tidal, waves and biomass sources

Hydrogen This is the great hope for the future. Obtainable through water, its only by-product is water vapour. By 2050 it is hoped that hydrogen-fuel cells will power emission-free cars, buses and lorries (replacing oil) and will heat and light factories and homes. Governments are keen on its development, as are the major car companies, all of which are spending huge sums on trying to advance the technology.

Tidal Like the Rance in Brittany (Figure 8.20), several British rivers have a large tidal range that could generate electricity. However, schemes such as the proposed Severn and Solway Barrages have been dismissed as being too expensive to build and they would destroy important wildlife habitats.

Waves Waves, especially during storms, have very high energy levels. At present there are two experimental schemes off the Scottish coast. The first, LIMPET, creates an oscillating water column on the Isle of Islay (Figure 8.21); the second is the Pelamis sea-snake which is a hinged device located on Orkney.

Biomass Fermenting dung gives off methane gas which can be used in developing countries instead of fuelwood (Figure 8.22). Although this energy is cheap to produce, it means the dung can no longer be used as a fertiliser, while methane is a greenhouse gas (Figure 13.20). Biomass can also be used to produce **biofuels**, the first being sugar cane which was used to power cars in Brazil. Now oil palm in Malaysia (page 103), oilseed rape in the EU and maize in the USA are being grown for use as a cheaper, cleaner fuel.

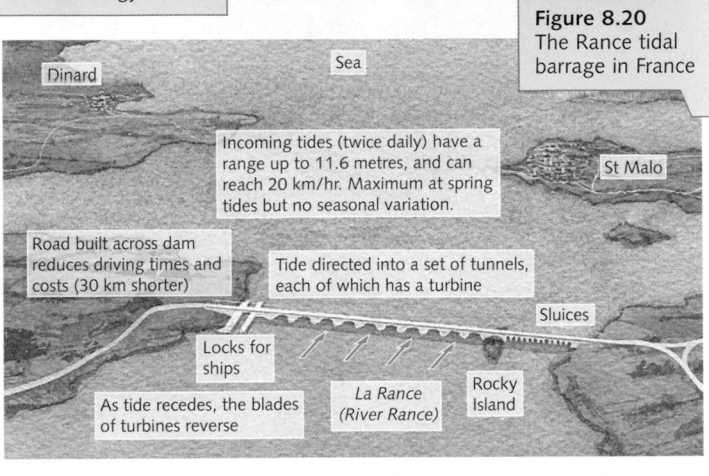

Figure 8.19
A solar energy station

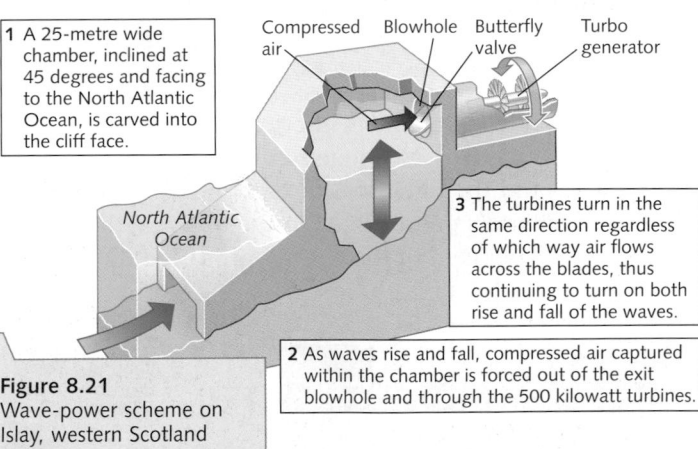

Figure 8.20
The Rance tidal barrage in France

Dinard

Sea

St Malo

Incoming tides (twice daily) have a range up to 11.6 metres, and can reach 20 km/hr. Maximum at spring tides but no seasonal variation.

Road built across dam reduces driving times and costs (30 km shorter)

Tide directed into a set of tunnels, each of which has a turbine

Sluices

Locks for ships

As tide recedes, the blades of turbines reverse

La Rance (River Rance)

Rocky Island

1 A 25-metre wide chamber, inclined at 45 degrees and facing to the North Atlantic Ocean, is carved into the cliff face.

Compressed air   Blowhole   Butterfly valve   Turbo generator

North Atlantic Ocean

3 The turbines turn in the same direction regardless of which way air flows across the blades, thus continuing to turn on both rise and fall of the waves.

2 As waves rise and fall, compressed air captured within the chamber is forced out of the exit blowhole and through the 500 kilowatt turbines.

Figure 8.21
Wave-power scheme on Islay, western Scotland

Figure 8.22
Biomass – fermentation of animal dung

World energy

Figure 8.23 shows how much energy per capita (per person) each country in the world produced in 2008. It also shows the main areas where fossil fuels are exploited (remember that fossil fuels still account for over 85 per cent of the world's energy – Figure 8.3b).

In both cases, it is the industrialised countries of North America, Western Europe (including the UK) and Russia that are best endowed, together with the oil-producing countries that form OPEC (Figure 11.14).

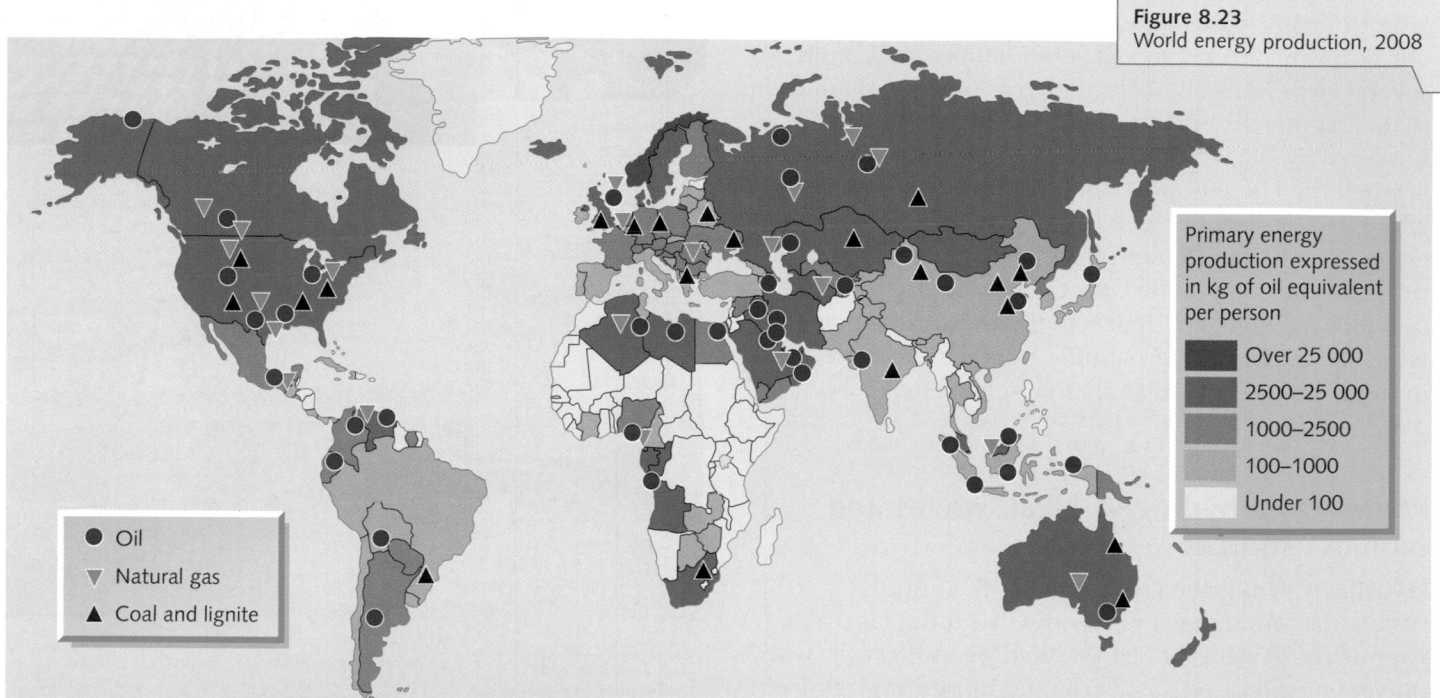

Figure 8.23
World energy production, 2008

Primary energy production expressed in kg of oil equivalent per person

	Over 25 000
	2500–25 000
	1000–2500
	100–1000
	Under 100

- ● Oil
- ▽ Natural gas
- ▲ Coal and lignite

Figure 8.24 shows how much energy per capita each country consumed in 2008 together with the gap between the MEDCs and the LEDCs. Although in 2008 only 16 per cent of the world's people lived in the MEDCs, they consumed 43 per cent of the energy produced. Of the remainder, the emerging countries of China and India, with 37 per cent of the world's population, consumed 14 per cent of the energy produced. (The remaining 47 per cent living in the LEDCs consumed 43 per cent.)

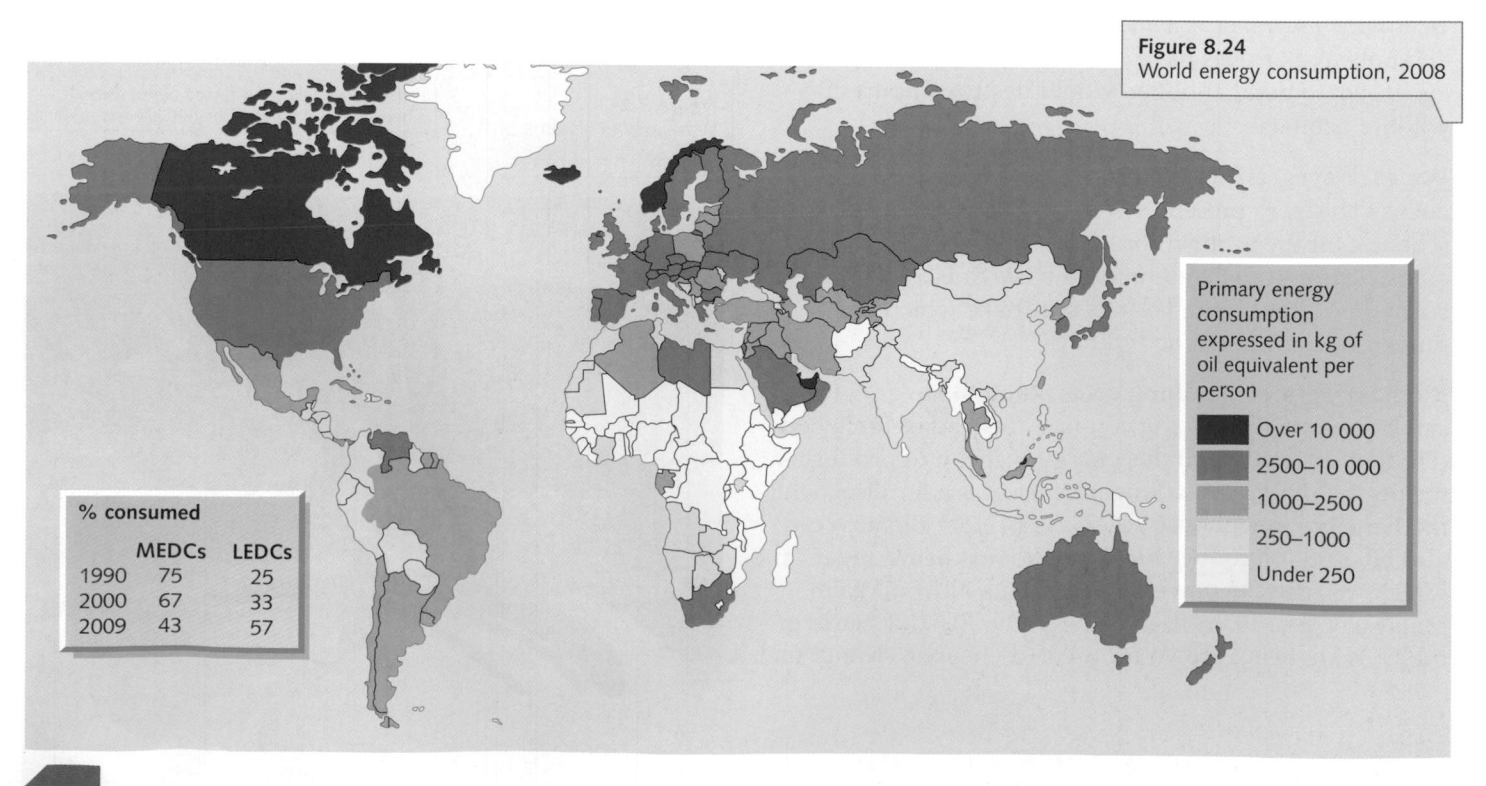

Figure 8.24
World energy consumption, 2008

Primary energy consumption expressed in kg of oil equivalent per person

	Over 10 000
	2500–10 000
	1000–2500
	250–1000
	Under 250

% consumed	MEDCs	LEDCs
1990	75	25
2000	67	33
2009	43	57

It is accepted that one measure of development (page 181) is the link between the amount of energy consumed per capita in a country and that country's level of development (Figure 8.25). The more developed a country is, the more energy it needs for industry, transport and the home. However, since 1990 there has been a slow-down in the consumption of energy by most MEDCs, partly due to industrial decline and growing environmental concerns, and an increase in consumption by many LEDCs with their growth in population and aspirations for a higher standard of living (Figure 8.24). Recent international environmental conferences have seen a clash between:

- the USA, which does not wish to reduce its energy consumption in case it results in a falling standard of living for its inhabitants
- most of the remaining MEDCs, which see the necessity to reduce energy consumption, especially of fossil fuels, if resources are to be conserved and the environment protected
- the LEDCs, which blame the MEDCs for most of the world's pollution and depletion of resources and which feel it is now their turn to use energy resources in order to develop economically and to improve their way of life
- China and India, which are both rapidly becoming industrialised.

	Country	Barrels of oil equivalent per person	Energy resources
MEDCs	USA	7.79	Coal, oil, natural gas, nuclear
	Germany	4.51	Coal, nuclear
	France	4.22	Nuclear, HEP
	Japan	4.04	HEP, nuclear
	UK	3.92	Oil, natural gas, nuclear
NICs	Korea	4.35	Nuclear
	Malaysia	2.32	HEP, oil, natural gas
	Mexico	1.53	Oil
	Thailand	1.41	Wood
	Brazil	1.07	Oil, natural gas, HEP, wood
Emerging	China	1.14	Coal, HEP, oil
	India	0.51	Coal, HEP
LEDCs with resources	Nigeria	0.78	Oil, natural gas, wood
	Egypt	0.76	HEP, oil
Poorer LEDCs	Kenya	0.48	Wood
	Ethiopia	0.28	Wood
	Bangladesh	0.16	Natural gas
	Nepal	0.08	Wood

World average = 1.75 barrels of oil equivalent

Figure 8.25
Sources and consumption of world energy, 2008

Appropriate technology

This is the introduction and use of a technology that is both appropriate to the needs, skills, knowledge and wealth of local people, and suitable to the environment in which they live (e.g. Nepal, page 150). Practical Action is a British charitable organisation that works with people in several developing countries including Nepal (page 150) and Kenya (page 151). It believes that energy provision is fundamental to improving living conditions for many people living in developing countries, especially in rural areas (Figure 8.26). One example was the improvement of cooking stoves which has reduced the use of fuelwood and improved the health of people in Kenya (Figures 8.27 and 9.38).

Reliable, accessible and affordable energy supplies can play an important role in improving living conditions in the developing world. They provide light and heat for homes, and power workshops that create jobs and generate wealth. Poor people in developing countries face particular problems in securing energy for their daily needs. This is a pressing issue in rural areas where most people live. More than half the world's population relies on biomass fuel (usually wood but also charcoal, crop residues or animal dung). Poor people cannot afford alternative fuels such as gas or kerosene. National grids mainly serve urban areas, or large industrial operations. It is prohibitively expensive to extend them far into the countryside. In places where there are renewable energy resources such as the sun, wind and water, communities often lack the knowledge, expertise and capital needed to install the appropriate system.

Practical Action

Figure 8.26
The need for energy in developing countries

Figure 8.27
Using an improved cooking stove, Kenya

Energy and the environment – oil in Alaska

Oil exploration in the harsh arctic climate of northern Alaska began in the 1950s. In 1968, two years after exploration began there, North America's largest oilfield was discovered at Prudhoe Bay (Figure 8.28). The field contains one-third of the USA's known oil reserves and 12 per cent of its known gas reserves. However, before the oil could be used, a route and a method of transport had to be decided by which the oil could be moved south. Two routes were suggested.

1 By giant oiltankers through the Arctic Ocean and around Alaska. This route was rejected as being too dangerous (the northern ocean is ice-covered for most of the year), and not economically practicable, despite two trial runs by a supertanker.

2 By pipeline, 1242 km in length, southwards across Alaska to the ice-free port of Valdez. This route faced such enormous physical difficulties and environmental opposition that it took until November 1973 before the USA government finally gave the go-ahead for the pipeline to be constructed, and it was June 1977 before the first crude oil was pumped along it.

Physical problems and environmental concerns

Many of the physical problems facing the oil companies and the concerns expressed by numerous environmental groups in 1968 are summarised in Figure 8.29.

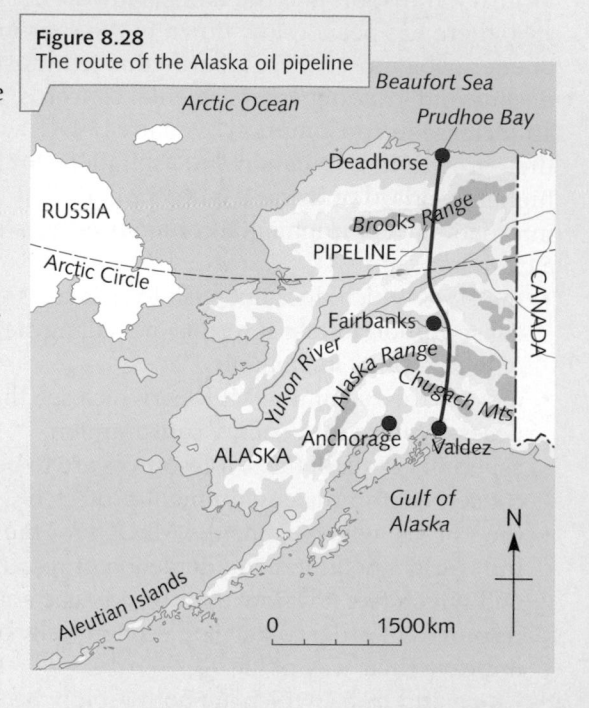

Figure 8.28
The route of the Alaska oil pipeline

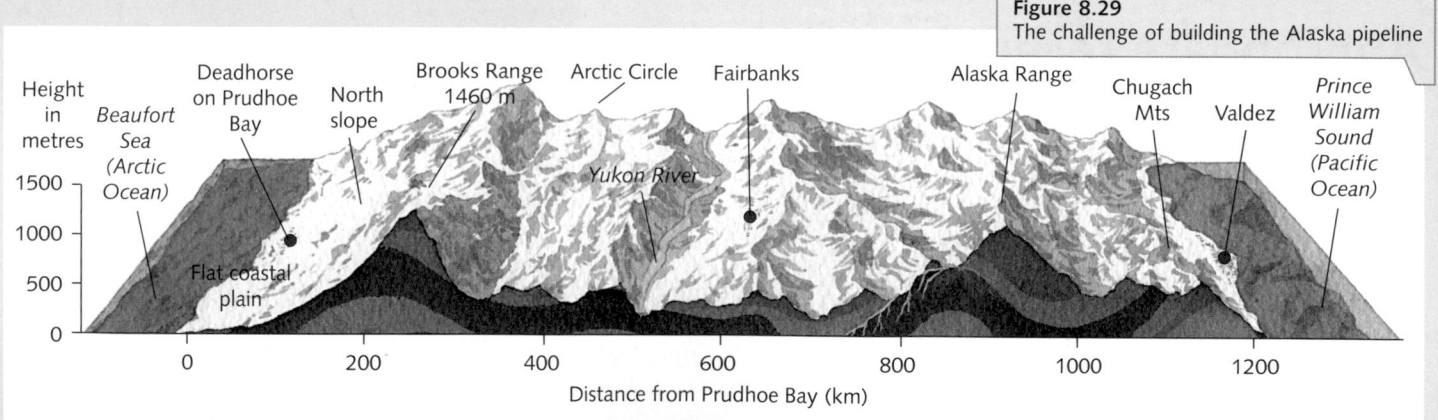

Figure 8.29
The challenge of building the Alaska pipeline

Physical problems		Environmental concerns	
Isolation	No road in 1968	Vegetation	Tundra ecosystem very fragile – concerns that once destroyed it could take decades to regenerate
Climate	Temperatures of –50°C in winter. Problems of keeping oil flowing, and human working conditions		
	Heavy snowfall: 1–2 m at Valdez	Wildlife	Crosses caribou migration route
	Fog, gales, icebergs in Prince William Sound and Gulf of Alaska		Near caribou breeding grounds
Soil	Permafrost over much of Alaska meant pipeline could not go underground		Might affect habitats of bears, wolves and moose
	Three months when surface thaws in summer, ground becomes unstable and pipeline could move	Wilderness	Crosses, or is adjacent to, National Parks and wildlife refuges
	Buildings would raise temperature and thaw the permafrost	Local people	May be forced to sell their land and to move
	Pipeline had to go over three mountain ranges		
Relief	Southern area is a major earthquake zone – tsunamis on coast	Oil spills	Major threat
	Over 350 rivers to cross (though many are frozen in winter)		

Some attempted solutions

Figures 8.30 to 8.33, taken in mid-1996, show some of the attempted solutions.

- **Figure 8.30** The first task, completed even before the government's go-ahead, was to extend the existing road, from just north of Fairbanks to Prudhoe Bay – the Dalton Highway. It was built on gravel pads to allow drainage and to limit the formation of ice. The pipeline was built on stilts 3 metres high to avoid melting the permafrost, and raised to 6 metres on caribou migration routes, so that animals can pass below the pipe. A total of 12 pumping stations (not shown) means that, in the event of a break, the flow of oil can be stopped. Notice how, after 20 years, the vegetation near the pipeline has still not fully recovered.
- **Figure 8.31** The pipeline supports are sunk into the permafrost for stability. The pipe itself, which has a diameter of 1.22 m, rests on 'sliding shoes'. These can move horizontally by one or two metres in the event of an earthquake. The oil is pumped through at a

Figure 8.30
The Dalton Highway and Alaska pipeline north of the Yukon River

Figure 8.31
Winter at Prudhoe Bay

Figure 8.32
Approaching the Brooks Range

Figure 8.33
In the Atigun Pass

temperature of about 80°C to prevent it freezing.
- **Figure 8.32** shows the pipeline as it approaches the Brooks Range. The first snow of winter had just fallen – on 3 August!

- **Figure 8.33** The Atigun Pass, in the Brooks Range, is the highest point that the pipeline reaches (1460 m). It has recently been sunk underground in river gravels (an environmental improvement) and encased in concrete (to prevent the oil from freezing and the ground from thawing).

Figure 8.34
Cross-section of the route of the pipeline across Alaska

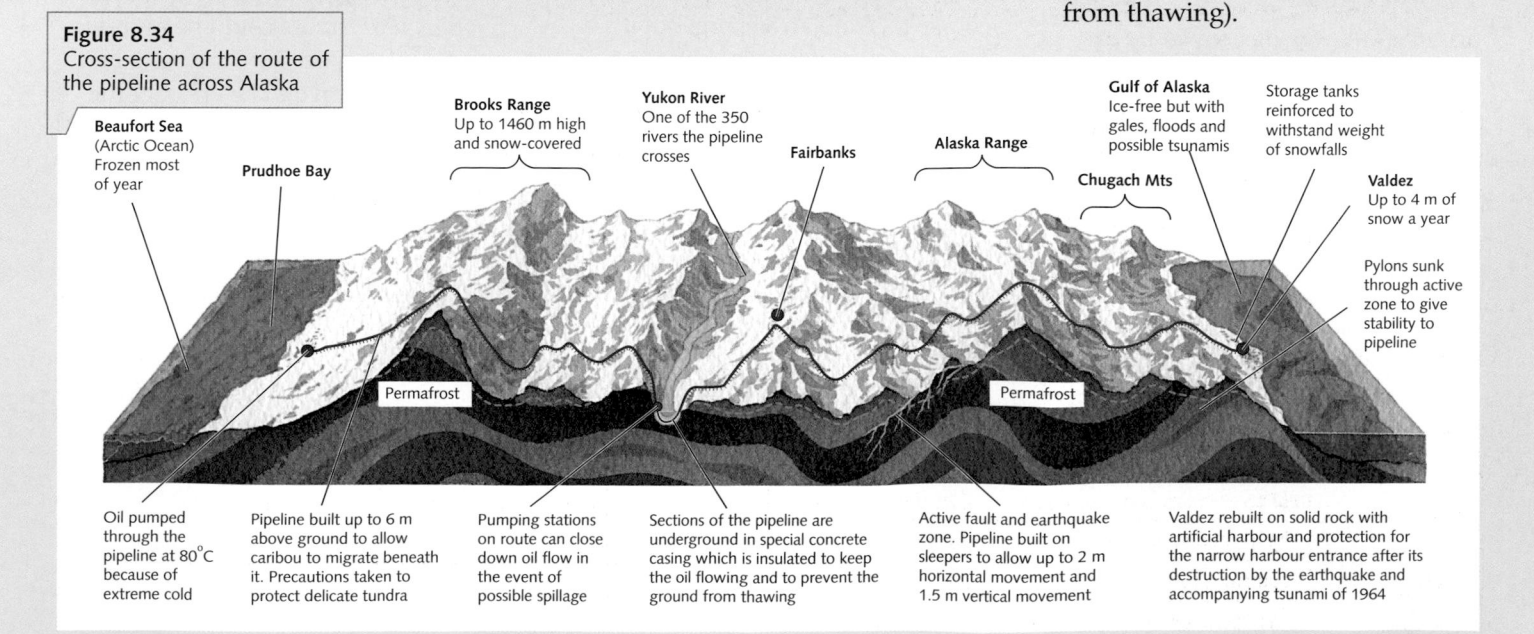

Beaufort Sea (Arctic Ocean) Frozen most of year

Prudhoe Bay

Brooks Range Up to 1460 m high and snow-covered

Yukon River One of the 350 rivers the pipeline crosses

Fairbanks

Alaska Range

Gulf of Alaska Ice-free but with gales, floods and possible tsunamis

Chugach Mts

Storage tanks reinforced to withstand weight of snowfalls

Valdez Up to 4 m of snow a year

Pylons sunk through active zone to give stability to pipeline

Permafrost

Permafrost

Oil pumped through the pipeline at 80°C because of extreme cold

Pipeline built up to 6 m above ground to allow caribou to migrate beneath it. Precautions taken to protect delicate tundra

Pumping stations on route can close down oil flow in the event of possible spillage

Sections of the pipeline are underground in special concrete casing which is insulated to keep the oil flowing and to prevent the ground from thawing

Active fault and earthquake zone. Pipeline built on sleepers to allow up to 2 m horizontal movement and 1.5 m vertical movement

Valdez rebuilt on solid rock with artificial harbour and protection for the narrow harbour entrance after its destruction by the earthquake and accompanying tsunami of 1964

Figure 8.35
Valdez oil terminal

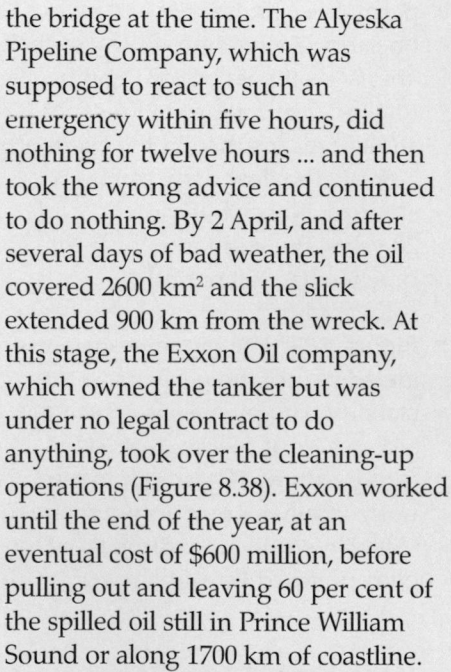

Figure 8.36
The Valdez Arm

Up to 1990 there had been some 300 reports of relatively minor oil leaks. The oil companies have, so far, been able to act quickly to seal these leaks. However, their efforts tend to be in restoring the flow (economic gain) rather than in cleaning up the spillage (environmental loss) – although efforts to clean up often *increase* the damage to the ecosystem.

The *Exxon Valdez* oil spill, 1989

Valdez is the ice-free port at the end of the Alaskan pipeline (Figure 8.35). The port was re-sited after the original one was destroyed by an earthquake in 1964. To reach the open sea from Valdez, oiltankers have no option but to follow a potentially dangerous route which takes them through Prince William Sound (Figure 8.36) and then through a narrow gap between two islands (Figure 8.37).

When the long-feared disaster occurred on 24 March 1989, it was not due to any of the physical dangers shown on Figure 8.37 but to human negligence and incompetence. Shortly after midnight the supertanker *Exxon Valdez*, 40 km out of Valdez and carrying 50 million tonnes of crude oil, ran aground in near-perfect weather conditions on Bligh Reef. It was first assumed that the ship had veered off course to avoid an iceberg, but it was later discovered that the captain was drunk and no senior officer was on the bridge at the time. The Alyeska Pipeline Company, which was supposed to react to such an emergency within five hours, did nothing for twelve hours ... and then took the wrong advice and continued to do nothing. By 2 April, and after several days of bad weather, the oil covered 2600 km^2 and the slick extended 900 km from the wreck. At this stage, the Exxon Oil company, which owned the tanker but was under no legal contract to do anything, took over the cleaning-up operations (Figure 8.38). Exxon worked until the end of the year, at an eventual cost of $600 million, before pulling out and leaving 60 per cent of the spilled oil still in Prince William Sound or along 1700 km of coastline. Even in 1997, traces could be found in remote inlets, and the ecosystem had still to fully recover. The effect of the

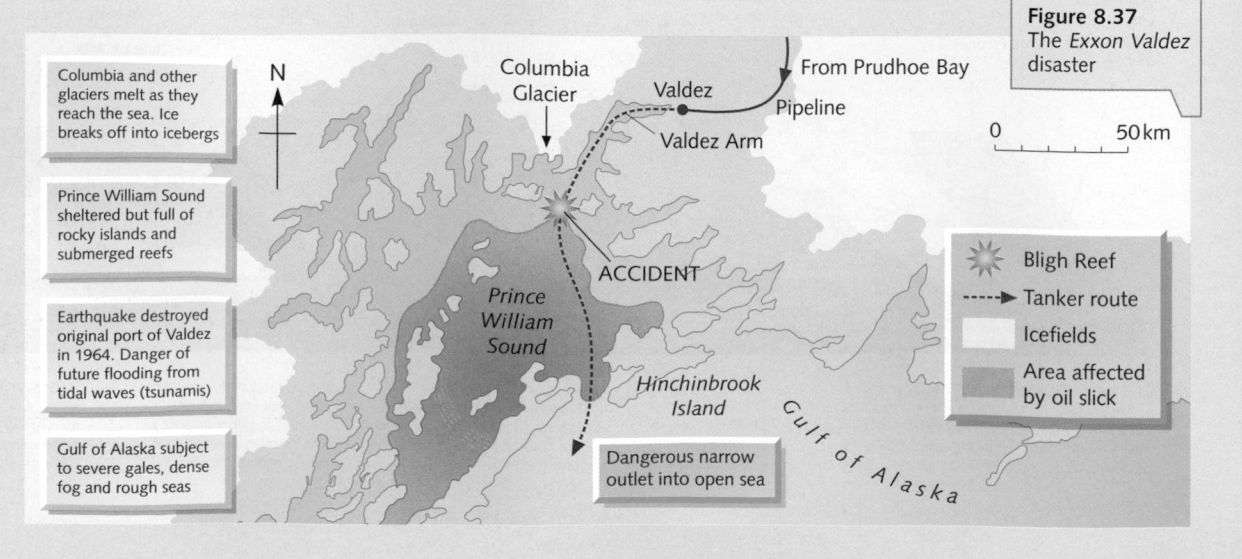

Figure 8.37
The *Exxon Valdez* disaster

Columbia and other glaciers melt as they reach the sea. Ice breaks off into icebergs

Prince William Sound sheltered but full of rocky islands and submerged reefs

Earthquake destroyed original port of Valdez in 1964. Danger of future flooding from tidal waves (tsunamis)

Gulf of Alaska subject to severe gales, dense fog and rough seas

N

Columbia Glacier

Valdez

From Prudhoe Bay
Pipeline

Valdez Arm

0   50 km

ACCIDENT

Prince William Sound

Hinchinbrook Island

Dangerous narrow outlet into open sea

Gulf of Alaska

☀ Bligh Reef
⇢ Tanker route
☐ Icefields
☐ Area affected by oil slick

spillage on wildlife and the marine ecosystem was enormous (Figure 8.39). The bodies of 35 000 seabirds and 3000 sea-otters were recovered (thousands more must have perished, while the bodies of dead seals sink). Likewise the local economy, which relied heavily upon fishing, was badly hit as salmon hatcheries were polluted and deep-sea fishing ruined.

Today and the future

For the last three decades oil has accounted for 85 per cent of Alaska's income. During that time the state has invested £20 billion of this income into a communal fund that pays every person in the state, including children, an annual dividend (sometimes the

Figure 8.38 Clean-up operations

Figure 8.39 Some effects of the oil slick on Prince William Sound

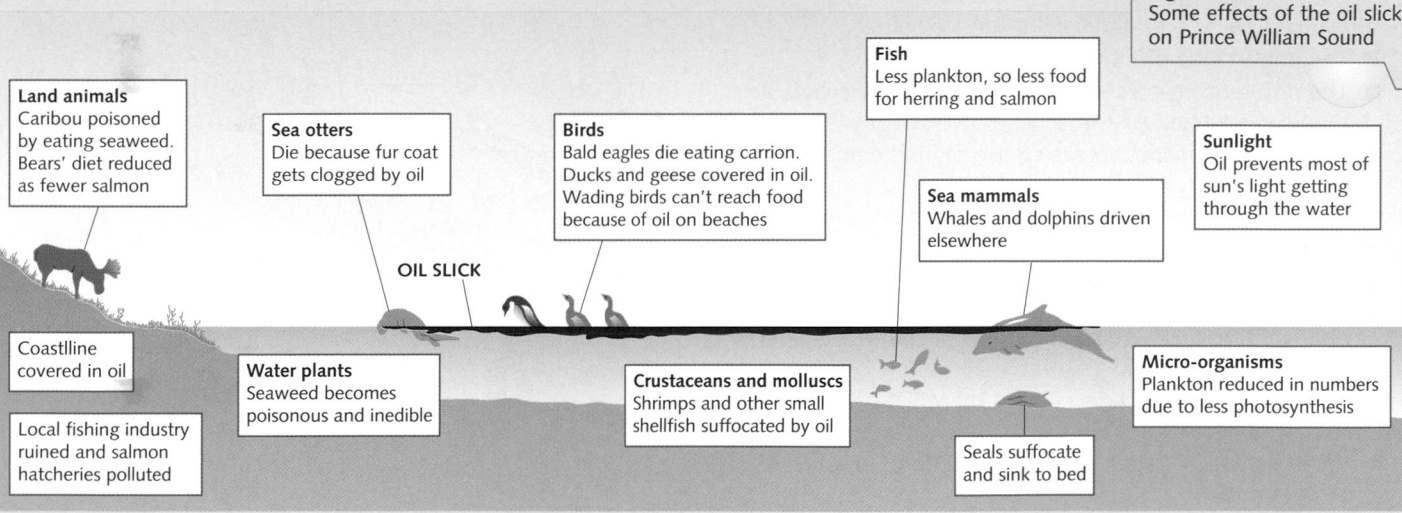

Land animals Caribou poisoned by eating seaweed. Bears' diet reduced as fewer salmon

Sea otters Die because fur coat gets clogged by oil

Birds Bald eagles die eating carrion. Ducks and geese covered in oil. Wading birds can't reach food because of oil on beaches

Fish Less plankton, so less food for herring and salmon

Sea mammals Whales and dolphins driven elsewhere

Sunlight Oil prevents most of sun's light getting through the water

OIL SLICK

Coastlline covered in oil

Water plants Seaweed becomes poisonous and inedible

Crustaceans and molluscs Shrimps and other small shellfish suffocated by oil

Micro-organisms Plankton reduced in numbers due to less photosynthesis

Local fishing industry ruined and salmon hatcheries polluted

Seals suffocate and sink to bed

equivalent of over £1000 per person). However, oil production is declining rapidly. In 1996, when it was almost 45 per cent down on its 1987 peak, the oil companies sought permission to explore new areas. One such area was the Arctic National Wildlife Refuge, a wilderness area on the North Slope where caribou, snowgeese and other wildlife are protected. As always there are groups of people in favour of and against the proposal. Some of their arguments are presented in Figure 8.40. After over a decade of debate, the US government did finally give the go-ahead in 2008 for further exploration to take place.

For (mainly economic and social)
- Most Alaskans see oil as their main source of jobs and wealth.
- The income from oil has been used within Alaska to build schools, hospitals and public buildings, and to improve water supplies and roads.
- Many American statesmen appreciate the importance of Alaskan oil, which still provides nearly one-third of the USA's total oil reserves and one-eighth of its total natural gas reserves.
- Alaskans see little point in protecting 'a herd of flea-ridden caribou seen by only a few hundred humans a year'.

Figure 8.40 Arguments for and against further exploration

Against (mainly environmental)
- Environmental groups, mainly outside Alaska, want the fragile Arctic National Wildlife Refuge to be protected from all forms of development.
- Some Native American groups claim they would prefer to live in their traditional way rather than benefit from oil income.
- There are real fears of a major oil spillage from either a pipeline breakage or another tanker accident.
- A criminal enquiry is investigating charges that one oil company has been illegally dumping toxic waste into its northern Alaskan field.

Resources (energy)

Key Words and Terms

a You should know the meaning of the following terms:
- natural resources • finite resources • recycling • fossil fuels
- greenhouses gases • nuclear power • fuelwood • hydro-electricity
- geothermal • solar energy • hydrogen cell • biomass/biofuels
- sustainable development • appropriate technology.

b You should know the difference between:
- renewable and non-renewable resources
- fossil fuels and sustainable resources
- wave and tidal power.

Key Ideas

You should know and understand the following:
- ■ The importance of natural resources.
- ■ The difference between renewable and non-renewable (finite) resources.
- ■ The reasons for the increasing use of, and demand for, resources.
- ■ The social, economic and environmental advantages and disadvantages (or issues) in the use of non-renewable energy resources.
- ■ The social, economic and environmental advantages and disadvantages (or issues) in the use of renewable energy resources.
- ■ That the amount of energy used by MEDCs is far greater than that used by LEDCs.
- ■ Ways by which a sustainable use of energy resources may be achieved.
- ■ How the development of energy resources in one place can affect people and the environment in other places.

Skills, Theories and Models

- ■ Classification of resource types
- ■ Interpret and use diagrams
- ■ Understand the idea of sustainable development
- ■ Investigate the impact of resource development
- ■ Appreciate other people's values and attitudes
- ■ Appreciate the need to plan and manage resource development

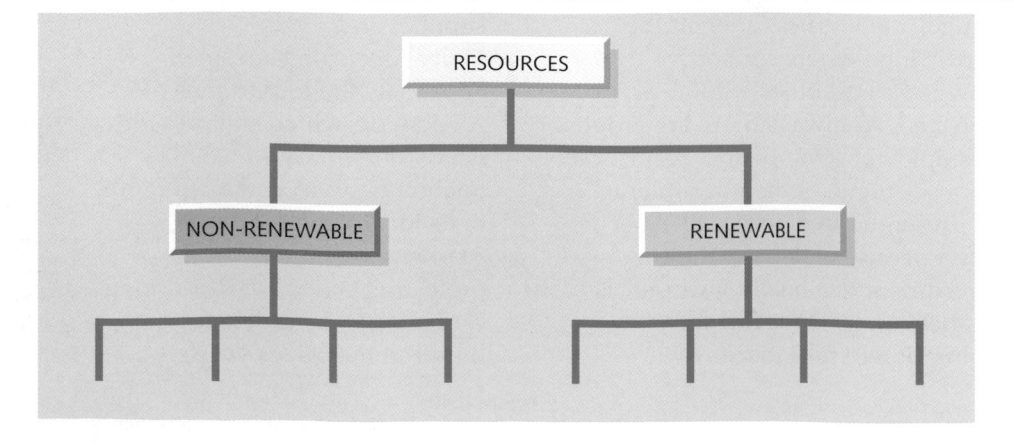

1 (Page 118)

a Make a larger copy of the diagram. Complete it by using examples from the following list:
- clean water
- fossil fuels
- landscapes
- minerals
- soils
- solar
- tides and waves
- trees
- wind. (9)

b Give three reasons why world demand for resources is increasing. (3)

c Describe five ways by which the Earth's resources may be protected. (5)

d Why is it important to protect the Earth's resources? (2)

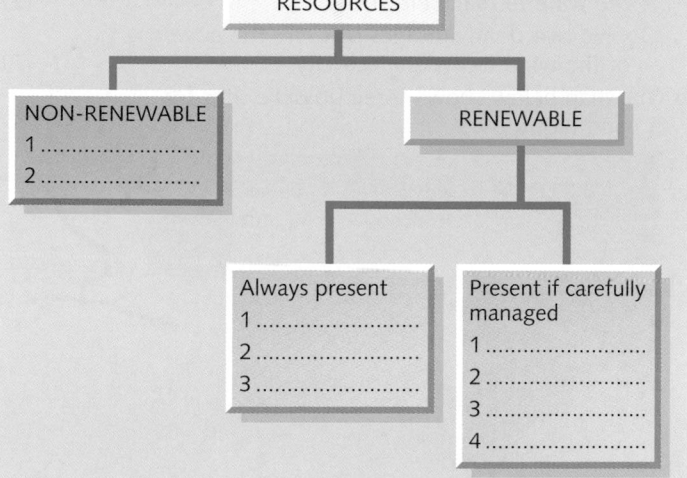

RESOURCES

NON-RENEWABLE
1 ....
2 ....

RENEWABLE

Always present
1 ....
2 ....
3 ....

Present if carefully managed
1 ....
2 ....
3 ....
4 ....

2 (Page 119)

Look at the six drawings showing different types of energy.

a Name the energy source in each drawing. (6)

b Which of the energy sources shown are renewable and which are non-renewable? (6)

c Name two other energy sources **not** shown here that are:
- renewable
- non-renewable. (4)

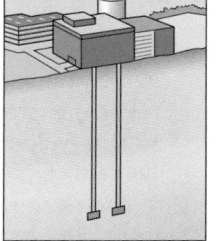
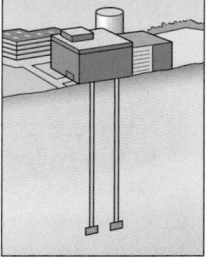

3 (Pages 120 to 125)

a Give two advantages and two disadvantages of using each of the following types of energy:
- coal
- oil
- natural gas
- nuclear. (4 × 4)

b Give three reasons why fuelwood is used more in less economically developed countries than any of the types of energy named in part (**a**). (3)

c i) Draw an annotated (labelled) sketch to show what makes the ideal site for a hydro-electric power station. (4)

ii) For Itaipù (Brazil) or any hydro-electric power scheme that you have studied, list three advantages and three disadvantages of the scheme. (6)

d i) Describe how geothermal energy is harnessed. (3)

ii) Why is geothermal energy important in a country like Iceland or New Zealand? (2)

e i) List three advantages of solar energy. (3)

ii) Give three reasons why Britain uses relatively little solar energy. (3)

iii) Why do relatively few tropical countries use solar energy? (2)

f i) Other than hydro-electricity, geothermal and solar energy, name two other types of renewable energy. (2)

ii) For each type, describe what it is and explain why it is not yet important in the UK. (2 × 2)

4 (Page 124)

a i) Give two advantages of using the power of the wind to make electricity. (2)

ii) Give two disadvantages of using the power of the wind to make electricity. (2)

b The map below shows three possible sites for a wind farm.

i) Which of the three sites X, Y or Z would you choose for a wind farm? Give two reasons for your answer. (2)

ii) For each of the two sites that you rejected, give one reason why you did so. (2 × 1)

Direction of prevailing wind

Sea

- ▬▬ Principal road
- ■ Town
- Village or hamlet
- Existing industry
- Areas of protected wildlife, nature reserves, etc.
- X Y Z Possible wind farm sites

N

0   0.5 km

Read the newspaper article.

c i) The farmer suggests that wind farms ruin the landscape. Describe two ways in which they might do this. (2)

ii) Name one group of people that is likely to agree with the farmer. (1)

iii) Name one group of people, other than the Electricity Board, that might disagree with the farmer. (1)

d Give two reasons why the number of wind farms is likely to increase in the next few years. (2)

Mid-Wales Farmer Faces Blight from Wind Farm

A farmer told our reporter today that plans to build a huge wind farm on Mynydd Carn should be scrapped. Mr Hywel Davies and his wife Mair, of Ty Coch farm, say they would be able to see the 60 metre tall turbines from every window of their house. 'They ruin the landscape,' said Mrs Davies 'We don't want them here in mid-Wales, and lots of other people agree with us.' An Electricity Board official defended the proposals for the wind farm. 'We cannot keep using fossil fuels forever – the local farmer might not see the sense of a wind farm, but plenty of others do.'

5 (Pages 126 and 127)

a According to Figure 8.23, which three continents produce most energy per person? (3)

b i) According to Figure 8.24, which three continents consume most energy per person? (3)

ii) Give two reasons for your answer to part (i). (2)

iii) According to Figure 8.24, which three continents consume least energy? (3)

iv) Give two reasons for your answer to part (iii). (2)

c With reference to Figure 8.25, describe any possible links between energy consumption per person and level of development. (3)

d According to Figure 8.26, 'Reliable, accessible and affordable energy supplies can play an important role in improving living conditions in the developing world'.

i) Describe three ways in which energy supplies can improve living standards. (3)

ii) Why is it difficult for less economically developed countries to improve their energy supplies? (3)

iii) How can appropriate technology help countries improve their energy supplies? (3)

6 *(Pages 120 to 125)*

a The map shows the location and distribution of electricity generating stations in the UK. Give two reasons for the location of each of the following:
  - coal-fired
  - oil-fired
  - gas-fired
  - nuclear
  - hydro-electric (HEP). *(10)*

b The table below shows percentage changes in the sources of energy in the UK between 1960 and 2008.
  i) Which source of energy has decreased the most? Give two reasons for your answer. *(3)*
  ii) Which source of energy has increased the most? Give two reasons for your answer. *(3)*
  iii) Suggest three likely changes in Britain's sources of energy by 2020. *(3)*

	1960	1970	1980	1990	2000	2008
Coal	76.5	50.7	35.1	30.6	17.7	18.1
Oil	22.6	42.7	39.7	34.7	36.4	33.4
Natural gas	0.1	2.7	20.1	25.9	33.7	39.8
Nuclear	0.2	3.3	4.5	8.3	11.3	6.4
Hydro-electric	0.6	0.6	0.6	0.4	0.2	0.4
Renewable	0	0	0	0.1	0.7	1.9

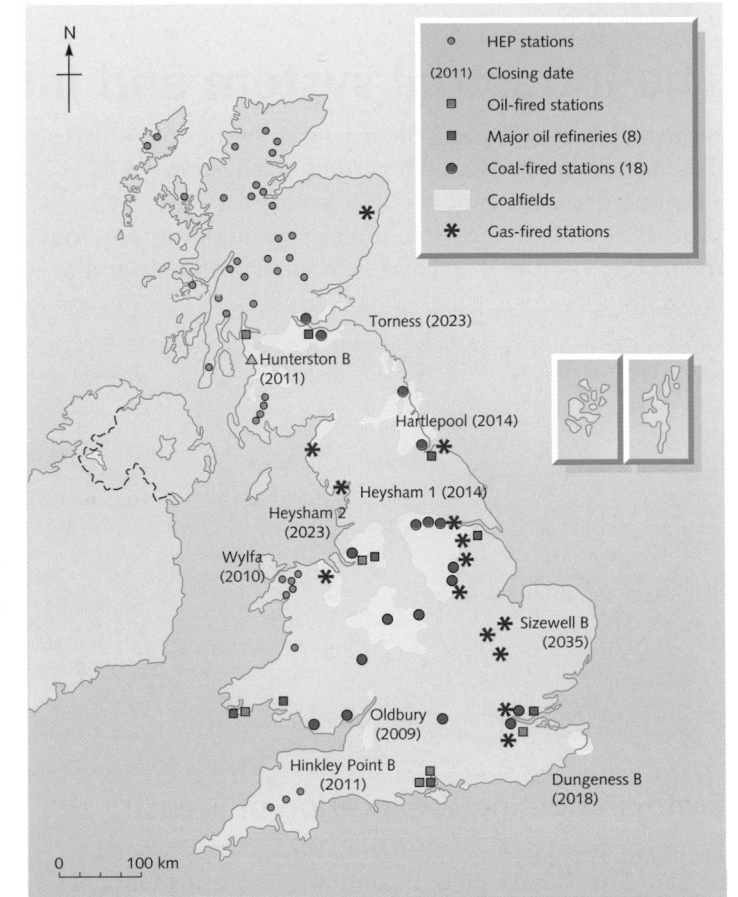

7 *(Pages 128 to 131)*

a Match up the following with the correct letter given on the cross-section of Alaska:
  - Alaska Range
  - Arctic Circle
  - Arctic Ocean
  - Brooks Range
  - Chugach Mountains
  - Prince William Sound
  - Prudhoe Bay
  - Valdez
  - Yukon River. *(9)*

b The following list gives some of the problems that had to be overcome before oil could be transported from Alaska to the rest of the USA. For each of the problems, describe the attempts made to solve it.
  - The ground surface thaws in summer but underneath remains frozen all year
  - Pipeline crosses an earthquake belt
  - Pipeline crosses caribou migration route
  - An underground pipe might thaw the ground
  - Beaufort Sea and Arctic Ocean frozen most of year
  - Air temperatures down to −50°C
  - The oilfield could only be reached by air. *(7)*

c Give four dangers faced by oiltankers using Prince William Sound. *(4)*

d Describe how the running aground of the *Exxon Valdez* affected:
  - fish
  - sea mammals
  - birds
  - land animals
  - plants (vegetation). *(5)*

e Do you think that the oil companies should be allowed to explore new areas for oil? Give three reasons for your answer. *(3)*

The industrial system and industrial location

Industry as a whole, or a factory as an individual unit, can, like farming (page 96), be regarded as a **system**. At its simplest, there are **inputs** into a factory (or industry), **processes** that take place in the factory, and **outputs** from the factory (Figure 9.1). For a firm to be profitable and to remain in business, the value of its outputs must be greater than the cost of its inputs. Some of the profit should then be re-invested, e.g. in modernising the factory and introducing new technology.

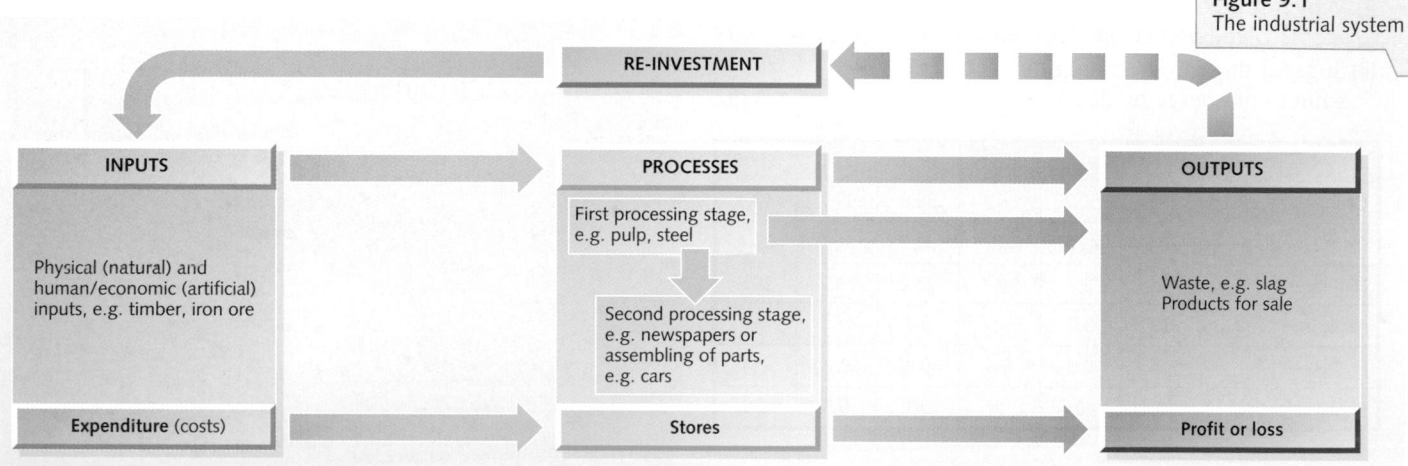

Figure 9.1
The industrial system

Factors affecting the location of industry

Factories may be built by an individual entrepreneur, a private firm, a nationalised company or a transnational corporation. Before the factory is built, however, decisions have to be made as to which will be the best site for its location. It is unlikely that any site will have all the factors that are listed as advantages in Figure 9.2. Where several sites are available, the individual or company must decide which is likely to provide the best location. In many cases this decision is determined by predicting which site will give the greatest profit. This will be where the costs of raw materials, land, energy, labour and transport are minimised and where there is a large market for the product. The decision may also be affected by government policies.

Physical factors
Raw materials The bulkier and heavier these are to transport, the nearer the factory should be located to the raw materials. This was even more important in times when transport was less developed.

Power – energy This is needed to work the machines in the factory. Early industry needed to be sited near to fast-flowing rivers or coal reserves, but today electricity can be transported long distances.

Natural routes River valleys and flat areas were essential in the days before the railway, car or lorry.

Site and land Although early industry did not at first take up much space, it did need flat land. As the size of plant increased (e.g. steelworks), more land was needed. Ideally such sites should be on low-quality farmland where the cost of purchase is lower. Last century many sites were in today's 'inner city' areas whereas now they tend to be on edge-of-city 'greenfield' locations.

> In the nineteenth century it was physical factors such as the source of raw materials (e.g. iron ore) and sources of energy (e.g. coal) which determined industrial locations.

Human and economic factors
Labour This includes both quantity (large numbers in nineteenth-century factories) and quality (some areas demand special skills as technology develops).

Capital (money) Early industry depended on wealthy entrepreneurs. Now banks and governments may provide the money.

Markets The size and location of markets have become more important than the source of raw materials.

Transport Costs increase when items moved are bulky, fragile, heavy or perishable.

Economies of scale Small units may become unprofitable and so merge with, or are taken over by, other firms.

Government policies As governments tend to control most wealth, they can influence industrial location.

Improved technology Examples are facsimile (fax) machines and electronic mail.

Leisure facilities Both within the town and the surrounding countryside, leisure activities are becoming more desirable.

Figure 9.2
Factors affecting the location of industry in the UK

> By the late twentieth century, the three main factors deciding industrial location were more likely to be the nearness to a large market, the availability of skilled labour, and government policies.

Location of industry in the UK

Figure 9.3 shows the location and distribution of Britain's traditional heavy industries. Most of these industries were established in the nineteenth century. Their growth was based on the use of coal, the development of technology to process local and imported raw materials, the creativity of the people and the ability to export manufactured goods. Consequently the major industrial areas were either on Britain's coalfields or in coastal ports located on deep-water estuaries.

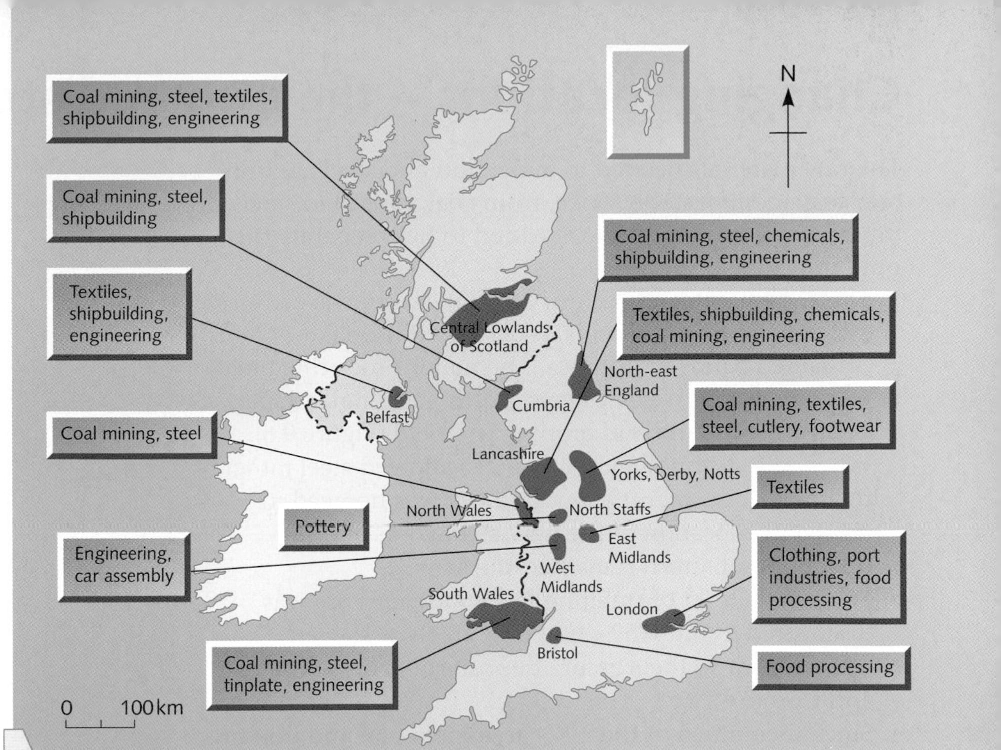

The location, distribution and type of Britain's present-day manufacturing industry has changed considerably (Figure 9.4). For a variety of reasons (Figure 9.5) coal mines began to close in the 1920s, textile mills in the 1960s, shipyards in the 1970s and steelworks in the 1980s. Modern replacement industries, many of which are high-tech and connected with electronics (page 140), employ fewer people and are often located well away from the traditional manufacturing areas. They are said to be **footloose** as, not being tied to raw materials, they have a relatively free choice of location.

These newer industries have:
- either opted for a more pleasant working environment near to large markets and major transport routes, or
- especially in the case of foreign companies, been tempted by government policies to locate in former industrial areas which often had higher levels of unemployment.

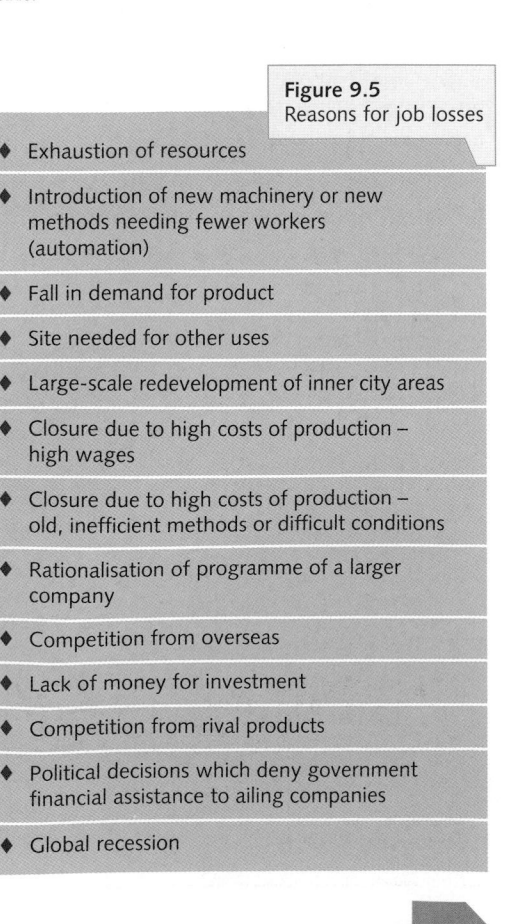

Figure 9.3
Traditional industrial areas in the UK (before 1970)

Coal mining, steel, textiles, shipbuilding, engineering

Coal mining, steel, shipbuilding

Coal mining, steel, chemicals, shipbuilding, engineering

Textiles, shipbuilding, chemicals, coal mining, engineering

Textiles, shipbuilding, engineering

Central Lowlands of Scotland

Belfast

North-east England

Cumbria

Coal mining, textiles, steel, cutlery, footwear

Coal mining, steel

Lancashire

Yorks, Derby, Notts

Textiles

North Wales

North Staffs

Pottery

East Midlands

Clothing, port industries, food processing

Engineering, car assembly

West Midlands

South Wales

London

Bristol

Coal mining, steel, tinplate, engineering

Food processing

0   100 km

Figure 9.4
Location of present-day industries in the UK

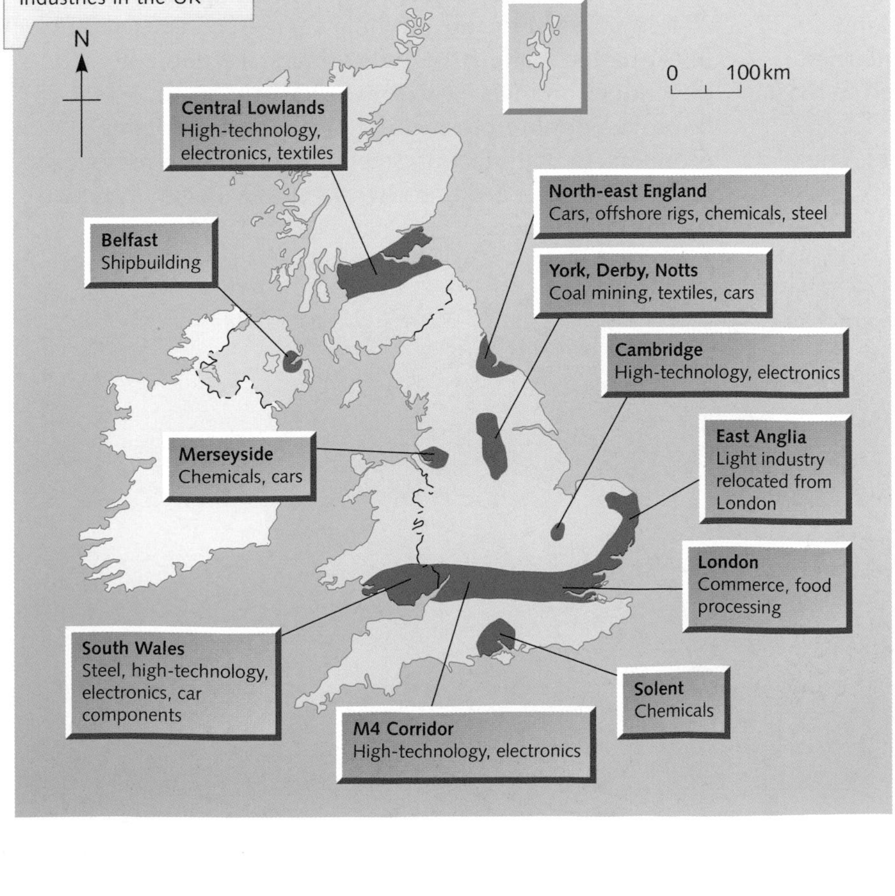

Central Lowlands
High-technology, electronics, textiles

Belfast
Shipbuilding

North-east England
Cars, offshore rigs, chemicals, steel

York, Derby, Notts
Coal mining, textiles, cars

Cambridge
High-technology, electronics

East Anglia
Light industry relocated from London

Merseyside
Chemicals, cars

London
Commerce, food processing

South Wales
Steel, high-technology, electronics, car components

M4 Corridor
High-technology, electronics

Solent
Chemicals

0   100 km

Figure 9.5
Reasons for job losses

- Exhaustion of resources
- Introduction of new machinery or new methods needing fewer workers (automation)
- Fall in demand for product
- Site needed for other uses
- Large-scale redevelopment of inner city areas
- Closure due to high costs of production – high wages
- Closure due to high costs of production – old, inefficient methods or difficult conditions
- Rationalisation of programme of a larger company
- Competition from overseas
- Lack of money for investment
- Competition from rival products
- Political decisions which deny government financial assistance to ailing companies
- Global recession

Changing locations – the iron and steel industry

The raw materials needed to make iron and steel are iron ore, coal and limestone. Coke, from coal, is used to smelt the iron ore, and limestone is added to help separate the pure iron from impurities.

- The early nineteenth-century iron industry was concentrated on coalfields, especially those that contained bands of iron ore. The major producing areas were the South Wales valleys, north-east England, parts of central Scotland and around Sheffield (Figure 9.6). When later technological advances allowed steel rather than iron to be manufactured, there was no need to change the location of the new steelworks and the distribution pattern remained the same.
- When a method of smelting low-grade iron ore was discovered in the 1950s, two new steelworks were opened near to the iron ore deposits at Scunthorpe and Corby.
- Since then much of the UK's accessible coal and iron ore have been used up. As these two raw materials had to be imported, three large steelworks were built on the coast at Port Talbot, Llanwern and Redcar.
- Recently, competition from overseas and a drop in steel prices have forced all but three of Britain's steelworks to close (Figure 9.6). These could be further threatened by the global recession, which saw Redcar mothballed in 2009.

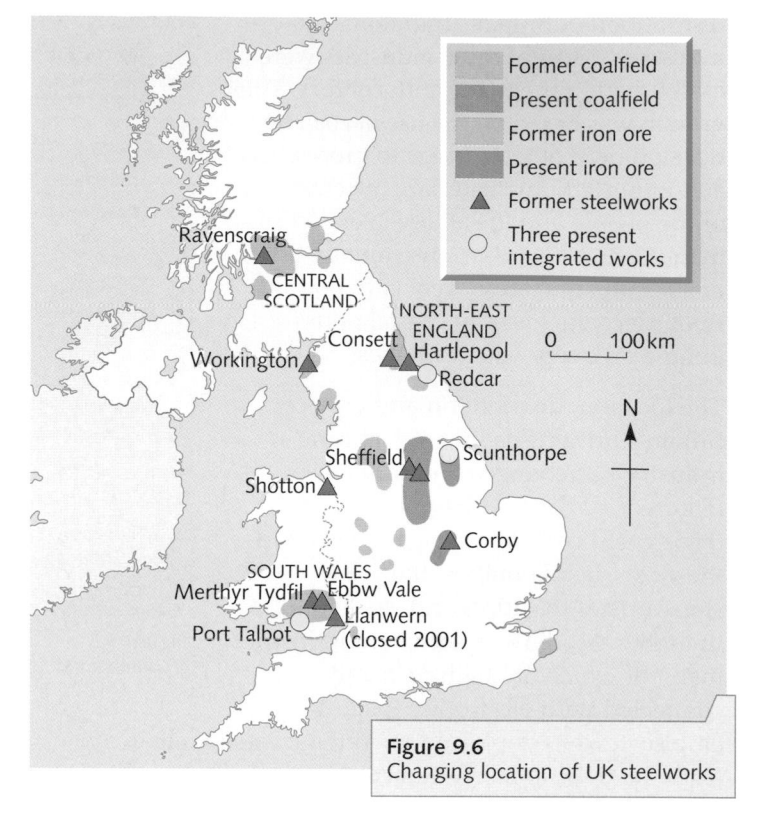

Figure 9.6
Changing location of UK steelworks

Iron and steel in South Wales

The valleys of South Wales were an ideal location for iron-making. Coal and iron ore were often found together on valley sides and limestone was quarried nearby (Figure 9.7). The valleys themselves led to coastal ports from where iron products and surplus coal were exported to many parts of the world. The industry became centred on places like Ebbw Vale

and Merthyr Tydfil (Figure 9.8a). By 1850 there were 35 ironworks in the area. Whole villages, constructed in linear patterns along the valley floors, were totally dependent on their local ironworks. In 1856 an improvement in iron smelting meant that it became economic to manufacture steel rather than the previously brittle iron. After 1860, steelworks slowly began to replace the iron foundries.

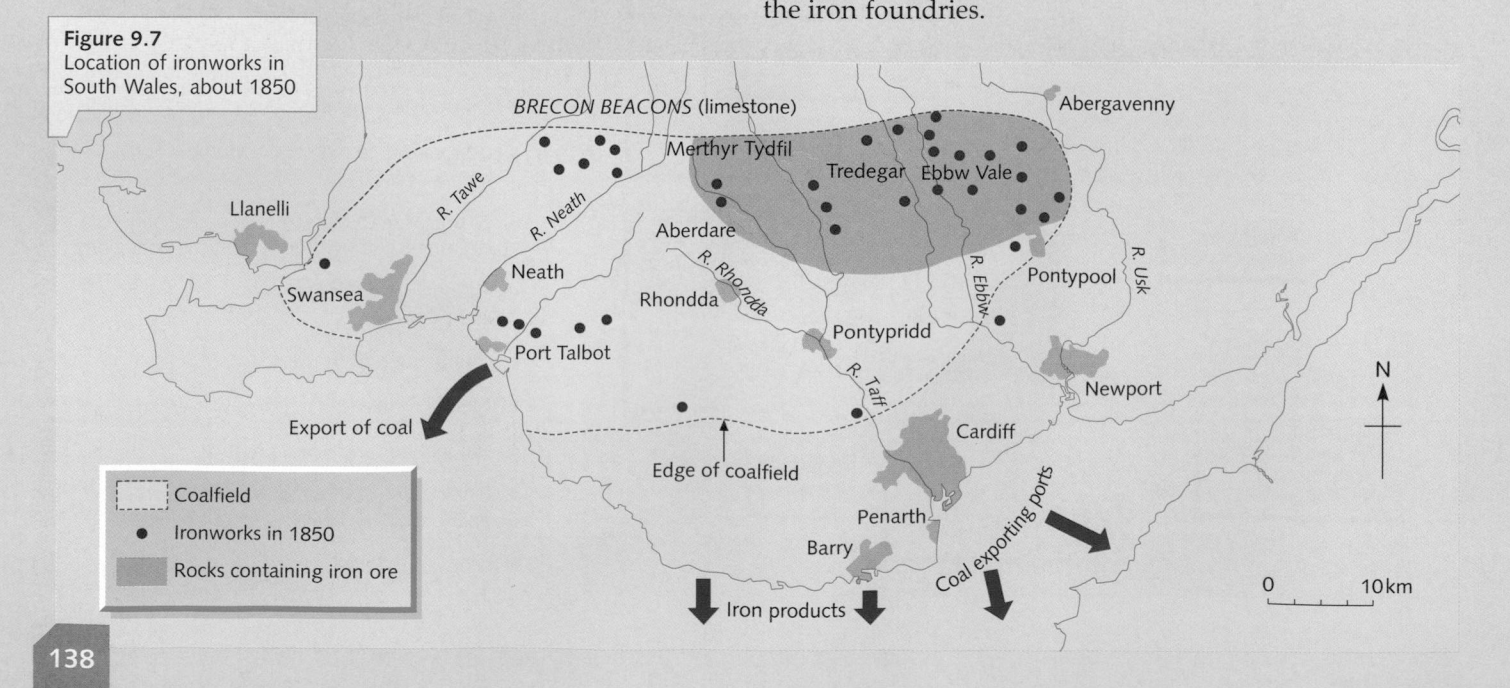

Figure 9.7
Location of ironworks in South Wales, about 1850

PLACES

Period of time			(a) Location of early 19th-century foundries in South Wales (i.e. Ebbw Vale)	(b) Location of integrated steelworks at Port Talbot
Physical	Raw materials	Coal	Mined locally in valleys	Only one coal mine open, most of coal imported
		Iron ore	Found within the Coal Measures	Imported from North Africa and North America
		Limestone	Found locally	Found locally
		Water	For power and effluent – local rivers	For cooling – coastal site
	Energy – fuel		Charcoal for early smelting, later rivers to drive machinery and then coal	Electricity from national grid (using coal, oil, natural gas and nuclear)
	Natural routes		Materials mainly on hand. 'Export' routes via the valleys	Coastal sites
	Site and land		Small valley-floor locations	Large areas of flat, low-capacity farmland
Human and economic	Labour		Large number of unskilled workers	Still relatively large numbers but with a higher level of skill. Fewer needed because high-tech
	Capital		Local entrepreneurs	Government, EU
	Markets		Local	The car industry
	Transport		Little needed, some canals	M4. Purpose-built ports
	Economies of scale		Not applicable	Two large steelworks more economical than numerous small iron foundries
	Government policy		Not applicable	Government can determine locations and closures
	Technology		Small scale – mainly manual	High-technology – computers, lasers, etc.

Figure 9.8
Reasons for changes in the location of the iron and steel industry in South Wales

By the 1970s there were only two steelworks left in South Wales. These were not, however, located in the valleys but on the coast at Port Talbot and Llanwern (Figure 9.9). This was because many of the initial advantages for steelmaking no longer existed (Figure 9.8b). Only a handful of coal mines remained open and the iron ore had long since been exhausted. As both these raw materials needed to be imported, it was logical to build any new modern steelworks on the coast at **break of bulk** locations. Break of bulk is when a product has to be transferred from one form of transport to another – a process that takes up both time and money. It was easier and cheaper, therefore, to have the new steelworks where the imported raw materials were unloaded, rather than transporting them to the older, inland works. Added to this, it was a government decision, as they were helping financially, to locate the new sites on the coast.

Port Talbot had its own harbour and docks which allowed the import of coal and iron ore. These, together with limestone, fed into a blast furnace where the ore is smelted and most of the impurities are removed. The resultant pig iron has oxygen added to it, reducing the carbon content (which makes iron brittle) to give steel. The steel is then rolled into thin sheets which can be used to make, among other things, car bodies. Port Talbot is one of Britain's three remaining **integrated** works and it uses the latest technology ('integrated' means all the stages in the manufacture take place on the same site). The Llanwern works was closed in 2001 due to overseas competition, global overproduction and a fall in the price of steel. Whether Port Talbot remains open or not is likely to depend on the EU, as political decisions are now more important than the physical advantages of the nineteenth century.

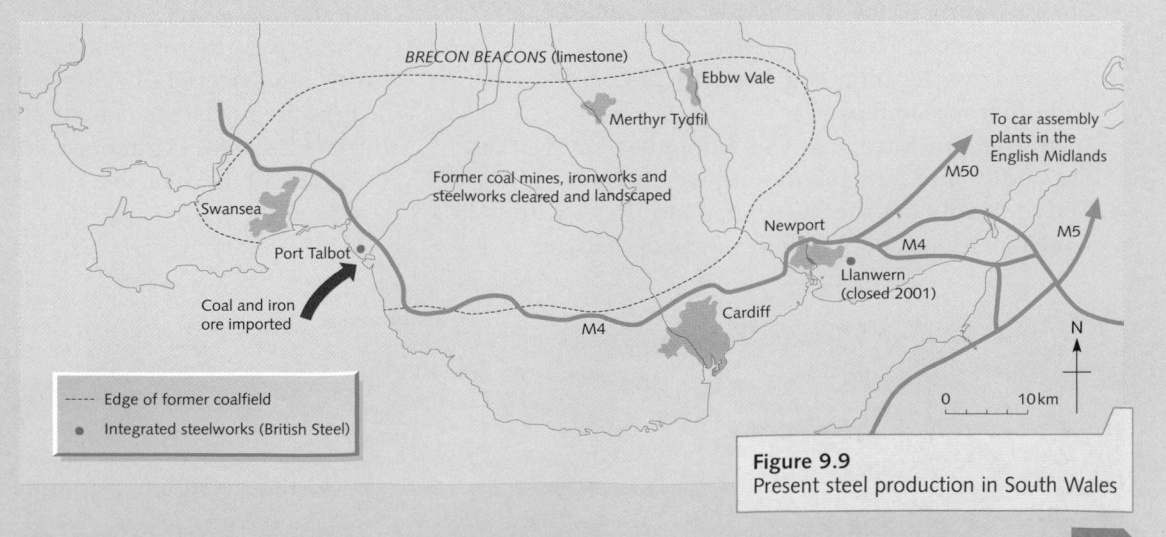

Figure 9.9
Present steel production in South Wales

Changing location – high-technology industries

The term **high-technology industry** (or high-tech) refers, usually, to industries developed within the last 25 years and whose processing techniques often involve micro-electronics. These industries, many of which are footloose (page 137), have been the 'growth industries' of recent years though unfortunately they employ few people in comparison with the older, declining heavy industries. Two possible subdivisions of high-tech industries are:

1 The 'sunrise industries' which have a high-technology base.
2 Information technology industries involving computers, telecommunications and micro-electronics.

As a highly skilled, inventive, intelligent workforce is essential, and as access to raw materials is relatively unimportant, these high-tech footloose industries tend to become attracted to areas which the researchers and operators find attractive – from a climatic, scenic, health and social point of view. Such areas include:

• Silicon Glen in central Scotland
• Silicon Valley in California
• Sunrise Strip which follows the route of the M4 from London westwards towards Newbury (locally known as Video Valley), Bristol (Aztec West) and into South Wales (Figure 9.10)
• south of France behind Nice.

M4 Corridor

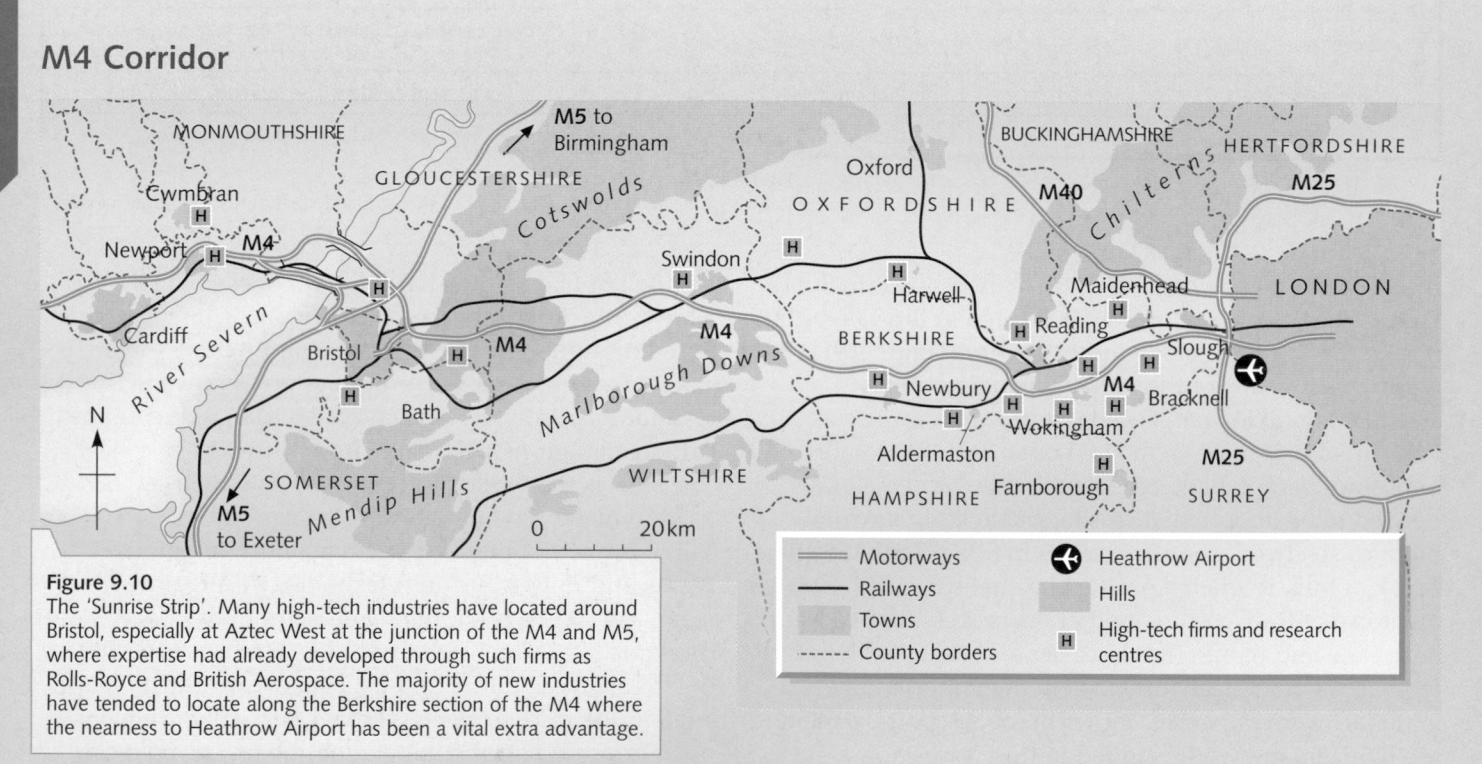

Figure 9.10
The 'Sunrise Strip'. Many high-tech industries have located around Bristol, especially at Aztec West at the junction of the M4 and M5, where expertise had already developed through such firms as Rolls-Royce and British Aerospace. The majority of new industries have tended to locate along the Berkshire section of the M4 where the nearness to Heathrow Airport has been a vital extra advantage.

The advantages of this area for spontaneous, unplanned growth of micro-electronics industries include:
• The proximity of the M4 and mainline railways.
• The presence of Heathrow Airport.
• The previous location and existence of government and other research centres.
• A large labour force, many of whom have moved out of London into new towns and overspill towns.
• The proximity of other associated industries with which ideas and information can be exchanged.

• Nearness to universities with expertise and research facilities available. High-tech firms can work closely with the university campus.
• An attractive environment. Figure 9.10 names the Cotswolds, Mendips, Chilterns and Marlborough Downs. Nearby are the North Downs, three National Parks (Brecon Beacons, Dartmoor and Exmoor) and, through the centre of the area, the Thames Valley.

Figure 9.11
Windmill Hill Business Park, Swindon. In all, three large business parks have been developed, each offering facilities and accommodation in a landscaped, parkland environment.

Business and science parks

Most business parks have grown up on edge-of-city greenfield sites, the remainder as part of inner city redevelopment schemes. The major attractions of greenfield sites are the relatively low cost of land and a pleasant working environment with a low density of buildings. Usually over 70 per cent of the land in business parks is left under grass and trees or converted into ornamental gardens and lakes (Figures 9.11 and 9.13). Business parks form an ideal location for high-tech industries such as electronics, and research institutions.

Science parks are similar but with the addition of direct links with universities (Figure 9.12). Some business parks include offices, hypermarkets and leisure complexes.

Why do similar industries locate together?

By locating near to each other, high-tech firms have the advantages of being able to exchange ideas and information with neighbouring companies, sharing maintenance and support services, sharing basic amenities such as connecting roads, and building up a pool of highly skilled, increasingly female labour.

Tsukuba Science City

Tsukuba Science City was specifically created to relieve the pressure on overcrowded, over-expensive Tokyo 40 km to the south-west (Figure 9.14). It has its own university and educational institutions, over 50 national research institutes and laboratories, and more than 150 private firms (Figure 9.12).

Tsukuba's population of 152 000 includes 7000 resident scientists, engineers and researchers, of whom nearly 2000 are non-Japanese. Not only has Tsukuba become a centre of science, it is also in the centre of a scenic environment, part of which has been declared a Quasi National Park (Figure 9.13).

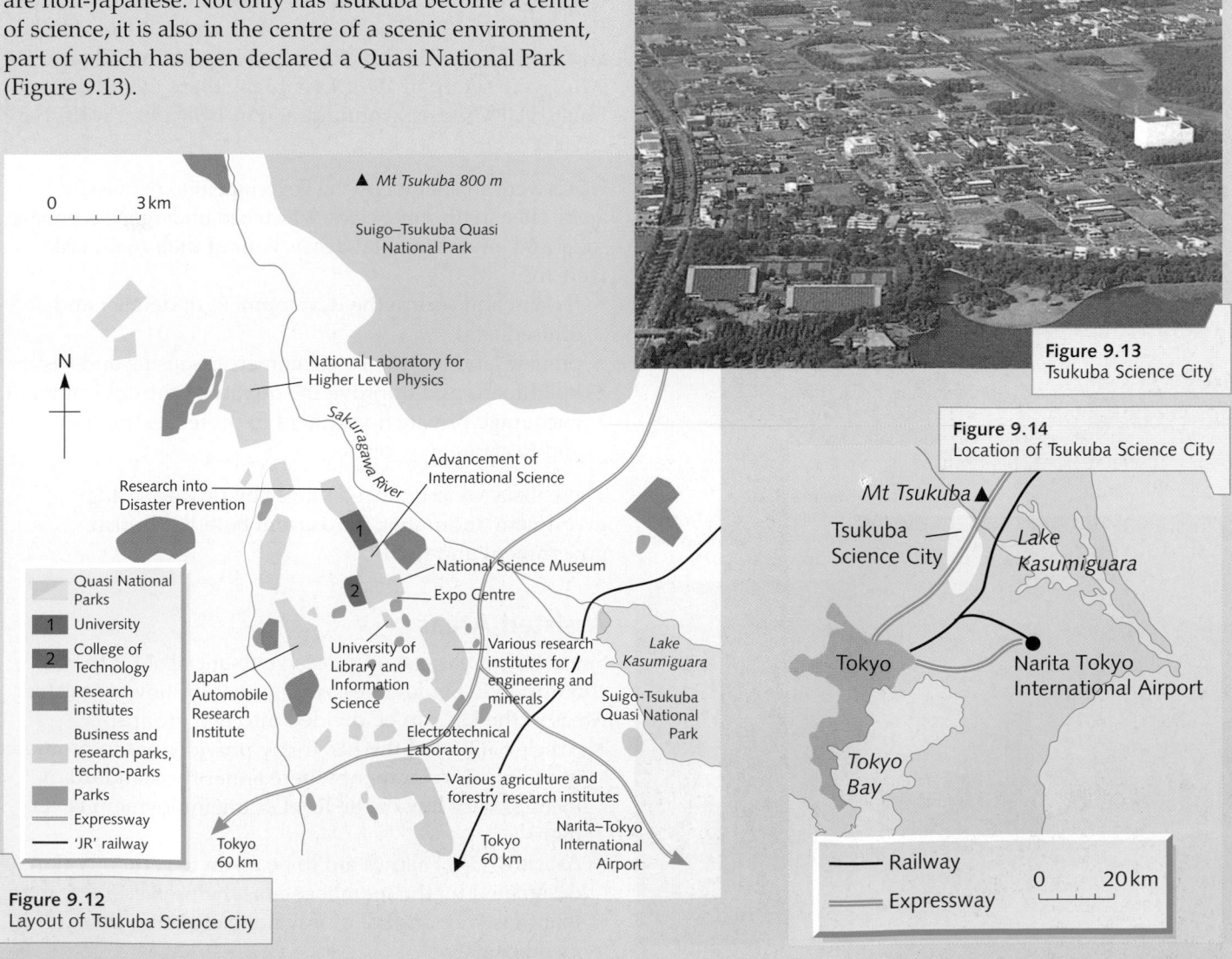

Figure 9.13
Tsukuba Science City

Figure 9.14
Location of Tsukuba Science City

Figure 9.12
Layout of Tsukuba Science City

141

Changing location – government policies in the UK

Successive British governments have tried, since 1945, to encourage industry to move to areas of high unemployment. The size and location of these areas of high unemployment have changed over time. Figure 9.15 shows the areas which were regarded as needing the most assistance in 2009. Over the years governments have tried to encourage new industries to reduce unemployment by:

- industrial development certificates which control where a firm can locate – these were first issued by the British government in 1947
- the creation of new towns in order to take work to the unemployed
- providing 'advanced factories' and industrial estates with services already present (e.g. roads, electricity)
- financial aid in the form of removal grants, rent-free periods, tax relief on new machinery, and reduced interest rates
- decentralising government offices
- improving communications and accessibility
- subsidies to keep firms going which otherwise would close down
- retraining schemes
- job creation, Manpower Creation Schemes (MCS) and Youth Training Schemes (YTS)
- Enterprise Zones
- assistance from the EU.

Urban Development Corporations (UDCs)

Urban Development Corporations were created by Act of Parliament in 1980. The first two, in London's Docklands and the Merseyside Development Corporation (Figure 9.16), were set up in 1981. Eventually there were 13. The English UDCs were wound down in 1998 and Cardiff Bay in 2000.

UDCs were an attempt by the government to rejuvenate areas, often in the inner cities, which had undergone economic, social and environmental decay. Four of their main tasks were to:

- reclaim and secure the development of derelict and unused land
- provide land for industry, commerce, housing and leisure
- build roads and improve the quality of the environment
- encourage private investment to protect existing jobs and to create new ones.

These tasks were financed jointly by private sector investment and public funds from both the British government and the EU.

Assisted Areas

Since 2007, areas can only receive financial aid if they conform to EU guidelines. As Figure 9.15 shows, Assisted Areas in the UK can be divided into two groups:

- Article 87(3)(a) of the EU Treaty provides aid to promote economic development where either the standard of living is very low or the level of unemployment is very high.
- Article 87(3)(c) allows aid to be given to certain regions designated by the member states themselves if it is felt that those regions are disadvantaged within that member state.

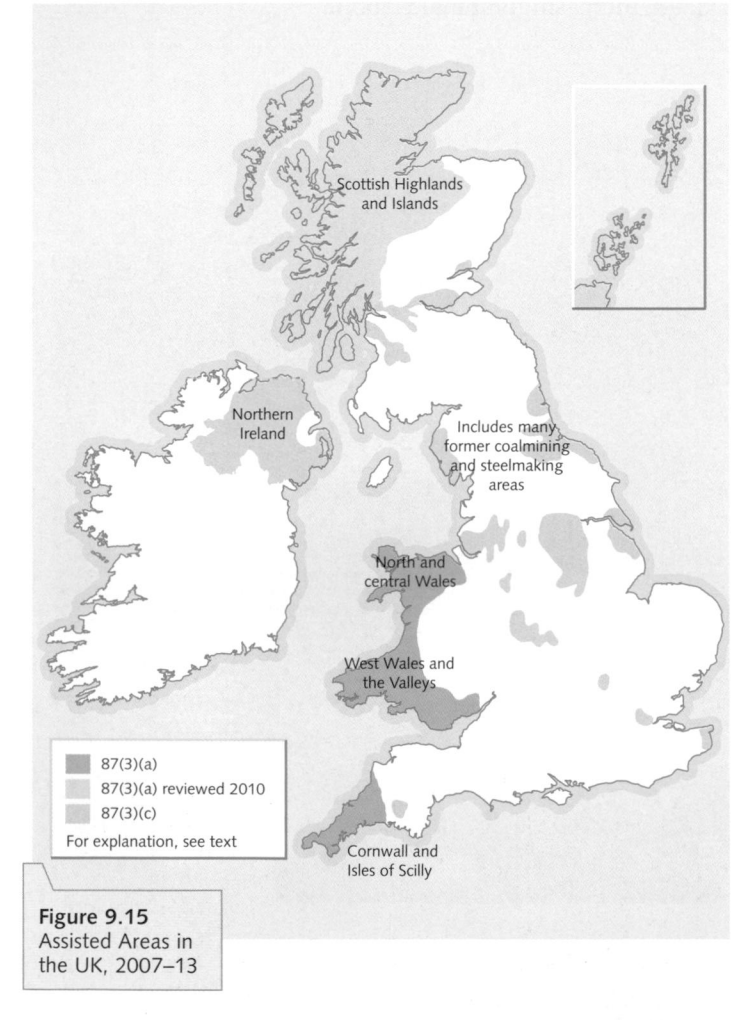

Scottish Highlands and Islands

Northern Ireland

Includes many former coalmining and steelmaking areas

North and central Wales

West Wales and the Valleys

Cornwall and Isles of Scilly

- 87(3)(a)
- 87(3)(a) reviewed 2010
- 87(3)(c)

For explanation, see text

Figure 9.15
Assisted Areas in the UK, 2007–13

Figure 9.16
Former Liverpool Docks, Merseyside Development Corporation

Trafford Park Development Corporation

During the early 1980s there was growing concern over the decline of Trafford Park, an area in Manchester which included Britain's first large industrial estate. Early attempts to improve the area included granting it Assisted Area status and, later, parts were designated an Enterprise Zone (page 142). Although these initiatives produced some results, they failed to tackle the underlying problems of an increasingly outdated infrastructure and the lack of attractive development opportunities to encourage investment. In 1987 Trafford Park, and the former steelworks site at Irlam 4 km to the south-west, became an Urban Development area. The Development Corporation identified four main development areas – Northbank Industrial Park, Village, Wharfside and Hadfield Street (Figure 9.17). Most recently opened (1998) is the Trafford Centre, a huge two-level 250-unit retail/recreation complex. The Corporation was wound up in March 1998 having exceeded in all eight of its targets for investment, jobs and new companies attracted.

Village
22 ha. Former housing area in centre of Trafford Park. Seen as the thriving focal point of the new development area.

Wharfside
39 ha. Alongside Manchester Ship Canal. Now a commercial development in an attractive waterside location.

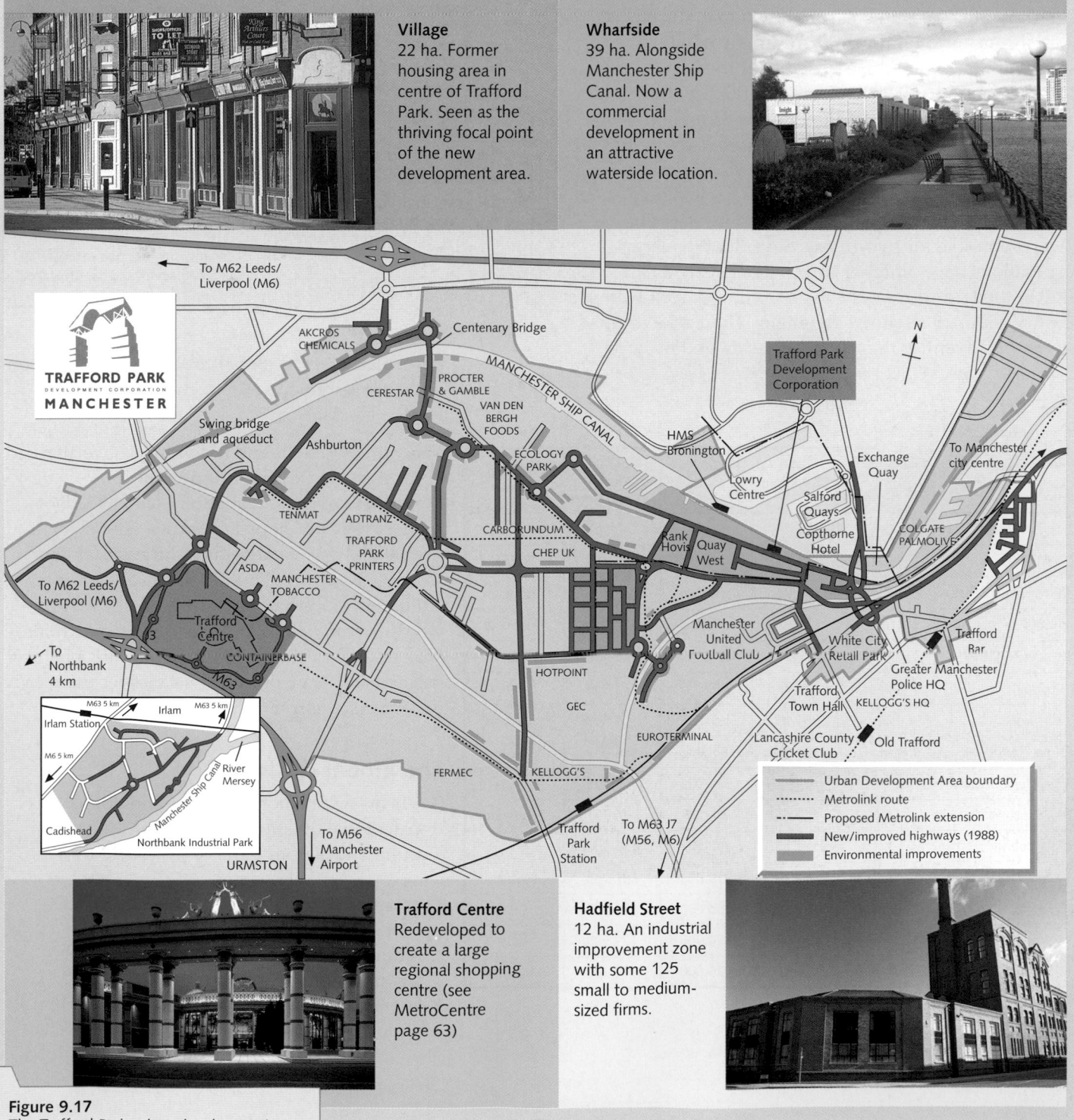

Trafford Centre
Redeveloped to create a large regional shopping centre (see MetroCentre page 63)

Hadfield Street
12 ha. An industrial improvement zone with some 125 small to medium-sized firms.

Figure 9.17
The Trafford Park urban development area

Transnational (or multinational) corporations

A transnational corporation (TNC), also referred to as a multinational company, is one that operates in many countries regardless of national boundaries, and sees the world – rather than the local area – as its supplier of labour, raw materials, component parts and its area of sales (globalisation, Chapter 20). Its headquarters and main factory are usually located in a more economically developed country and most of its manufacturing and assembling plants in those that are less economically developed. In 1970 there were 7000 TNCs, the largest being car manufacturers and oil corporations. This number had grown to 63 000 in 2000, mainly due to the emergence of electronics and high-tech firms, many of which were based in the newly industrialised countries of eastern Asia (page 146); and to 80 000 in 2009 as China continued to emerge. TNCs are now said to employ 82 million people worldwide, to indirectly influence an even larger number, and to control over 75 per cent of the world's trade. It has been estimated that whereas TNCs controlled only about one-fifth of the world's manufacturing in 1966, they are now responsible for two-thirds. Several of the largest corporations have a higher turnover than all of Africa's GDP in total – and ExxonMobil would rank 26th were it a country. Figure 9.18 shows the major TNCs in 2007 (just before the advent of the world recession and its impact

Advantages to the country	Disadvantages to the country
Brings work to the country and uses local labour	Numbers employed small in comparison with amount of investment
Local workforce receives a guaranteed income	Local labour force usually poorly paid
Improves the levels of education and technical skill of the people	Very few local skilled workers employed
Brings welcome investment and foreign currency to the country	Most of the profits go overseas (outflow of wealth)
Companies provide expensive machinery and modern technology	Mechanisation reduces the size of the labour force
Increased Gross Domestic Product/ personal income can lead to an increased demand for consumer goods and the growth of new industries	GDP grows less quickly than that of the parent company's headquarters, widening the gap between developed and developing countries
Leads to the development of mineral wealth and new energy resources	Minerals are usually exported rather than manufactured and energy costs may lead to a national debt
Improvements in roads, airports and services	Money possibly better spent on improving housing, diet and sanitation
Prestige value (e.g. Volta Project)	Big schemes can increase national debt (e.g. Brazil)
Widens economic base of country	Decisions are made outside the country, and the firm could pull out at any time
Some improvement in standards of production, health control, and recently in environmental control	Insufficient attention to safety and health factors and the protection of the environment

Figure 9.19
Advantages and disadvantages of transnational corporations

particularly on the global car industry), based on three measurements: sales, employees and foreign assets.

Many organisations and individuals have attacked TNCs as being exploiters of poor people, especially women and children, who live in less economically developed countries. Yet talking to several of these workers in countries as far apart as Brazil, Kenya, Sri Lanka and Malaysia, their attitude was: 'Perhaps, but it is the only way by which we can find full-time work.' Do you think, after studying Figure 9.19, that TNCs are, on balance, a blessing or a curse to a less economically developed country?

The global car industry

Car firms were amongst the first to opt for transnational operations. They found that by locating in different parts of the world they could:

- get around trade barriers which may have been erected to protect home markets
- reduce costs by gaining access to cheaper labour and/or raw materials
- be nearer to large markets (centres of population).

Figure 9.18
Dominance of transnational corporations, 2007

TNC	HQ country	Sales ($ million)	Employees	Foreign assets ($ million)	Industry
1 ExxonMobil	USA	365 467	82 000	245 467	Petroleum
2 Wal-Mart	USA	344 992	1 910 000	151 193	Retail
3 Royal Dutch/Shell	UK/Neth	318 845	108 000	235 276	Petroleum
4 British Petroleum	UK	270 600	97 000	217 600	Petroleum
5 General Motors	USA	207 349	280 000	186 192	Cars
6 Toyota	Japan	205 918	298 000	273 853	Cars
7 Chevron	USA	204 892	83 000	111 608	Cars
8 Total	France	192 950	95 000	138 579	Petroleum
9 Daimler-Chrysler	Germ/USA	190 196	360 000	250 259	Cars
10 Conoco-Phillips	USA	183 650	38 000	164 781	Petroleum
11 Mitsubishi	Japan	176 410	58 000	126 000	Cars
12 General Electric	USA	163 391	319 000	697 239	Electrical
13 Ford	USA	160 123	283 000	278 000	Cars
14 Volkswagen	Germany	131 571	324 000	179 900	Cars
15 Siemens	Germany	109 500	475 000	119 200	Electrical

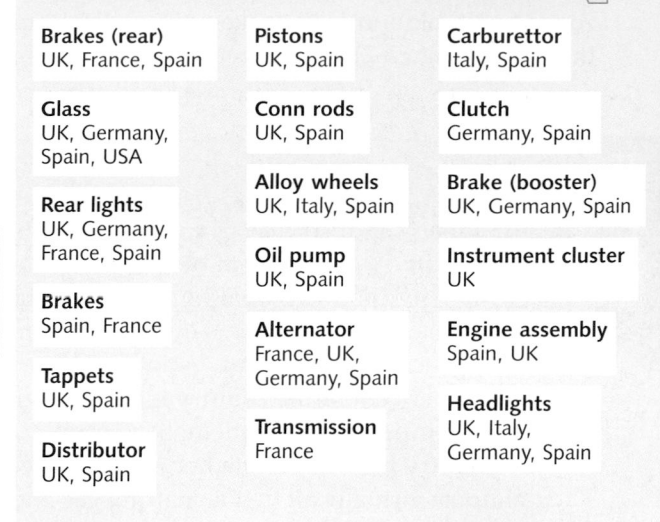

Figure 9.20
The worldwide production and assembly of Ford cars

Countries with Ford factories

Figure 9.21
Ford's 'European' car

Brakes (rear)
UK, France, Spain

Glass
UK, Germany, Spain, USA

Rear lights
UK, Germany, France, Spain

Brakes
Spain, France

Tappets
UK, Spain

Distributor
UK, Spain

Pistons
UK, Spain

Conn rods
UK, Spain

Alloy wheels
UK, Italy, Spain

Oil pump
UK, Spain

Alternator
France, UK, Germany, Spain

Transmission
France

Carburettor
Italy, Spain

Clutch
Germany, Spain

Brake (booster)
UK, Germany, Spain

Instrument cluster
UK

Engine assembly
Spain, UK

Headlights
UK, Italy, Germany, Spain

Ford – a global car corporation

The giant Ford Corporation originally located in Detroit (see Places below). By the late 1990s, a century later, it was:

- manufacturing and/or assembling its cars worldwide, although the bulk of the parts were still produced in the more industrialised parts of North America, Japan and the EU (Figure 9.20)
- increasingly locating its new factories, which either manufactured cars (e.g. São Paulo) or assembled parts made elsewhere (e.g. Malaysia, the Philippines), in less economically developed countries
- increasingly making parts in several countries (reducing the risk of strikes) so that each particular model is no longer made in one country – Figure 9.21 shows that a Ford car sold in the UK may have been assembled here or elsewhere in the EU, and that its parts may have come from several European countries

- facing increased competition, especially from Japanese manufacturers, at a time of economic recession – any resultant factory closures are more likely in the less economically developed countries where the local market is smaller and where redundancies are of less concern to the parent company
- working in Detroit with its previous rivals, Chrysler and General Motors, to produce a car that will use less fuel, cause less pollution and challenge the dominance of Japanese and, increasingly, Korean cars.

Detroit – motor city

Henry Ford, a local man, saw Detroit as an ideal location for what was to be the world's first mass-production line. He built his factory on flat land next to the Detroit River (Figures 9.22 and 9.23) at the heart of the Great Lakes waterway system. Steel was produced on an adjacent site using relatively local iron ore (brought by ship) and coal (brought by train). Ford developed a large local market by paying his workers $5 a day, when the national average was $9 a week, enabling them to buy their own cars.

The high wages attracted workers from all over the world, especially from the south-east of the USA. Later, America's two other car giants, Chrysler and General Motors, located their main factories at Detroit. After a serious depression in the 1970s (due to the world oil crisis and competition from Japan), Detroit's car industry thrived again until the disastrous global recession that began in 2008, leading to General Motors declaring itself bankrupt.

Figure 9.22
Industry along the banks of the Detroit River, USA

Figure 9.23
Ford's car plant, Detroit

The Pacific Rim

The Pacific Rim includes all those countries in Australasia, Asia, North America and Latin America that border the Pacific Ocean. The term was originally used to describe a zone of volcanic and earthquake activity which surrounds the Pacific (page 262). More recently it acknowledges the shift in the location of the world's manufacturing industry away from countries bordering the North Atlantic (i.e. in North America and Western Europe) and the emergence of **newly industrialised countries** (**NICs**), especially in eastern Asia. This industrialisation began in Japan in the 1950s.

Japan

In 1945 Japan's industry lay in ruins after the Second World War. By 1990 Japan had become, after the USA, the world's second most wealthy and industrialised country. This achievement was even more remarkable because:

- only 17 per cent of the country is flat enough for farming, industry and settlement
- the country has very limited energy resources and has to import virtually all the oil, natural gas and coal it needs
- the country lacks most of the basic raw materials needed by industry. It no longer has workable supplies of iron ore and coking coal (yet it is the world's second largest steel producer), nor has minerals of any significance.

How was Japan's economic miracle achieved?

- Post-war demilitarisation meant money was invested into the economy rather than spent on armaments.
- It had political stability and a government committed to industrialisation.
- Modern machinery and technology was introduced and the profits re-invested into research.
- The workforce were prepared to work long hours, to become better educated and trained, to work as a team, and to give total loyalty to their company.
- The country had, especially along its Pacific coastline, many deep and sheltered harbours which facilitated the import of energy and raw materials and the export of manufactured goods (Figure 9.24).
- The development of transport is based on a first-class rail network, motorway expressways and the use of coastal shipping.
- The domestic demand (market) for high-quality goods increased rapidly as Japan's population and standard of living increased.
- Japan had the wealth and the technology to reclaim land from the sea (Figure 9.25).

These factors have led to the concentration of industry around five main coastal areas (Figure 9.24). The Japanese themselves consider the most important industrial location factors to be the distribution of lowland, the distribution of people, and the availability of sheltered deep-water ports.

Japan began to improve its economy by using its limited supplies of iron ore and coal to produce steel. The steel was used to make ships specifically designed to carry the necessary raw materials (oil tankers and ore carriers). Attention was then turned to making cars and, later, developing electronics and high-technology industries. Although most of the world's largest car and electronic transnationals are Japanese, the vast majority of local firms are still small family units, many of which make component parts for their transnational neighbours.

Key

	5 main industrial areas
	Other industrial areas
	Mountains
	Exports
	Imports

0   200km

Sea of Japan

Pacific Ocean

N

HOKKAIDO

Sapporo

Sendai

•Niigata

Hanshin (Osaka–Kobe)

Setouchi (Hiroshima)   Mizushima

HONSHU

Kyoto

Kitakyushu (Kitakyushu)

Keihin (Tokyo–Kawasaki–Yokohama)

Chukyo (Nagoya)

Nagasaki

SHIKOKU

KYUSHU

Imports of energy Coal, oil, natural gas

Imports for industry Iron ore, coking coal, bauxite and minerals

Exports Cars, steel, videos, electronics, computers

Figure 9.24 Industry in Japan

Figure 9.25 The Mizushima works of the Kawasaki Steel Corporation, built on land reclaimed from the sea

146

The NICs in East Asia

Encouraged by Japan's success, other governments in eastern Asia set out to improve their standards of living, initially by investing in manufacturing industry before, later, concentrating on high-tech industries (page 140). Manufacturing output rose most rapidly in South Korea, Taiwan, Hong Kong and Singapore – four countries that collectively became known as the 'four tigers'. Like Japan, these countries lacked raw materials, had governments committed to long-term planning and had a dedicated workforce that was reliable and, at first, was prepared to work long hours for relatively little pay. Economic growth in these NICs continued during the 1980s and 1990s when it was slowing down in the more developed economies and world manufacturing was declining.

Since the 1980s, Malaysia (the most successful), Thailand, Indonesia and, to a lesser extent, the Philippines have formed a 'second-wave' of 'Asian tigers'. The latest to emerge, and likely to be the largest, is China (Figure 9.26).

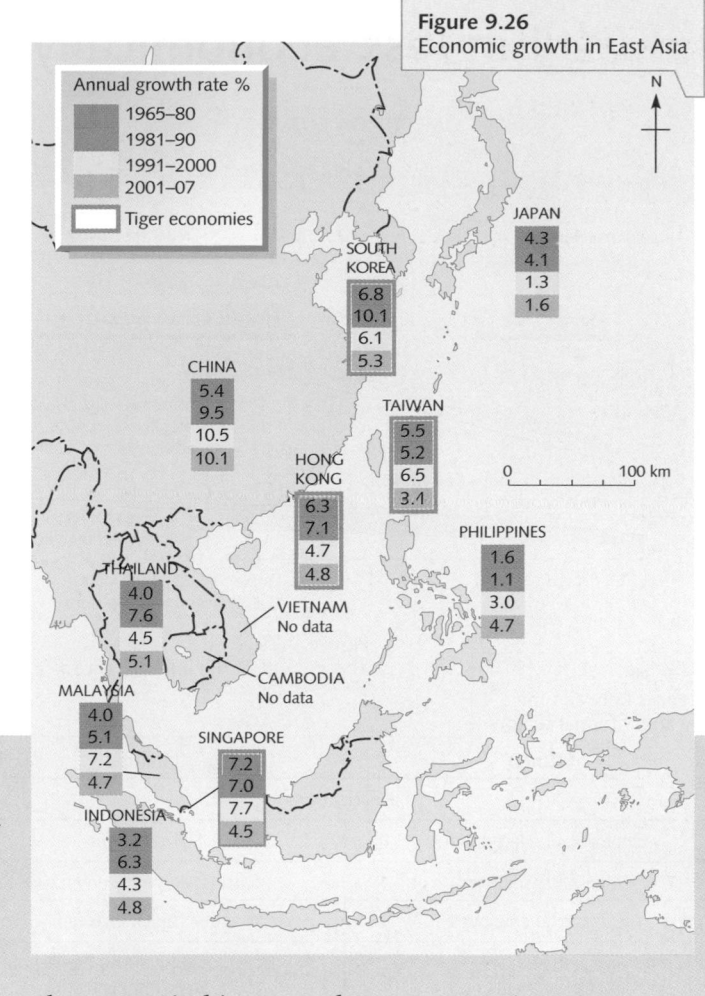

Figure 9.26
Economic growth in East Asia

Malaysia

Until the 1980s, Malaysia's economy was based on primary products such as rubber and palm oil (page 103), tin, timber (Case Study 14) and oil. The government then proclaimed its vision of Malaysia becoming fully developed and industrialised by 2020 and the country emerged as the leader of the second wave of Asian 'tiger economies'. Its success was based on its hub position as a gateway to ASEAN (Figure 11.14) and the encouragement, by tax incentives, for TNCs to locate there. The country's industrial strategy emphasised the development of high-value goods for both the domestic market and export (e.g. cars) and the encouragement of high-tech industries (e.g. electronics, Figure 9.27).

The Second Industrial Plan (1996 to 2005) focused on the manufacturing sector, research and development, and the integration of support industries. It concentrated on the production of electrical and electronic goods (including ICT and multimedia), oleochemicals (from palm oil, rubber and timber), chemicals (petrochemicals and pharmaceuticals), cars and transport equipment, machinery and high-value textiles and clothing.

In its latest Master Plan (2006 to 2020) the government has targeted the development of:

- six non-resource-based manufacturing industries (electrical and electronics, medical devices, textiles, machinery and equipment, metals and transport equipment)
- six resource-based manufacturing industries (wood-based, rubber-based, oil palm-based, food processing, petrochemicals and pharmaceuticals)
- eight service-sector-based activities (logistics, business and professional, ICT, distributive trade, construction, education and training, health care and tourism).

Figure 9.27
High-tech industry in Penang ('Silicon Island')

Figure 9.28
Petronas Towers, Kuala Lumpur

Industry in less economically developed countries

COMPETITION FOR JOBS

Inhabitants of the city	Migrants into the city
Perhaps some education and skills	Usually no education and very few skills
May get one of the few relatively regular paid jobs, possibly with a transnational company	Unlikely to get a regular or a paid job – they need to create their own jobs in order to survive
Formal sector	**Increase in the informal sector**

Increase in population of a city →

Rapid growth of population in working age group

Proportion of population working in the informal sector   (Recession)

Manufacturing employment

Time →

Informal sector usually employs over 60% of working population

Formal sector usually employs less than 40%

Employment in army, police, civil services and professions

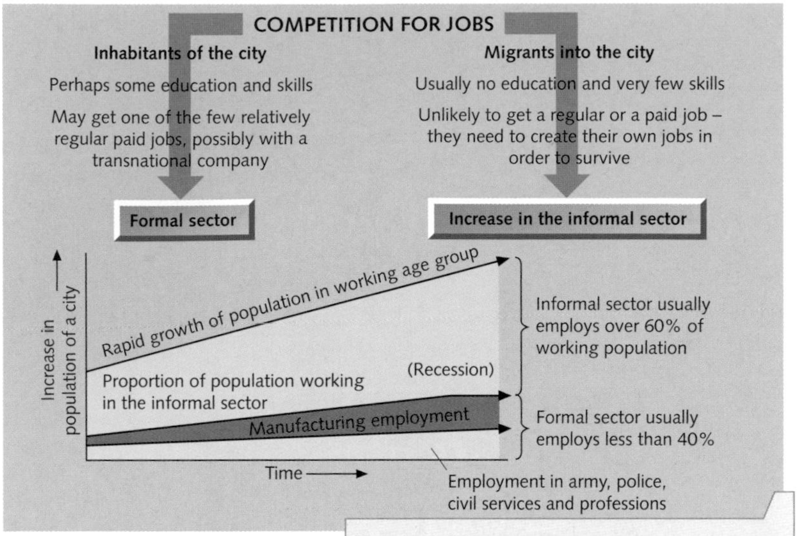

Figure 9.29
The growth of the informal sector of industry

Figure 9.30
Differences between the formal and informal sectors

Formal	Informal
Description	
Employee of a large firm	Self-employed
Often a transnational	Small-scale/family enterprise
Much capital employed	Little capital involved
Capital-intensive with relatively few workers. Mechanised	Labour-intensive with the use of very few tools
Expensive raw materials	Using cheap or recycled waste materials
A guaranteed standard in the final product	Often a low standard in quality of goods
Regular hours (often long) and wages (often low)	Irregular hours and uncertain wages
Fixed prices	Prices rarely fixed and so negotiable (bartering)
Jobs done in factories	Jobs often done in the home (cottage industry) or on the streets
Government and multinational help	No government assistance
Legal	Often outside the law (illegal)
Usually males	Often children and females
Type of job	
Manufacturing – both local and transnational industries	Distributive, e.g. street pedlars and small stalls
Government-created jobs such as the police, army and civil service	Services, e.g. shoecleaners, selling clothes and fruit
	Small-scale industry, e.g. food processing, dress repairs, furniture repairs
Advantages	
Uses some skilled and many unskilled workers	Employs many thousands of unskilled workers
Provides permanent jobs and regular wages	Jobs may provide some training and skills which might lead to better jobs in the future
Produces goods for the more wealthy (cars, food) within their *own* country so that profits may remain within the country	Any profit will be used within the city. The products will be for local use by the lower-paid people
Waste materials provide raw materials for the informal sector	Uses local and waste materials

Formal and informal sectors

In cities in less economically developed countries, the number of inhabitants greatly outweighs the number of jobs available. With the rapid growth of these cities the job situation is continually worsening. An increasing number of people have to find work for themselves and thus enter the informal sector of employment (Figure 9.29), as opposed to the formal sector (i.e. the professions, offices, shops and organised modern industry). The differences between the formal and informal sectors are given in Figure 9.30.

In many cities there are now publicly and privately promoted schemes to support these self-help efforts. In Nairobi, for example, there are several **jua kali** (meaning 'under the hot sun') with metal workshops. In one area little more than the size of three football pitches, 1000 workers hammer scrap metal into an assortment of products (Figure 9.31).

Figure 9.31
Workers in the informal sector

Role of children

Children, many of whom are under 10 years old, make up a large proportion of the informal sector workers. Very few of them have schools to go to, and from an early age they go out onto the streets to try to supplement the family income (Figure 9.32). One such 'worker', who has become well known in British schools, was a shoe-shine boy called Mauru who lived in São Paulo, and was seen in a TV programme *Skyscrapers and Slums*. He would try to earn money during the day, and study to become an airline pilot, at night.

'Our small group of 11 tourists from Britain was staying at the Rio Palace, a five-star hotel overlooking the famous Copacabana beach in Rio de Janeiro. Walking along the beach one afternoon, I noticed numerous beach vendors. These form part of Rio's large informal sector. Several vendors carried large umbrellas from which dangled an assortment of sun hats, suntan lotion or *tangas* (bikinis). One vendor carried pineapples in a basket perched on his head and, in his hand, a large knife with which to cut the fruit. Some vendors carried cool-boxes in which were ice cream, Coca-Cola, coconut water and other drinks, while others carried large metal drums which contained *maté* (a local drink). These people drew attention to themselves by shouting, blowing whistles, beating metal drums or whirling a metal ratchet that clattered loudly. On the pavement next to the beach were small children trying to sell sweets and chocolate, and numerous kiosks with fruit and drinks available.

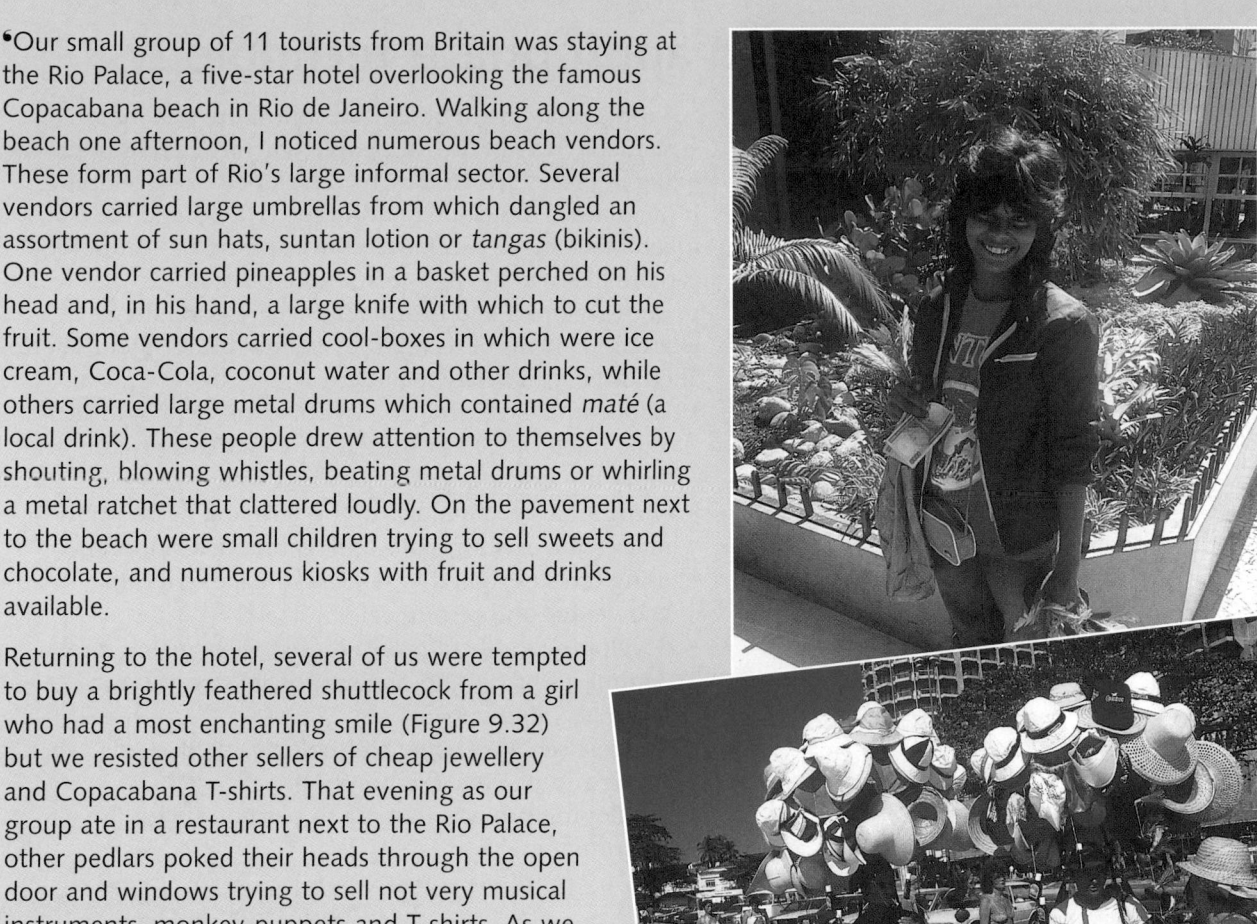

Returning to the hotel, several of us were tempted to buy a brightly feathered shuttlecock from a girl who had a most enchanting smile (Figure 9.32) but we resisted other sellers of cheap jewellery and Copacabana T-shirts. That evening as our group ate in a restaurant next to the Rio Palace, other pedlars poked their heads through the open door and windows trying to sell not very musical instruments, monkey-puppets and T-shirts. As we left the restaurant at 9.30 pm to visit one of Rio's famous Samba shows, we saw a small boy of four or five years with sad, appealing eyes trying to sell roses individually wrapped in cellophane. How could one fail to buy a rose which, he indicated by his fingers, were only 15 crusadas (about 50p)? However, by the time the money was produced, he had raised his price to 20 crusadas – and shown us that his eyes had been trained and his brain sharpened for business. When we returned at 1.00 am, he was still outside the hotel with the remainder of his roses.'

David Waugh

Figure 9.32
The informal sector
Main photo: Copacabana beach and Sugarloaf Mountain
Insets: Girl selling shuttlecocks in Rio de Janeiro; beach vendors on Copacabana beach

Sustainable development and appropriate technology

In most less economically developed countries, not only are high-tech industries too expensive to develop, they are usually inappropriate to the needs of the local people and to the environment in which they live. An **appropriate technology** is exactly what it says – a technology appropriate or suitable to the place in which it is used.

An appropriate technology can contribute to a more sustainable way of life for people who are rich or poor, living in places which may be developed or developing. If the place is developed and industrialised and its inhabitants are well-off, then the appropriate technology is more likely to be high-tech. If the place is underdeveloped and its inhabitants are poor, then alternative forms of technology should be adopted. These alternative forms may include:

- labour-intensive projects – with so many people likely to be unemployed or underemployed, it is of little value replacing existing workers by machines
- encouraging technology that is sustainable and fully utilises the existing skills and techniques of local people
- using tools designed to take advantage of local knowledge and resources
- developing local crafts and industries by using local natural resources and, where possible, recycling materials (Figure 9.33)
- low-cost schemes using technologies that people can afford and manage (Figure 9.34)
- developing projects that are in harmony with the environment.

Figure 9.33
Recycling materials in Nairobi

Figure 9.34
Appropriate technology in Nepal, one of the poorest countries in the world

BEFORE

For women, most tasks are labour-intensive, time-consuming and have to be done by hand

AFTER

Villagers can now hull their rice mechanically with this 3 kW mill driven by a micro-hydro turbine – time is saved and quality and productivity increased

BEFORE

Grinding enough corn to feed a family for just 3 days takes 15 hours when it is done by hand

AFTER

By taking corn to the grinder in the mill-house – usually a popular meeting-place for villagers – 3 days' worth of corn can be ground in just 15 minutes

Practical Action

Practical Action is a British charitable organisation which works with people in developing countries, especially in rural areas (Figure 9.35). It helps them to acquire the tools and techniques needed if they are to work themselves out of poverty. Practical Action helps people to meet their needs of food, clothing, housing, farm and industrial equipment, energy and employment. Practical Action uses, and adds to, local knowledge by providing technical advice, training, basic equipment and financial support so that people can become more self-sufficient and independent. Ideally the aim is for people to create a small surplus which can then be invested into their small businesses and communities.

KENYA
Stoves
Wind pumps
Rainwater harvesting
Sugar processing
Fibre-concrete roof tiles
Fish smoking
Animal husbandry

SUDAN
Stoves for refugee camps
Agricultural equipment

NEPAL
Stoves
Micro-hydro

BANGLADESH
Sugar processing
Textiles
Food processing

SRI LANKA
Stoves
Food processing
Building materials
Solar timber kiln
Micro-hydro

PERU
Micro-hydro
Food processing
Textiles
Agriculture and water
Building materials

PRACTICAL ACTION
Technology challenging poverty

ZIMBABWE
Small-scale mining
Fibre-concrete roof tiles
Carpenters' hand tools

Figure 9.35
Practical Action: project activities

Practical Action projects in Kenya

Most Practical Action projects in Kenya, as elsewhere in the world, are in response to requests from local groups, often women, who need assistance in one or more stages of their planned work. For example:

1 **The Maasai Housing Project – towards sustainable rural housing improvement** (Figure 9.36) 'The traditional Maasai house, the *enkang*, was made from posts, a soil/cowdung mixture, and grass (Figure 10.33). A narrow opening served as an entrance and source of ventilation. Occupants slept and cooked in the dwelling, usually along with their sheep and goats. *Enkangs* were often characterised by low, leaking roofs; smoky, dark rooms; cramped space; a stench of animal odour; termite-infested posts; no water; and a lack of permanence. They posed health risks, such as eye and respiratory ailments; they were uncomfortable; and they needed constant repair. The community-based Maasai Housing Project is being helped by Practical Action to find sustainable and affordable ways to upgrade and make *enkangs* more permanent. Fibre-cement skin roofs consist of a polythene sheet laid between a twigs/grass base and a thin covering of cement. Water is collected from the now-impermeable roof, saving the women the daily chore of walking several kilometres to the nearest river. Ferro-cement panel walls are made by nailing wire mesh to termite-proof posts, while larger openings and a chimney cowl make the inside of the dwelling lighter, less smoky and more healthy.'

2 **Roofing materials** (Figure 9.37) Improved urban building materials include roofing tiles which are made by adding natural fibres (such as sisal) and lime (or cement) to soil. The resultant blocks are dried, and prove to be far less expensive than commercially produced tiles.

3 **Energy** (Figures 8.27 and 9.38) Potters are trained in the production of two types of fuel-efficient cooking stoves, both of which drastically reduce the amounts of fuel needed – wood for the *mandaleo* in rural areas and charcoal for the *jiko* in urban areas. The stoves are made from recycled scrap (Figure 9.31) and have a ceramic

Figure 9.36
The Maasai Housing Project

Figure 9.37
Making low-cost tiles

Figure 9.38
A fuel-efficient cooking stove (*jiko*)

lining. They save people time in collecting wood (rural) and money buying charcoal (urban) as well improving women's health by reducing the amount of smoke produced.

Industry in a city in a developed country – Osaka–Kobe

The Osaka–Kobe conurbation is one of Japan's major industrial areas. The many reasons for this industrial growth include the following:

- It is a natural harbour situated within Osaka Bay, an inlet of the Inland Sea (Figure 4.28). The Inland Sea itself is protected from the worst of the typhoon winds by the island of Shikoku (page 73).
- Osaka imports many of the raw materials needed by Japan. Many of these raw materials are processed within the port area, e.g. oil, coal and iron ore. The port is also a major outlet for Japanese exports (Figure 4.30).
- The land around Osaka Bay is one of the relatively few areas of flat land in a country that is 83 per cent too mountainous for development. Even so, Osaka–Kobe has spread outwards as far as the highland allows (Figure 9.39).
- Land in Osaka Bay has been reclaimed for port development, new industries and an international airport (Figures 4.32 and 4.34).
- It has a population of over 9 million which provides a highly skilled, dedicated workforce and a large, wealthy domestic market.
- High-tech industries have grown rapidly as a consequence of the inventiveness of the Japanese.
- It is the centre of many banks and large Japanese companies (Figure 4.29).

Japanese industry is highly efficient. It is based on hard work, loyalty and trust. The Japanese compare their industrial organisation to a pyramid (Figure 9.40). At the base are thousands of small, often family-run firms. Small firms, usually employing fewer than six people, account for 90 per cent of Japanese companies.

Figure 9.39
Osaka

Above them are medium-sized companies which, together with the small firms, produce 60 per cent of Japan's manufactured goods. At the top of the pyramid are the large corporations, many of which, like Toyota, Nissan, Sharp, Panasonic and Mitsubishi, have become major transnationals.

From oil to camcorders

The port petrochemical industry processes oil into many by-products, one of which is plastic. Some of this plastic is taken to the workshop district of Osaka which, for convenience, is located near to the motorway which skirts the east of the city (it can be identified on Figure 4.31). Within this district are numerous small work units, including the one run in a converted garage by Mr and Mrs Yamasaki. Mr and Mrs Yamasaki produce plastic parts which they send daily to a nearby medium-sized company run by the Kawasaki brothers. The Kawasakis, who employ 18 people, produce a range of products which include, for example, the tops of car batteries and safety goggles. Recently they have won orders to

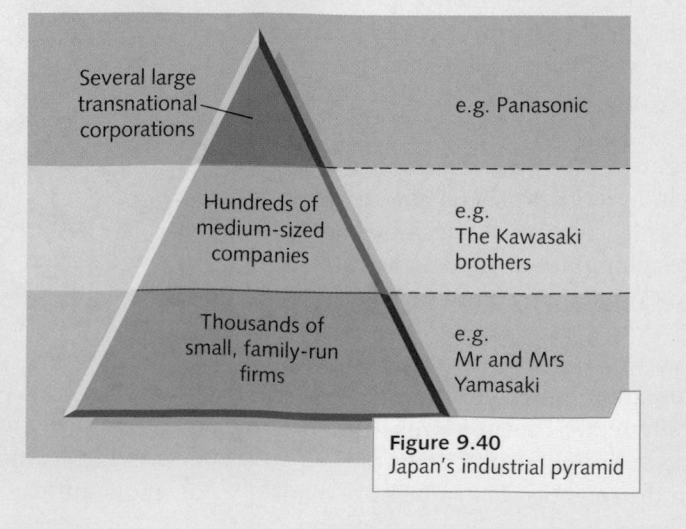

Several large transnational corporations — e.g. Panasonic

Hundreds of medium-sized companies — e.g. The Kawasaki brothers

Thousands of small, family-run firms — e.g. Mr and Mrs Yamasaki

Figure 9.40
Japan's industrial pyramid

produce sunglasses for Italy, and 100 000 lenses for Panasonic to use in its camcorders. The lenses, when finished, are sent 20 km across the city to the huge Panasonic plant. Panasonic is a transnational corporation.

Panasonic assembles the component parts made by many small and medium-sized firms. Although the assembly is highly automated with robots doing many very technical jobs, Panasonic still employs several hundred workers. Mrs Yamagami, who supervises eight robots, points out that she feels very much part of the company, despite its large size. Along with Japanese companies worldwide, Panasonic encourages close ties between management and staff. Mrs Yamagami, like other employees, is encouraged to suggest ways of improving both production and working conditions. The completed camcorders, along with other Panasonic products which include TVs and videos, are exported through the port at Osaka to 160 countries worldwide (Figure 9.42).

The major problem facing Panasonic at present is competition from the NICs, especially those in the Pacific Rim of South-east Asia, whose labour and other production costs are cheaper (page 147). Panasonic is trying to meet this challenge by: investing in and setting up assembly in other countries; research and development so as to keep ahead in technology; and seeking to develop new products.

Japanese industry does not stand still. Two recent developments include:
• demolishing the 100-year-old Nippon steelworks, located alongside Osaka Bay, and using former employees to create and develop new forms of technology and automation (Japanese industrialists have little sentiment for anything that has outlived its usefulness)
• building up to 11 new science parks, some the size of a small city, in pleasant environments beyond the present urban limits (page 141).

Figure 9.41
The working district of Osaka

The science parks, with their research centres paid for by Japan's leading corporations, are being developed to 'make sure that Japan's high-tech stays ahead of the rest of the world'.

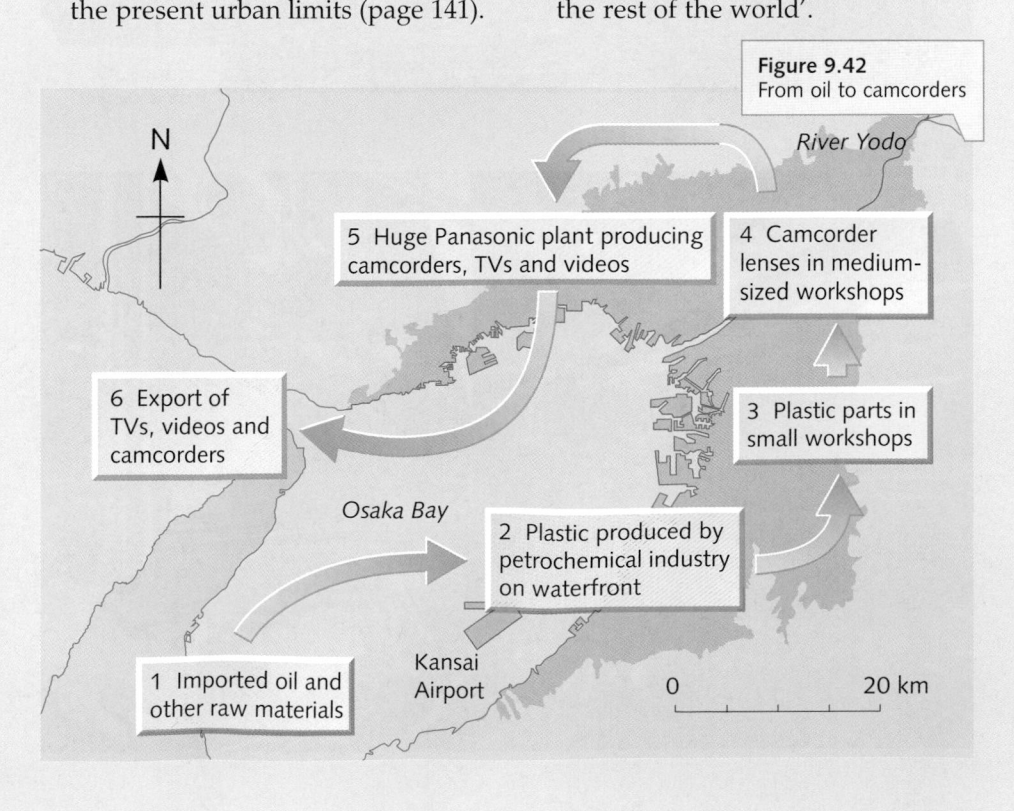

Figure 9.42
From oil to camcorders

River Yodo

5 Huge Panasonic plant producing camcorders, TVs and videos

4 Camcorder lenses in medium-sized workshops

6 Export of TVs, videos and camcorders

3 Plastic parts in small workshops

Osaka Bay

2 Plastic produced by petrochemical industry on waterfront

Kansai Airport

1 Imported oil and other raw materials

0   20 km

Industry in a city in a developing country – São Paulo

Brazil is the most industrialised of the world's economically developing countries. Industrialisation is not, however, evenly spread across the country but is mainly concentrated in and around São Paulo (Figure 1.24). During the nineteenth century São Paulo was only a small town situated in the centre of a major coffee-growing area. Its rapid urbanisation and industrialisation took place during the so-called 'economic miracle' of the 1960s and 1970s. At this time up to half a million people a year (as many as live in present-day Leeds or Sheffield) migrated to the city looking for work (rural–urban migration). Today the city has an estimated population of 18 million living in an urban area that extends across 100 km.

The region around São Paulo was well endowed with minerals, including iron ore, and had access to energy resources. This led to the development of the iron and steel and engineering industries (Figure 9.43) and the manufacture of machinery, aircraft and cars (Figure 9.44). Brazil is the world's ninth largest producer of cars. Four large car **transnational** corporations – Ford, Volkswagen, General Motors and Mercedes – all have assembly plants in São Paulo (Figure 9.44). They produce cars in sufficiently high numbers to satisfy the domestic market and for export to the rest of South America and even to the USA and the Middle East. Brazilian car workers may earn more than twice the average Brazilian wage, but to international manufacturers they are cheaper to employ than car workers in developed countries.

Figure 9.43
Iron and steel works, São Paulo

Figure 9.44
A car plant in São Paulo

Figure 9.45
Formal jobs in the CBD

Industrialisation brought employment and created problems, especially in and around downtown São Paulo (Figure 9.45). Heavy industry and traffic have caused air pollution; the great number of cars has led to gridlock; and the location of commercial buildings (e.g. banks) and offices (headquarters of large companies) has created a 'sky-scraper jungle', high land prices and a lack of open space. One result has been the movement of industry to new towns. These have been created, mainly to attract industry, by building major roads and locating in a cleaner, less congested environment. One such town, Jundaia, is 100 km from São Paulo's city centre.

Figure 9.46
Pepsi plant at Jundaia

Figure 9.47
The informal sector

Jundaia provides employment in the **formal sector** (page 148). Several transnational companies have already located there, including Pepsi (Figure 9.46). Although the plant is heavily automated, it still employs 350 people.

Pepsi's workers not only have a regular job with regular pay, they also get free lunches and free medical care. Many people living and working in Jundaia are enjoying a rising standard of living – although others still have to seek low-paid jobs as cleaners or shop assistants.

Nearer the CBD, some of São Paulo's increasing wealth is being used to turn run-down areas into modern business, retail and leisure centres. Unfortunately, the development of these new centres with their ultra-modern buildings can only take place by clearing existing *favelas* such as the one at Edith Gardens (Figure 5.12).

People who live in Edith Gardens favela usually only find work in the **informal sector**, as indeed do over one-third (about 4 million) of São Paulo's working population. The jobs in the informal sector include recycling materials, repairing goods and processing and/or selling food (Figure 9.47). Most of the

residents living in Edith Gardens *favela* are not enjoying a rising standard of living.

Alcione Florencia is one such person. Alcione, who had no schooling, moved to São Paulo with her two small children. For the first three years she was unemployed. She then turned the front of her house into a bar, selling soft drinks to builders working on local construction sites, and took in washing. Now the council wish to 'improve' Edith

Gardens (that is, to demolish and redevelop it) and its residents are being offered £1100 to move out. Although Alcione wants to return to her home area, she feels the money is insufficient as it would take more than that to restart a business elsewhere. To those in Edith Gardens who are unemployed, £1100 (if it is ever paid) is an inducement to dismantle their often flimsy homes and re-erect them in other *favelas*.

Industry

Key Words and Terms

a You should know the meaning of the following terms:
  - industrial system • raw materials • site • labour • capital • markets
  - economies of scale • government policies • footloose • break of bulk
  - integrated works • high-tech industries • science/business parks
  - Enterprise Zones (EZs) • Urban Development Corporations (UDCs)
  - transnationals • globalisation • Pacific Rim
  - newly industrialised countries (NICs).

b You should know the difference between:
  - the formal and the informal sectors of industry
  - sustainable development and appropriate technology.

Key Ideas

You should know and understand the following:
- Industry is a system, with inputs, processes and outputs.
- The industrial company or entrepreneur is a decision maker.
- The location of industry is affected by physical, human and economic factors.
- These location factors can change over a period of time.
- Physical factors were usually the more important for the location of nineteenth-century industries, and human and economic factors for the location of present-day industries.
- Over a period of time, some areas and/or industries experience a decline, while others experience growth.
- Government policies can affect industrial location, growth and decline.
- Transnational corporations have both advantages and disadvantages and have become key players in the global economy.
- Some countries in eastern Asia and Latin America have developed to become newly industrialised countries (NICs).
- Rapid urbanisation in LEDCs leads to a growth in the informal sector of the economy.
- Contemporary solutions to industrial problems in many LEDCs depend on sustainable development and the use of appropriate technology.
- There is a considerable difference in the types of jobs available in cities in the MEDCs and the LEDCs.

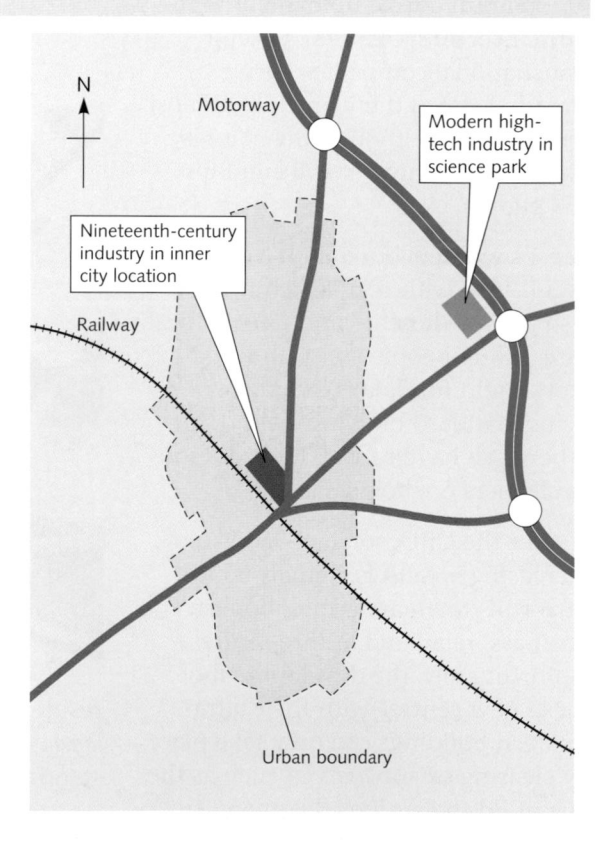

Skills, Theories and Models

- Interpret and use diagrams and maps
- Identify factors affecting the location of industry
- Understand the causes and effects of change
- Understand the idea of sustainable development
- Recognise the distinctive character of places
- The industrial system

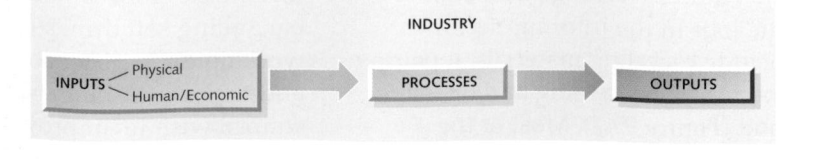

1 *(Page 136)*

a Choose and name a manufacturing industry you have studied. Copy and complete the top diagram (right) to show your chosen industry as a system. *(6)*

b Complete the star diagram to show some of the factors affecting the location of industry. The following list should help you:
- capital
- energy supplies
- labour
- natural routes
- raw materials
- relief
- transport
- markets. *(8)*

c Choose **two** physical and **two** human and economic factors, and explain how they can influence the location of a factory. *(2 × 4)*

d i) Give three reasons why physical factors were more important than human and economic factors in the location of nineteenth-century industry. *(3)*

ii) Give three reasons why human and economic factors are usually more important than physical factors in the location of a modern industry. *(3)*

e Describe how political factors can affect the location of a factory. *(3)*

INPUTS   PROCESSES   OUTPUTS

Physical factors | Human and economic factors

1 2 3 4   Factory   1 2 3 4

2 *(Pages 138 and 139)*

a i) Give four reasons why town X became important for the manufacture of iron in the early nineteenth century. *(4)*

ii) Why did town Y become an important exporting port? *(2)*

b i) Give four reasons why a large modern steelworks has been built at town A. *(4)*

ii) Why was the modern steelworks not built at town B? *(2)*

Fast-flowing rivers in narrow valleys
Attractive coastline
Y
B
Deep, sheltered bay
Good farmland
A
Coalfield
Motorway
X
Coalfield with iron ore deposits
Limestone hills

3 *(Pages 62, 137 and 140)*

Many of Britain's newer industries are said to be 'footloose'.

a i) Give three examples of footloose industries. *(3)*

ii) Give three reasons why footloose industries locate on greenfield sites on the edges of cities. *(3)*

b The sketch (right) has been drawn from the photo above it which shows new industries located on the edge of a city. Copy and complete the sketch, choosing the correct six labels from the following list:
- nearby motorway
- near to city centre
- cheap land on edge of city
- nearby housing estate for workforce
- attractive greenfield site
- nearby city provides a large market
- space for car parking and future expansion. *(6)*

QUESTIONS

4 (Pages 136 to 141)

For any industry that you have studied:

a Name an area where it is located. (1)

b Draw a labelled sketch map to show its location. (4)

c Give three reasons why that industry has either grown or declined in the area named. (3)

d What effects has this growth or decline had on the local economy? (3)

5 (Page 140)

If a town wants to attract new industry it has to advertise its advantages.

a i) Give four advantages listed in the advertisement for locating a new factory in Swindon. (4)

ii) Explain why each would be an advantage for a firm seeking a new industrial site. (4)

b Give three other advantages of locating in Swindon that are not listed in the advertisement. (3)

SWINDON ENTERPRISE

THE PROFIT BASE.

££'s LOWER OVERHEADS – AROUND ONE-FIFTH CENTRAL LONDON'S ✈ HEATHROW 60 MINS – GATWICK 1 HR 30 MINS ✈ LONDON 50 MINS BY HIGH-SPEED TRAIN 🚗 PRIME M4 CORRIDOR LOCATION – LONDON 90 MINS 🚇 EXCELLENT ADVANCED COMMUNICATIONS 🌳 OUTSTANDING QUALITY OF LIFE 👥 YOUNG, DYNAMIC AND VERSATILE WORKFORCE

THAMESDOWN BOROUGH COUNCIL HAS A RANGE OF SITES AVAILABLE

7 (Pages 142 and 143)

a i) Describe the location of the Assisted Areas in the UK. (3)

ii) Give three reasons why these areas need government and EU help. (3)

b i) Where is Trafford Park? (1)

ii) What were the main problems in Trafford Park in the 1980s? (3)

c Describe the improvements made by the Trafford Park Development Corporation, under the headings:

• transport

• employment

• the environment. (3 × 3)

6 (Pages 140 and 141)

In recent years there has been a rapid growth of high-tech industries in the area between London and Bristol.

a Give two examples of high-tech industries. (2)

b How have the following encouraged the location of high-tech industries between London and Bristol?

• Accessibility • Labour supply
• Universities • Attractive countryside
• Cultural and social attractions. (5)

c What advantages does the photo of Tsukuba Science City (Figure 9.13) show that is likely to attract high-tech industry? (3)

d Study the information for the Cambridge Science Park below.

i) How many companies employ fewer than 10 people? (1)

ii) Why do most companies only employ a few people? (1)

iii) What type of people are likely to be employed in a science park? (2)

e From the map, give two pieces of evidence which suggest that most employees travel to the science park by car. (2)

f Describe the layout of the park under the headings:
i) buildings ii) road pattern iii) landscaped areas. (3 × 2)

CAMBRIDGE SCIENCE PARK

To M11 London
A14 northern by-pass
A10 Milton Road
0 100 metres
N
Railway

Buildings
Landscaped area
Trees
Car park
Lake

8 *(Page 146)*

a Japan's industry developed despite three major physical disadvantages. What were these three disadvantages? (3)

b Give five reasons why Japan has become a rich industrial country. (5)

c The photo shows the Mazda car plant in Hiroshima, Japan.
  i) Why has Mazda built its factory on land reclaimed from the sea? (3)
  ii) Give five other reasons why Hiroshima was a good site for a car plant. (5)

A	Flat land on a river delta	**E**	Port for imports and exports
B & C	Reclaimed land	**F**	Population of 1.2 million
D	Deep, sheltered inlet of inland sea	**G**	Bullet train route
		H	Attractive environment

9 *(Pages 144, 145 and 147)*

a i) What is a transnational corporation (TNC)? (1)
  ii) Name two TNCs connected with each of:
    • petroleum • cars • electronics. (3)
b i) Give three possible advantages that a TNC can bring to a less economically developed country. (3)
  ii) Give three possible disadvantages that a TNC can bring to a less economically developed country. (3)

c i) Name the city and country where Ford has its headquarters. (2)
  ii) Why are Ford cars partly made and assembled in several other countries? (2)
d i) Apart from Malaysia, name four other NICs. (4)
  ii) Give three reasons why Malaysia has become a successful NIC. (3)
  iii) How is Malaysia hoping to continue being successful in the future? (3)

10 *(Pages 148 and 149)*

a Rearrange the jobs listed below, which were noted during a visit to a country in the less economically developed world, into formal and informal jobs.
  • shoe-shiner
  • maker of sandals from old wood and tyres
  • policeman • bus driver
  • tin smith using old beer cans • soldier
  • self-employed fisherman with fish stall
  • car factory worker • snake charmer
  • bottler at a Coca-Cola works
  • hotel waitress • tour guide
  • smallholder with a few goats
  • cool-drinks vendor • gardener in a city park. (8)

Formal	Informal

b Give two reasons why the number of informal jobs is growing so rapidly in many less economically developed countries. (2)

11 *(Pages 150 and 151)*

a Give five ways by which appropriate technology can provide a sustainable way of living in one of the poorer less economically developed countries. (5)
b i) What are the main aims of Practical Action? (3)
  ii) Describe three of Practical Action's projects in Kenya. How do they help local people? (6)

12 *(Pages 152 to 155)*

a Why has industry grown so rapidly in:
  • Osaka • São Paulo? (3 × 2)
b i) What is meant by Japan's 'industrial pyramid'? (3)
  ii) How does it work in the case of the Panasonic corporation? (6)
c i) In which parts of São Paulo are most:
    • formal • informal jobs to be found? (2)
  ii) Describe the importance of formal and informal jobs in São Paulo. (4)

10 Tourism

Recent trends and changing patterns

Tourism has become the world's fastest-growing and largest industry. It is an important factor in the economy of most developed countries and is seen by many developing countries as the one possible way to obtain income, create jobs and to improve their standard of living. It has grown rapidly in the last 60 years, mainly as a result of **globalisation** – globalisation in this context meaning the easier movement of people around the world. In 1950, 25 million 'international arrivals' were recorded worldwide, a figure that had risen to 225 million by 1975 and 903 million by 2008. Even so, this figure is small compared with the number of 'domestic' tourists who travel within their own countries. Travel for a privileged minority has been replaced by **mass tourism** – an industry that employs 8.4 per cent of the world's workforce and produces 9.9 per cent of its GDP.

Figure 10.1 shows that the more economically developed countries receive just over 70 per cent of all international tourist arrivals. However, their share has fallen from 82 per cent in 1980 as destinations in less economically developed countries in Africa, South-east Asia, the Pacific and, lately, the Middle East have become more popular and accessible.

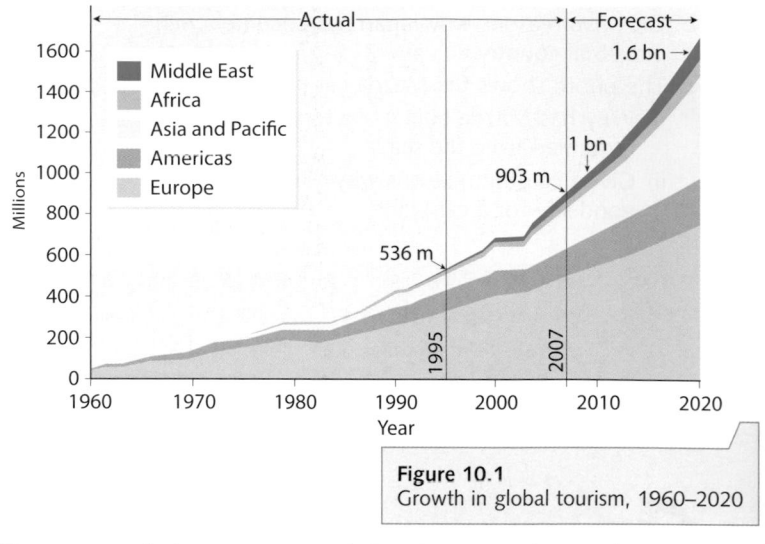

Figure 10.1
Growth in global tourism, 1960–2020

Recent trends in tourism, and the changing demands made by tourists, have resulted from a wide range of factors. These factors, which are summarised in Figure 10.2, include:

- an increase in wealth (affluence), mobility, accessibility and leisure time
- improvements in transport and technology
- changes in lifestyles, interests and demands
- a greater awareness of places due to coverage by the media and TV.

Figure 10.2
Recent trends in tourism in the UK

Greater affluence	• People who have employment within the UK usually earn a high salary – certainly higher than several decades ago. • People in full-time employment also receive holiday with pay. This means that they can take more than one holiday a year and can afford to travel further.
Greater mobility	• The increase in car ownership has given people greater freedom to choose where and when they go for the day, or for a longer period. In 1951, only 1 UK family in 20 had a car. By 2000, 76 per cent had at least one car. • Chartered aircraft have reduced the costs of overseas travel.
Improved accessibility and transport facilities	• Improvements in roads, especially motorways and urban by-passes, have reduced driving times between places and encourage people to travel more frequently and greater distances. • Improved and enlarged airports (although many are still congested at peak periods). Reduced air fares. package holidays, computerised reservation systems. • Computerised reservation systems and use of the internet. • Organisation of mass tourism by transnational companies.
More leisure time	• Shorter working week (although still the longest in the EU) and longer paid holidays (on average 3 weeks a year, compared with 1 week in the USA). • Flexitime, more people working from home, and more firms (especially retailing) employing part-time workers. • An ageing population, many of whom are still active (page 14).
Changing lifestyles	• People are retiring early and are able to take advantage of their greater fitness. • People at work need longer/more frequent rest periods as pressure of work seems to increase. • Changing fashions, e.g. health resorts, fitness holidays, winter sun.
Changing recreational activities	• Slight decline in the 'beach holiday' – partly due to the threat of skin cancer. • Increase in active holidays (skiing, water sports) and in self-catering. • Most rapid growth in mid-1990s has been in 'cruise holidays' followed by city breaks. • Importance of theme parks.
Advertising and TV programmes	• Holiday programmes, magazines and brochures promote new and different places and activities.
'Green' or sustainable tourism	• Need to benefit local economy, environment and people without spoiling the attractiveness and amenities of the places visited (page 168).

The UK

In the eighteenth century, spa towns such as Bath developed so that the wealthy could 'take the waters', while in the early nineteenth century George IV popularised the coastal resort of Brighton. By the end of that century, many industrial workers enjoyed a day, or even a few days, by 'the seaside' at places such as Blackpool. Since then, the annual holiday has become part of most British families' way of life, with an increasing number:

- taking more than one holiday a year, including **short breaks**
- travelling abroad and to increasingly distant destinations (Figures 10.3 and 10.4).

Within the UK itself, the most popular tourist region is the South West with its warmer climate and wide range of attractive scenery (Figure 10.5). Abroad, Spain remains the major destination (pages 164–5), although its importance has declined in the last twenty years as more people are prepared to take **long-haul** holidays to more distant, exotic and adventurous locations. Figure 10.6 summarises, with examples, some of the factors that have influenced the development of tourism in both the UK and the world.

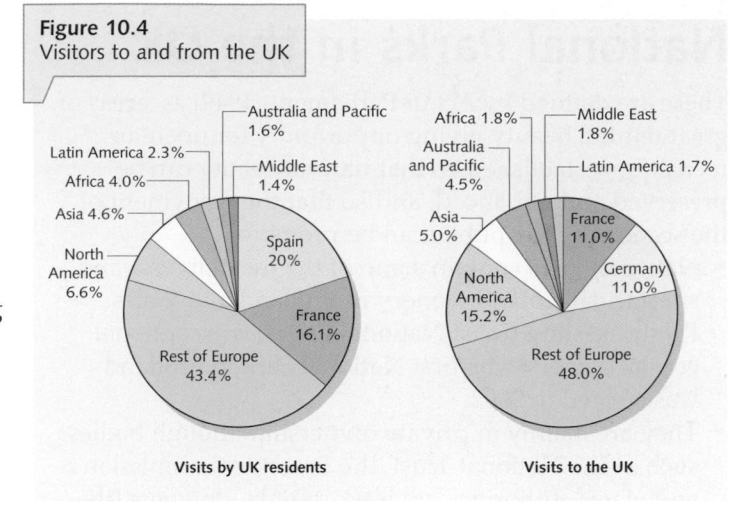

Figure 10.4
Visitors to and from the UK

Visits by UK residents

Visits to the UK

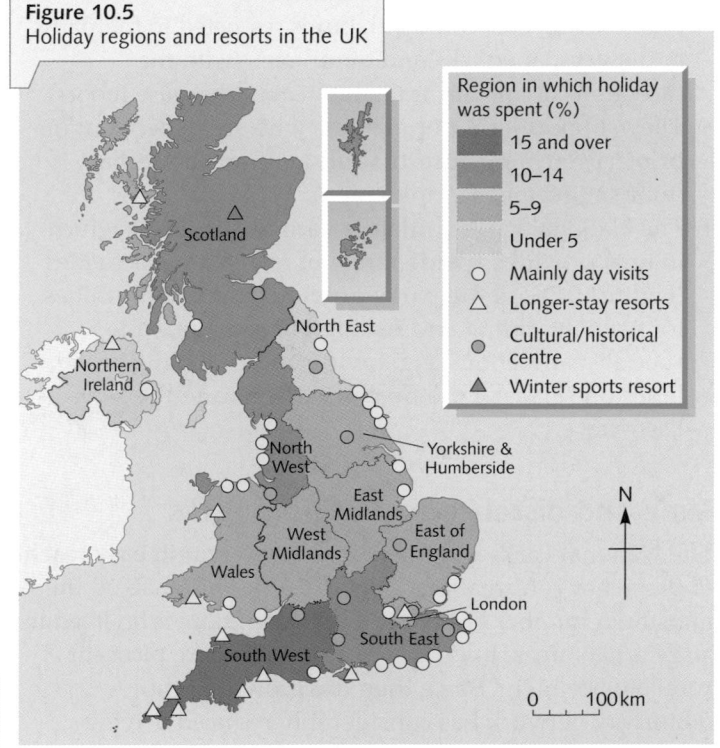

Figure 10.5
Holiday regions and resorts in the UK

Region in which holiday was spent (%)
- 15 and over
- 10–14
- 5–9
- Under 5
- ○ Mainly day visits
- △ Longer-stay resorts
- ◐ Cultural/historical centre
- ▲ Winter sports resort

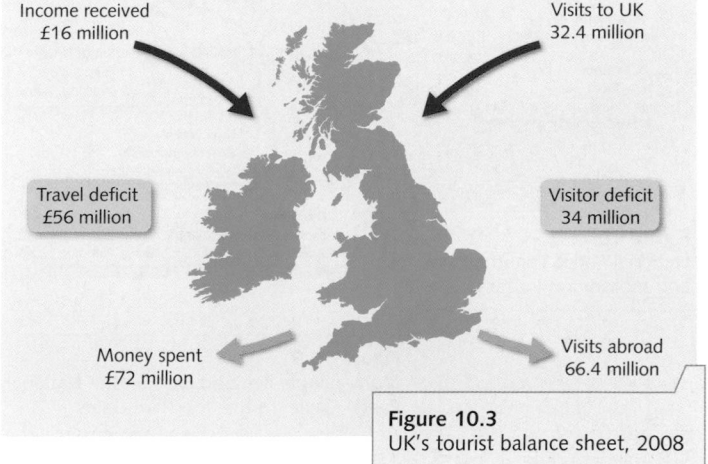

Income received £16 million

Visits to UK 32.4 million

Travel deficit £56 million

Visitor deficit 34 million

Money spent £72 million

Visits abroad 66.4 million

Figure 10.3
UK's tourist balance sheet, 2008

Figure 10.6
Factors affecting the growth of tourism in the UK and the world

Factor	Specific examples	UK examples of area or resort	World examples
1 Transport and accessibility	• Early resorts (stage coach), spa towns • Water transport (18th century) • Railways • Car and coach • Plane • Cruise ships	Bath Margate Blackpool, Brighton Cornwall, Scottish Highlands Channel Islands Scottish Islands	Baden-Baden Amsterdam Paris, Istanbul (Orient Express) Rhinelands, Alps Florida, Uluru (Ayers Rock) Norwegian Fiords, Caribbean
2 Scenery	• Sandy beaches • Coasts of outstanding beauty • Mountains, lakes and rivers	Margate, Blackpool Pembroke, Antrim, Cornwall Lake District, Snowdonia	Benidorm, Rio de Janeiro Barrier Reef Nepal, Switzerland, Rhine
3 Weather	• Hot, dry, sunny summers • Snow	Margate Aviemore	Costa del Sol, Greece Alps, Rockies
4 Accommodation	• Hotels and boarding house resorts • Holiday camps • Caravan parks and campsites	Margate, Blackpool Minehead, Pwllheli National and Forest Parks	Benidorm, Costa del Sol Caribbean French Riviera
5 Amenities	• Culture and history (castles, cathedrals) • Active amenities (sailing, golf, water-skiing) • Passive amenities (shops, cinemas) • Theme parks • Theatres, museums, music	York, Edinburgh Kielder, St Andrews Most resorts Alton Towers, Longleat London	Athens, Florence, Paris Costa del Sol Most resorts Disney World Paris, Vienna
6 Ecotourism and sustainability	• Nature reserves and game parks • World Heritage Sites	Slimbridge Hadrian's Wall	Kenya Cairo, Venice, Grand Canyon

National Parks in the UK

These are defined by Act of Parliament (1949) as 'areas of great natural beauty giving opportunity for open-air recreation, established so that natural beauty can be preserved and enhanced, and so that the enjoyment of the scenery by the public can be promoted'.

- National Parks contain some of the most diverse and spectacular upland scenery in England and Wales. Pembrokeshire Coast National Park has exceptional coastal scenery. The first National Park in Scotland was created in 2002.
- They are mainly in private ownership, though bodies such as the National Trust, the Forestry Commission and water authorities are important landowners (the Parks are not owned by the nation).
- Public access is encouraged, but is restricted to footpaths, bridleways, open fells and mountains (with the exceptions of military training areas and grouse moors).
- They support local populations who are dependent on primary (farming, forestry and mining) and tertiary (tourism) forms of employment.
- The National Parks contain a variety of scenery which in turn provides a wide range of recreational activities (Figure 10.7). All the parks provide basic opportunities for walking, riding and fishing but some provide specialist attractions, e.g. caving and potholing in the limestone areas of the Brecon Beacons and the Peak District.

Time and distance to National Parks

The National Parks were usually located within easy reach of the major conurbations (Figure 10.7). This enabled the maximum number of people, including those who lived in large urban areas, to escape to a quieter, more pleasant rural environment. Since then the growth of the motorway network has considerably reduced driving times and, in effect, has reduced distances between the conurbations and the National Parks (Figure 10.8).

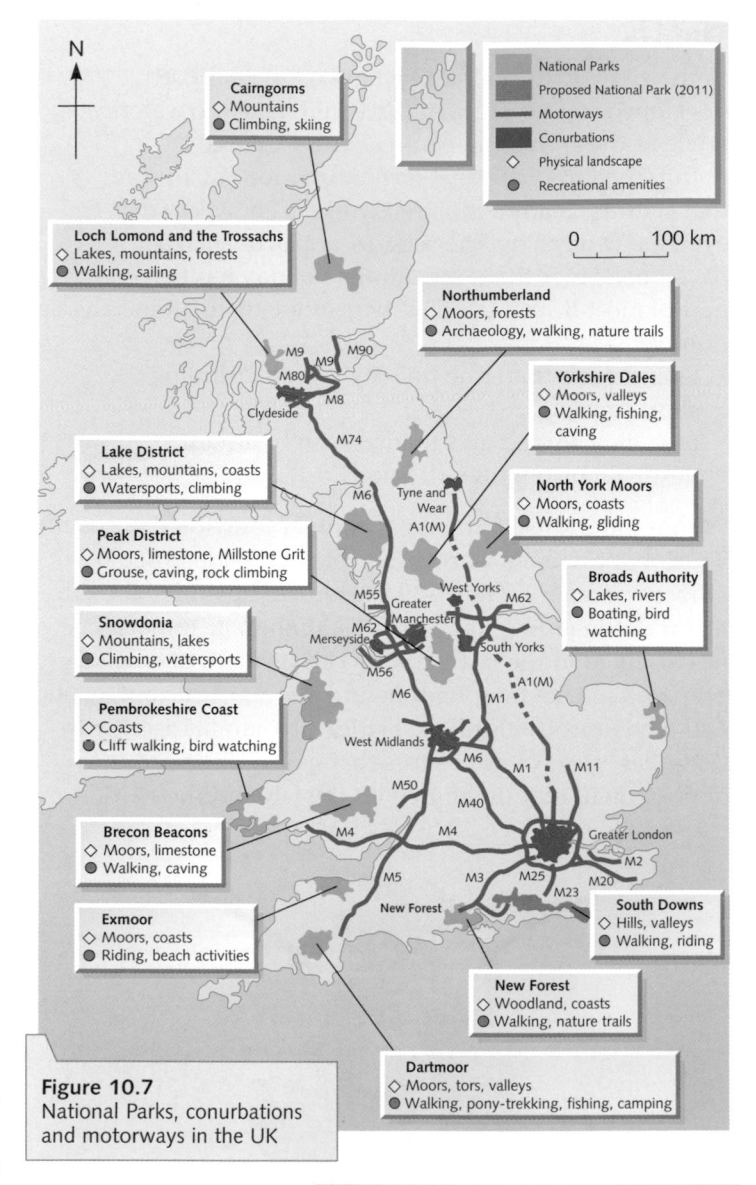

Figure 10.7
National Parks, conurbations and motorways in the UK

Figure 10.9
Many people live and work in the National Parks: Gayle, in the Yorkshire Dales

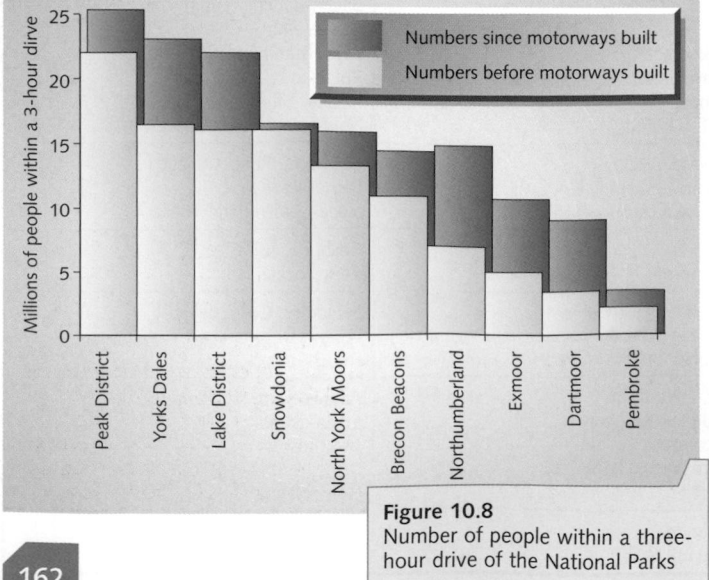

Figure 10.8
Number of people within a three-hour drive of the National Parks

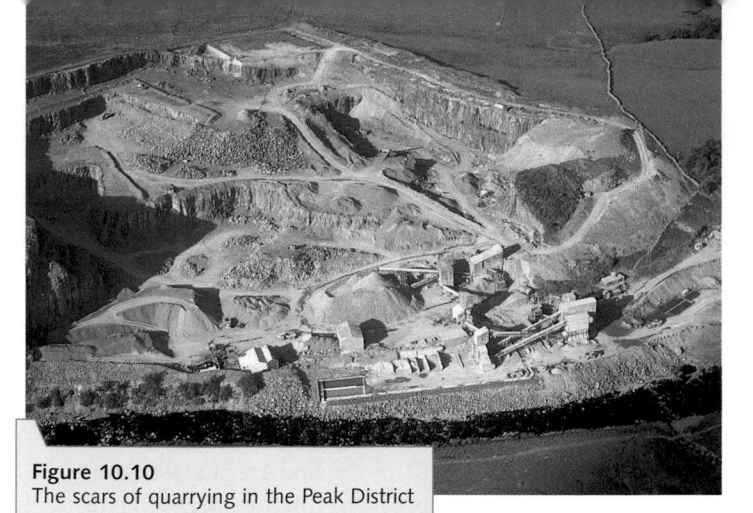

Figure 10.10
The scars of quarrying in the Peak District

Who owns the National Parks?

In total, 81 per cent of the land is owned privately, mainly by farmers, with 6 per cent belonging to the Forestry Commission, 5 per cent to the National Trust (a charitable organisation earning revenue from membership, admission fees and souvenirs), 3 per cent to water authorities, 3 per cent to the Ministry of Defence, 1 per cent to county councils and 1 per cent to the National Parks themselves. It should be remembered that many people live and work within National Parks (Figure 10.9).

Conflict between land users in National Parks

With over two people per hectare, the UK is one of the most densely populated countries in the world, and so there is considerable competition for land. This competition is also seen within the National Parks.

- Town dwellers wish to use the countryside for recreation and relaxation.
- Farmers wish to protect their land and in areas such as Exmoor are ploughing to a higher altitude because they receive government grants.
- The Forestry Commission has planted many hectares of trees in the poorer soils of Northumberland, the North York Moors and the Snowdonia Parks.
- The mining and quarrying of slate (Lake District and Snowdonia) and limestone (Peak District) creates local jobs but ruins the environment (Figure 10.10 and page 252).

Figure 10.11
A badly eroded footpath being repaired

- Water authorities have created reservoirs in the Lake District and Peak District Parks.
- The Ministry of Defence owns nearly a quarter of the Northumberland Park.
- Walkers and climbers wish for free access to all parts of the Parks, and campers and caravanners seek more sites for accommodation.
- Despite planning controls, the demand for housing has led to an increased suburbanisation of villages (page 64) and the use of property as 'second homes' for town dwellers.
- Nature lovers wish to create nature reserves and to protect birds, animals and plants from the invading tourists.

Honeypots

The National Parks include many of the nation's **honeypots** – areas of attractive scenery (Malham Cove in the Yorkshire Dales), or of historic interest (the Roman Wall in the Northumberland Park), to which tourists swarm in large numbers. The problem is how to preserve the honeypots' natural beauty and their unspoilt quality (the essence of their appeal), while providing facilities for the hordes who arrive at peak summer periods (Figure 10.12). At Malham Cove steps have been cut into the limestone to safeguard paths. It is estimated that £1.5 million is needed to repair the six paths leading to the top of Snowdon where, on a summer's day, 2500 people might reach the summit. Parts of the 400 km Pennine Way have had to be laid with artificial surfaces as the tracks of walkers have penetrated over a metre into the peat in certain places. The footpaths on the Roman Wall are being eroded and, as soil is washed away, the foundations of the wall are being exposed.

Figure 10.12
How planning in a National Park can help solve problems such as over-use, congestion and conflicts of use

Problems	Attempted solutions
Footpaths worn away (see Figure 10.11)	New routes planned; signposted routes; artificial surfaces laid
Destruction of vegetation, erosion of footpaths	Areas fenced off; education of visitors; landscaping
Litter, vandalism, trespassing	Provision of picnic areas with litter bins; park wardens
Cars parked on grass verges or in narrow lanes	Car parks; one-way systems; park and ride schemes
Congestion on narrow roads	Roads closed to traffic in tourist season/at weekends; park and ride; encouragement to use minibuses, to cycle or to walk
Heavy lorries, local traffic and tourist traffic	Scenic routes separating local and tourist traffic
'Honeypots' (views, cafés) cause crowding	Develop alternative honeypots, direct visitors to other attractions
Conflict of users, e.g. a) between local farmers and tourists b) between tourists	Restricting tourist access to footpaths and bridleways. separating activities, e.g. water skiing and angling
Unsightly new cafés, car parks and caravan parks	Screened behind trees. Only certain natural colours allowed in paint schemes

Coastal resorts

The Costa del Sol

The Costa del Sol (the sun coast) is the most southerly of Spain's many tourist coasts (*costas*). It faces the sun, the Mediterranean Sea and North Africa (Figure 10.13). In the 1950s the area was important only for farming and fishing. Since then, both the landscape and the lives of local people have been transformed by tourism (Figure 10.14). In summer in the main resorts of Torremolinos and Marbella it is more usual to hear English being spoken than Spanish. Why has this become such an important tourist region?

Climate (Figure 10.15) Summers are hot, sunny and dry. Although winters are wet, it rarely rains all day and it is usually mild enough for people to sit out of doors.

Landscape There are long stretches of sandy beach beside the warm, blue Mediterranean Sea. Some beaches consist of shingle; others are artificial. Inland are the spectacular Sierra Nevada Mountains.

Accommodation Torremolinos consists of high-density, low priced, high-rise hotels and apartments (Figure 10.16). Fuengirola's hotels provide the cheapest accommodation along the coast. Marbella has the most modern and luxurious of the hotels, and the hills behind are dotted with many time-share apartments. There are numerous campsites near to the N340 road.

Nightlife and shopping Illuminated shops attract tourists after dark. Numerous restaurants, cafés and bars provide flamenco and disco music, wine and beer, and Spanish, British and other European food. Most resorts have nightclubs. Shops range from the cheaper local bazaars and souvenir shops (mainly Torremolinos area) to chic boutiques with designer clothes (more likely in Marbella). Leather goods, ceramics and perfume can be bought in most places.

Things to do Many activities are linked to the sea, e.g. water sports and Aquapark at Torremolinos, and yachting marinas and harbours at Benalmadena, Puerto Banus and Estepona (Figure 10.17). There are also many golf courses, especially near Marbella. Day visits can be made to the whitewashed village of Mijas (perhaps the only 'real Spain' seen by tourists), to Ronda in the higher mountains, or to the historic centres of Granada and Seville.

Figure 10.13
Location of the Costa del Sol

"Sun-drenched southern Spain has always been one of the most popular holiday destinations and there's a very good reason for it. This wonderful stretch of coastline, fringed by long sandy beaches and backed by dramatic mountain ranges, has so much to offer. Variety is the key to the coast's success. Dedicate your days to that all-important tan, taking an occasional, refreshing dip in the western Mediterranean. Explore this fascinating part of Spain by visiting romantic Seville, British Gibraltar, stylish Puerto Banus and historic Granada. Once the sun goes down, enjoy a range of nightspots, bars and restaurants from the sophisticated to the informal."

Figure 10.14
Extract from a Thomson travel brochure

Average hours of sunshine per day

	London	Malaga
J	2	6
F	3	7
M	4	7
A	5	9
M	6	10
J	7	11
J	6	12
A	6	11
S	5	9
O	3	7
N	2	6
D	2	6

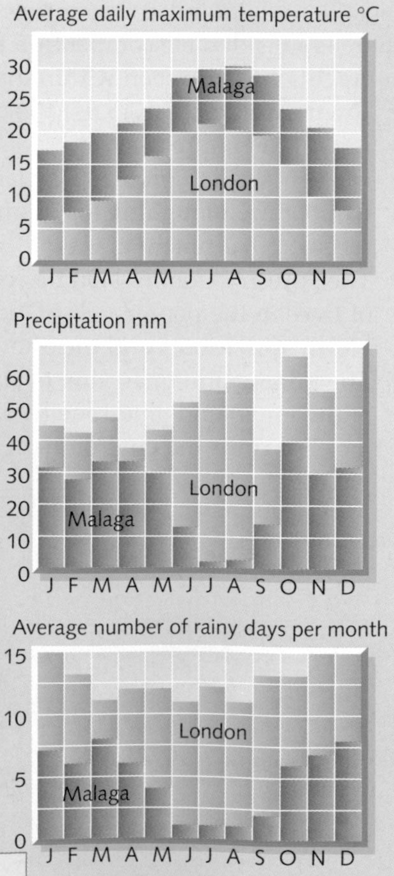

Figure 10.15
Climate graphs for the Costa del Sol and London

Figure 10.16
Torremolinos

Figure 10.17
Yachting marina at
Puerto Banus

	Stage 1 The traditional society	Stage 2 Take-off and development	Stage 3 Peak growth	Stage 4 Stagnation	Stage 5 Rejuvenation or decline?

Growth in tourism ↑

Date	1960s	1970s	1980s	1990s	2000s
Social effects					
Tourists from UK to Spain	1960 = 0.4 million	1971 = 3.0 million	1984 = 6.2 million 1988 = 7.5 million	1990 = 7.0 million	2008 = 4.5 million
State of, and changes in, tourism	Very few tourists.	Rapid increase in tourism. Government encouragement.	Carrying capacity reached – tourists outstrip resources, e.g. water supply and sewerage.	Decline – world recession and prices too high – cheaper upper-market hotels elsewhere. Beach boredom.	Further decline. Theme parks.
Holiday accommodation	Limited accommodation, very few hotels and apartments, some holiday cottages.	Large hotels built (using breeze-blocks and concrete), more apartment blocks and villas.	More large hotels built, also apartments and time-share, luxury villas.	Older hotels looking dirty and run-down. Fall in house prices. Only high-class hotels allowed to be built.	Refurbishment of older hotels.
Economic effects					
Local employment	Mainly in farming and fishing.	Construction workers. Jobs in hotels, cafés, shops. Decline in farming and fishing.	Many in tourism – up to 70% in some areas.	Unemployment increases as tourism declines (30%). Farmers use irrigation.	Some diversification including high-tech.
Infrastructure (amenities and activities)	Limited access and few amenities. Poor roads. Limited street lighting and electricity.	Some road improvements but congestion in towns. Bars, discos, restaurants and shops added.	E340 opened – 'The highway of death'. More congestion in towns. Marinas and golf courses built.	Bars/cafés closing, Malaga by-pass and new air terminal opened.	
Environmental effects					
Landscape and environment	Clean, unspoilt beaches. Warm sea with relatively little pollution. Pleasant villages. Quiet. Little visual pollution.	Farmland built upon. Wildlife frightened away. Beaches and seas less clean.	Mountains hidden behind hotels. Litter on beaches. Polluted seas (sewage). Crime (drugs, vandalism and mugging). Noise from traffic and tourism.	Attempts to clean up beaches and seas (EU blue flag beaches). New public parks and gardens opened. Nature reserves.	Opening of historic and cultural sites.

Figure 10.18
Social, economic and environmental effects of tourism

Changes in tourism and to the environment

Places with pleasant climates and spectacular scenery attract tourists. Tourists demand amenities to make their visit more comfortable (e.g. accommodation, car parks) and activities to fill their leisure time (e.g. water sports). As more amenities are added and more leisure activities become available, the environment which first attracted people to these areas either deteriorates or becomes congested. Tourists will therefore seek alternative places to visit. This is happening on the Costa del Sol (Figure 10.18).

Role of the Spanish government

The Spanish government saw tourism as one way to provide jobs and to raise the country's standard of living. In places like the Costa del Sol it encouraged the construction of new hotels and apartment blocks and the provision of leisure amenities such as swimming pools, marinas and entertainment. More recently it has reduced VAT to 6 per cent in luxury hotels to try to maintain cheap holidays. It has also introduced stricter controls to improve the quality of the environment, which includes cleaner beaches and reducing sea pollution (Spain now has the most 'Blue Flag' beaches in the EU).

Mountain resorts

Courmayeur (Italy)

The mountaineering and winter sports resort of Courmayeur is located near the head of the scenic Val d'Aosta in the extreme north-west of Italy (Figure 10.19). It lies at the foot of Mont Blanc (Figure 10.20), the highest mountain in the Alps, close to where the main road linking Turin and France passes through the Mont Blanc tunnel.

Figure 10.19
Courmayeur in the Alps

Figure 10.20
Noticeboard of the Mont Blanc massif

Figure 10.21
Courmayeur ski resort

Ski facts
High-altitude skiing and snowboarding

Max. skiers permitted	8400
Snow cannons (artificial snow)	316
Length of piste	100 km
Longest run	20 km
Easy runs	6 km
Medium runs	12 km
Difficult runs	4 km
Artificial piste	16 km
Cross-country	20 km
No. of lifts	19
Types of lift	
Chair	8
Cable-car/gondolas	8
Drag	3
Slopes face	N, NW, NE, E
Mountain restaurants	27

Skiing The resort area rises from 1224 m to 3470 m. The main skiing area, with 100 km of marked piste, is reached by cable car which begins just 150 m from the Hotel Pavilion. Skiing is excellent for intermediates and advanced. Apart from the famous Vallée Blanche run to Chamonix, there are numerous runs under Mont Blanc (Figure 10.20).

Après-ski In the evening you can sample local Aosta specialities such as fontina cheese or beef steak Valdostan, and enjoy the many wines from northern Italy. Check out the best bars – Roma or Steve's cocktail bar among them – then go dancing at one of the clubs, like the Clochard or Abat Jour.

The resort Courmayeur (Figure 10.21) is a traditional town, full of character and a marvellous Italian atmosphere. It has narrow, twisting cobblestone streets, a range of accommodation, a bustling weekly market, great cuisine, varied shopping, a lively nightlife, an ice rink and a swimming pool.

Nearby There are opportunities for walking, climbing and horse-riding as well as buses to Chamonix (40 minutes away in France) and Aosta. You can even take a cable car to the Cresta d'Arp and cross the Mont Blanc massif.

While the increase in tourism has brought many social and economic benefits to mountain resorts such as Courmayeur, it has also created many social, economic and environmental problems. These are summarised in Figure 10.22.

Benefits
- More and better paid jobs, especially for younger people.
- Younger people no longer have to leave the area (out-migration) to find work.
- Improved accessibility due to better roads.
- Improved services, including the provision of electricity, a reliable water supply and sewage disposal.
- Leisure amenities added which can be used by local people, e.g. ice-rink, swimming pool.
- Improved shopping.
- The **multiplier effect**, which is when the success of one type of industry – in this case tourism – attracts other forms of economic development and creates more jobs, for example in shops, hotels, restaurants and bars, and also as guides and instructors.

Disadvantages
- At peak times, tourists outnumber the local population.
- The traditional village is swamped by new buildings, e.g. hotels, chalets, souvenir shops, restaurants, car parks.
- Many of these new buildings, together with ski lifts and ski runs, are unsightly and create visual pollution. As eye-sores, they spoil the beautiful scenery that first attracted tourists.
- The traditional way of life is changed and the local culture is likely to be lost.
- Farmers have lost land and their traditional jobs.
- House prices have risen and become too expensive for local people.
- Unemployment may be seasonal, as many of the jobs are linked to the winter skiing season, e.g. ski instructors.
- Hillsides are deforested to create new and longer ski runs. This destroys the fragile alpine ecosystem (vegetation is damaged, wildlife is frightened away) and increases the risk of soil erosion and avalanches (page 321).
- The increase in traffic has been blamed for the increase in acid rain that is killing local vegetation (page 221).

Tourism in developing countries

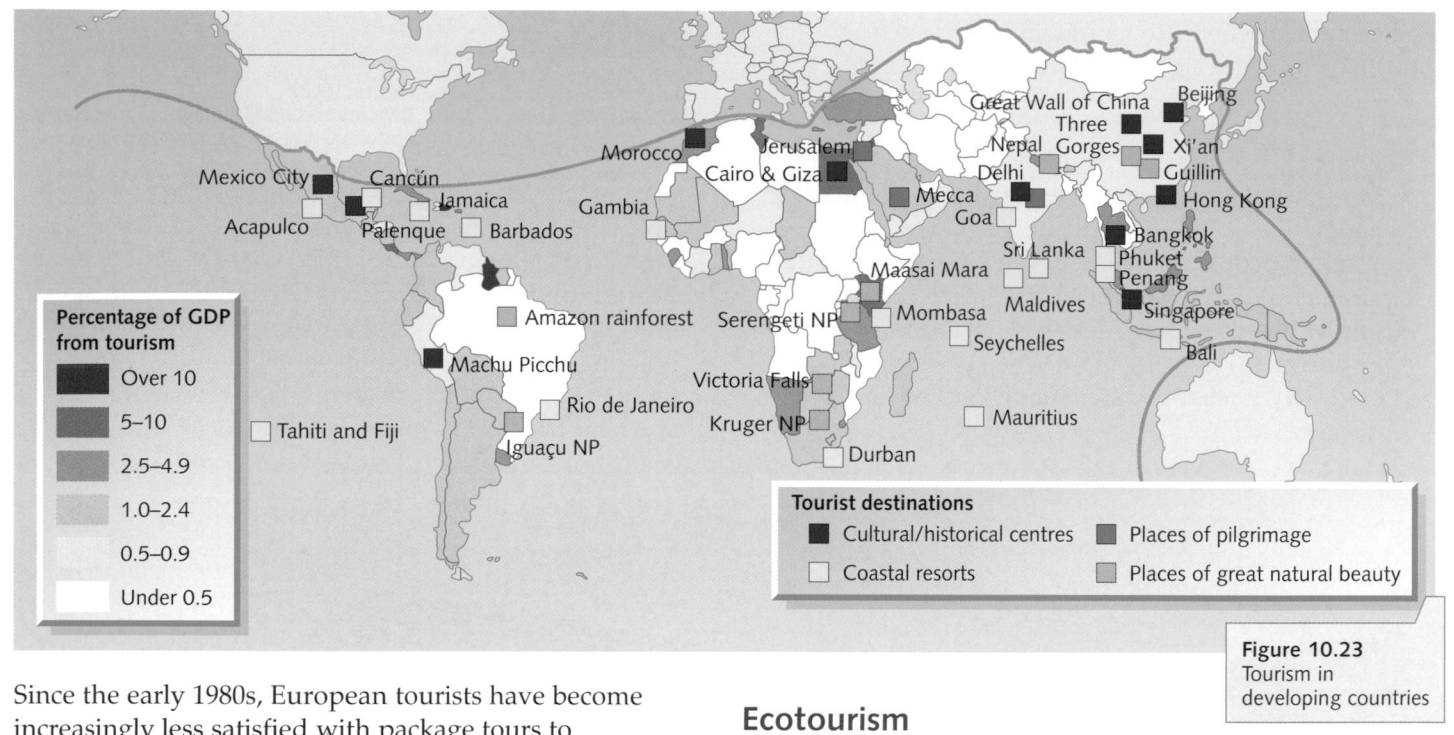

Figure 10.23
Tourism in developing countries

Since the early 1980s, European tourists have become increasingly less satisfied with package tours to Mediterranean coastal resorts. Like tourists from North America and Japan, they have sought holidays in places farther afield where the environment and culture is very different from their own. The chief beneficiaries of this change in holiday fashion and demand have been several developing countries located within, or near to, the tropics – places such as Kenya (page 170), Egypt, Sri Lanka, Thailand, Malaysia and in the West Indies (Figure 10.23). The attraction of earning money from tourism is considerable to less economically developed countries, many of which see it as the only possible way to raise their standard of living. However, only a limited number of LEDCs have the potential to develop a successful tourist industry, and even then the damage to their culture and environment can, at times, outweigh the benefits (Figure 10.24).

Ecotourism

Ecotourism, sometimes known as **green tourism**, is a sustainable form of tourism that is more appropriate to developing countries than the mass tourism associated with places like Florida and the Spanish costas (page 164) in developed countries. Ecotourism includes:

- visiting places in order to appreciate their scenery and wildlife and to understand their culture
- creating economic opportunities (jobs) in an area while at the same time protecting natural resources (scenery and wildlife) and the local way of life (culture).

Compared with mass tourists, ecotourists usually travel in small groups and share specialist interests (bird watching, photography). They are more likely to merge and live with local communities and to appreciate local cultures rather than 'stop, take a photo and move on'. They often visit National Parks and game reserves (page 170) where the scenery (e.g. sandy beaches, coral islands, forests and waterfalls) and wildlife (whales, elephants, polar bears, giant pandas and mountain gorillas) that attracted them there in the first place, are carefully protected and managed (Figure 10.24). Even so, ecotourists usually pay for their holiday in advance (meaning they spend relatively little in the developing country); are not all environmentally educated or concerned; can cause land prices to rise; congregate at prime sites (honeypots); and may still cause conflict with local people. There is a real danger that tour operators, by adding 'eco' as a prefix, give certain holidays unwarranted respectability.

a

b

Figure 10.24
Tourist sites:
(a) Bali
(b) Machu Picchu

A West Indies beach village

Attractions of the West Indies

- Winters (25°C) are much warmer than those in North America and Europe. Summers are hot (28°C) but not oppressive.
- Most days have more than eight hours of sunshine.
- The scenery is attractive, usually either volcanic mountains covered in forest or coral islands with sandy beaches.
- The warm, clear blue seas are ideal for water sports such as sailing, water-skiing, scuba diving and snorkelling.
- There is varied wildlife – plants, birds, fish and animals.
- Customs are different – calypsos, steel bands, food, festivals and carnivals.
- There are many cultural and historic resorts.
- The region is situated only a relatively short flight-time from North America.

The **beach village** is a recent attempt to try to disperse accommodation and amenities so that they merge with the natural environment, reduce overcrowding at other places, and avoid spoiling the physical advantages that originally attracted tourists to the islands (Figures 10.25 and 10.26). The advantages and disadvantages of tourism to islands in the West Indies are summarised in Figure 10.27.

Disadvantages of tourism

- Hotels, airports and roads spoil the visual appearance and create noise, air pollution and litter.
- Usually only 10–20% of the income received from tourists stays in the country. Most hotels are foreign-owned and profits go overseas. Tourists spend most of their money in the hotels.
- Much employment is seasonal. Overseas labour may be brought in to fill the better-paid jobs.
- Local craft industries may be destroyed in order to provide mass-produced, cheap souvenirs.
- Farming economy is damaged as land is sold to developers. Much of the food eaten by tourists is imported either because local production is insufficient or to meet the demands for European-style foods (but sold at the developing country's prices).
- Local people cannot afford tourist facilities.
- Borrowed money increases national debt.
- Tourists expect unlimited water – up to 500 litres a day or ten times that used by local people. Many areas may be short of water for domestic and farming use.
- Local cultures and traditions are destroyed. New social problems of prostitution, crime, drugs and drunkenness. Lack of respect for local customs and religious beliefs (e.g. semi-naked tourists in mosques and temples).
- The building of hotels means that local people lose their homes, land and traditional means of livelihood (e.g. fishermen, as hotels are built next to beaches) and become dependent on serving wealthy tourists.

... Tourism is a form of economic colonialism

Advantages of tourism

- The natural environment (sun, sand, sea and scenery) is used to attract tourists and their much-needed money.
- Income from tourism is usually greater than the income from the export of a few raw materials.
- Creates domestic employment, e.g. hotels, entertainment and guides. It is labour intensive.
- Encourages the production of souvenirs.
- Creates a market for local farm produce.
- Local people can use tourist facilities.
- Overseas investment in airports, roads and hotels.
- Profits can be used to improve local housing, schools, hospitals, electricity and water supplies.
- Increased cultural links with foreign countries, and the preservation of local customs and heritage.
- Reduces migration.

... Tourism raises the standard of living

Figure 10.25 A beach village merging with the natural environment

Figure 10.26 A beach village in Barbados

Figure 10.27 The disadvantages and advantages of tourism

Tourism and the environment

Kenya

Few countries in the world can offer the traveller the variety of landscapes that Kenya can. It has mountains, grassy plains, sandy beaches, coral reefs and an abundance of wildlife. Kenya appreciates these natural resources and has set up over 50 National Parks and game reserves to protect and manage its environment. Tourism has become Kenya's major source of overseas income. **Safaris**, meaning 'journeys', are organised so that tourists can be driven around, usually in seven- or nine-seater minibuses with adjustable roofs to allow easier viewing. Unlike on early safaris, today's tourists are only allowed to shoot with cameras. An advertisement for one safari is given in Figure 10.28.

Day 1 Nairobi/Samburu (310 km)
After breakfast drive north, cross the Equator and pass Mt Kenya, to Samburu Lodge. After lunch there will be a game drive when you should see elephant, buffalo, lion, reticulated giraffe, zebra, crocodile and many bird species.

Day 2 Samburu
Early morning game drive. Relax at midday around the swimming pool or watch the Samburu perform traditional dances. Late afternoon game drive.

Day 3 Samburu/Treetops (200 km)
Drive south for lunch at the Outspan Hotel. A short journey takes you into the Aberdare Mountains where you will spend the night at Treetops, the world-famous tree hotel. As evening approaches, buffalo, elephant and rhino join other animals at the waterhole.

Day 4 Treetops/Nakuru/Naivasha (240 km)
Transfer to Outspan for breakfast. Drive to the Thomson's Falls, and down into the Rift Valley to Nakuru for lunch. A short drive will let you see vast flocks of flamingos and the endangered Rothschild giraffe. Continue to Lake Naivasha Hotel for the night.

Day 5 Naivasha/Maasai Mara (240 km)
Leisurely morning by the lakeside. After lunch, drive to Keekorok Lodge in the Maasai Mara.

Day 6 Maasai Mara
The huge Mara plain provides some of the best game-viewing in East Africa. During early morning and late afternoon game drives you are likely to see huge herds of wildebeest and zebra, as well as lion, elephant, cheetah, leopard, Maasai giraffe, and hippo. An option is the early morning balloon safari.

Day 7 Maasai Mara/Nairobi (260 km)
Early morning departure arriving at Nairobi for lunch. Afternoon flight to Mombasa to continue your holiday at a beach hotel.

SAFARIWISE

Safari lodges in Kenya provide all modern comforts. While simple in design, your room will have bath or shower (except Shimba Hills and Treetops where shared facilities are provided) and most lodges have a pool. Cuisine, though not *cordon bleu*, is of good standard and sometimes includes game meat.

We use the best available vehicles – 7- or 9-seater safari cruisers with roof hatches and sliding windows for easy game viewing and photography. Journeys, particularly between game reserves, can be long, dusty and tiring but the excitement of seeing wildlife in its natural habitat usually makes it all worthwhile. The occasional change in routeing and/or hotels/lodges may be necessary due to weather conditions or shortage of accommodation.

Tented accommodation is sometimes included at Samburu or Keekorok – but do not be alarmed! The tents have stand-up room, are heavy-duty and erected on a concrete base under an awning, while to the rear (direct access from tent) are simple but private shower and toilet facilities.

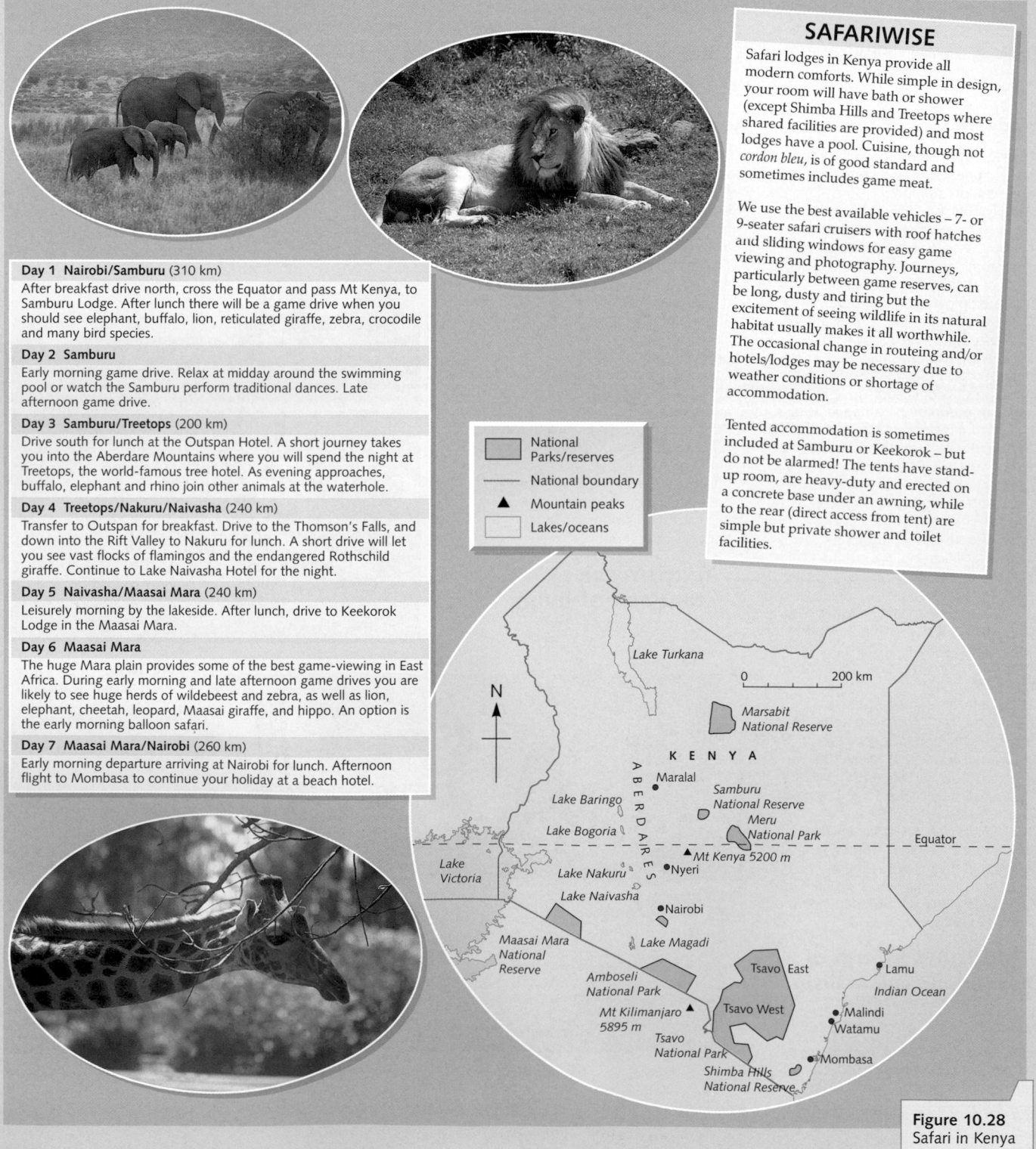

Legend:
- National Parks/reserves
- National boundary
- ▲ Mountain peaks
- Lakes/oceans

Figure 10.28
Safari in Kenya

Kenya appreciates the value of its landscape and wildlife. They are seen as a major source of income in a country desperately short of money. Money from tourism can be used to pay for improving services, building more houses and roads, and creating new jobs. But tourism also has its problems. Large numbers of people going on safaris can damage the very environment which attracted them there in the first place. Pressure is put upon wildlife and local people, as well as on Kenya's many fragile environments.

The environment Safari minibuses are meant to keep to well-defined tracks in National Parks and game reserves. However, drivers often form new routes, either to enable their passengers to get as close as possible to wildlife, or to avoid wet season marshy areas. Minibuses can get stuck in the mud, ruining vegetation (Figure 10.29), or widening existing tracks (Figure 10.30). In Amboseli, as in other parks, the wind, minibuses and herds of animals all cause mini dust storms which increase the rate of soil erosion (Figure 10.31).

Wildlife Minibuses are not meant to go within 25 m of animals, but their drivers often ignore this as they are unlikely to get good tips from their passengers if the best close-up views of wildlife are not obtained. Animals may be prevented from mating, making a kill, or forced to move to less favourable areas. Balloon safaris (Figure 10.32) cause controversy as conservationists claim that the intermittent release of hot air and the shadow of passing balloons disturb wildlife.

People Today, apart from employees at safari lodges, nobody is allowed to live in National Parks. Even game reserves only permit a limited number of herders and their cattle. The setting up of National Parks meant that nomadic tribes, such as the Maasai, had to be moved away from their traditional grazing grounds. Many now have to live a more permanent life, earning money by selling small artefacts to, or performing traditional dances for, the tourists (Figure 10.33). Recently the government has begun to work with the Maasai, allocating them a share of the wealth obtained from tourism to help improve their education, housing and water supply (page 151).

Figure 10.29
Erosion by minibuses

Figure 10.32
Balloon safari

Figure 10.30
Dust track in Amboseli

Figure 10.31
A dust storm causing soil erosion

Figure 10.33
Maasai people selling artefacts to tourists

The Lake District National Park

Each year, 12 million people visit the Lake District. They visit the National Park because of its mountain, valley and lake scenery and its centuries of human history (Figure 10.34). Some of the early tourists came for spiritual refreshment, while most of the present-day visitors seek a range of recreational activities. What attracts the tourist also attracts second-home owners and retired people. These people, together with the long-established local population, add up to a population of over 40 000 living, and often working, within the National Park.

The Lake District National Park Plan identifies several special qualities (Figure 10.35).

- The mixture of natural and farmed landscapes provide a release from the pressures of urban living.
- There is a diversity of landscapes – sandy coasts and clear rivers, rugged mountains and deep lakes, woodlands and moorland, wildlife and historic remains.
- Its most spectacular scenery results from glaciation (page 317). The work of frost and ice has created landforms which include tarns (corries), narrow ridges (arêtes), and U-shaped valleys (glacial troughs) with waterfalls (hanging valleys), cliff-like sides (truncated spurs) and lakes (ribbon lakes).
- The wide range of ecosystems (Chapter 14) include freshwater habitats, ancient broad-leaved woodland, limestone grassland, heath, arctic–alpine communities, heathland, coastal marsh and estuary. The National Park includes 101 Sites of Special Scientific Interest (SSSIs), six national nature reserves and eight areas of protected limestone pavement (page 249).

Figure 10.34
View in the Lake District National Park

Figure 10.35
The Lake District National Park

N

0   5 km

Land below 150 metres
150 m–300 m
300 m–500 m
500 m–700 m
700 m–900 m
Above 900 metres
— ·— National Park boundary
Major honeypots

Caldbeck

Cockermouth
A66
Workington

Bassenthwaite Lake
Skiddaw 931 m
A66
Penrith

Keswick
Crummock Water
Derwent Water
Thirlmere
Ullswater
Helvellyn 950 m
Shap

Buttermere
Borrowdale
Haweswater

Ennerdale Water
A591
M6

Scafell Pike 978 m
Wast Water
Grasmere
Gt Langdale
Ambleside
Kentmere

Windermere

Coniston
Bowness
Windermere
Kendal

Coniston Water

Irish Sea

- The 3200 km of footpaths, bridleways and green lanes give access to large areas and create a sense of freedom and discovery.
- There are 20 larger settlements and 35 villages, each with its own architecture and character, as well as 6000 archaeological sites and monuments and 1740 listed buildings. The area, with its human history extending back for 12 000 years, includes Neolithic sites, Roman remains and more recent mining and industrial activities. Stone-walled fields have developed since the twelfth century, and parklands from medieval times.

- The rural character of Lake District life with its dialect, sports and links with literature and artistic movements, has created a strong local culture.

Figure 10.36 gives information about the Lake District National Park taken from several surveys conducted over the last few years. Although visitor numbers to the Lake District have not increased significantly in recent years, the nature of demand has changed in several ways:
- Short breaks, especially of three nights or fewer, continue to increase while longer stays have shown a decrease.

- There has been a significant shift from serviced holiday (hotels and B&Bs) to self-catering accommodation (holiday flats and camping).
- Day visits make up the greater proportion (69 per cent) but represent a relatively small proportion of money spent.
- Visitors, certainly before the 2008 recession, had increasing levels of disposable income which meant that they expected higher levels of quality, and standards, which some tourism businesses found hard to meet.

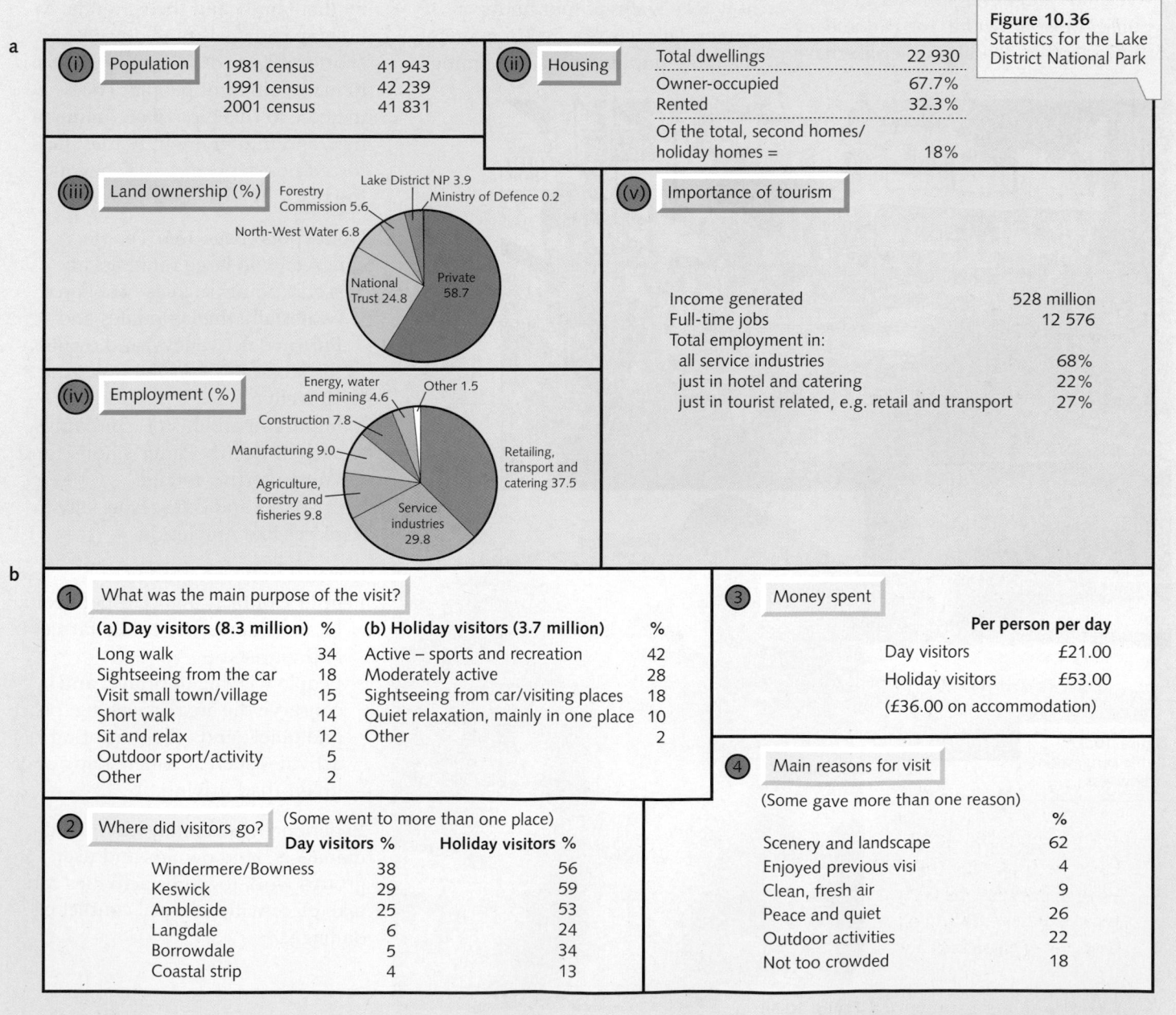

Figure 10.36
Statistics for the Lake District National Park

a

(i) Population

1981 census	41 943
1991 census	42 239
2001 census	41 831

(ii) Housing

Total dwellings	22 930
Owner-occupied	67.7%
Rented	32.3%
Of the total, second homes/ holiday homes =	18%

(iii) Land ownership (%)

- Forestry Commission 5.6
- Lake District NP 3.9
- Ministry of Defence 0.2
- North-West Water 6.8
- National Trust 24.8
- Private 58.7

(iv) Employment (%)

- Energy, water and mining 4.6
- Other 1.5
- Construction 7.8
- Manufacturing 9.0
- Agriculture, forestry and fisheries 9.8
- Service industries 29.8
- Retailing, transport and catering 37.5

(v) Importance of tourism

Income generated	528 million
Full-time jobs	12 576
Total employment in:	
all service industries	68%
just in hotel and catering	22%
just in tourist related, e.g. retail and transport	27%

b

(1) What was the main purpose of the visit?

(a) Day visitors (8.3 million)	%	(b) Holiday visitors (3.7 million)	%
Long walk	34	Active – sports and recreation	42
Sightseeing from the car	18	Moderately active	28
Visit small town/village	15	Sightseeing from car/visiting places	18
Short walk	14	Quiet relaxation, mainly in one place	10
Sit and relax	12	Other	2
Outdoor sport/activity	5		
Other	2		

(2) Where did visitors go? (Some went to more than one place)

	Day visitors %	Holiday visitors %
Windermere/Bowness	38	56
Keswick	29	59
Ambleside	25	53
Langdale	6	24
Borrowdale	5	34
Coastal strip	4	13

(3) Money spent

	Per person per day
Day visitors	£21.00
Holiday visitors	£53.00
(£36.00 on accommodation)	

(4) Main reasons for visit (Some gave more than one reason)

	%
Scenery and landscape	62
Enjoyed previous visi	4
Clean, fresh air	9
Peace and quiet	26
Outdoor activities	22
Not too crowded	18

Problems created by visitors to the Lake District National Park

The 12 million annual visitors to the Lake District National Park are bound to cause problems both to the environment and for the 42 000 local residents. These problems include the following.

Traffic Although 83 per cent of visitors arrive by car, traffic levels do appear to have levelled off. Even so, congestion is often heavy on approach roads to the Park, especially at weekends and in the summer. Limited public transport within the Park means that visitors and local people alike have to rely on the car. This creates periodic congestion in bottlenecks such as Bowness and Ambleside (Figure 10.37), parking problems, and noise and air pollution. Many roads outside the built-up areas are narrow, steep and winding, making them unsuitable for large volumes of traffic and especially for tourist buses. A more recent problem is recreational green-road driving where four-wheel drives and motorbikes can cause noise, pollution, erosion and conflict.

Footpath erosion Latest surveys show that 4 million visitors a year now take walks of four hours or longer. This has resulted in many of the more popular routes becoming severely eroded and ugly scars being created on hillsides (Figure 10.38). A constant stream of new guidebooks is attracting more people to previously remote areas.

Second homes and holiday homes One in six properties in the Lake District (16 per cent) are either second homes or holiday cottages, and in places in the north-east and south-east (nearer to M6 interchanges) the proportion is more than one in four. A second home is one that is owned by a family who live elsewhere, and is used primarily by that family and their friends. A holiday cottage is rented out on short-term lets on a commercial basis to many different people. The increase in this type of accommodation, and in retirement homes, has forced house prices to rise beyond the reach of most local people.

Honeypots (page 163) People congregate in large numbers at:
- scenic points such as Aira Force (waterfall), the Langdales and Borrowdale (valleys) and even on the summits of Scafell Pike and Helvellyn
- important and historic buildings such as Beatrix Potter's house and Wordsworth's cottage
- car parks and cafés, especially in places like Ambleside.

Conflict between groups of people Conflict can occur between:
- local residents (especially farmers) and tourists
- people wishing to protect and conserve the area or seeking 'peace and quiet', and others wanting activities such as water-skiing and green-road driving.

Experience shows that when land managers, land owners and user groups work together, activities can take place with minimal conflict or damage.

Figure 10.37
Traffic congestion in Bowness

Figure 10.38
Footpath erosion on Catbells

Possible solutions to the problems

The latest management plan put forward by the Lake District National Park Authority sees, as one of its main aims, the need to balance the needs of local residents with those of visitors. It must ensure that 'the Lake District remains a beautiful place, where a range of landscapes retain their character and give shelter to a diversity of flora and fauna. It must continue to be an area for recreation and a source of relaxation and inspiration. The future must see the landscape maintained, visitors catered for and the local community flourishing.' Two important aspects of the plan are landscaping and traffic management.

Landscaping
This includes:
- repairing eroded footpaths and damaged stone walls (Figure 10.39)
- planting trees to screen, wherever possible, car parks and quarries (page 251).

The Lake District Traffic Management Initiative
The latest, and revised, main objectives of this initiative are to:
- improve access and help everyone to move around easily, safely and enjoyably by train, bus, boat, bicycle, on horseback, by wheelchair and on foot
- encourage people to choose to leave their private vehicles behind in favour of these more sustainable modes of travel
- ensure that traffic is tailored to the character of the roads, not the other way round
- make sure that traffic management, cycling, pedestrian, equestrian or limited mobility measures reflect the special qualities of the National Park.

a Road hierarchy It was felt that the current roads classification did not reflect the ability to accommodate traffic and gave the wrong impression to motorists. It has been proposed that all the roads within the Park, including feeder roads, be reclassified according to their type and strategic importance. A six-layer hierarchy has been proposed (Figure 10.40) with design and traffic control standards (e.g. speed and weight limits) based on the volume and type of traffic and the destinations served. A 40 mph (65 km/hr) restriction is proposed for all

but the trunk roads, and many minor roads would be downgraded to cater for the needs and safety of walkers, cyclists and horse-riders using roads.

b Integration of rail, bus and lake steamer transport

c Cycle routes (Figure 10.41)

Between 1996 and 2006, the number of visitors travelling to the Lake District fell from 90 per cent to 81 per cent and those travelling around the area from 86 per cent to 73 per cent.

Figure 10.39
Repairing a wall

Figure 10.40
Proposed road hierarchy

- **Trunk roads** – intended principally to carry long-distance/through traffic.
- **County strategic roads** – linking the main population and activity centres to the motorway and trunk road network.
- **Local distributor roads** – carrying traffic between the main centres and linking all other centres of activity. Local distributors would be subdivided into grade 1, grade 2 or grade 3, depending on their function, and with regard to their characteristics (particularly width and visibility standards) and the amount of traffic they would be expected to carry.
- **Local access roads** – remainder of the network, which would be relatively traffic-free routes for walkers, cyclists and horse riders.

Town and village streets, other than the main through routes, are omitted from these designations.

Figure 10.41
Cycle routes

Cycle routes have become increasingly important for both utility and recreational journeys. Most of the existing network still uses roads, including minor roads, where it is possible, safe and practical. There is also a growing number of purpose-made cycle tracks. The strategic cycle routes in the National Park include a section of the coast-to-coast route (C2C linking the Irish Sea and the North Sea), a link between Kendal and Keswick and the Cumbria Cycle Way which, together, form part of the National Cycle Network (NCN).

Tourism

Key Words and Terms

a You should know the meaning of the following terms:
- tourism • globalisation • mass tourism
- National Parks • honeypots • multiplier effect
- ecotourism • green tourism • beach village
- safaris • game parks/reserves • second homes.

b You should know the difference between:
- short-break and long-haul holidays
- mass tourism and sustainable tourism
- coastal resorts and mountain resorts.

Key Ideas

You should know and understand the following:
- Why tourism is the world's fastest-growing industry.
- That the growth of tourism is partly the result of globalisation.
- Recent trends in tourism.
- Recent changes in the patterns of tourism.
- That tourists are attracted by contrasting scenic areas and purpose-built amenities.
- The location and aims of National Parks in England and Wales.
- That there are conflicts between different land users in National Parks.
- Alternative strategies relating to management and conservation in National Parks.
- The social, economic and environmental impacts of tourism.
- That tourism can benefit and harm local communities, different groups of people and local environments, whether they are in a National Park, a coastal area, a mountainous area, a wildlife/game reserve or a beach village.
- That tourism is seen by several LEDCs as the sole way of improving their standard of living yet at the same time it creates problems.
- The need for conservation of scenic areas and fragile ecosystems through ecotourism and sustainable development.

Skills, Theories and Models

- Interpret and use graphs
- Analyse and use statistical data
- Identify the advantages and disadvantages of tourism
- Understand how development can cause conflict
- Identify ways to plan and manage tourist environments
- Understand the concept of sustainable tourism
- Understand the causes and effects of change
- Appreciate other people's values and attitudes
- Recognise the distinctive character of places

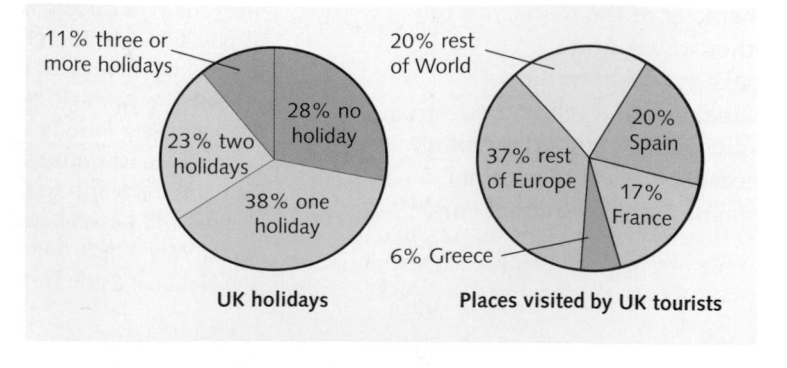

1 (Pages 160 and 161)

a How has tourism been affected by changes in each of the following?
- Affluence (wealth)
- Mobility
- Transport facilities
- Retirement age
- Leisure time
- Advertising and TV programmes. (6)

b Using Figure 10.6, describe how each of the following has affected the development of tourism. For each factor give one example of a place in the UK and one outside the UK.
- Sandy beaches
- Mountains
- The weather
- Accommodation
- Purpose-built amenities (10)

2 (Pages 162 and 163)

a
i) Which UK conurbation is not near to a National Park? (1)
ii) Which National Park is surrounded by the most motorways and conurbations? (1)
iii) Which two National Parks have been least affected by the construction of motorways? (2)

b Many more people now visit National Parks than in 1950. Give four reasons (other than the building of motorways) for the increase in numbers. (4)

c The pie graph shows different land owners in the National Parks. Match up each segment of the graph with its correct land owner. (6)

d National Parks contain a variety of scenery which in turn provides a wide range of recreational activities. Explain this on a completed copy of the two tables below. (10)

e Visitors to a National Park may come into conflict with both local residents and other visitors. Suggest two conflicts that might occur between each of:
- visitors and local residents
- different groups of visitors. (2 × 2)

f
i) Give three problems that may result from the increase in number of visitors to a honeypot. (3)
ii) Give three ways by which pressure on a honeypot may be reduced. (3)
iii) Name two famous buildings and two places of natural beauty in the UK that have become honeypots. (4)

a Scenery	National Park
Coasts	Pembroke
Forests	
Lakes	
Limestone	
Moors	
Mountains	

b Activity	National Park
Archaeology	Northumberland
Bird-watching	
Caving	
Climbing	
Gliding	
Riding	

3 (Pages 166 and 167)

a Courmayeur, in the Italian Alps, has become increasingly dependent on tourism.
i) Describe its:
- natural scenery
- climate
- traditional way of life. (6)
ii) On a larger copy of the table (right), list ways in which tourism has either harmed or benefited the local economy, the local community and the natural environment. (6)

b There is a proposal to build a new ski resort in a village in the Alps. Explain why some villagers may be:
- in favour of the proposal
- alarmed by the proposal. (2)

	Harmed	Benefited
Local economy		
Local community		
Natural environment		

4 (Pages 164 and 165)

Study the map below showing the number of tourists visiting Spain.

a i) Name the area marked A. (1)
　 ii) Name the sea area marked B. (1)
　 iii) From which country do most tourists come? (1)
　 iv) From which of the named countries do fewest visitors come? (1)

b i) Why do many people from the UK visit Spain? (3)
　 ii) Why do relatively few people from Portugal and Italy visit Spain? (3)

c Look back at Figure 10.15.
　 i) How many hours of sunshine do Malaga and London get on an average day in:
　 • July • January? (4)
　 ii) Apart from hours of sunshine, give two other differences in the weather between Malaga and London in summer. (2)

d Study the landsketch below which shows part of the southern coast of Spain.
　 i) Apart from the weather, what other physical attractions does this coast possess? (3)
　 ii) Name five amenities added for the benefit of tourists. (5)
　 iii) It is proposed to build a new hotel and leisure complex at location A on the sketch. Why might this development be:
　 • opposed by conservationists and some local residents
　 • welcomed by many local residents? (4)

e i) How did the growth of tourism between 1960 and 2000 affect:
　 • the local environment • the local community? (6)
　 ii) What problems are facing the tourist industry in this part of Spain today? (3)

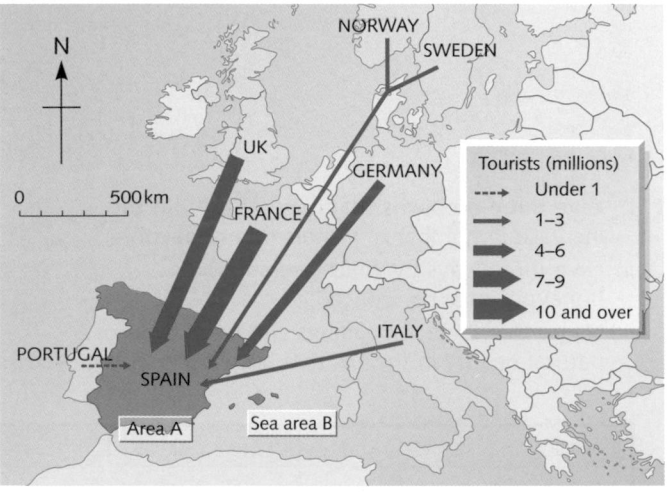

5 (Pages 168 and 171)

a Refer to Figure 10.23 on page 168.
　 i) Which less economically developed country obtains most of its income (GDP) from tourism? (1)
　 ii) Make a larger copy of, and then complete, the table by naming two coastal resorts, two cultural/historic centres and two areas of natural beauty in each of South America, Africa and Asia. (3 × 6)
　 iii) Why are an increasing number of tourists from Europe and North America visiting LEDCs? (3)

b For a named LEDC (except Kenya) or beach village that you have studied:
　 i) Describe the advantages that tourism has brought. (3)
　 ii) Describe the problems that tourism has created. (3)

c i) Give three natural advantages that attract tourists to Kenya. (3)
　 ii) How has the growth of safari holidays in Kenya affected:
　 • the landscape • wildlife • local people? (6)

d i) Give three advantages of ecotourism. (3)
　 ii) Why can ecotourism still create problems? (2)

		Coastal resort	Area of natural beauty	Historical/cultural centre
South America	1			
	2			
Africa	1			
	2			
Asia	1			
	2			

6 (Page 169)

a Using the climate graph below of Montego Bay, Jamaica (West Indies):
  i) What is the maximum temperature? *(1)*
  ii) What is the annual range of temperature? *(1)*
  iii) Which month has both the lowest rainfall and most hours of sunshine? *(1)*
  iv) Which month has both the highest rainfall and the fewest hours of sunshine? *(1)*

b Draw a star diagram to show the main features of climate and scenery in the West Indies. *(6)*

c Study the map (right).
  i) At which two airports in Jamaica do tourists arrive? *(2)*
  ii) Do most visitors to Jamaica stay in hotels or resort cottages? *(1)*
  iii) Name three physical attractions of the area around Montego Bay. *(3)*

d It is proposed to build a beach village between Montego Bay and Negril.
  i) What is a beach village? *(1)*
  ii) How may the building of this beach village benefit and harm the natural environment, the local economy, and the local community? *(6)*
  iii) Who else might benefit from the building of the beach village? Give a reason for your answer. *(2)*

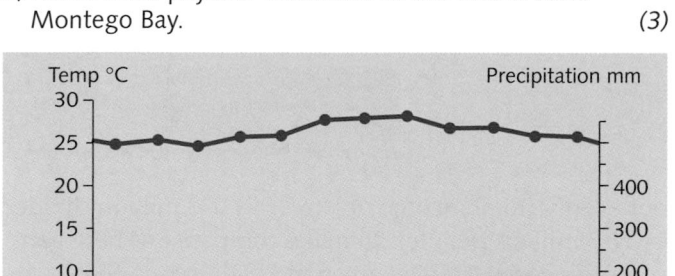

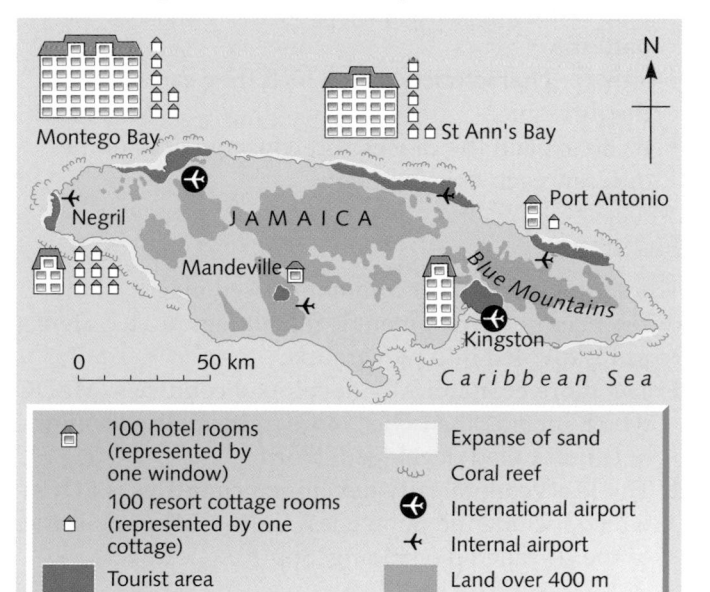

7 (Pages 172 to 175)

a With reference to the map of the Lake District National Park, name the following:
  i) Three of the lakes numbered 1 to 5. *(3)*
  ii) Three of the towns numbered 6 to 10. *(3)*
  iii) Three mountains lettered A to C. *(3)*

b The latest Lake District Plan identifies several special qualities of the National Park. Describe these under the following headings:
  • Landscape (scenery) • Vegetation and wildlife (ecosystems)
  • Human settlement and culture. *(9)*

c What, according to Figure 10.36a, is the:
  i) resident population
  ii) percentage of dwellings used as second homes
  iii) major type of land ownership
  iv) two most important types of employment? *(5)*

d According to Figure 10.36b:
  i) What is the difference between a day visitor and a holiday visitor? *(1)*
  ii) What were the two main purposes for:
    • day visitors • holiday visitors? *(2)*
  iii) Did most tourists visit small towns or valleys? *(1)*

e For *three* of the following, describe how visitors to the Lake District have added to problems of:
  • traffic congestion • footpath erosion • second homes
  • honeypots • conflict between different groups of people. *(9)*

f i) What are the main objectives of the Lake District Traffic Management Initiative? *(3)*
  ii) Describe why a new road hierarchy has been proposed and explain how it would work. *(5)*
  iii) Describe the increasing importance of cycle routes. *(3)*

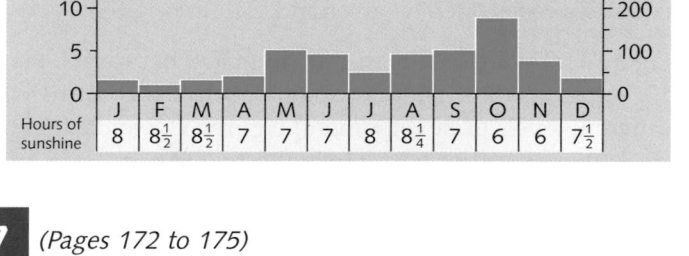

179

World development and interdependence

Patterns and characteristics of development

Geographers are interested in differences in levels of development and rates of growth either between countries or within a country. They can:

- map these differences to see if there are recognisable **patterns**
- look for **characteristics** by which they can measure the differences
- try to explain the **causes** and **consequences** of the differences in development.

Patterns

The simplest and most commonly used map to show differences in development is that of Figure 11.1. Here, the world is divided into two groups:

1 The **more economically developed countries** (**MEDCs**) which include the richer, more industrialised countries of the so-called developed 'North'.
2 The **less economically developed countries** (**LEDCs**) which include the poorer, less industrialised countries of the so-called developing 'South'.

Characteristics

Geographers have suggested several methods by which they can measure development.

1 Economic wealth

This book has, so far, referred to just one measure of development, i.e. economic development. This is because the traditional, and arguably the easiest, way in which to compare places is to measure their wealth. The wealth of a country is measured by its **gross domestic product per capita**, i.e. its **GDP per person**. The GDP per person is the total value of goods and services produced by a country in a year, divided by the total number of people living in that country. In a recent year in the UK, for example, every person living there would have received US$18 800 had the UK's wealth been shared

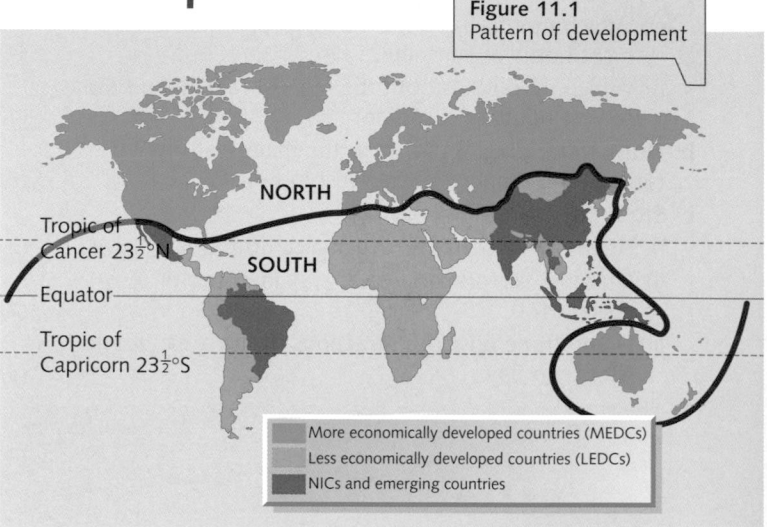

Figure 11.1
Pattern of development

out evenly (total income of US$1 094 000 million divided by 58.3 million people). To make comparisons between countries easier, GDP is given in US dollars (US$).

Figure 11.2 is a world map showing GDP per capita. Figure 11.3 includes the GDP per capita for selected countries at different levels of development. Figures 11.2 and 11.3 both show that countries with the highest GDP per capita are those said to be the most economically developed. Indeed, to many people living in a Western, industrialised country such as the UK, economic development tends to be solely associated with a growth in wealth. The disadvantage of GDP, however, is that it does not show differences in wealth between people and places in a country.

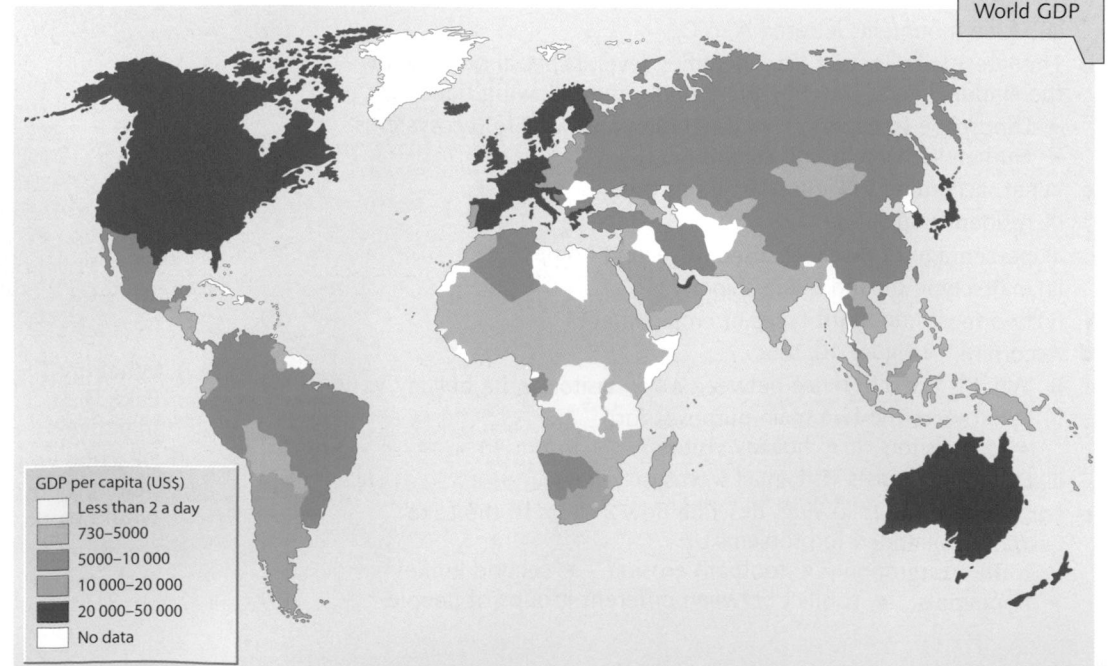

Figure 11.2
World GDP

GDP per capita (US$)

- Less than 2 a day
- 730–5000
- 5000–10 000
- 10 000–20 000
- 20 000–50 000
- No data

Figure 11.3 Indicators (measures) of development, 2008	Country	Economic wealth	Social						Others			
			Population			Health			Literacy	Employment	Urbanisation	Energy consumption
		GDP per capita (US$)	Birth rate	Death rate	Total fertility rate	Infant mortality rate	Life expectancy	% below poverty line	% adult	% in agriculture	% in urban areas	Tonnes oil equiv./person/yr
High income (MEDCs)	USA	45 800	14	8	2.1	6	78	12	99	2	82	8.1
	UK	35 100	11	10	1.7	5	79	11	99	1	90	3.9
	Japan	33 600	8	9	1.2	3	82	6	99	2	66	4.1
	Italy	30 400	8	11	1.3	6	80	13	98	2	68	3.1
Middle income (NICs)	Malaysia	13 300	22	5	2.9	16	73	15	89	13	58	2.8
	Brazil	9 700	19	6	2.2	23	72	31	88	20	86	1.1
Emerging countries	China	5 300	14	7	1.8	20	74	8	91	42	43	1.3
	India	2 700	22	7	2.7	30	70	25	61	18	29	0.6
Low income (LEDCs)	Kenya	1 700	38	10	4.6	55	58	51	85	75	34	0.5
	Bangladesh	1 300	29	8	2.8	59	60	45	43	63	27	0.3
	Ethiopia	800	45	12	6.2	81	54	39	43	80	17	0.1

2 Social indicators

Although, as we have seen, economic development to people living in a Western society often means a growth in wealth, other indicators have also been suggested. Figures 11.3 and 11.4 show possible links between development and a range of social measures.

a Population

In general, the MEDCs have lower birth rates and a slower natural increase than do LEDCs. Population structures (pages 8 and 9) show that the MEDCs have a smaller proportion of children aged under 15 (page 12) and a higher proportion of people aged over 65 (page 14) than do developing countries.

b Health

Similarly, MEDCs have a lower infant mortality rate, a longer life expectancy and fewer people per doctor than the LEDCs.

3 Other indicators

These include measures related to adult literacy, diet, employment structures and energy consumption (Figure 11.3). Notice, however, that many of these criteria are themselves related to the wealth of a country. For example, the more wealthy and, therefore, economically developed a country is, the more it can spend on health care, education, energy provision and providing other services. This suggests that a country has to increase its GDP if it is to improve the standard of living and quality of life of its inhabitants.

Indicators/measures		MEDCs	LEDCs
Economic (GDP)		Majority over US$5000 per capita. 80% of the world's income.	Majority under US$2000 per capita. 20% of the world's total income.
Social	Population	Low, steady birth rate partly due to family planning. Slow natural increase – takes 100 years to double. 18% of world's population.	High birth rate, partly due to limited family planning, but falling. Relatively high death rate but falling except in sub-Saharan Africa. Rapid natural increase – doubling population in 30 years. 82% of world's population.
	Health	Low infant mortality rate, long and increasing life expectancy. Mainly due to vaccines, large number of doctors and hospitals, and high standard of health care.	High, though falling, infant mortality rate. Shorter life expectancy, especially in poorest countries. Relatively few doctors and hospitals. Little money available.
Others	Education	Majority have full-time education, many have 16+ education. Most adults, including women, are literate.	Limited numbers get full-time education, and few go on after 16. Low adult literacy rate, and women are disadvantaged.
	Diet	Balanced. Several meals per day. High protein intake.	Unbalanced. 35% of children under 5 are underfed. Low protein intake.
	Employment	Few jobs in primary sector, more in secondary, most in tertiary sector. Mechanisation and technology.	Most in primary sector, few in secondary and tertiary sectors. Limited mechanisation and technology.
	Energy	High levels of consumption – mainly coal, oil, natural gas and nuclear power. Use 70% of world's energy.	Low levels of consumption – often only fuelwood. Only use 30% of world's energy.
	Trade	Large volume and value – mainly manufactured goods.	Small volume and value – mainly unprocessed raw materials.

Figure 11.4
Differences between MEDCs and LEDCs

Inequalities in world development

Human Development Index (HDI)

It is now widely accepted that the term **development** means more than just wealth. In 1990, the United Nations replaced GDP as their measure of development with the **Human Development Index** (**HDI**). Since then the UN has published an annual report in which it ranks countries according to the **quality of life** of its inhabitants rather than using traditional economic figures. The HDI is a social welfare index measuring three variables: life expectancy (health), adult literacy (education), and real GDP per capita (standard of living).

- **Life expectancy** is regarded by the UN as the best measure of the health and safety of a country.
- **Education attainment** is obtained by combining adult literacy rates and the primary, secondary and tertiary student enrolment rate.
- **Real GDP per capita** is the GDP adjusted to the actual purchasing power (PPP in US$), i.e. what the average income will actually buy in a country.

Each variable is given a score ranging from 1.000 (the best) to 0.000 (the poorest). Figure 11.5, apart from the final HDI score, gives the data for a country before it has been converted (this should be more meaningful to you). The HDI, which is the average of the three scores, in 2008 put Iceland and Norway top with a score of 0.968 and Sierra Leone bottom with a score of 0.336. Even so, the world map showing HDI (Figure 11.6) shows that:

- countries with scores exceeding 0.900 correspond very closely with the MEDCs in the 'North' (Figure 11.1) which have a high GDP per capita (Figure 11.2)
- countries with scores below 0.900 equate closely with the LEDCs in the 'South' with their lower GDPs
- with the exception of Afghanistan, for which there was no data, the 24 countries with a score of below 0.510 were all in sub-Saharan Africa.

Figure 11.5
Human Development Index, 2008

	HDI rank order	Country	Life expectancy at birth (years)	Adult literacy rate (%)	School enrolment rate[1] (%)	Real GDP per capita (PPP US$)	HDI
Top 4	1	Iceland	80.7	99.0	104	39 900	0.968
	2	Norway	79.9	99.0	97	55 200	0.968
	3	Australia	81.6	99.0	116	38 100	0.962
	4	Canada	81.2	99.0	97	39 300	0.961
Selected countries	8	Japan	82.1	99.0	82	34 200	0.953
	10	France	80.9	99.0	94	32 700	0.952
	12	USA	78.1	99.0	95	47 000	0.951
	13	Spain	80.0	97.9	95	34 600	0.949
	16	UK	79.0	99.0	106	36 600	0.946
	20	Italy	80.2	98.4	84	31 000	0.941
	52	Mexico	76.1	91.0	71	14 200	0.829
	63	Malaysia	73.3	88.7	66	15 300	0.811
	70	Brazil	72.0	88.6	80	10 100	0.801
	81	China	73.5	90.9	73	6 000	0.777
	128	India	69.9	61.0	56	2 800	0.619
	140	Bangladesh	60.2	43.1	37	1 500	0.547
	148	Kenya	57.8	85.1	51	1 600	0.521
	169	Ethiopia	55.4	42.7	27	800	0.406
Bottom 3	175	Niger	52.6	28.7	16	700	0.374
	176	Burkina Faso	52.9	21.8	14	1 200	0.370
	177	Sierra Leone	41.2	35.1	27	700	0.336

[1]The gross enrolment ratio is the number of students enrolled in a level of education, regardless of age, as a percentage of the population of official school age for that level.

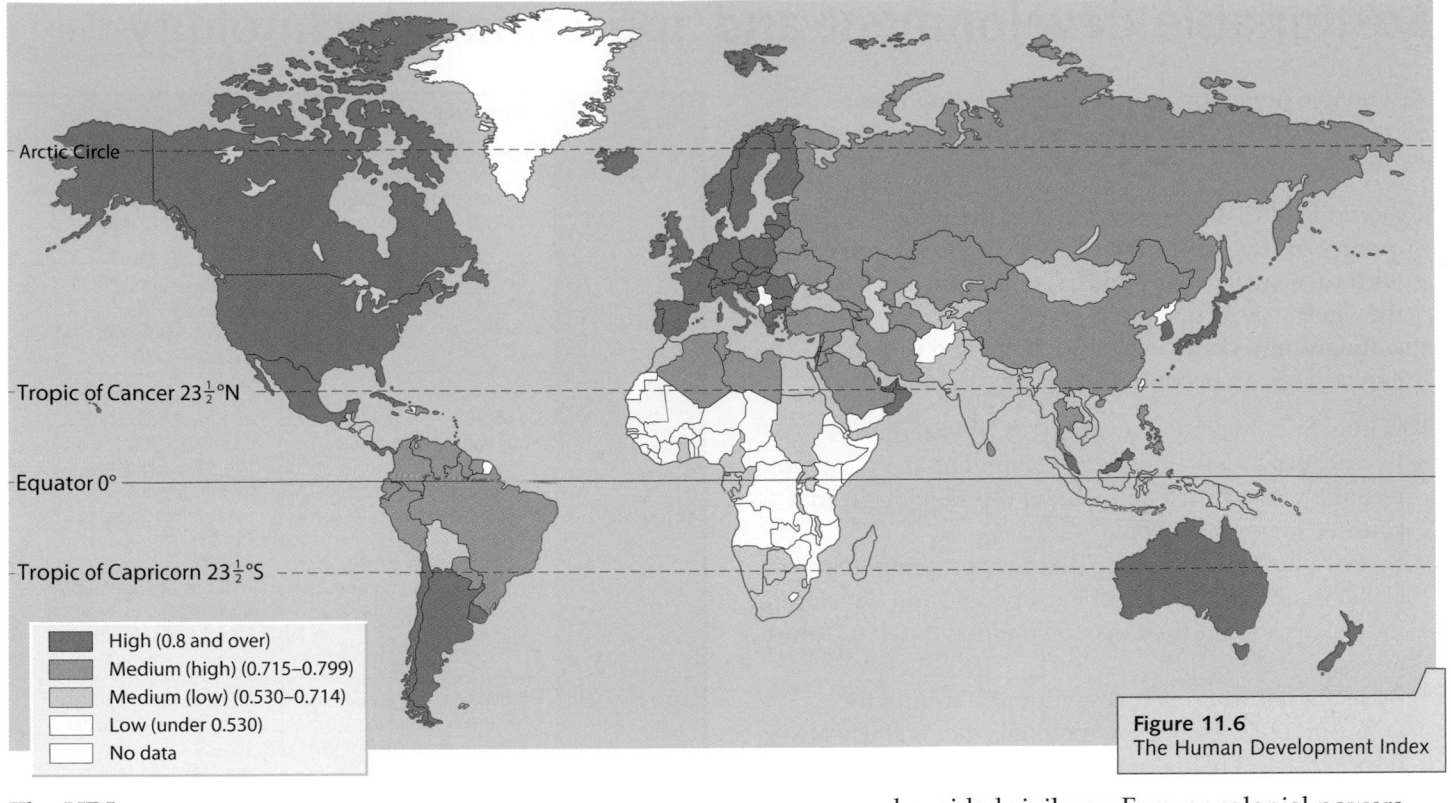

Figure 11.6
The Human Development Index

Legend:
- High (0.8 and over)
- Medium (high) (0.715–0.799)
- Medium (low) (0.530–0.714)
- Low (under 0.530)
- No data

Arctic Circle
Tropic of Cancer 23½°N
Equator 0°
Tropic of Capricorn 23½°S

The HDI can:

- measure differences within a country (unlike GDP) as well as between countries
- highlight where poverty is worst, both within a country and between countries
- act as a measure to show how far a country has developed and whether there are improvements in its level and rate of development
- help a country to set targets that can lead to improvements in the quality of life of its citizens, e.g. better health care to improve life expectancy and better education to improve literacy rates.

Causes of inequality

There are numerous factors that may, individually or collectively, lead to inequalities in development. For example:

Economic Countries with mineral deposits (e.g. iron ore) and energy resources (e.g. coal) were more likely to have been the first to develop industrially. As they accumulated wealth through the sale of manufactured goods, they were able to buy resources either because their own became exhausted or because they did not possess them. They were able to provide numerous jobs, improve transport systems and develop new technologies as well as imposing trade restrictions on less developed countries.

Social The more a country develops, the more money it has to spend on education, health and better-quality housing.

Political Countries were likely to develop more quickly if they had a government that was stable, invested in the economy and avoided civil war. Former colonial powers developed at the expense of their colonies.

Environmental Countries were also likely to develop more quickly if they were unaffected by frequent and severe natural hazards (e.g. drought, flooding, earthquakes, volcanic eruptions and tropical storms), disease and illness (malaria, AIDS), pests (locusts), and if they were able to provide a reliable supply of clean water (page 225).

Consequences of inequality

Inequalities and imbalances in development have led to differences in standards of living and quality of life. As summarised in Figure 11.4, the LEDCs, when compared with the MEDCs, usually have:

- higher birth rates and infant mortality rates, a greater natural increase in population, a higher proportion of their population aged under 15, and a shorter life expectancy (pages 9–15)
- poorer education facilities (lower levels of literacy, less time spent at school), poorer health care (fewer hospitals and doctors per size of population), a poorer diet and a poorer provision of services (less energy/electricity available, unreliable or polluted water supplies, limited transport provision)
- more jobs in the primary and informal sectors and fewer in the secondary, tertiary and formal sectors (pages 94 and 148)
- a smaller volume of trade, a likely trade deficit and a need to seek aid (pages 186–91)
- less purchasing power per capita (page 182).

Sustainable development and appropriate technology

Sustainable development should, according to the UN:
'... meet the needs of the present without compromising the ability of future generations to meet their own needs.'

It should lead to an improvement in people's:
- **quality of life** – allowing them to become more content with their way of life and the environment in which they live
- **standard of living** – enabling them, and future generations, to become better off economically.

This may be achieved in a variety of ways (Figure 11.7):
- By encouraging economic development at a pace that a country can afford and manage so as to avoid that country falling into debt.
- By developing technology that is appropriate to the skills, wealth and needs of local people and developing local skills so that they may be handed down to future generations.
- By using natural resources without harming the environment, developing materials that will use fewer resources, and using materials that will last for longer. When using resources, remember the three Rs: renew, recycle or replace.

Sustainable development needs careful planning and, as it involves a commitment to conservation, the co-operation of different groups of countries adopting global agreements. Figure 11.8 lists some of the contrasts between sustainable and non-sustainable development referred to in this book.

	Sustainable development	Non-sustainable development
Socio-economic	• Family planning, steady replacement rate (pages 12–13)	• High birth rate, rapid natural increase, ageing population (pages 12, 14–15)
	• Building materials (pages 83–151) • Controlled urban growth	• Rapid urbanisation (pages 84–86) • Loss of countryside (pages 47, 62)
People and resources	• Renewable energy (pages 118–19) • Soil conservation (page 255) • Re-afforestation (page 240) • Clean water supply (page 225) • Organic farming (page 109) • Recycling materials (Figure 9.31)	• Continued use of minerals and fossil fuels (pages 118–20) • Soil erosion due to overgrazing and overcultivation (page 254) • Deforestation (pages 236–7) • Pollution of water supplies (page 225) • Heavy use of fertiliser in farming (page 108)
People and the environment	• Ecotourism (page 168) • National Parks and game reserves (pages 162, 170–5) • Protecting scenery and wildlife habitats	• Mass tourism, especially at coastal resorts (pages 164–5) and mountain resorts (pages 166–7)

Figure 11.8
Differences between sustainable and non-sustainable development

Figure 11.7
Appropriate technology – a village pump in India

Encouraging economic development at a pace a country can afford so as to prevent it falling into debt.

Developing materials that use fewer resources and will last for longer.

Using tools designed to take advantage of local knowledge and resources.

Developing projects that are in harmony with the environment.

Encouraging technology that uses existing skills and techniques of local people and which can be handed down to future generations.

Developing local crafts and industries by using local natural resources and, where possible, recycling materials.

Introducing labour-intensive projects as, with so many people already likely to be seeking work, it is no use replacing existing workers by machines.

Adopting low-cost schemes and technologies that people can afford and can manage.

Figure 11.9 describes some of the ways in which appropriate technology has been adopted to help the inhabitants of Ladakh, and pages 150–1 show how it helps people living in Nepal and Kenya.

Sustainable development in MEDCs

It is often mistakenly believed that the concept of sustainable development and the use of appropriate technology is not applicable to MEDCs. This is not the case, as it is the MEDCs that, at present:

- consume most of the world's non-renewable resources, such as fossil fuels, minerals and timber
- are the greatest contributors to global pollution, for example acid rain and the release of greenhouse gases.

An appropriate technology can contribute to a more sustainable way of life for people who are rich or poor, living in places that are considered to be either more developed or less developed. The only difference is that for those living in MEDCs, the appropriate technology is likely to involve more capital (money) and to be high-tech.

Ladakh (northern India)

Ladakh, in the foothills of the Himalayas, is very mountainous. It also has an extreme climate with, on average, temperatures for over half the year below or near to 0°C, a growing season of less than 5 months, an annual precipitation of under 20 cm (making it a desert) most of which falls as snow, 325 days of sunshine, and strong winds sweeping down its valleys. People in Ladakh do not want development based on the Western model with, as

they see it, its resultant socio-cultural, urbanisation and environmental problems. Instead they are trying to 'promote a development policy based on ecological principles, with an emphasis on local conditions, resources and traditions. This should help protect our environment and preserve the foundations of our traditional culture' (Figure 11.9).

Leh, the capital, is 3500 m above sea-level. Like other settlements it is next to a river.

Water from summer snowmelt can be used for:
- irrigation
- micro-hydro power.

Strong winds and the sun are important renewable sources of energy.

Houses
- Solar power heats water and houses.
- Longest wall faces south to get most sunshine.
- Solar cookers and new locally-made metal braziers (compare Figure 9.38) avoid use of precious dung and fuelwood, and improve health as less smoke.
- Cavity walls filled with mud and stones to conserve heat.

Solar greenhouses allow vegetables to grow throughout the year.

Small, locally made hydraulic rams can pump water to a higher level, extending the area of cultivation and allowing Ladakh to remain self-sufficient in food.

The ecological development centre trains people in traditional handicrafts.

Small micro-hydro schemes generate electricity for lighting and for grinding corn (see Figure 9.34).

Figure 11.9
Features of Ladakh

World trade and interdependence

No country is self-sufficient in the full range of raw materials (food, minerals and energy) and manufactured goods that are needed by its inhabitants. To try to achieve this, countries must **trade** with one another. Trade is the flow of commodities from producers to consumers, and it is important in the development of a country. Countries that trade with other countries are said to be **interdependent**.

Raw materials, goods and services bought by a country are called **imports**, and those sold by a country are **exports**. The difference between a country's imports and exports is known as its **trade balance**. One way for a country to improve its standard of living and to grow more wealthy is to sell more goods than it buys. Unfortunately, if several countries export more than they import, then other countries will have to import more than they export. The result is that some countries will have a **trade surplus**, allowing them to become richer, while others will have a **trade deficit**, making them poorer and likely to fall into debt.

Patterns of world trade

There is a wide imbalance of trade between the LEDCs (the less economically developed countries) and the MEDCs (the more economically developed countries). This is mainly because:

- the LEDCs provide **primary goods** such as foodstuffs and raw materials – primary goods are usually sold to the MEDCs at low and often fluctuating prices
- the MEDCs process primary goods, which they either possess themselves or obtain from LEDCs, into **secondary** (or **manufactured**) **goods** – secondary goods are sold at high and usually steady prices.

Although the prices of primary goods have increased over the years, the prices of secondary goods have increased far more rapidly. This means (Figure 11.10):

Trade of less economically developed countries	Trade of more economically developed countries
A legacy of former colonial economies where a mineral once mined, or a crop once grown is exported in its 'raw state'. Most exports are primary products.	Mainly manufactured goods are traded, as these countries have become industrialised. Cereals also exported.
Often only two or three items are exported.	A wide range of items are exported.
Prices of, and demand for, these products fluctuate annually. Prices rise less quickly than for manufactured goods.	Prices of, and demand for, these products tend to be steady. Prices have risen considerably in comparison with raw materials.
The total trade of these countries is small.	The total trade of these countries is large.
Most exports come from transnational companies which tend to send profits back to the parent company.	Profits are retained by the exporting country.
Trade is hindered by poor internal transport networks.	Trade is helped by good internal transport networks.
Trade is severely hit at times of world economic recession.	Trade is badly affected at times of world economic recession.

Figure 11.10
Differences between trade of LEDCs and MEDCs

- The MEDCs that export manufactured goods earn increasingly more than the LEDCs which have only primary goods to sell. The result is (with the exception of the NICs – Figure 11.11 and page 147) a widening trade gap between the MEDCs and the LEDCs.
- Although only 14 per cent of the world's population lived in the MEDCs in 2008 (page 11), those countries still contributed 58 per cent of the world's trade.

Figure 11.12 gives some of the socio-economic, environmental and political advantages and disadvantages of this pattern of world trade to LEDCs and MEDCs.

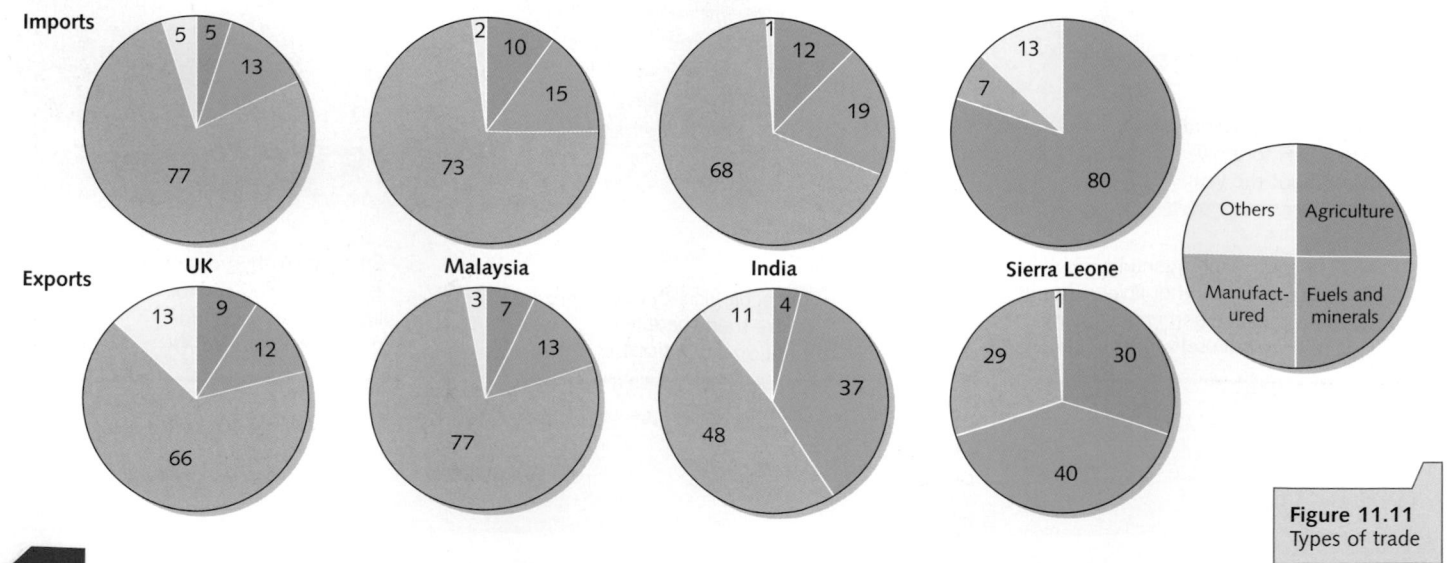

Figure 11.11
Types of trade

Figure 11.12
International trade

	Advantages	Disadvantages
Richer countries (MEDCs)		
Socio-economic	Cheap imports of foodstuffs and raw materials. Expensive exports of manufactured goods. Trade surplus.	Often obtainable from considerable distances: high transport costs.
Environmental	Limited mining and deforestation. Money to improve environment.	Manufacturing goods can create air, water, noise and visual pollution.
Political	Can exert pressure on LEDCs.	
Poorer countries (LEDCs)		
Socio-economic	Raw materials have a ready market in MEDCs. Source of work.	Limited range of exports. Trade deficit as imports cost more than exports.
Environmental		Problems created by mining, deforestation and overgrazing.
Political	May be able to obtain overseas aid.	Often tied to/dominated by MEDCs.

A further problem for many LEDCs, and especially those in Africa, is that they rely on just one or two major commodities for export (Figure 11.13). The price paid for these commodities is often fixed by the MEDCs. If there is a world recession, an overproduction of a crop or mineral, a change or a fall in the demand for a product, a crop failure, or the exhaustion of a mineral, then the economy of the producing country can be seriously affected. The MEDCs also try to depress the price of products that they have to buy from the LEDCs. They may also, in order to protect their own manufacturing industries, limit the import of processed goods. In many cases, it is large transnational corporations (page 144) that finance and organise the growing of crops, the mining of minerals and the processing of the raw materials. As the headquarters of these corporations are usually located in MEDCs, profits are therefore sent overseas and the LEDCs supplying the commodities do not get their fair share of the wealth.

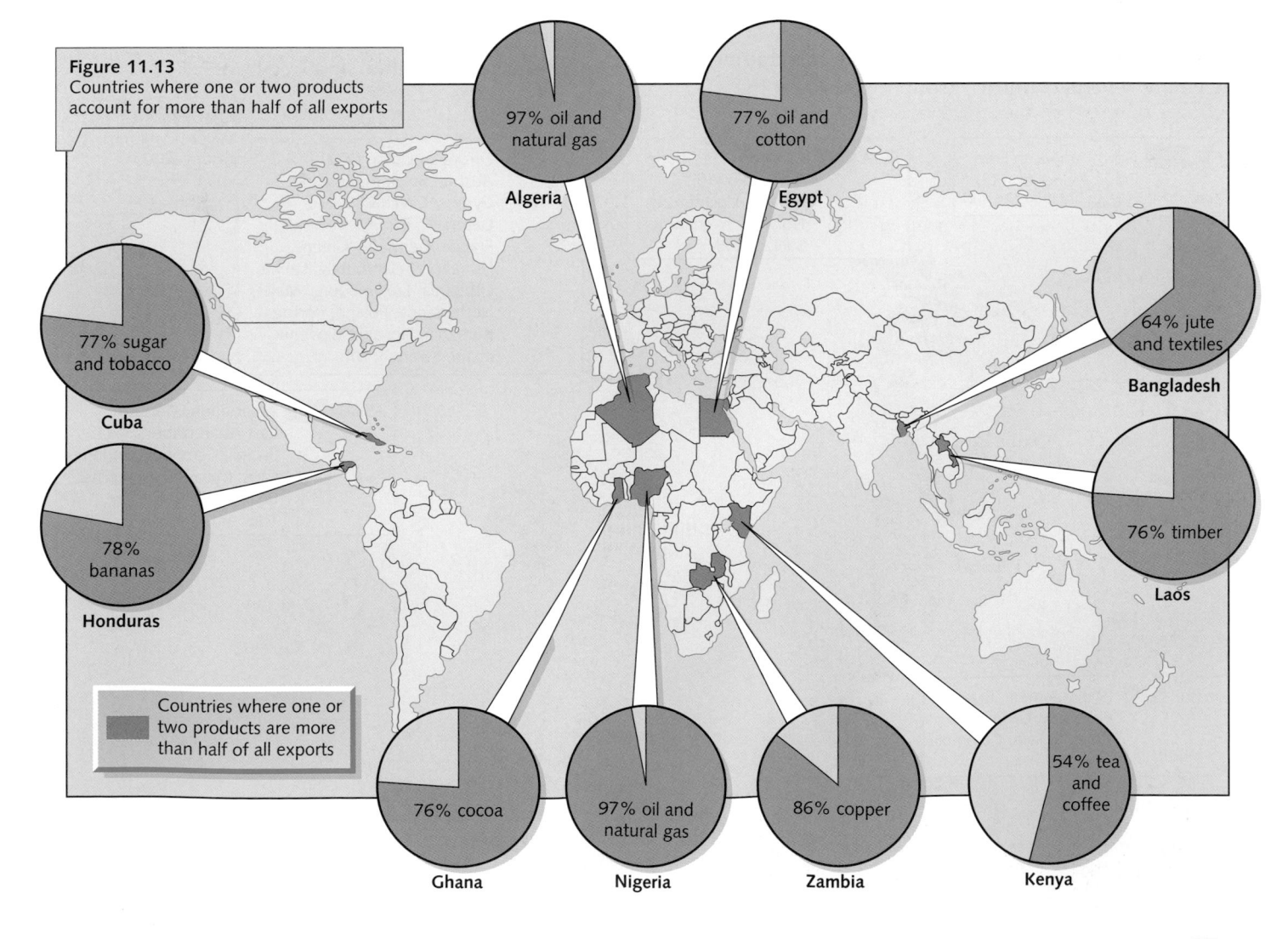

Figure 11.13
Countries where one or two products account for more than half of all exports

97% oil and natural gas — Algeria

77% oil and cotton — Egypt

77% sugar and tobacco — Cuba

64% jute and textiles — Bangladesh

78% bananas — Honduras

76% timber — Laos

76% cocoa — Ghana

97% oil and natural gas — Nigeria

86% copper — Zambia

54% tea and coffee — Kenya

Countries where one or two products are more than half of all exports

Free trade, tariffs and quotas

In an ideal world there should be **free trade** between all countries. Free trade is when governments neither restrict nor encourage trade. In the real world this rarely happens. Virtually all governments, and especially those of the MEDCs, have tried to regulate and control overseas trade. They have tried to achieve this by creating **trade barriers** which they hope will protect jobs and industries within their own country. There are two common ways of affecting the levels and patterns of international trade:

1 Tariffs are taxes or customs duties paid on imports. The exporter has to pay a percentage of the value of the goods to the importer. Importers may add tariffs just to raise money, but usually the idea is to increase the price of imported goods in order to make them more expensive, and less competitive, in comparison with home-produced goods. Tariffs, therefore, can either increase the cost of imports (helping the trade balance) or protect home-made products.

2 Quotas limit the amount of goods that can be imported. Quotas tend to be restricted to primary goods and so work against the LEDCs.

Trading groups

The European Union (EU) is one example of a **trading bloc** where several countries group together for the purpose of trying to increase the volume and value of their trade (Figure 11.14). The EU and NAFTA, each with about 500 million people in 2008, form the world's two largest internal markets. By eliminating customs duties (tariffs), the EU was able to reduce the cost of products sold between member countries. This meant, for example, that Germany could sell cars to the UK at a cheaper price than could Japan. Although this made the EU more competitive against its major rivals (the USA and Japan), it also created restrictions (trade barriers) which protected goods made in the EU from those produced in, and imported from, the LEDCs. This meant that developing countries found it even harder to sell their products in the world market, increasing the trade gap between themselves and the MEDCs.

As the number of EU members has grown (from the original six of 1957 to the present twenty-seven), so too has the size of its internal market. The larger the internal market, the greater the number of potential customers. This is even more important when, as in the EU, many of these potential customers are wealthy and can afford to buy numerous high-valued goods. The size of the EU, which has recently accepted new members from the former eastern Europe, and NAFTA, which is considering admitting countries from Latin America, poses the question: 'Do super-trading blocs increase or hinder the move towards globalisation and free trade?'

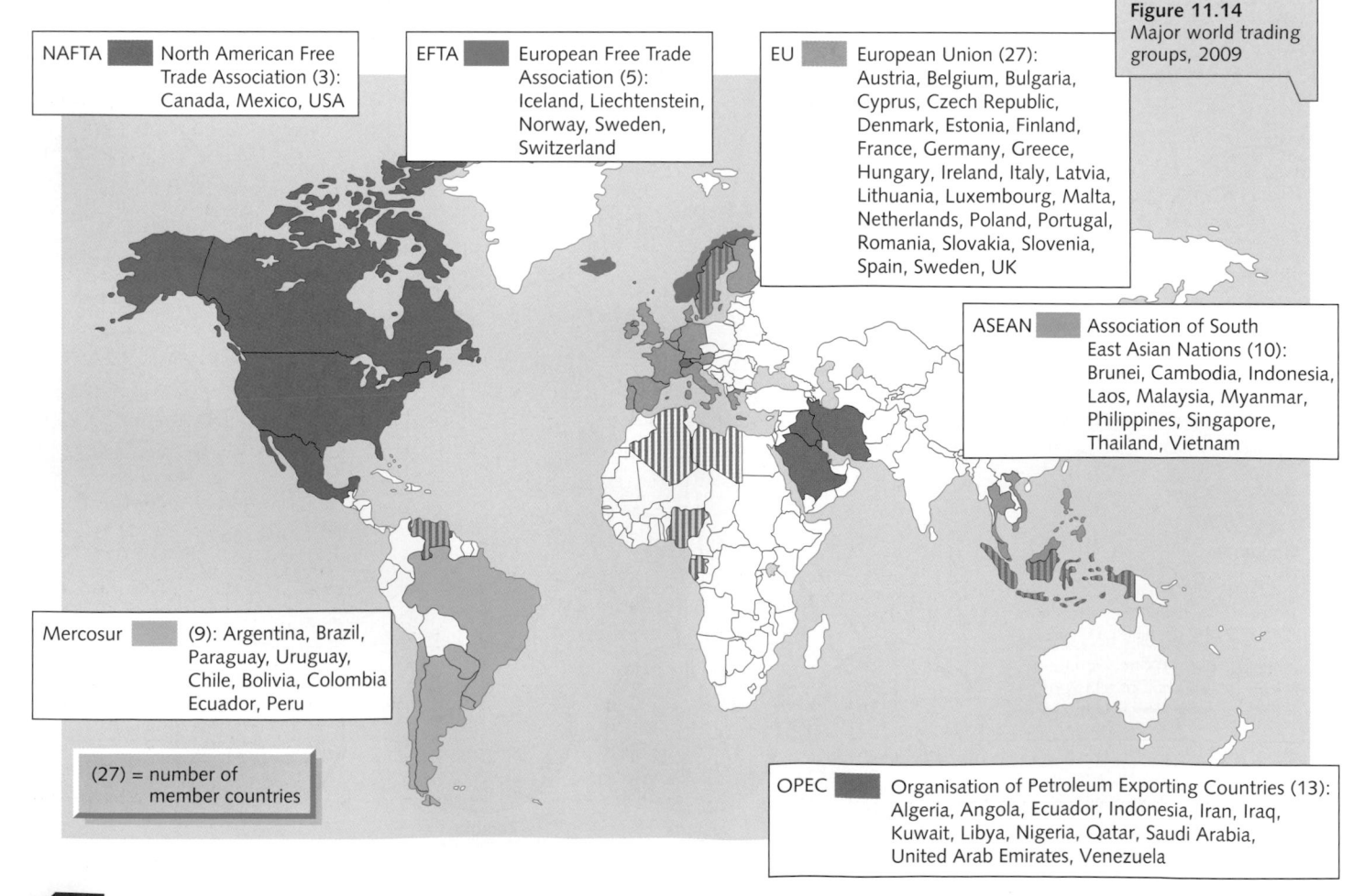

NAFTA North American Free Trade Association (3): Canada, Mexico, USA

EFTA European Free Trade Association (5): Iceland, Liechtenstein, Norway, Sweden, Switzerland

EU European Union (27): Austria, Belgium, Bulgaria, Cyprus, Czech Republic, Denmark, Estonia, Finland, France, Germany, Greece, Hungary, Ireland, Italy, Latvia, Lithuania, Luxembourg, Malta, Netherlands, Poland, Portugal, Romania, Slovakia, Slovenia, Spain, Sweden, UK

Figure 11.14 Major world trading groups, 2009

ASEAN Association of South East Asian Nations (10): Brunei, Cambodia, Indonesia, Laos, Malaysia, Myanmar, Philippines, Singapore, Thailand, Vietnam

Mercosur (9): Argentina, Brazil, Paraguay, Uruguay, Chile, Bolivia, Colombia Ecuador, Peru

(27) = number of member countries

OPEC Organisation of Petroleum Exporting Countries (13): Algeria, Angola, Ecuador, Indonesia, Iran, Iraq, Kuwait, Libya, Nigeria, Qatar, Saudi Arabia, United Arab Emirates, Venezuela

Direction of world trade

- Despite the growing number of trading groups, the world's trade is still dominated by MEDCs (Figure 11.15) in North America and the EU together with Japan – the so-called advanced market economies. However, their share of this trade fell from 72 per cent in 1990 to 68 per cent in 1998 and 58 per cent in 2008.
- Most countries tend to trade with one or more of their closest neighbours, e.g. Canada with the USA, South Korea with Japan and the UK with countries in north-west Europe.
- The advanced market economies trade mainly between themselves and have relatively little contact with developing countries. Where they do, they generally import low-value raw materials and export high-value manufactured goods.
- There is relatively little trade between the developing countries themselves, partly due to their low rates of economic development and partly their production of similar types of goods.
- Since the 1970s, the MEDCs have experienced increasing competition from the so-called newly industrialised countries (NICs, page 147) and, even more recently, from the emergence of China.
- Even so, world trade is no longer dominated by countries but rather by large and powerful transnational corporations (TNCs, page 144).

Trade – a measure of development

Figure 11.3 gives several indicators that are used to measure differences in levels of development between countries. Trade can also be added as a measure. Figure 11.16 shows that it is the MEDCs that have the highest trade per capita, followed by the NICs, the middle-income LEDCs and, finally, the lower-income LEDCs. Singapore is an anomaly (an exception to the rule) in that, although it is an NIC, it has the world's highest trade per capita.

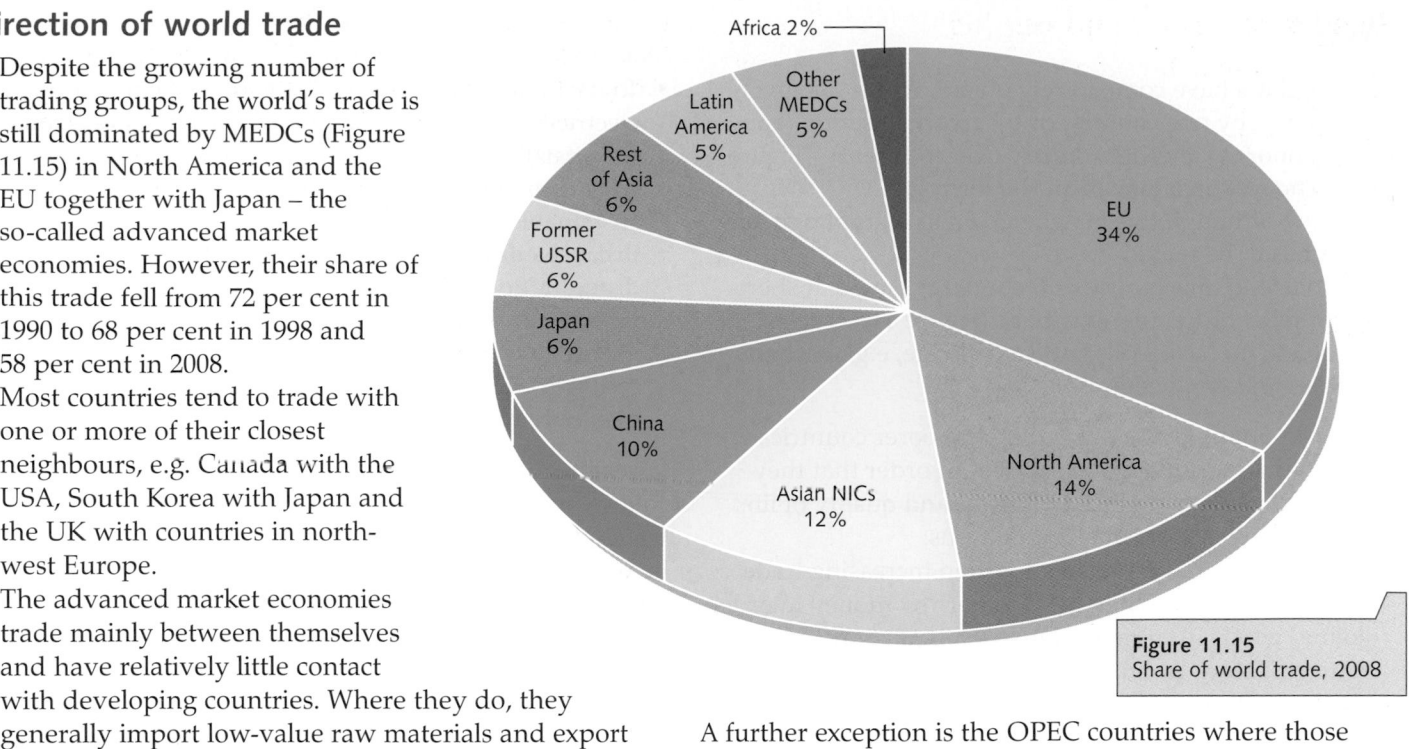

Figure 11.15
Share of world trade, 2008

A further exception is the OPEC countries where those located in the Middle East have an extremely high trade per capita, whereas it is often very low for countries located elsewhere.

Recent trends

The **WTO (World Trade Organization)** and, previously, **GATT (General Agreement on Trade and Tariffs)** have been trying for many years to get an international agreement that would stimulate world trade through the reduction of tariffs, quotas and other trade barriers set up by countries, usually the MEDCs, to protect their own products. In the UK, for example, we complain about cheap imports (e.g. textiles) that close factories in Britain, yet without the LEDCs being able to sell goods to the UK, how can they earn sufficient money to buy British goods?

During the 2000s, beginning with the slow-down in the North American economy, the trade of many MEDCs slowly declined whereas, in contrast, that of emerging countries and many of the LEDCs was barely affected. That was until 2008 when all economies were affected by the global recession.

Figure 11.16
Trade per capita (US$) for selected countries

MEDCs		NICs		Middle income		Emerging		OPEC		Low income	
Germany	26 400	Singapore	124 769	Chile	5 402	Brazil	1 234	UAE	39 288	Kenya	275
UK	21 389	Taiwan	19 399	Mexico	4 643	China	1 207	Qatar	38 276	Bangladesh	177
Australia	14 890	South Korea	17 608	Argentina	2 091	India	307	Kuwait	28 096	Nepal	141
USA	10 864	Malaysia	11 603					Angola	1 747	Sierra Leone	127
Japan	10 112	Thailand	4 053					Nigeria	447	Ethiopia	89

Aid

Many LEDCs have come to rely on aid. **Aid** is the giving of resources by one country, or by an organisation (known as the **donor**) to another country (the **recipient**) – Figure 11.18. The resource may be in the form of:

- money, although this may be given as a grant or a loan that has to be repaid
- goods, food, machinery or technology aimed at short-term relief or long-term benefit (Figure 11.17)
- people who have skills and knowledge, e.g. teachers, nurses and engineers.

The basic aim in giving aid is to help poorer countries develop their economy and services in order that they may improve their standard of living and quality of life. LEDCs require aid for different reasons:

- Because they have a large and often increasing trade deficit (page 186). They need to borrow money in order to buy goods from richer, industrialised countries. Unfortunately, by borrowing money, the LEDCs fall further into debt (Figure 11.18). This aid is often long-term.

- To try to improve their basic amenities (water supply, electricity) and infrastructure (transport, schools and hospitals).
- To encourage self-help schemes and to promote sustainable development.
- Because either they are prone to natural disasters (e.g. drought, flooding, tropical storms, earthquakes or volcanic eruptions) or they suffer as a result of human-induced disasters (e.g. desertification, page 256, and civil war). This aid is often needed in an emergency and may only be short-term.

In reality, the giving of aid is often complex and controversial as it does not always benefit the country to which it is given. Figure 11.17 describes five different ways by which aid may be given and gives the advantages and disadvantages of each to the recipient country. Figure 11.19 describes how the traditional and often unsatisfactory trade–aid cycle may be improved.

Figure 11.17 Advantages and disadvantages of aid

Type of aid	Definition	• Disadvantage • Advantage
Government (bilateral)	Given directly by a richer country (donor) to a poorer country (recipient) – often **tied** `with strings attached´	• `Tied´, meaning LEDC has to buy goods from the donor, e.g. arms, manufactured goods • Money often has to be spent on prestigious schemes such as dams (Figure 8.12) and international airports • Large schemes take up land belonging to local people • Aid often encourages corruption – money rarely reaches poorer people living in more remote areas • LEDC unable to repay money – gets further into debt • LEDC becomes increasingly dependent on donor country • Can provide grants for students to study in MEDCs
International organisations (multilateral)	Given by organisations such as the World Bank and the IMF (International Monetary Fund)	• Not meant to be tied, but less likely to be given to countries with unfavourable economic and political systems • Encourages farming and industry but products are sent to MEDCs rather than consumed in LEDCs • LEDCs become increasingly dependent on aid, and often fall increasingly into debt • Helps LEDCs to develop new crops, raw materials and industry (page 113)
Voluntary	Non-governmental organisations such as Oxfam and ActionAid which collect money and receive gifts for people in LEDCs	• Not tied • Deals with emergencies (page 272) • Encourages low-cost self-help schemes (pages 127, 151 and 185) • Money more likely to reach poorer people in more remote areas • Dependent on charity´s ability to collect money • Annual amounts uncertain – requires longer-term planning
Short-term/emergency	Needed to cope with the effects of environmental hazards such as earthquakes and tropical storms	• Immediate help – provides food, clothes, medical supplies and shelter • Goes to places and people most in need • Not tied, and less chance of corruption • Also helps refugees (Figure 2.5)
Long-term/sustainable	Organisations such as Practical Action which help people in LEDCs to support themselves	• Encourages development of local skills and use of local raw materials (page 151) • Trains local people to be teachers, nurses, health workers • Helps equip schools and development of local agriculture and small-scale industry • LEDCs do not fall into debt

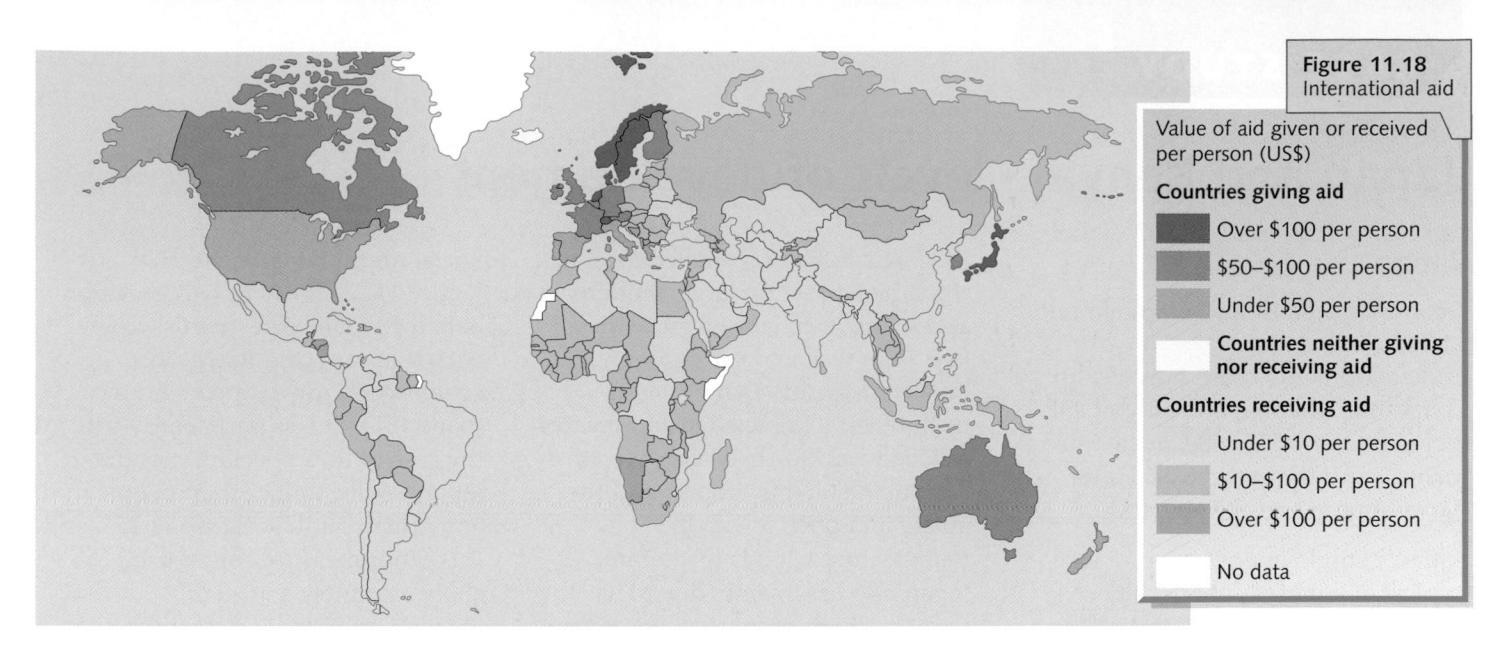

Value of aid given or received per person (US$)

Countries giving aid

Over $100 per person

$50–$100 per person

Under $50 per person

Countries neither giving nor receiving aid

Countries receiving aid

Under $10 per person

$10–$100 per person

Over $100 per person

No data

The advantages of aid to a recipient country

Short-term aid can include food, clothing, shelter and medical supplies needed after a natural disaster or civil war. Long-term aid should try to encourage poorer countries to become increasingly self-sufficient and independent. This might be achieved through improving education and health standards, growing higher-yielding crops for their own use rather than for export, developing small-scale sustainable industries using appropriate technology, and encouraging the MEDCs to buy more products rather than setting up tariffs. Many people believe that, as we all live in the same world, we must help each other and try to improve the quality of life for everyone – the concept of **global citizenship**.

The disadvantages of aid to a recipient country

Aid rarely reaches the poorest people, who tend to live in rural areas. Inefficient and corrupt officials direct it to themselves and the urban areas where most of them live. Aid can force countries to produce materials for the MEDCs rather than growing food or developing industries for themselves. In time, the poorest countries come to rely upon aid which, if it comes in the form of loans on which interest in payable, makes the LEDCs even poorer. Many people question whether aid should even be given, as they believe it discourages development within LEDCs and makes them dependent on richer countries and organisations (Figure 11.20).

Figure 11.19
The trade–aid cycle

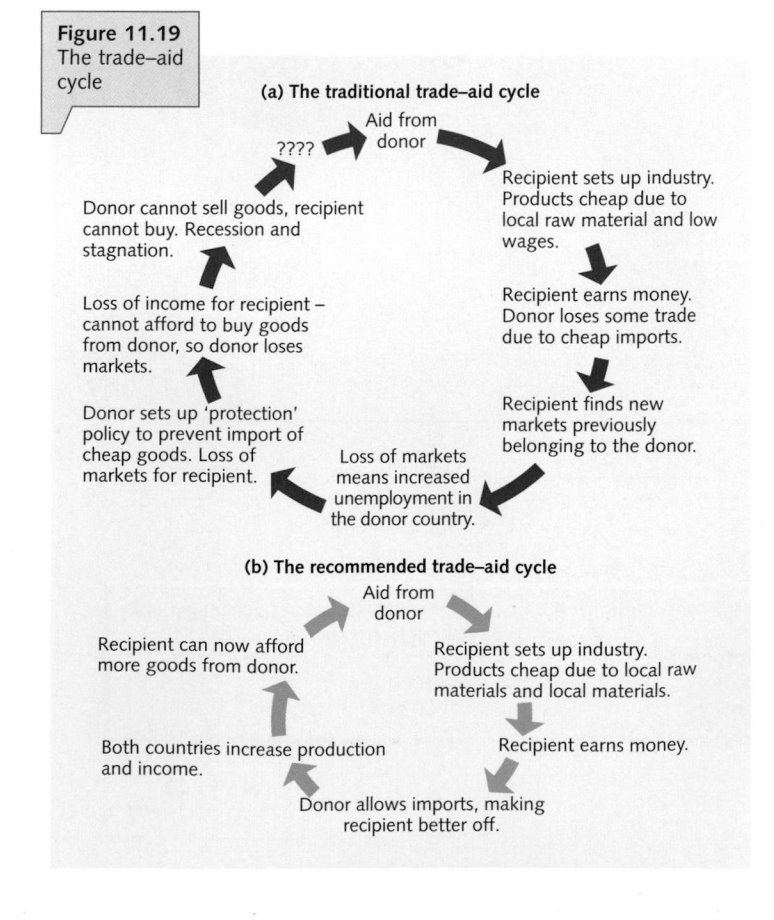

(a) The traditional trade–aid cycle

???? → Aid from donor

Recipient sets up industry. Products cheap due to local raw material and low wages.

Recipient earns money. Donor loses some trade due to cheap imports.

Recipient finds new markets previously belonging to the donor.

Loss of markets means increased unemployment in the donor country.

Donor sets up 'protection' policy to prevent import of cheap goods. Loss of markets for recipient.

Loss of income for recipient – cannot afford to buy goods from donor, so donor loses markets.

Donor cannot sell goods, recipient cannot buy. Recession and stagnation.

(b) The recommended trade–aid cycle

Aid from donor

Recipient sets up industry. Products cheap due to local raw materials and local materials.

Recipient earns money.

Donor allows imports, making recipient better off.

Both countries increase production and income.

Recipient can now afford more goods from donor.

There might have been some justification for the steady increases from year to year in external aid if there was evidence that the battle against poverty was being won. All the evidence, however, clearly demonstrates that the poor are getting poorer.

The reason why there is little effect on poverty is that most aid is spent on heavily capitalised infrastructure projects such as railways, bridges and roads. These may have some indirect effect on the lives of the poor people, but what people really want is better health and education services, improvements in village roads rather than highways, and access to proper credit facilities to help them improve agriculture and raise their incomes. Although there are some enlightened donors, most big Western governments and UN agencies find these types of project too small-scale and difficult to measure and administer; they would rather give money for power stations and fertiliser plants since such things can be seen and their performance measured. They also look impressive in glossy magazines telling the tax-payers back home how the government is spending their money.

Charity statement

Figure 11.20
The aid/debt relationship

THE CUMULATIVE effect of the way in which the developing world is portrayed by charitable organisations and the media is grossly misleading. And it results in deeply held public misconceptions that are ultimately damaging to the understanding they seek to promote.

Ninety per cent of people's knowledge and impressions of the developing world come from two sources: the news media and fund-raising agencies. Both are responsible for distorting the public's perception of the developing world – the one because it is in the business of reporting the exceptional; the other because it is in the business of raising money.

One of the worst examples of a distorted message came during the massive campaign mounted in response to the African emergency in the mid-eighties. The public in the industrialised world donated roughly half a billion dollars in one 12-month period – one of the greatest fund-raising responses of all time. At the same time, the industrialised world's governments gave $2bn in extra emergency aid to Africa. However, almost three times that sum was paid to the industrialised world by Africa in debt and interest repayments. The net flow of finance in that year was therefore *out of* Africa. This year the outflow is at least $10bn in interest repayments alone – more than Africa spends on its health and education services. *UNESCO*

Japan and Kenya – levels of development

Japan

Economic wealth and employment structures

If development is based on wealth (page 180), then Japan, with a GDP per capita of US$33 600, is the third richest country in the world after Luxembourg and Switzerland.

Japan's employment structure is typical of a more economically developed country (Figure 11.21 and page 94). There is only a relatively small proportion of the working population in the primary sector. This is because most young people prefer to live and work in urban areas rather than on farms; farming itself is highly mechanised; Japan has few minerals and so there is very little mining; and there is little forestry, as most domestic forests are protected. Japan, like other developed countries, has a high proportion of its workforce engaged in the secondary sector.

Despite a lack of raw materials, it has the capital to set up large, modern and highly mechanised industries (e.g. high-tech and car assembly); an education system that provides technological knowledge and creates a skilled workforce; a large, wealthy local market to buy goods; and the ability and ports to export manufactured goods (page 146). Japan also has, mainly due to its wealth, a high proportion employed in the tertiary sector, in health, education, commerce, transport and recreation.

Social measures and the HDI

Figure 11.22 shows how population data for Japan corresponds closely with that expected in a more economically developed country (page 181). This is particularly true in the case of life expectancy, as the Japanese live longer than people in any other country. Figure 11.22 also lists other measures often used to show a country's level of development. In 2008, Japan had an HDI of 0.953, making it eighth in the world rankings (second equal in 1995) – a ranking lower than that based on its GDP per capita (page 182).

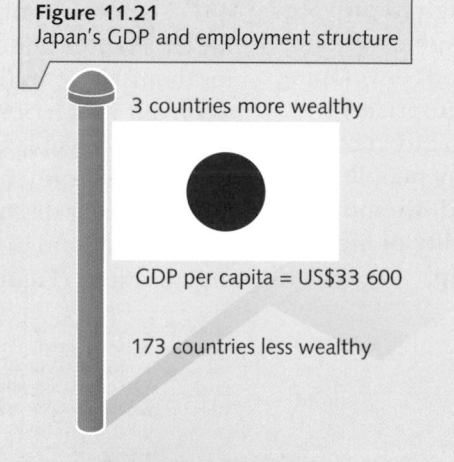

Figure 11.21
Japan's GDP and employment structure

3 countries more wealthy

GDP per capita = US$33 600

173 countries less wealthy

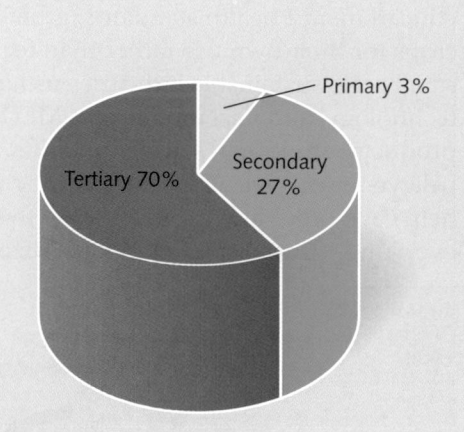

Primary 3%
Tertiary 70%
Secondary 27%

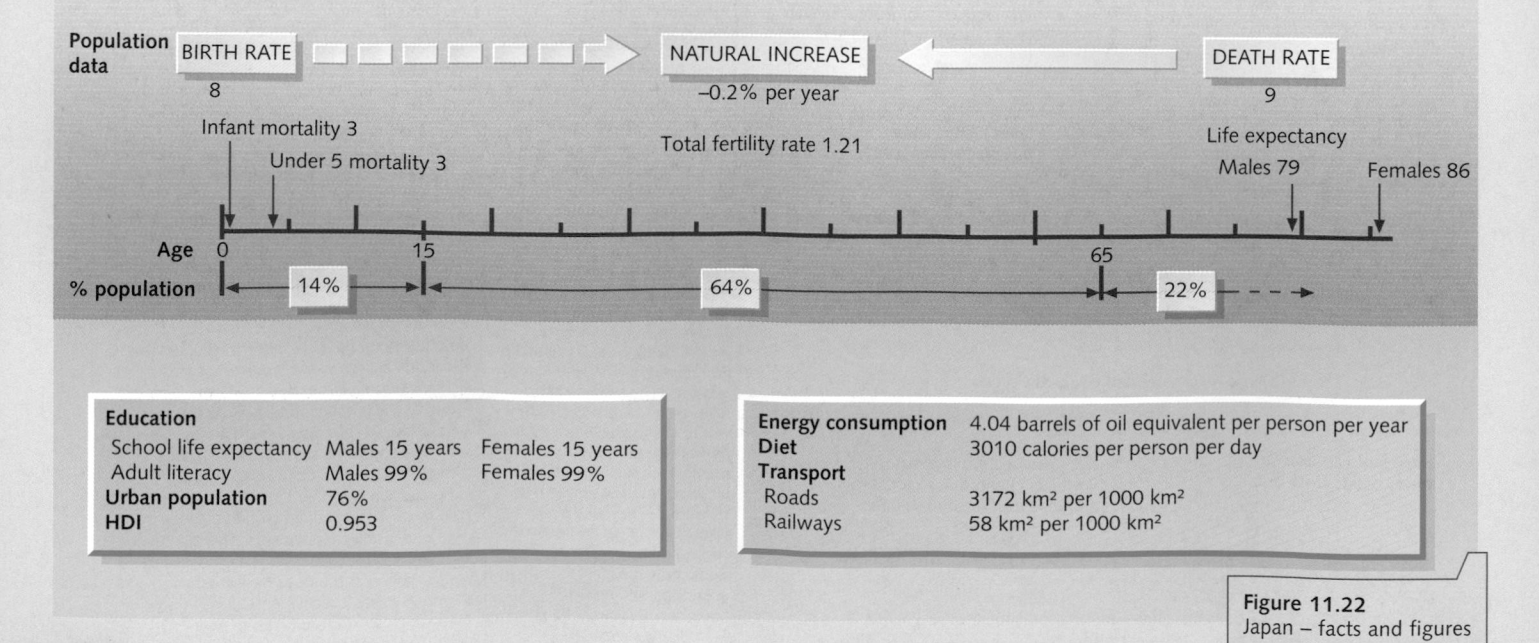

Population data

BIRTH RATE 8 → NATURAL INCREASE −0.2% per year ← DEATH RATE 9

Infant mortality 3
Under 5 mortality 3

Total fertility rate 1.21

Life expectancy
Males 79
Females 86

Age 0 ─ 15 ─ 65
% population 14% 64% 22%

Education		
School life expectancy	Males 15 years	Females 15 years
Adult literacy	Males 99%	Females 99%
Urban population	76%	
HDI	0.953	

Energy consumption	4.04 barrels of oil equivalent per person per year
Diet	3010 calories per person per day
Transport	
Roads	3172 km² per 1000 km²
Railways	58 km² per 1000 km²

Figure 11.22
Japan – facts and figures

Kenya

Economic wealth and employment structures

If development is based purely on wealth (page 180), then Kenya, with a GDP per capita of US$1240 (giving it ranking of 159 out 177), is one of the world's poorest and least economically developed countries.

Kenya's employment structure is typical of a less economically developed country (Figure 11.23 and page 94). There is a high proportion of the working population in the primary sector. Most of these people are engaged in farming, which is labour-intensive and often at a subsistence level (Places, page 104). Smaller numbers are employed in mining, forestry and fishing. Kenya, like other developing countries, has a low proportion engaged in the secondary sector. This is mainly due to a lack of capital, energy supplies and technical knowledge to establish industry; a limited education system leaving a less skilled workforce; the export of raw materials and agricultural produce; and a relatively small local market unable to afford to buy manufactured goods. Mainly due to its lack of wealth, Kenya has a relatively small proportion employed in the tertiary sector with limitations in its health, education, commerce and transport services. The exception is tourism (page 170), which is the country's main money earner. It should also be remembered that large numbers are employed in the informal sector (page 148).

Social measures and the HDI

Figure 11.24 shows how population data for Kenya corresponds closely with that expected in a less economically developed country (page 181). This is particularly true in the case of natural increase, especially during the early 1990s, when Kenya had the highest natural increase in the world. Figure 11.24 also lists other measures often used to show a country's level of development. In 2008, Kenya had an HDI of 0.521. Although this was a low world ranking of 148 (it was 151 in 1995), it was, nevertheless, higher than that based on its GDP per capita (page 182).

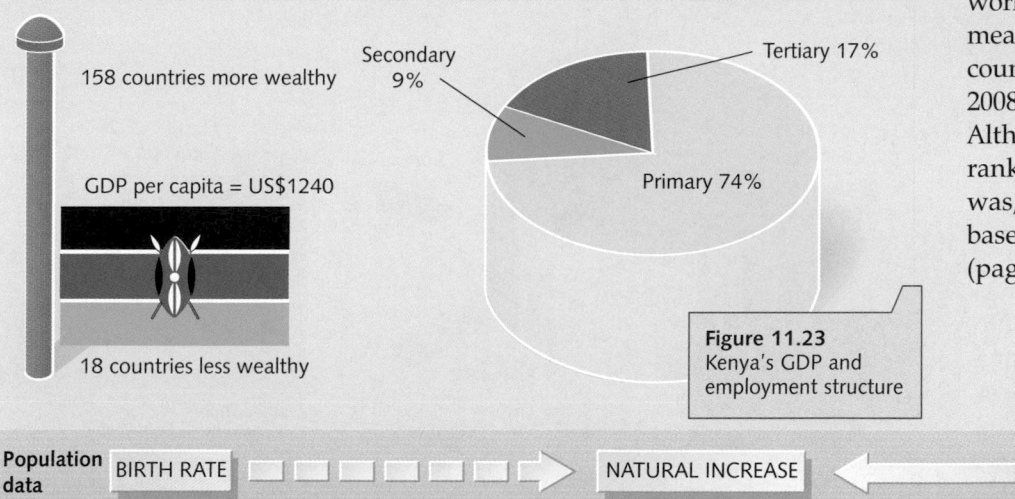

158 countries more wealthy

GDP per capita = US$1240

18 countries less wealthy

Secondary 9%

Tertiary 17%

Primary 74%

Figure 11.23
Kenya's GDP and employment structure

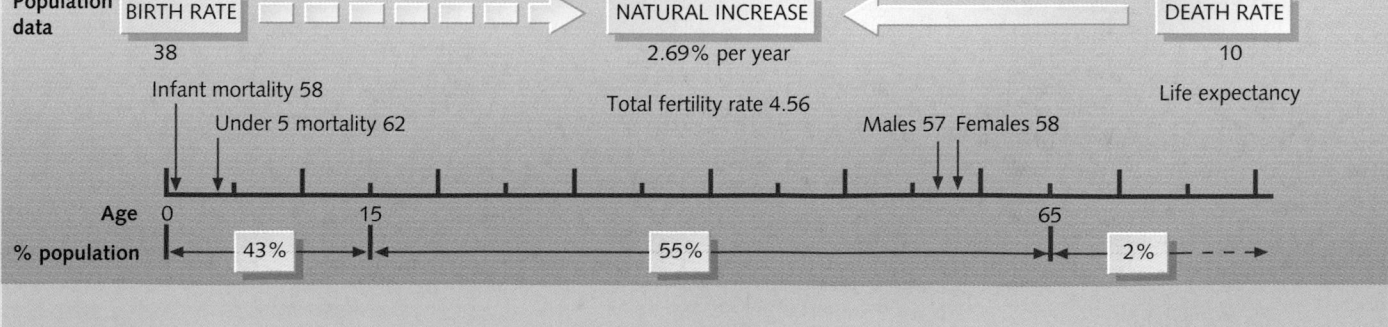

Population data

BIRTH RATE	NATURAL INCREASE	DEATH RATE
38	2.69% per year	10

Infant mortality 58
Under 5 mortality 62
Total fertility rate 4.56
Life expectancy
Males 57 Females 58

Age 0   15   65

% population 43% 55% 2%

Education
School life expectancy Males 10 years Females 8 years
Adult literacy Males 90% Females 79%
Urban population 22%
HDI 0.521

Energy consumption 0.48 barrels of oil equivalent per person per year
Diet 2150 calories per person per day
Transport
Roads 118 km² per 1000 km²
Railways 4 km² per 1000 km²

Figure 11.24
Kenya – facts and figures

Japan and Kenya – trade and interdependence

Japan

Japan has a large population, limited amounts of flat land and few natural resources of its own (page 146). It has, therefore, to import:

- virtually all of its energy supplies (e.g. oil and coal), which are expensive, as well as various raw materials (e.g. timber) and minerals (e.g. iron ore) needed by its industries (Figure 11.25)
- considerable amounts of foodstuffs because although Japanese farming is intensive and highly mechanised, there is insufficient space to grow enough for the country to be self-sufficient.

On the other hand, by working long hours, introducing modern machinery and developing high levels of technology, the Japanese are able to produce and export across the world a range of goods noted for their high quality and reliability (e.g. electrical goods, cars and high-tech products).

This has resulted in Japan becoming the world's fourth largest trader after the USA, the EU and, recently, China (Figure 11.26). Before the 1980s, most of Japan's trade was with the advanced market economies in North America and the EU, but since then there has been a large increase first with the Asian NICs and, since 2000, China. For many years between the 1980s and 2000, Japan had the world's largest trade surplus, a position it no longer holds. This healthy trade surplus was due to Japan:

- reducing its previously high energy bill by changing from expensive imports of oil to the more controversial use of nuclear power

- importing relatively cheap raw materials (often from poorer, developing countries) and exporting expensive processed goods (usually to richer, developed countries)
- protecting – until the 1990s – its domestic industries by imposing tariffs on imported goods, and gaining foreign markets by overseas investment and building

new factories abroad (e.g. Nissan, Toyota and Honda in the UK)
- financing scientific research projects and building projects in developing countries in return for their often non-renewable resources (e.g. Japanese logging companies are responsible for much deforestation in South-east Asia in order to protect Japan's own forests).

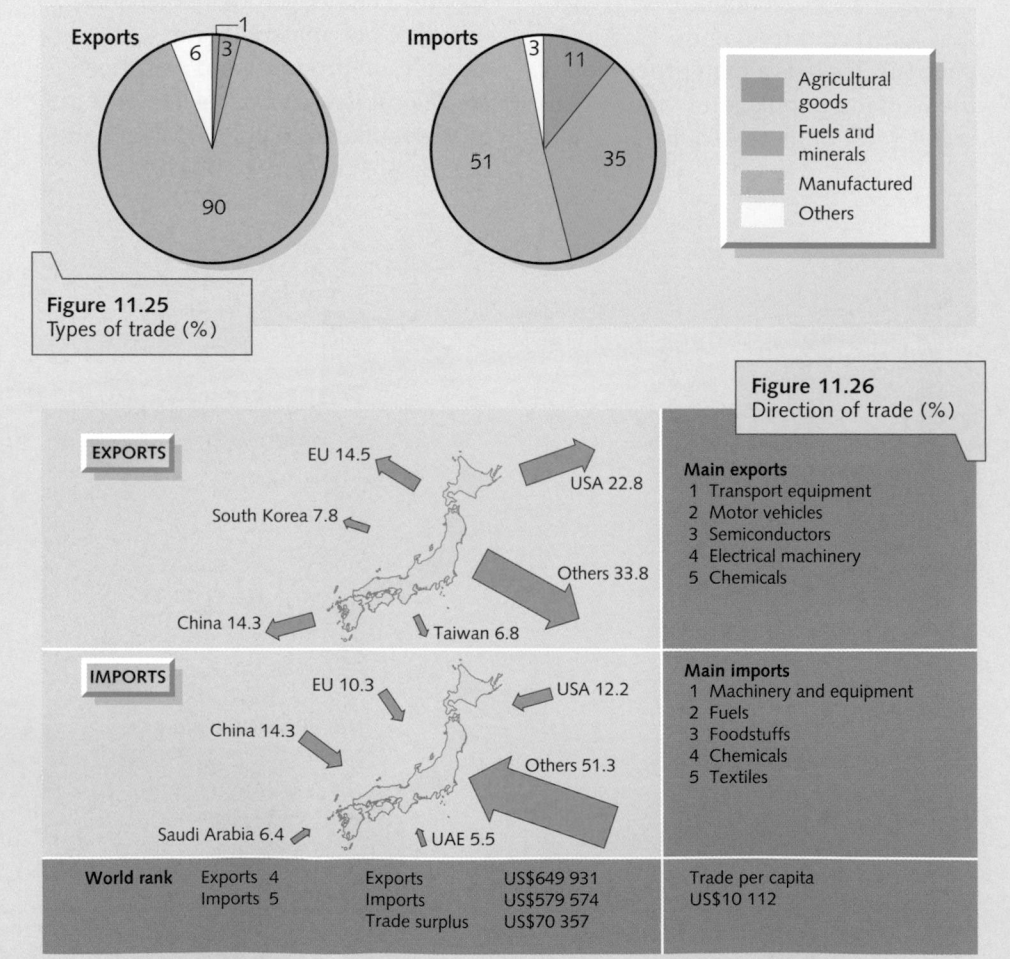

Figure 11.25
Types of trade (%)

Exports: 90, 6, 3, 1
Imports: 51, 35, 11, 3

Legend:
- Agricultural goods
- Fuels and minerals
- Manufactured
- Others

Figure 11.26
Direction of trade (%)

EXPORTS
- EU 14.5
- USA 22.8
- South Korea 7.8
- Others 33.8
- China 14.3
- Taiwan 6.8

Main exports
1 Transport equipment
2 Motor vehicles
3 Semiconductors
4 Electrical machinery
5 Chemicals

IMPORTS
- EU 10.3
- USA 12.2
- China 14.3
- Others 51.3
- Saudi Arabia 6.4
- UAE 5.5

Main imports
1 Machinery and equipment
2 Fuels
3 Foodstuffs
4 Chemicals
5 Textiles

World rank	Exports 4	Exports	US$649 931	Trade per capita
	Imports 5	Imports	US$579 574	US$10 112
		Trade surplus	US$70 357	

Figure 11.27
The busy port of Tokyo

Kenya

Although parts of northern Kenya are desert, many areas to the south are well suited to agriculture. Where the climate and soil are more favourable, subsistence farmers can, in a normal year, grow sufficient crops to be self-supporting. Where rainfall is abundant and the volcanic soils are fertile, crops such as tea, coffee and fruit can be grown commercially for export. Unfortunately, foodstuffs, and raw materials such as soda ash, are low in value and do not earn the country much money. In contrast, Kenya has little formal industry or energy reserves and so it has to import most of the manufactured goods, machinery, cars and oil it needs – all of them expensive to buy (Figure 11.28). The result is that Kenya has a large trade deficit.

Over 25 per cent of Kenya's trade is with the EU (Figure 11.29) and the UK remains, as in colonial times, both the most important single market for Kenyan exports and the country's main supplier. In both cases, however, the proportion of trade has declined. Four significant changes have occurred during and since the 1990s.

- Japan became Kenya's largest overseas investor, which has meant that, in return, Kenya has had to buy more Japanese goods ('Bilateral aid', page 190).
- Kenya has developed a trade surplus within Africa, exporting mainly cement and refined oil to neighbouring countries such as Uganda and Tanzania (Figure 11.29). Although Kenya earns four times more through exports to African countries than it pays out in imports, the volume of trade is too small to reduce significantly its balance of trade deficit (compare the amount of activity in the ports of Mombasa and Tokyo, Figures 11.27 and 11.30).
- Air freight is used to export perishable goods to Europe. Places near to Nairobi airport can cut flowers (e.g. carnations and roses, page 104) and pick vegetables (e.g. peas and beans) one day and have them sold in Europe (e.g. in London and Amsterdam) the next morning.
- Kenya has a large trade deficit with several Middle East countries from which it imports oil but a small trade surplus with neighbouring African states to which it sells refined oil.

Figure 11.28
Types of trade (%)

Exports

Imports

Legend:
- Agricultural goods
- Fuels and minerals
- Manufactured
- Others

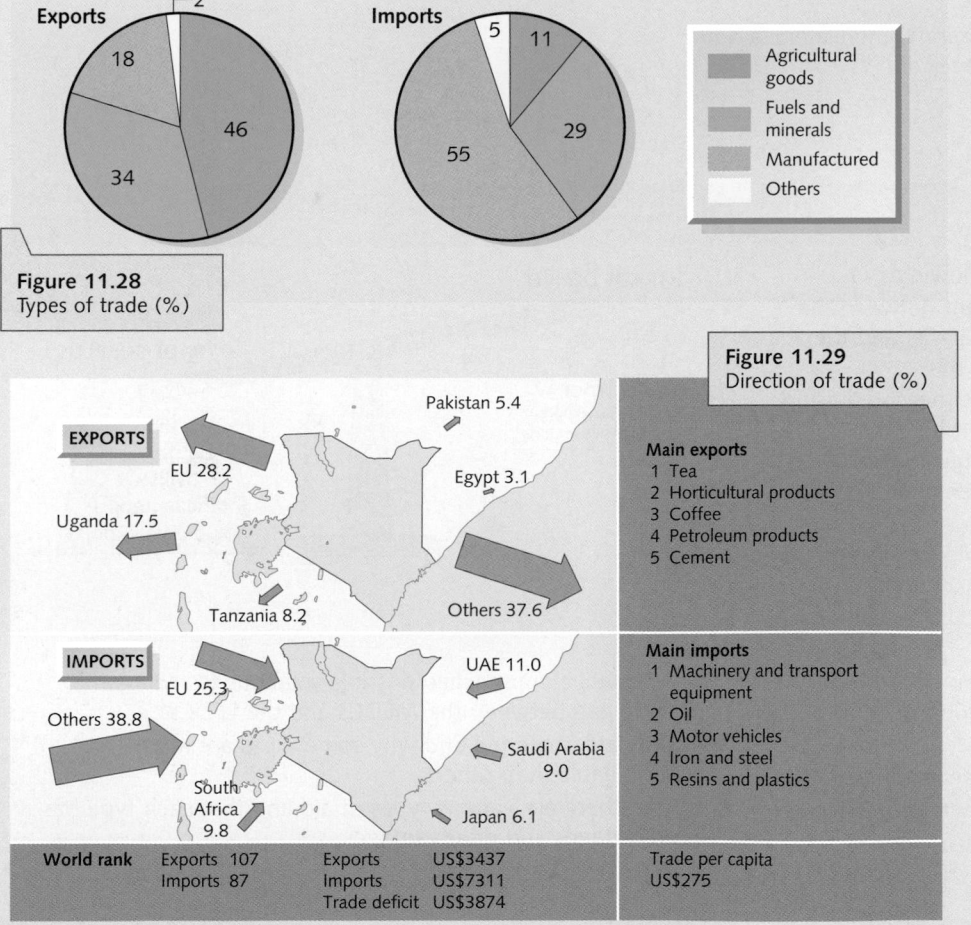

Figure 11.29
Direction of trade (%)

EXPORTS
Pakistan 5.4
EU 28.2
Egypt 3.1
Uganda 17.5
Tanzania 8.2
Others 37.6

Main exports
1 Tea
2 Horticultural products
3 Coffee
4 Petroleum products
5 Cement

IMPORTS
EU 25.3
UAE 11.0
Others 38.8
Saudi Arabia 9.0
South Africa 9.8
Japan 6.1

Main imports
1 Machinery and transport equipment
2 Oil
3 Motor vehicles
4 Iron and steel
5 Resins and plastics

World rank	Exports 107	Exports US$3437	Trade per capita
	Imports 87	Imports US$7311	US$275
		Trade deficit US$3874	

Figure 11.30
The quieter port of Mombasa

World development and interdependence

Key Words and Terms

a You should know the meaning of the following terms:
- more economically developed country (MEDC)
- less economically developed country (LEDC)
- gross domestic product (GDP) per capita
- birth rate • death rate • natural increase
- total fertility rate (TFR) • infant mortality
- life expectancy • adult literacy
- Human Development Index (HDI)
- actual purchasing power (PPP)
- sustainable development • appropriate technology
- quality of life • standard of living • interdependence
- trade balance • free trade • tariffs • quotas • aid
- trading groups/blocs • World Trade Organization (WTO)
- government/bilateral aid • international/multilateral aid
- charitable/voluntary aid • emergency aid
- globalisation.

b You should know the difference between:
- the 'North' and the 'South'
- economic and social indicators/measures
- imports and exports
- a trade surplus and a trade deficit
- primary goods and secondary goods
- short-term aid and long-term aid
- aid donors and aid recipients.

Key Ideas

You should know and understand the following:
- That there are contrasts in development between the richer countries of the 'North' (the MEDCs) and the poorer countries of the 'South' (LEDCs).
- That contrasts in development may be measured using economic and social indicators.
- That differences in development and quality of life are related to economic, social, environmental and political factors.
- The causes and consequences of imbalances in the development/quality of life between countries.
- How appropriate technology can encourage economic/sustainable development.
- Recent patterns/directions and trends in world trade.
- The causes and consequences of the imbalance of trade between MEDCs and LEDCs.
- The need for interdependence between trading countries.
- The advantages and disadvantages of creating trading blocs/groups.

TRADE DEFICIT

TRADE SURPLUS

LEDCs
Cheap exports
Expensive imports

MEDCs
Cheap imports
Expensive exports

- Alternative approaches/possible solutions to reduce the trade gap between the MEDCs and the LEDCs.
- Why aid is needed and why some types are more sustainable than others.
- That there are various types of aid and that each type has advantages and disadvantages.

Skills, Theories and Models

- Analyse and evaluate statistical data
- Interpret and use pie graphs and choropleth maps
- Use census data
- Use case studies to compare places
- Identify inequalities in development
- Understand the idea of sustainable development

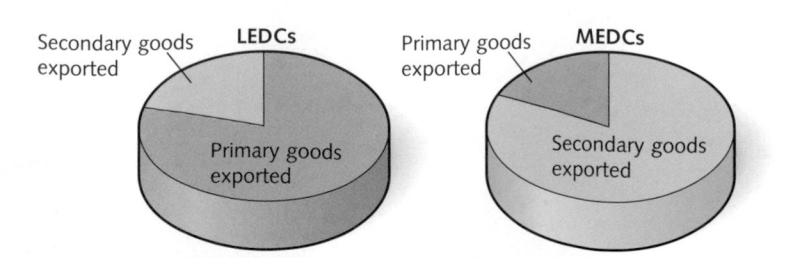

Secondary goods exported **LEDCs** Primary goods exported

Primary goods exported **MEDCs** Secondary goods exported

1 (Pages 180 and 183)

a Name three continents in the 'richer North' and three in the 'poorer South'. (6)

b i) State how the five indicators on the map show that Africa is economically the least well-off. (5)

ii) Suggest three other indicators that might be used to measure differences in development between countries. (3)

c Make a copy of the table headings below. In the correct columns add the information below the table, to show differences between MEDCs and LEDCs. (7)

d i) Describe the three variables used to measure the HDI. (3)

ii) Why is the HDI considered to be a better way than GDP of measuring levels of development? (2)

iii) Name the three countries with the highest HDI and the three countries with the lowest HDI in 2008. (2)

e For each of the following, describe two factors that can cause inequalities in development:
• economic   • social   • political
• environmental. (8)

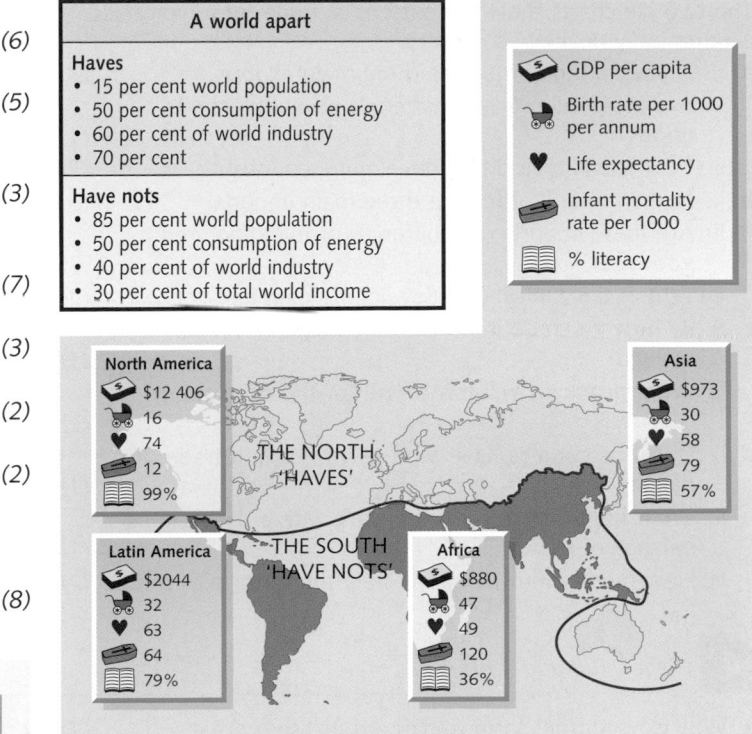

MEDCs	LEDCs

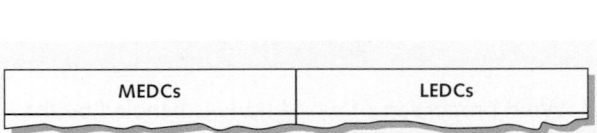

• High GDP • Low GDP • High birth rate • Low birth rate • Long life expectancy • Short life expectancy
• High percentage in agriculture • Low percentage in agriculture • High percentage in services
• Low percentage in services • High percentage of urban dwellers • Low percentage of urban dwellers
• High literacy levels • Low literacy levels

2 (Pages 184 and 185)

a How can sustainable development lead to an improvement in people's:
• quality of life   • standard of living? (2)

b Using the headings on the star diagram, and Figures 11.8 and 11.9, describe how sustainable development might be achieved. (10)

c Name one scheme or project in an LEDC that you have studied that uses appropriate technology in an attempt to achieve sustainable development.

i) Describe the scheme/project. (5)

ii) Say how successful you think it has been. (2)

d i) Why is it important for MEDCs to adopt methods of sustainable development? (2)

ii) How might appropriate technology differ between an MEDC and an LEDC? (1)

3 *(Page 186)*

The two pie charts show the pattern of trade for a typical less economically developed country.

a i) Name, in rank order, the three main exports. *(3)*

ii) Are these exports raw materials or manufactured goods? *(1)*

iii) Why is this typical of a developing country? *(2)*

b i) Name, in rank order, the three main imports. *(3)*

ii) Are these imports raw materials or manufactured goods? *(1)*

iii) Why is this typical of a developing country? *(2)*

c i) By how much do imports exceed exports (in US dollars)? *(1)*

ii) What problems will this create for the developing country? *(2)*

d i) Does the country have a trade surplus or a trade deficit? *(1)*

ii) What might happen to a country with a deficit in its balance of trade? *(1)*

iii) How might industrialisation help the balance of trade? *(2)*

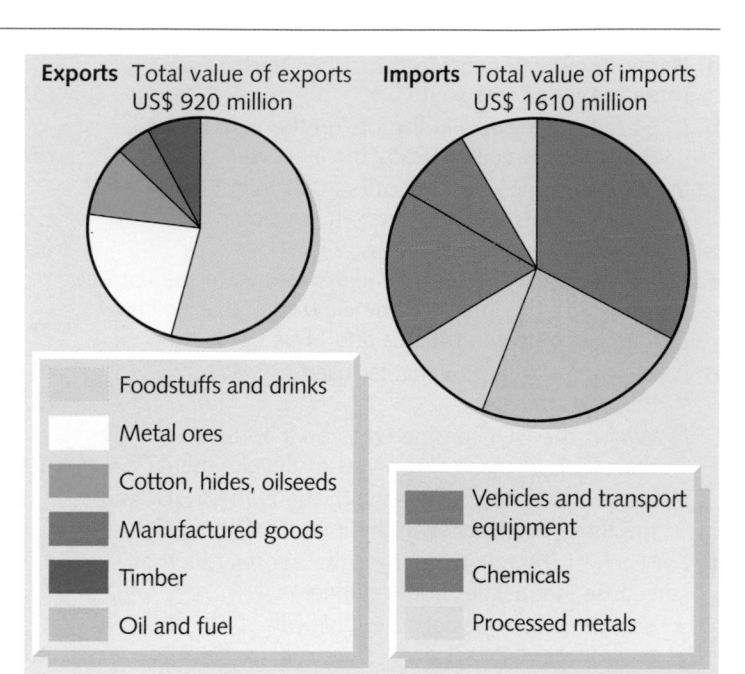

Exports Total value of exports US$ 920 million

Imports Total value of imports US$ 1610 million

- Foodstuffs and drinks
- Metal ores
- Cotton, hides, oilseeds
- Manufactured goods
- Timber
- Oil and fuel
- Vehicles and transport equipment
- Chemicals
- Processed metals

4 *(Pages 188 and 189)*

a Why do countries form trading groups (blocs)? *(1)*

b i) Name one trading group in each of:
- North America • Europe • Asia. *(3)*

ii) Name three countries in each of the trading groups that you have just named. *(9)*

iii) Name the trading group that deals in oil. *(1)*

iv) Name two important trading countries – one an MEDC and the other an LEDC – that are not members of a trading group. *(2)*

c i) What proportion of world trade is handled by the MEDCs? *(1)*

ii) Which group of LEDCs have been successful in developing their trade? *(1)*

d i) What are the main aims of the WTO? *(2)*

ii) Give one advantage and one disadvantage of:
- tariffs • quotas. *(4)*

iii) Why are the LEDCs keen to see the removal of tariffs? *(1)*

5 *(Pages 188 and 189)*

a Using the data for Britain's trade in 1973 (right):

i) Name Britain's main trading partner. *(1)*

ii) What proportion of Britain's trade was with developed countries? *(1)*

iii) Why did Britain trade with OPEC? *(1)*

iv) Why was relatively little of Britain's trade with developing countries? *(1)*

b Comparing the data for Britain's trade in 2008 with that of 1973:

i) With which trading partner did Britain's trade increase the most? *(1)*

ii) Why has there been an increase in trade with this partner? *(1)*

iii) Which country did Britain trade with in 2008 that was not named on the 1973 graph? *(1)*

iv) Give two reasons for your last answer. *(2)*

v) Which group of countries did Britain trade with in 2008 but not in 1973? *(1)*

vi) Which trade group did Britain trade with in 1973 but was not named in 2008? *(1)*

vii) Give one reason for your last answer. *(1)*

(1973 pie chart): Centrally planned economies (USSR and eastern Europe) 4%; Other developing countries 15%; OPEC 7%; Other developed countries 10%; North America 16%; EU 32%; Rest of Europe 16%

(2008 pie chart): Other developing countries 5%; China 6%; Asian 'tiger' economies (NICs) 6%; Other developed countries 5%; Japan 3%; North America 12%; EU 57%; Rest of Europe and former USSR 6%

6 *(Pages 190 and 191)*

a i) Name five types of aid. *(5)*

ii) What are the main advantages and disadvantages of each of these types of aid? *(10)*

iii) Which of these types of aid do you think is most helpful to the recipient country? Give two reasons for your answer. *(2)*

b i) Give four reasons why some countries need aid. *(4)*

ii) Name four countries that have recently needed aid urgently. For each give a reason why that aid was needed. *(8)*

iii) Suggest three forms of aid that might be given in the short term. *(3)*

iv) Suggest three forms of aid that might be given in the long term. *(3)*

v) Give three problems that might arise in a country as a result of receiving aid. *(3)*

c The graphs below show the major aid donors and major aid recipients in 2000.

i) The UN recommends that rich countries give 0.7 per cent of their GDP as aid to poorer countries. How many countries reached this target in 2000? *(1)*

ii) How much aid did the UK give? *(1)*

iii) Name the four countries that gave most aid. *(4)*

iv) Name the four countries that received most aid. *(4)*

d Some people criticise Britain for giving relatively little aid. Others say that if we stopped giving aid then everyone in the UK could be £100 a year better off. Do you think we should give more or less aid? Give reasons for your answer. *(3)*

2000

Rank	Country	Aid as % of GDP
20	USA	0.20
18	Japan	0.22
13	UK	0.26
11	Germany	0.28
6	Canada	0.34
5	France	0.45
4	Sweden	0.79
3	Netherlands	0.81
2	Norway	0.86
1	Denmark	0.97

UN target of 0.70% (only 4 countries exceeded this figure)

Aid as % of GDP — scale 0, 0.25, 0.5, 0.75, 1.0

Rank	Country	Debt as % of GDP
10	Chad	20
9	Rwanda	21
8	Burundi	22
7	Malawi	23
6	Bhutan	25
5	Tanzania	36
4	Guinea-Bissau	41
3	Bosnia-Herzegovina	42
2	Nicaragua	44
1	Mozambique	58

Debt as % of GDP — scale 0, 20, 40, 60

7 *(Pages 192 and 195)*

a Complete a larger copy of the table to show differences in development between Japan (an MEDC) and Kenya (an LEDC). *(16 × 3)*

b Describe, and give reasons for, differences in trade between Japan (an MEDC) and Kenya (an LEDC) under the following headings:

- Main types of imports
- Main types of exports
- Main trading partners
- Value of imports and exports
- Balance of trade. *(10)*

Measure of development		Japan	Kenya	Reasons
Wealth	GDP			
Employment structures	% primary			
	% secondary			
	% tertiary			
Population data	Birth rate			
	Death rate			
	Total fertility rate (TFR)			
	Infant mortality			
	Lif expectancy			
Population structure	% under 15			
	% over 60			
Other measures	Education			
	Urban population			
	HDI			
	Energy consumption			
	Diet			

12 Britain's weather and climate

Weather and climate

Weather is the hour-to-hour, day-to-day state of the atmosphere. It includes temperature, sunshine, cloud cover, precipitation, atmospheric pressure, wind speed and wind direction. It is short-term and can be localised in relatively small areas.

Climate is the average weather conditions of a place taken over a period of time, often 30 years. It is the expected, rather than the actual, conditions. It is long-term and is often applied to sizeable parts of the globe, e.g. the equatorial and the Mediterranean climates.

Britain has:
* a variable climate, which means that, unlike the equatorial climate (page 212), the weather changes from day to day, which makes it difficult to forecast
* an equable climate, which means that extremes of heat or cold, drought or prolonged rainfall, are rarely experienced.

Air masses

Britain is affected by four main **air masses**, though at different times. An air mass forms when air remains stationary over a place for several days. During this time the air is likely to assume the temperature and humidity properties of that area (Figure 12.1). Air masses are classified according to:

* the latitude in which they originate, as this determines their temperature, i.e. giving either cold polar air (**P**) if to the north, or warm tropical air (**T**) if to the south
* the surface over which they develop, as this affects their humidity and precipitation, i.e. giving either wet maritime air (**m**) if over the sea, or dry continental air (**c**) if over the land.

When these different air masses move towards the British Isles, they bring with them their own characteristic type of weather (Figure 12.1):
* **Polar maritime (Pm)** It occurs frequently and tends to bring with it cool and wet weather.
* **Polar continental (Pc)** This is not very frequent and gives cool, drier weather.
* **Tropical maritime (Tm)** This occurs most frequently and brings warm, wet conditions.
* **Tropical continental (Tc)** It is the least frequent and gives warm, dry conditions.

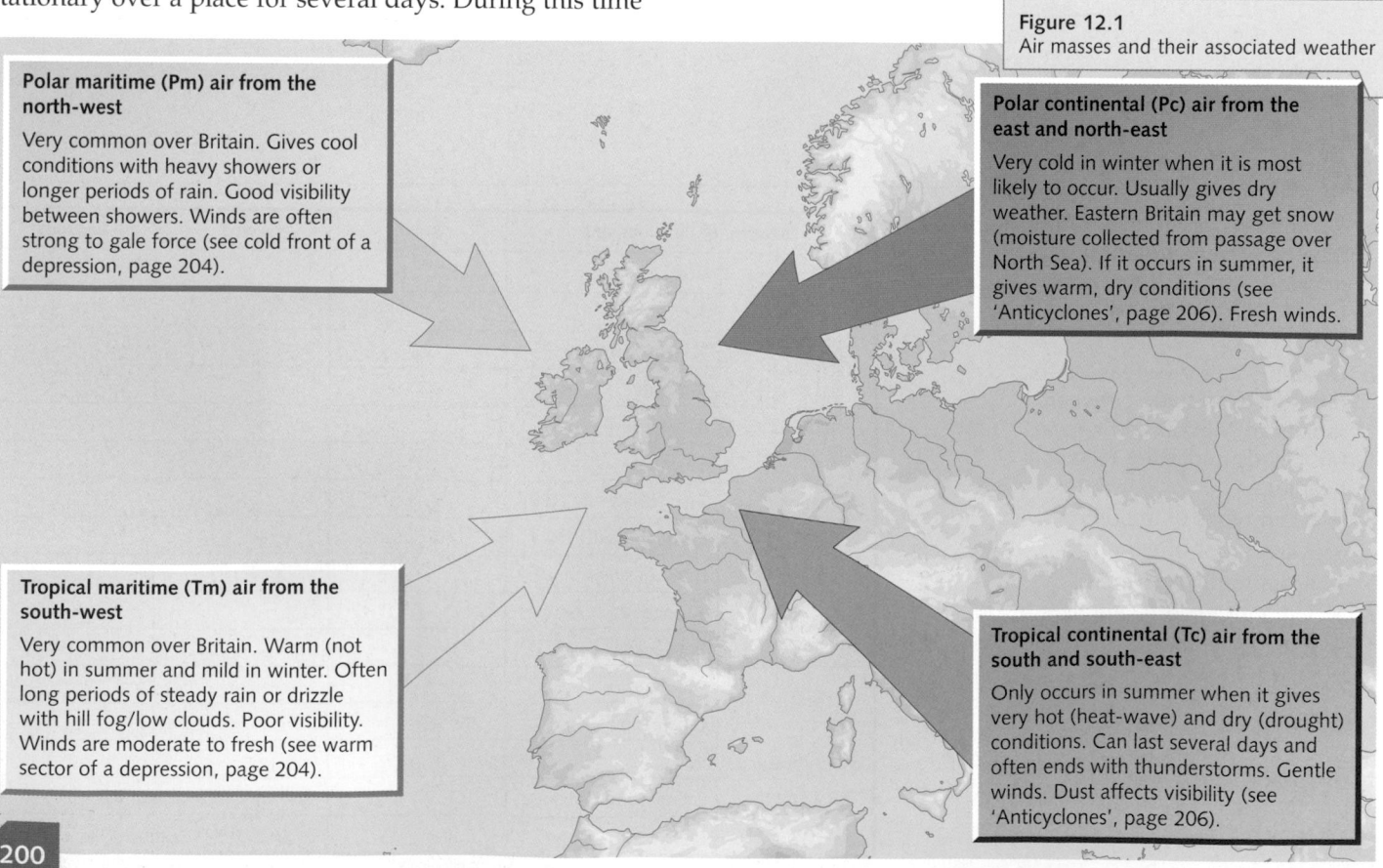

Figure 12.1
Air masses and their associated weather

Polar maritime (Pm) air from the north-west

Very common over Britain. Gives cool conditions with heavy showers or longer periods of rain. Good visibility between showers. Winds are often strong to gale force (see cold front of a depression, page 204).

Polar continental (Pc) air from the east and north-east

Very cold in winter when it is most likely to occur. Usually gives dry weather. Eastern Britain may get snow (moisture collected from passage over North Sea). If it occurs in summer, it gives warm, dry conditions (see 'Anticyclones', page 206). Fresh winds.

Tropical maritime (Tm) air from the south-west

Very common over Britain. Warm (not hot) in summer and mild in winter. Often long periods of steady rain or drizzle with hill fog/low clouds. Poor visibility. Winds are moderate to fresh (see warm sector of a depression, page 204).

Tropical continental (Tc) air from the south and south-east

Only occurs in summer when it gives very hot (heat-wave) and dry (drought) conditions. Can last several days and often ends with thunderstorms. Gentle winds. Dust affects visibility (see 'Anticyclones', page 206).

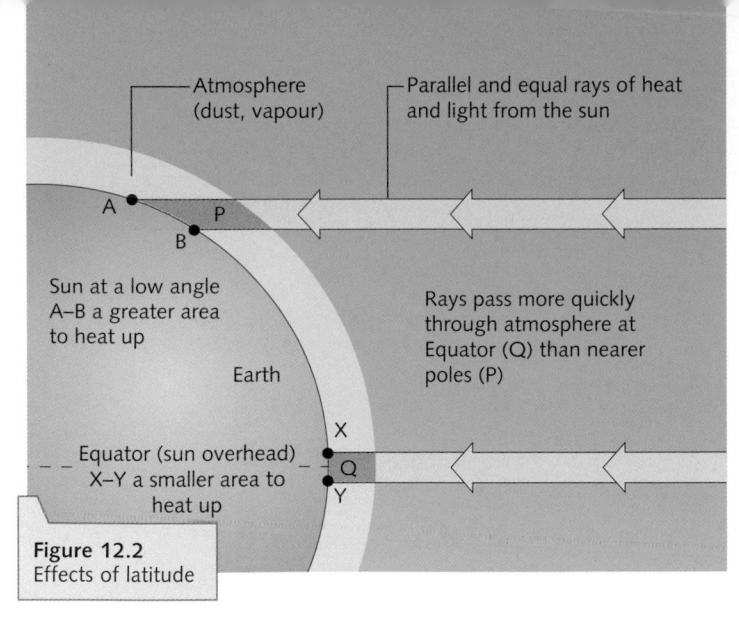

Figure 12.2
Effects of latitude

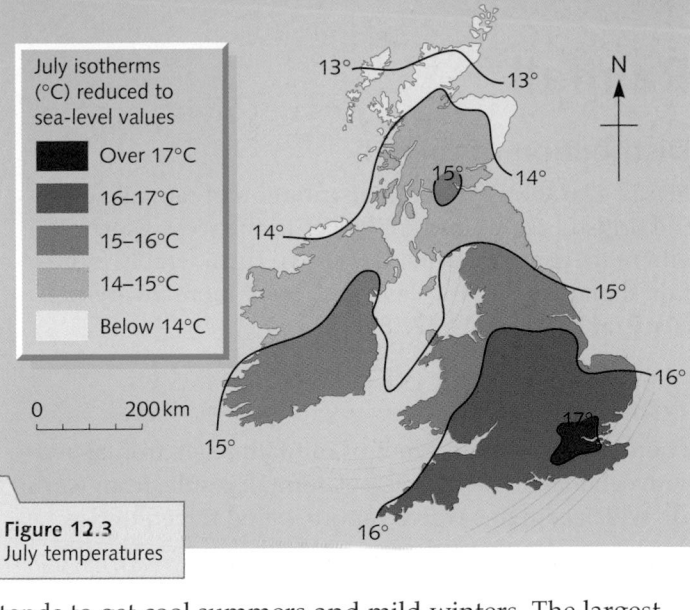

Figure 12.3
July temperatures

Factors affecting temperature

Latitude Places nearer to the Equator are much warmer than places nearer to the poles. This is due to the curvature of the Earth and the angle of the sun (Figure 12.2). At the Equator the sun is always high in the sky. When it is overhead it shines directly downwards, concentrating its heat into a small area which will become very hot. In contrast, the sun is always low in the sky towards the poles. This means that its heat is spread over a wide area, and so temperatures remain lower. Notice also that the lower the angle of the sun, the greater the amount of atmosphere through which the rays have to pass. This means that more heat will be lost to gases, dust and cloud in the atmosphere. This is why places in the south of Britain can expect to be warmer, especially in summer, than places further north (Figure 12.3).

Distance from the sea The sea (a liquid) is less dense than the land (a solid) and can be heated to a greater depth. This means that the sea takes much longer to heat up in summer than does the land. Once warmed, however, the sea retains its heat for much longer, and cools down more slowly than the land in winter. This is why places that are inland are warmer in summer but colder in winter than places on the coast. As Britain is surrounded by the sea, it tends to get cool summers and mild winters. The largest reservoir of heat in winter is the Atlantic Ocean, even though it is still cold enough to die from hypothermia within a few minutes should you fall into it. This explains why western parts of Britain are warmer than places to the east in winter (Figure 12.4).

Prevailing winds Prevailing winds will bring warm weather if they pass over warm surfaces (the land in summer, the sea in winter) and cold weather if they blow across cold surfaces (the land in winter, the sea in summer). As Britain's prevailing winds are from the south-west, they are cool in summer but warm (mild) in winter.

Ocean currents Many coastal areas are affected by ocean currents. The North Atlantic Drift is a warm current of water which originates in the Gulf of Mexico. It keeps the west coast of Britain much warmer in winter than other places in similar latitudes.

Altitude Temperatures decrease, on average, by 1°C for every 100 m in height. As many parts of the Scottish Highlands are over 1000 m, they will be at least 10°C cooler than coastal places. In fact, the windchill factor will make them even colder, and enables snow to lie for long periods during winter (Figure 12.5).

Figure 12.4
January temperatures

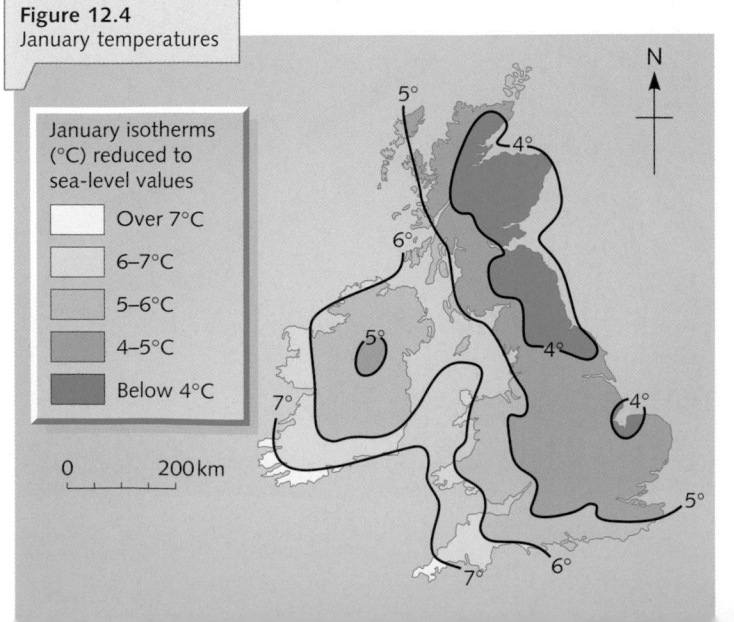

Figure 12.5
Influence of altitude on length of snow cover in winter

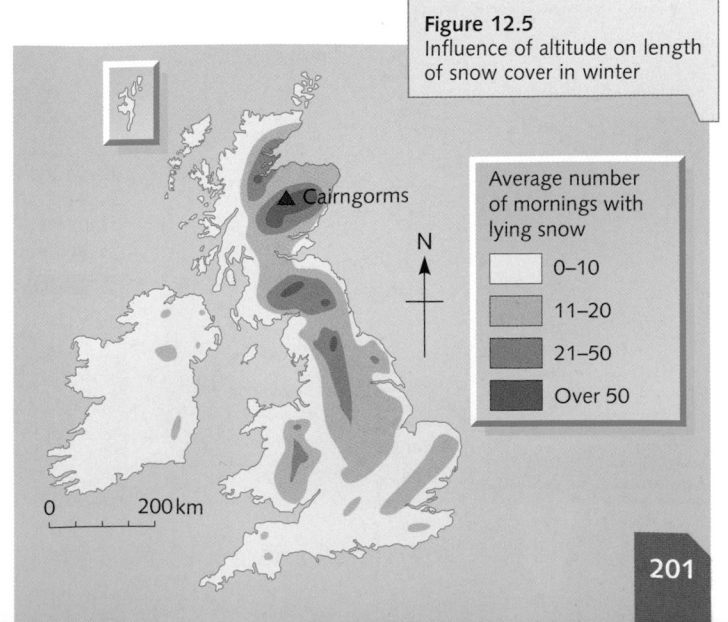

Rainfall

Distribution of rainfall

Figure 12.6 shows that the distribution of rainfall across the British Isles is uneven. Places to the west, especially where there are mountains, receive considerably more rain than places to the east, which are more low-lying (see graphs in Figure 12.18).

Types of rainfall

There are three main types of rainfall: relief, frontal and convectional. In all three cases rainfall results from warm air, which contains water vapour, being forced to rise until it cools sufficiently for condensation to take place (Figure 12.7). Condensation can only occur when two conditions are met:

i)   Cold air cannot hold as much moisture as warm air. As the warm air and water vapour rises, it cools until a critical temperature is reached, at which point the air becomes saturated. This critical temperature is called the **dew point**. If air continues to rise and cool, some of the water vapour in it condenses back into minute droplets of water.

ii) Condensation requires the presence of large numbers of microscopic particles known as **hygroscopic nuclei**. This is because condensation can only take place on solid surfaces such as volcanic dust, salt or smoke (or on windows and walls in a bathroom or kitchen).

The difference between the three types of rainfall is the condition that forces the warm air to rise in the first place.

Figure 12.7
Formation of rain

If cooling continues, increasing condensation will form rain (or snow if temperatures are below 0°C)

Further cooling causes condensation where water vapour is turned back into minute water droplets, which are visible as clouds

As it rises, it cools (effect of altitude)

Warm air and water vapour is forced to rise

Warm air causes water on the ground, in lakes and the sea, and from vegetation, to evaporate

Sea   Land   Lake   Vegetation

1   Relief rainfall (Figure 12.8)

Relief rain occurs when warm, almost saturated air from the sea is blown inland by the wind. Where there is a coastal mountain barrier, the air will be forced to rise over it. The rising air will cool and, if dew point is reached, condensation will take place. Once over the mountains the air will descend, warm and, therefore, the rain is likely to stop.

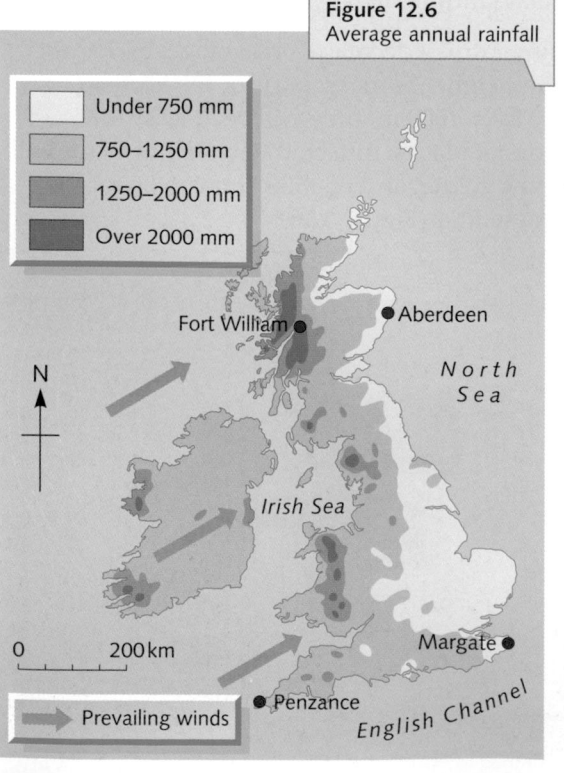

Figure 12.6
Average annual rainfall

- Under 750 mm
- 750–1250 mm
- 1250–2000 mm
- Over 2000 mm

Fort William
Aberdeen
North Sea
Irish Sea
N
Margate
0   200km
Penzance
English Channel
Prevailing winds

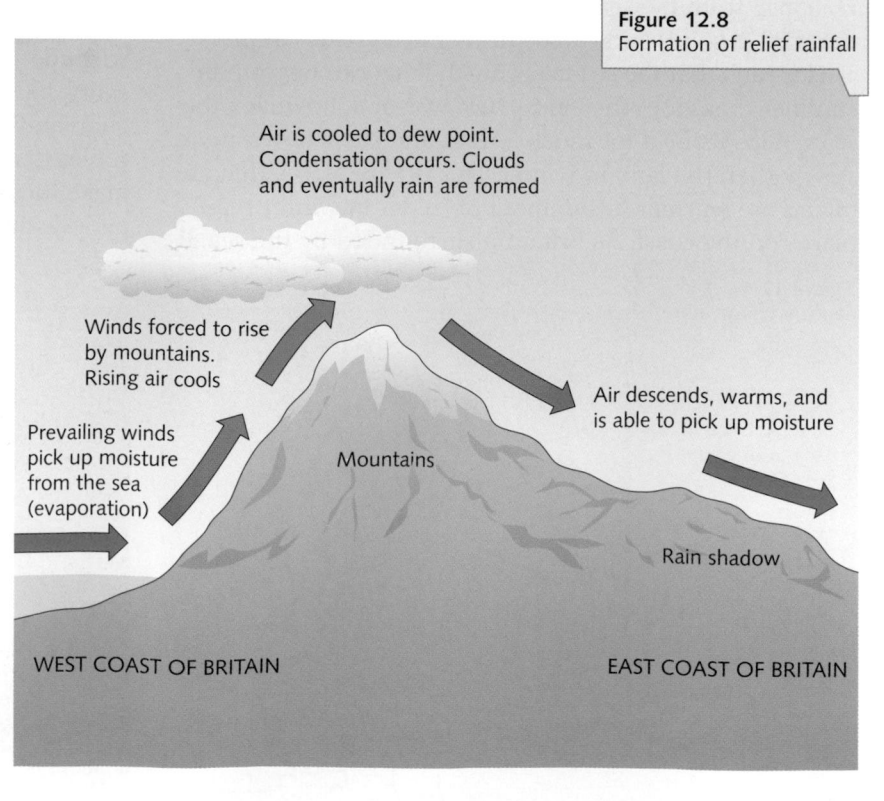

Figure 12.8
Formation of relief rainfall

Air is cooled to dew point. Condensation occurs. Clouds and eventually rain are formed

Winds forced to rise by mountains. Rising air cools

Prevailing winds pick up moisture from the sea (evaporation)

Mountains

Air descends, warms, and is able to pick up moisture

Rain shadow

WEST COAST OF BRITAIN   EAST COAST OF BRITAIN

The protected side of a mountain range is the **rain shadow**. In Britain the prevailing winds come from the south-west, collecting moisture as they cross the Atlantic Ocean. They bring heavy rainfall to western parts as they cross the mountains of Scotland, Wales and northern England. Eastern areas receive much less rain as they are in the rain shadow area. Places like Fort William and Penzance get heavy rainfall in late autumn when the sea is at its warmest, and winds blowing over it can pick up most moisture.

2 Frontal rainfall (Figure 12.9)

Frontal rain is associated with depressions (page 204) and results from warm, moist air from the tropics meeting colder, drier air from polar areas. As the two air masses have different densities, they cannot merge. Instead the warmer, moister and lighter air is forced to rise over the colder, denser air, setting the condensation process into motion. The boundary between the warm and cold air is called a **front**. Most depressions have two fronts, a warm and a cold front, giving two periods of rainfall (Figure 12.11). Britain receives many depressions and their associated fronts each year. Depressions usually come from the Atlantic Ocean, increasing rainfall on the west coasts. Depressions are more common in winter, as illustrated by the winter rainfall maximum.

3 Convectional rainfall (Figure 12.10)

Convectional rain occurs where the ground surface is heated by the sun. As the air adjacent to the ground is heated, it expands and begins to rise. If the ground surface is wet and heavily vegetated, as in equatorial areas (page 212), there will be rapid evaporation. As the air rises, it cools and water vapour condenses to form towering cumulonimbus clouds and, later, heavy thunderstorms. Equatorial areas, where the sun is constantly at a high angle in the sky, experience convectional storms most afternoons. Convectional rain is less frequent in cooler Britain, and is most likely in south-east England in summer when temperatures are at their highest. This also accounts for the summer rainfall maximum in this region.

Figure 12.10
Formation of convectional rainfall

a Morning

Warm air and water vapour rise

By midday, sun shines vertically giving maximum heat

Ground wet after previous day's rain

Sun's rays heat the Earth's surface

b Early afternoon

10 000 m

Water vapour cools and condenses into water droplets

Cumulus clouds

5000 m

Convection currents

Ground

c Late afternoon (4 pm)

10 000 m

Cumulonimbus clouds give heavy rain, usually with thunder and lightning

5000 m

Ground

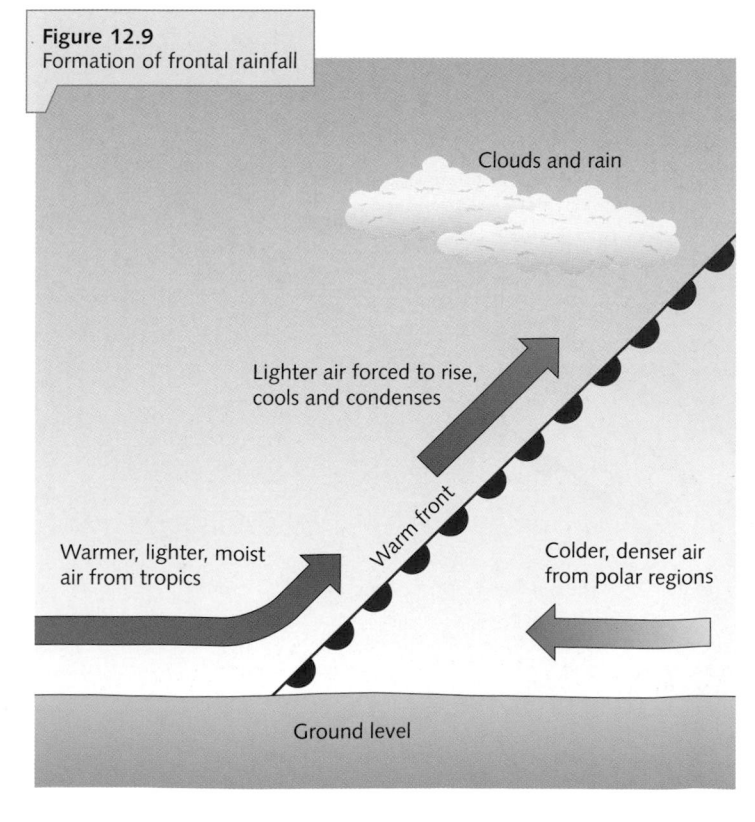

Figure 12.9
Formation of frontal rainfall

Clouds and rain

Lighter air forced to rise, cools and condenses

Warm front

Warmer, lighter, moist air from tropics

Colder, denser air from polar regions

Ground level

Depressions

Britain's weather changes from day to day. For much of the year our climate is dominated by the passing of depressions. **Depressions** are areas of low pressure which bring rain, cloud and wind. They form over the Atlantic Ocean when a mass of warm, moist tropical air from the south meets a mass of colder, drier, heavier polar air from the north. The two masses of air do not easily mix due to differences in temperature and density. The boundary between two air masses is called a **front**. When lighter, warmer air moves towards denser, colder air, it is forced to rise over the cold air at a warm front (Figure 12.11). When denser, colder air moves towards warm air, it undercuts the warm air forcing it to rise at a cold front. In both cases the rising warm air is cooled and some of its water vapour content condenses, producing cloud and frontal rain (page 203). The cold front travels faster than the warm front, catching it up to form an **occluded front**. Although each depression is unique, the weather they bring to Britain as they travel eastwards tends to have an easily recognisable pattern.

As a warm front approaches clouds begin to form. They get lower, and thicken (Figure 12.11). Winds blow from the south-east, in an anticlockwise direction, and slowly increase in strength. As the air rises, atmospheric pressure drops. The passing of the warm front is usually characterised by a lengthy period of steady rainfall, low cloud and strong winds. As the warm front passes there is a sudden rise in temperature and the wind turns to a south-westerly direction. The **warm sector** of a depression is usually a time of low and sometimes broken cloud, decreasing winds, and drizzle or even dry weather. As a cold front passes the weather deteriorates rapidly. Winds often reach gale force and swing round to the north-west. Rainfall is very heavy, though of relatively short duration, and temperatures fall rapidly. After the cold front passes, the weather slowly improves as pressure increases. The heavy rain gives way to heavy showers and eventually to sunny intervals. Winds are cold and slowly moderate, but still come from the north-west. Most depressions take between one and three days to pass over the British Isles.

Depressions can be seen on satellite images as masses of swirling cloud (Figure 12.12). **Satellite images** are photos taken from space and sent back to Earth. They are invaluable when trying to produce a weather forecast or predicting short-term changes in the weather. The state of the weather at any one given time is shown on a **synoptic chart** (a weather map). The three synoptic charts in Figure 12.13 are meant to match the satellite images in Figure 12.12.

Notice in Figures 12.12 and 12.13:

a how the cloud cover (white) and the two fronts are over Ireland

b twelve hours later, how the cloud and fronts have moved eastwards and are over East Anglia and the North Sea

c another twelve hours later, how both have moved further east to lie across Germany, Denmark and Scandinavia.

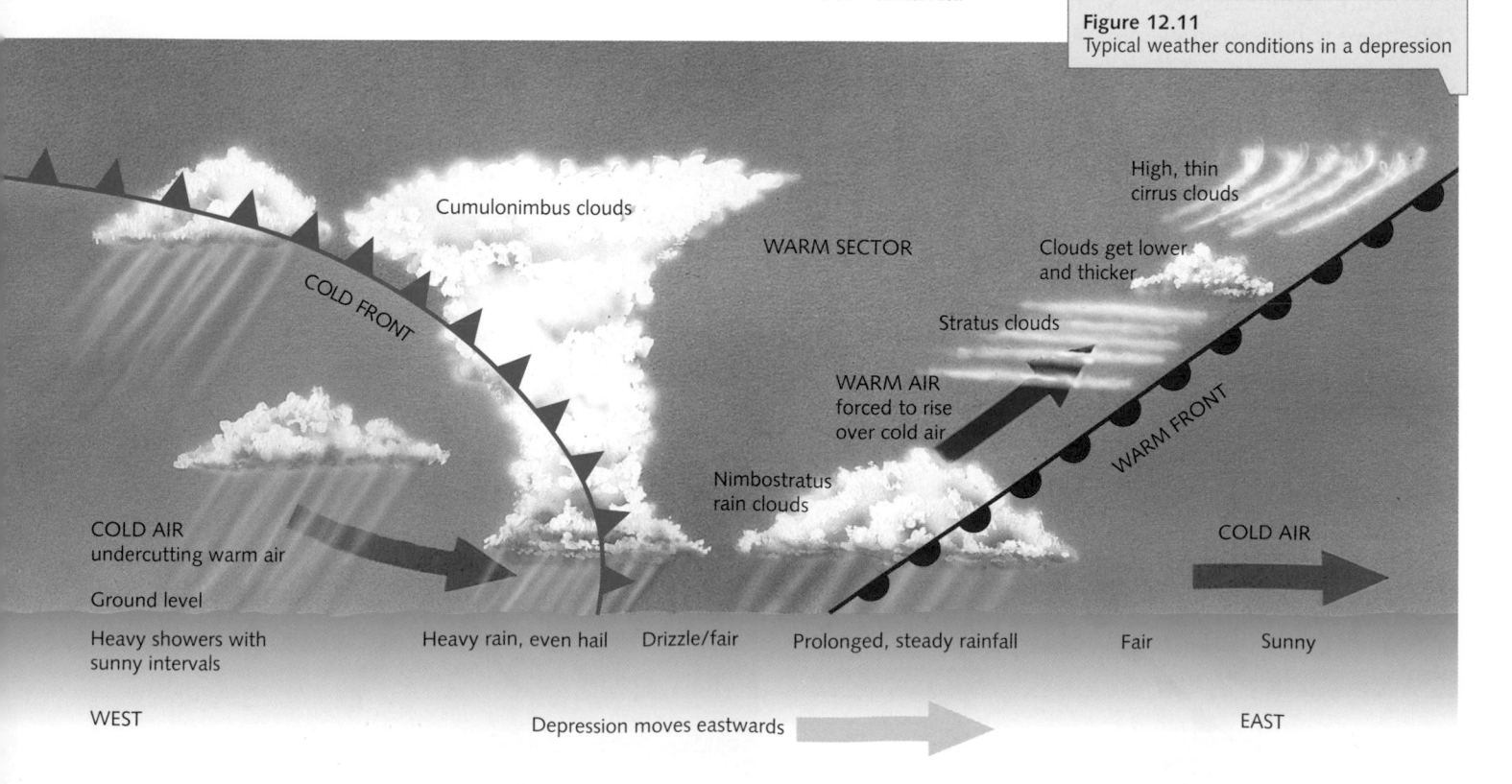

Figure 12.11
Typical weather conditions in a depression

High, thin cirrus clouds

Cumulonimbus clouds

WARM SECTOR

Clouds get lower and thicker

COLD FRONT

Stratus clouds

WARM AIR forced to rise over cold air

WARM FRONT

Nimbostratus rain clouds

COLD AIR undercutting warm air

COLD AIR

Ground level

Heavy showers with sunny intervals | Heavy rain, even hail | Drizzle/fair | Prolonged, steady rainfall | Fair | Sunny

WEST

Depression moves eastwards

EAST

a 0600 hours 4 January

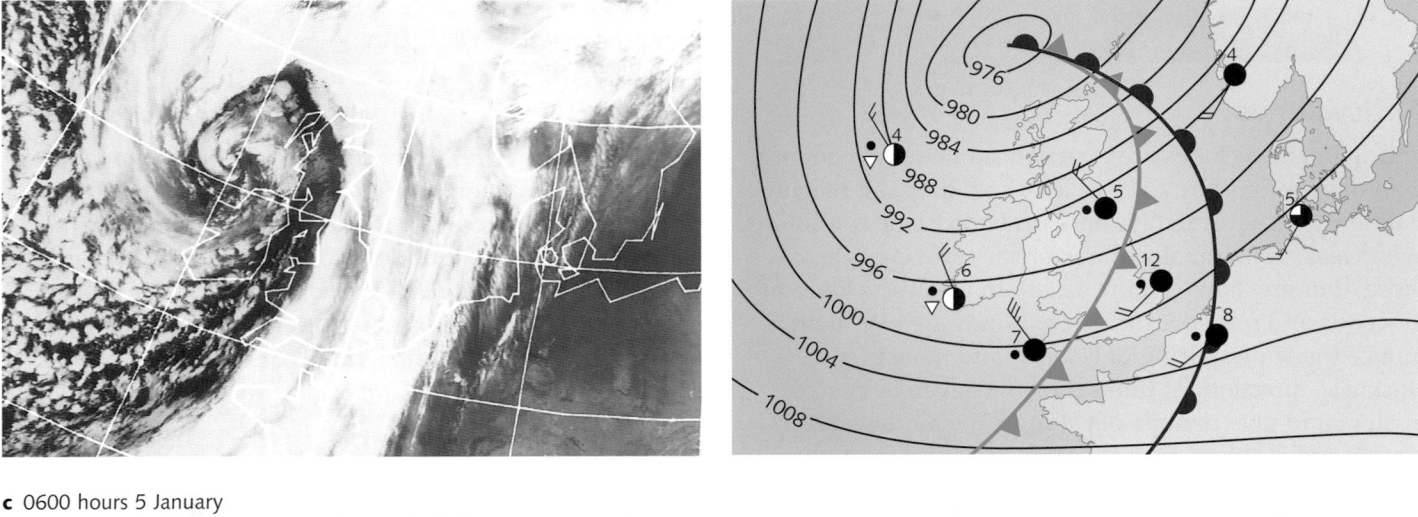

b 1800 hours 4 January

c 0600 hours 5 January

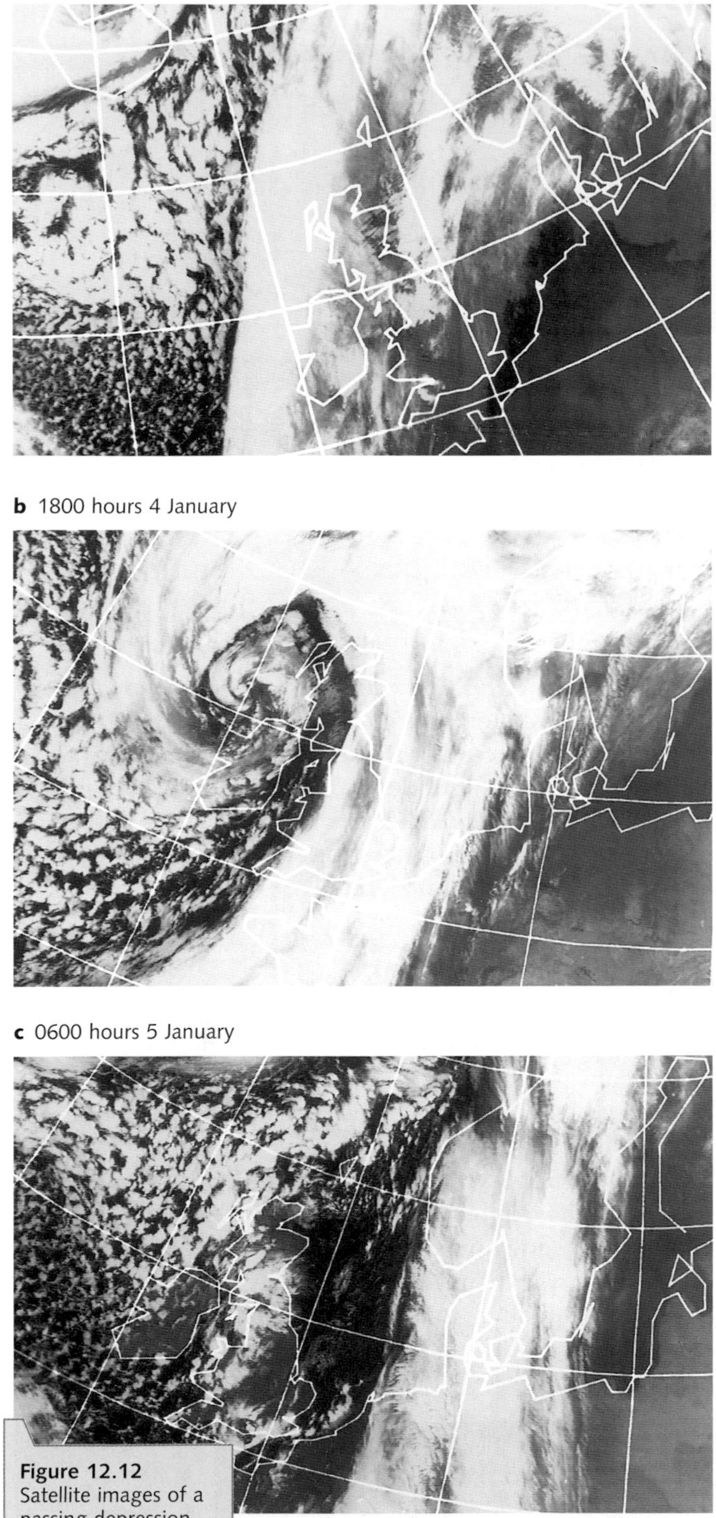

Figure 12.12
Satellite images of a passing depression

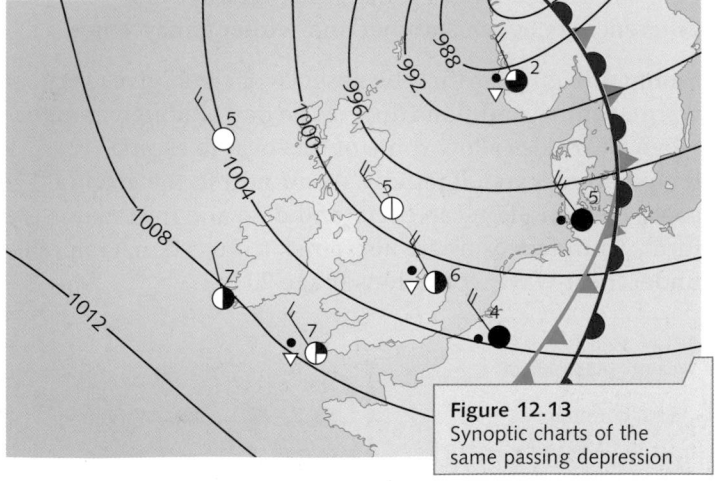
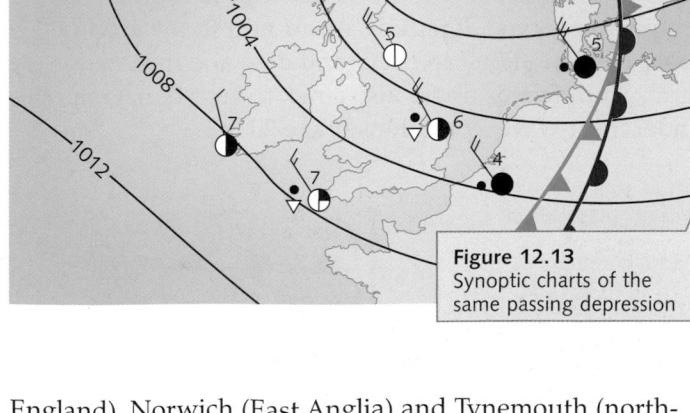

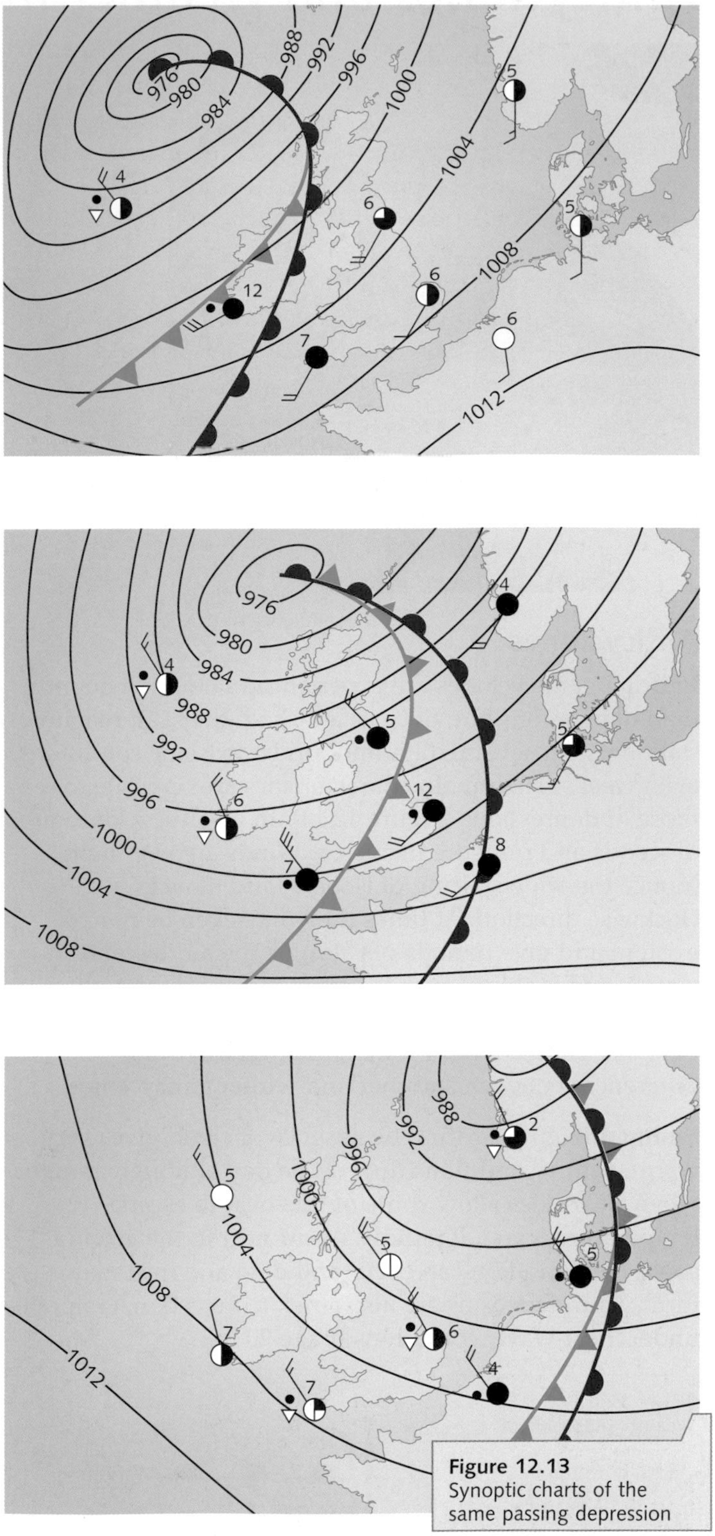

Figure 12.13
Synoptic charts of the same passing depression

The synoptic charts in Figure 12.13 show various **weather symbols** (Figure 12.17) for selected **weather stations**. The four weather stations in the British Isles are at Valencia (south-west Ireland), the Isles of Scilly (south-west England), Norwich (East Anglia) and Tynemouth (north-east England). The weather station to the north-west of Ireland is on a weather ship.

205

Anticyclones and weather maps

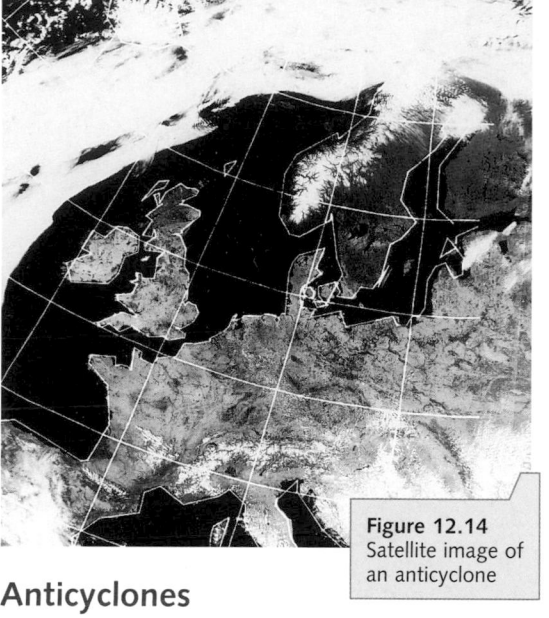

Figure 12.14
Satellite image of an anticyclone

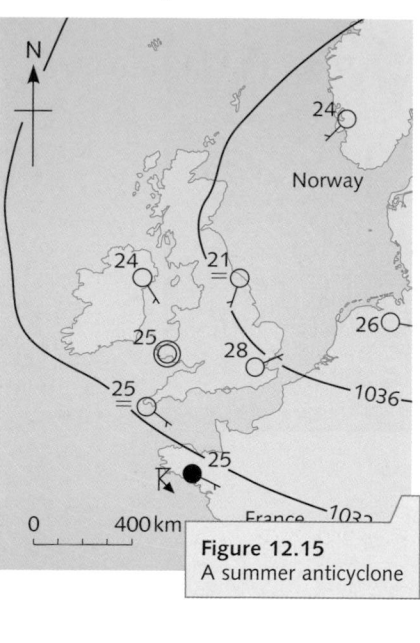

Figure 12.15
A summer anticyclone

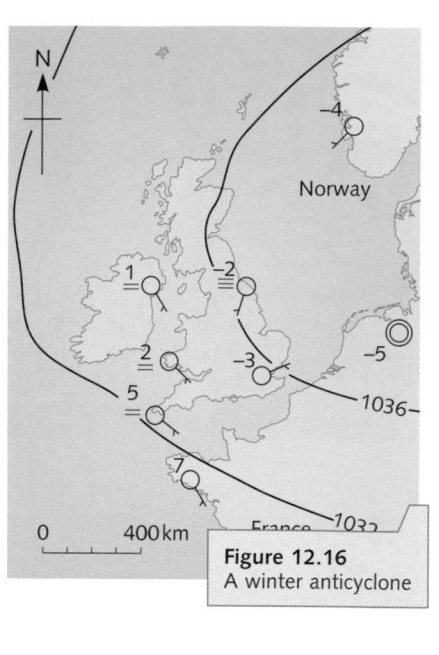

Figure 12.16
A winter anticyclone

Anticyclones

In Britain anticyclones are experienced far less frequently than depressions but, once established, they can remain stationary for several days and, under extreme conditions, even weeks. Their main characteristics are opposite to those of depressions (Figure 12.14). In an anticyclone, air descends and pressure increases. Winds are very light (notice the wide spacing of isobars) and blow in a clockwise direction. At times they may even be non-existent and give periods of calm. As the air descends it warms and is able to pick up more moisture through evaporation. This usually results in settled conditions with clear skies and a lack of rain. However, there are differences between summer and winter anticyclones.

Summer (Figure 12.15) The absence of cloud gives very warm, sunny conditions during the day although at night, when clear skies allow some of this heat to escape, temperatures can fall rapidly. As air next to the ground cools, condensation can occur and dew and mist may form. Thunderstorms are also a risk (convectional rainfall) under 'heat-wave' conditions (page 203).

Winter (Figure 12.16) Although temperatures remain low during the day, due to the sun's low angle in the sky (Figure 12.2), the weather is likely to be dry and bright. The rapid loss of heat under the clear evening skies means that nights can be very cold. Condensation near to the ground can produce frost and fog which may, due to the sun's lack of heat, persist all day.

The daily weather map

The daily weather map, as shown on television or in a newspaper, aims to give a clear but simplified forecast. This means that both media use symbols that are easily recognised by the average viewer or reader. Synoptic charts (Figure 12.13) produced by the Meteorological Office use official symbols to show conditions at specific weather stations (Figure 12.17). The weather stations on Figure 12.13 show five weather elements: temperature, wind speed, wind direction, amount of cloud cover, and type of precipitation, while a sixth, atmospheric pressure, can be obtained by interpreting the isobars.

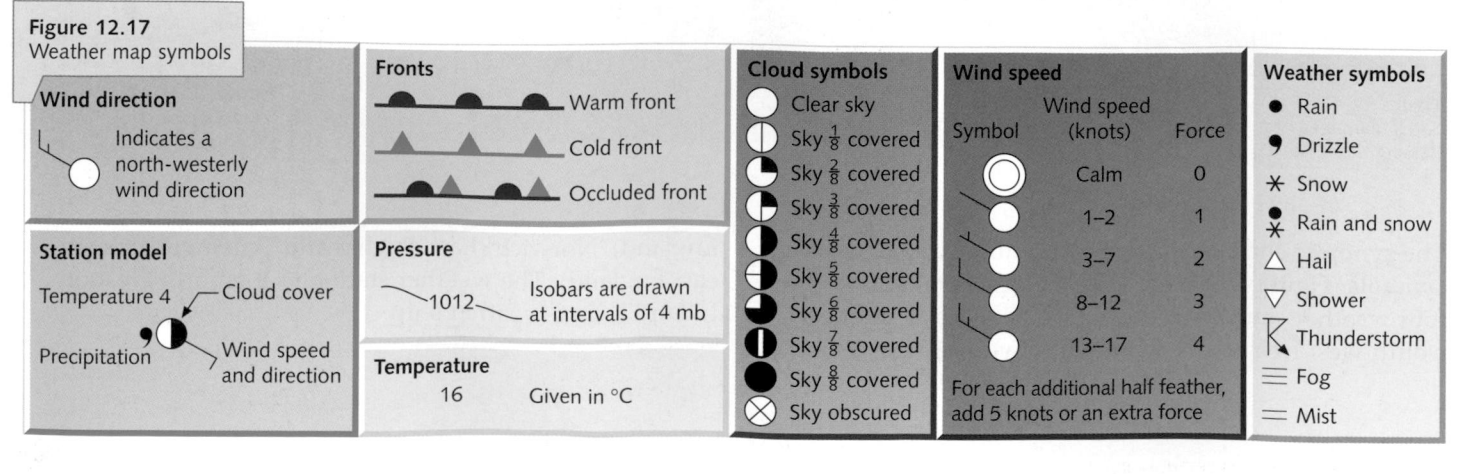

Figure 12.17
Weather map symbols

Seasonal and regional weather patterns

If we wish to generalise about Britain's climate, we can say that it has cool summers, mild winters and a steady, reliable rainfall that is spread evenly throughout the year. However, even across an area as small as the British Isles, there are significant differences:

- seasonally, between summer and winter
- regionally, between places in the extreme north and south, and places on the east and west coasts (Figure 12.18).

You should be able to see from Figure 12.18 that:

- Places in the south are warmer and sunnier than places in the north in summer. This is mainly due to the influence of latitude (page 201).
- Places in the west are milder and cloudier than places to the east in winter. This is due to the prevailing winds from the south-west and the influence of the North Atlantic Drift (page 201).
- Places in the west have a lower range of temperature, due to the moderating influence of the sea, than those to the east (page 201), which are more likely to be affected by more extremes of weather from the continent.
- Places in the north-west have lower temperatures throughout the year, and more snow in winter, than places in the south and east, due to the influence of altitude (page 201).
- Places in the west are wetter than places to the east due to the prevailing winds bringing frontal rain and giving relief rainfall as they pass over the mountains (pages 202–3).
- Places in the west receive most rain in winter, when depressions are most frequent, whereas those to the east have a summer maximum, due to convectional rainfall (page 203).

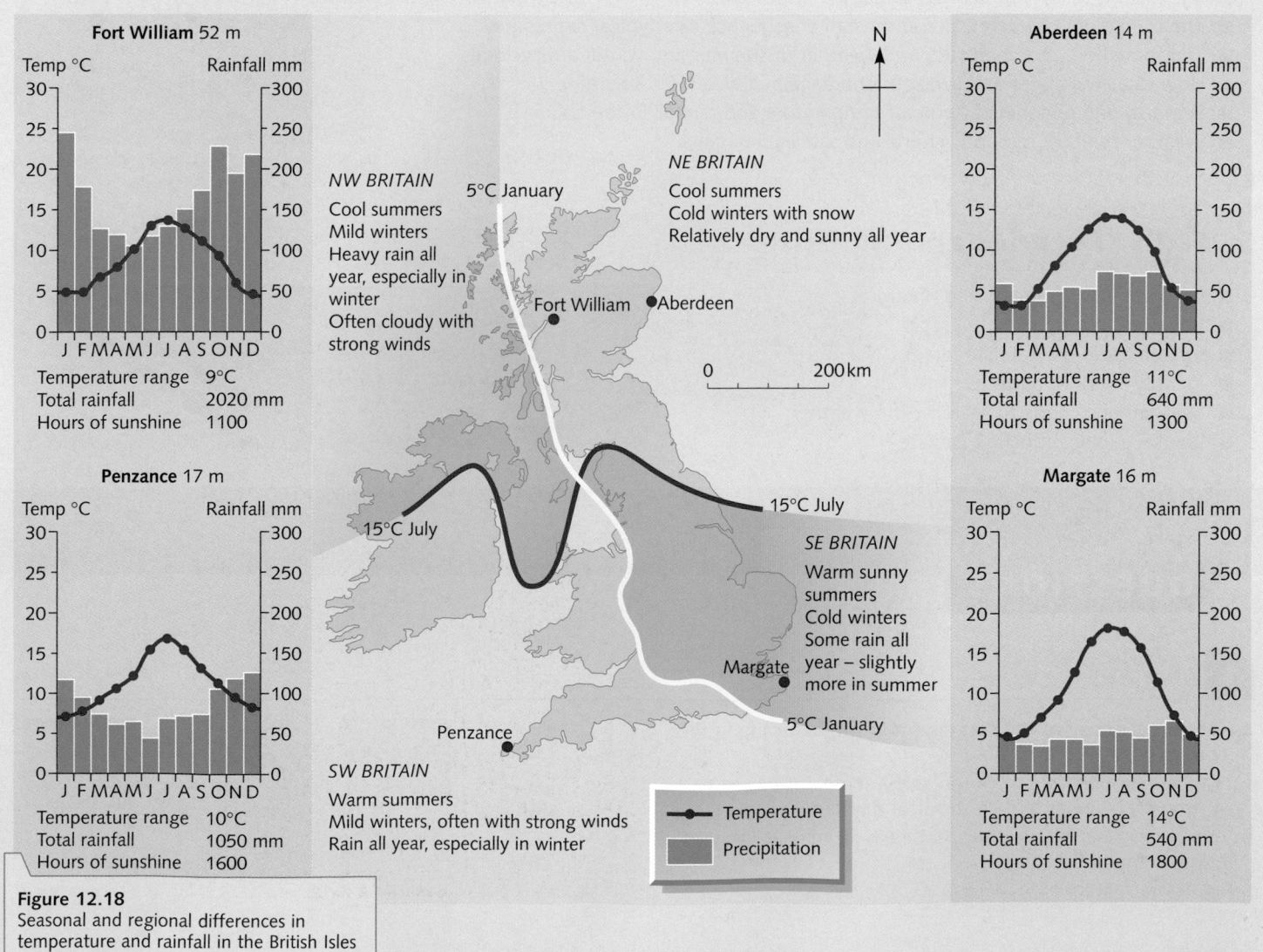

Figure 12.18
Seasonal and regional differences in temperature and rainfall in the British Isles

Britain's weather and climate

Key Words and Terms

a You should know the meaning of the following terms:
- air masses • prevailing winds • dew point • relief rainfall • frontal rainfall
- convectional rainfall • depressions • warm front • cold front
- occluded front • warm sector • weather symbols • weather station
- anticyclones.

b You should know the difference between the following:
- weather and climate
- maritime and continental air
- polar and tropical air
- satellite images and synoptic charts.

Key Ideas

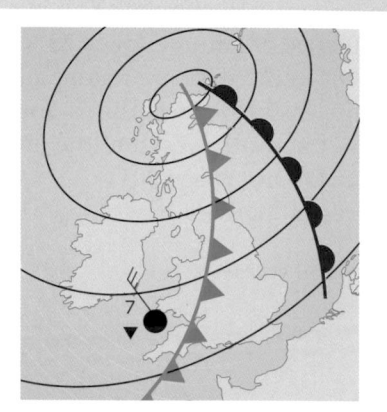

You should know and understand the following:
- The difference between weather and climate.
- The temperature and rainfall characteristics of the four main air masses that affect the UK.
- The effect on temperatures of latitude, distance from the sea, prevailing winds, ocean currents and altitude.
- The main types of rainfall that affect the British Isles.
- The formation, characteristics and weather sequence of a typical depression.
- The formation, characteristics and weather of summer and winter anticyclones.
- How satellite images and synoptic charts help in weather forecasting.
- Seasonal and regional patterns of temperature and rainfall in the UK.
- Weather symbols, synoptic charts and satellite images.

Skills, Theories and Models

- Interpret and use maps and diagrams
- Interpret and use climate graphs
- Interpret and use satellite images
- Identify weather patterns
- Use synoptic charts to forecast the weather
- Identify factors affecting climate

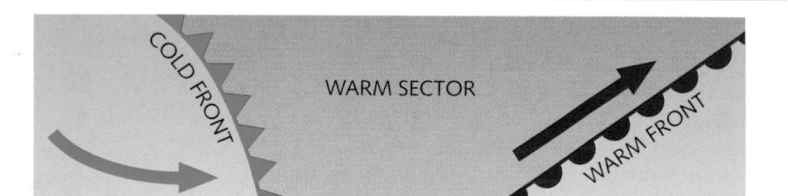

QUESTIONS

1 *(Page 200)*

a On a larger copy of the diagram, add the following in the correct boxes:
- polar continental
- polar maritime
- tropical continental
- tropical maritime. (4)

b Briefly describe the weather that each air mass is likely to bring to the British Isles. (8)

2 *(Pages 201 to 203 and 207)*

Five places are labelled A to E on the map below.

a With the help of a diagram, explain why place B is cooler than place A in summer. (3)

b Give two reasons why place C is warmer during winter than place D. (4)

c Why does place E have snow lying for over 50 days in an average year? (2)

d Why does place C receive more rainfall throughout the year than place D? (3)

e Draw labelled diagrams to show the formation of:
- relief rainfall   • frontal rainfall
- convectional rainfall. (9)

f Describe and account for the distribution of rainfall over the British Isles. (4)

4 *(Pages 205 and 206)*

Refer to the weather map below.

a i) Name the weather symbols labelled A and B. (2)
   ii) Is this an area of high pressure or low pressure? (1)

b What is the temperature, cloud cover, precipitation, wind speed, wind direction and pressure at:
- Belfast   • Manchester   • Aberdeen? (6 × 3)

c Draw a weather station model for London to show:
- Temperature: 15°C   • Wind speed: force 2 (6 knots)
- Wind direction: SSW
- Cloud cover: sky covered (8/8)
- Present weather: drizzle. (5)

d Give the likely forecast for London for the next few hours. (3)

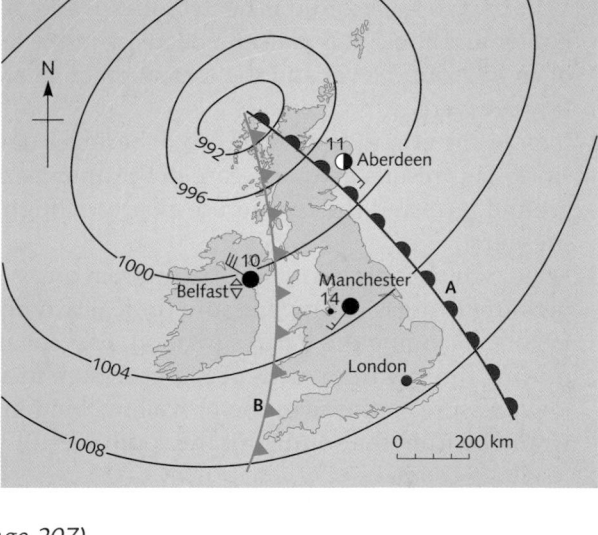

5 *(Page 207)*

Four climate graphs are shown in Figure 12.18.

a What is the: • maximum temperature
- minimum temperature
- annual range of temperature for Fort William? (3)

b Which of the four places has the:
- coolest summer   • warmest summer
- coldest winter   • mildest (warmest) winter
- largest range of temperature   • smallest range of temperature   • most sunshine in a year
- least hours of sunshine in a year
- most rainfall in a year   • least rainfall in a year? (9)

c i) What is the average rainfall for Penzance? (1)
   ii) Which is the wettest:
- month   • season at Penzance? (2)

3 *(Pages 204 to 206)*

a Why do depressions form over the Atlantic Ocean? (2)

b On an enlarged copy of table A, describe the likely weather conditions:
- as a warm front passes   • in a warm sector
- after a cold front passes. (18)

c Using table B as a guide, list six differences between the expected **summer** weather conditions between a depression and an anticyclone. (6 × 2)

d Give three likely differences in weather between a summer anticyclone and a winter anticyclone. (3)

e How are British people's lives affected by the weather in: • a depression   • an anticyclone? (6)

Weather conditions	A As a warm front passes	In a warm sector	Just after a cold front passes	B Depression in summer	Anticyclone in summer
Temperature					
Cloud cover					
Precipitation					
Wind speed					
Wind direction					
Pressure					

World climate

Atmospheric circulation and world climates

Although it may not be obvious at first glance, Figure 13.2 does show distinct patterns as to the location of the main climatic types. Take the British type as an example. Notice that it is located on the west coast of continents and between latitudes 40° and 60° north and south of the Equator. Can you, at this stage, identify patterns for the other five named climatic types? Any pattern that you could identify is mainly due to the circulation of air in the atmosphere (Figure 13.1). In order to understand this circulation, you need to be aware of three processes:

1 If air next to the ground is heated, it expands, gets lighter and rises. This causes a decrease in the amount of air at ground level and the formation of an area of **low pressure**.

2 If air in the atmosphere is cooled, it becomes denser and sinks. This results in an increase in the amount of air at ground level and the creation of an area of **high pressure**.

3 Wind, which is air in motion, blows from areas of high pressure to areas of low pressure (in reality other factors, including the Earth's rotation, prevent air from moving directly from areas of high pressure to areas of low pressure – otherwise the prevailing wind in Britain would be from the south, not the south-west).

As the air becomes denser, it descends to form a high pressure belt and, as it gets warmer and drier, creates an area of desert. Some of the descending air, on reaching the ground, returns to the Equator as the trade winds. The remainder moves away from the tropics (the westerlies) towards places like Britain where, as warm, moist tropical air, it meets with colder polar air and is forced to rise, creating low pressure and depressions (page 204).

This explanation is an oversimplified one (if it was not, then weather forecasting would be quite easy!). Due to the Earth's tilt and rotation, the position of the overhead sun appears to change, giving us our seasons. The northern hemisphere is hottest when the sun appears to be above the Tropic of Cancer (21 June). At this time, the pressure and wind belts move northwards. In contrast, the southern hemisphere is hottest when the sun appears to be over the Tropic of Capricorn (21 December), by which time the pressure and wind belts have moved south. The resultant changes in pressure and wind are responsible for the seasonal contrasts in climates, such as the tropical continental (page 213) and the monsoon (page 214) climate types.

Figure 13.1 shows the major areas of high and low pressure, and the general circulation of the atmosphere. The circulation is controlled by the build-up of heat at the Equator, which causes air to rise (convection currents) and low pressure to form. As the rising air cools, it condenses to give thunderstorms (page 203) before spreading outwards from the Equator.

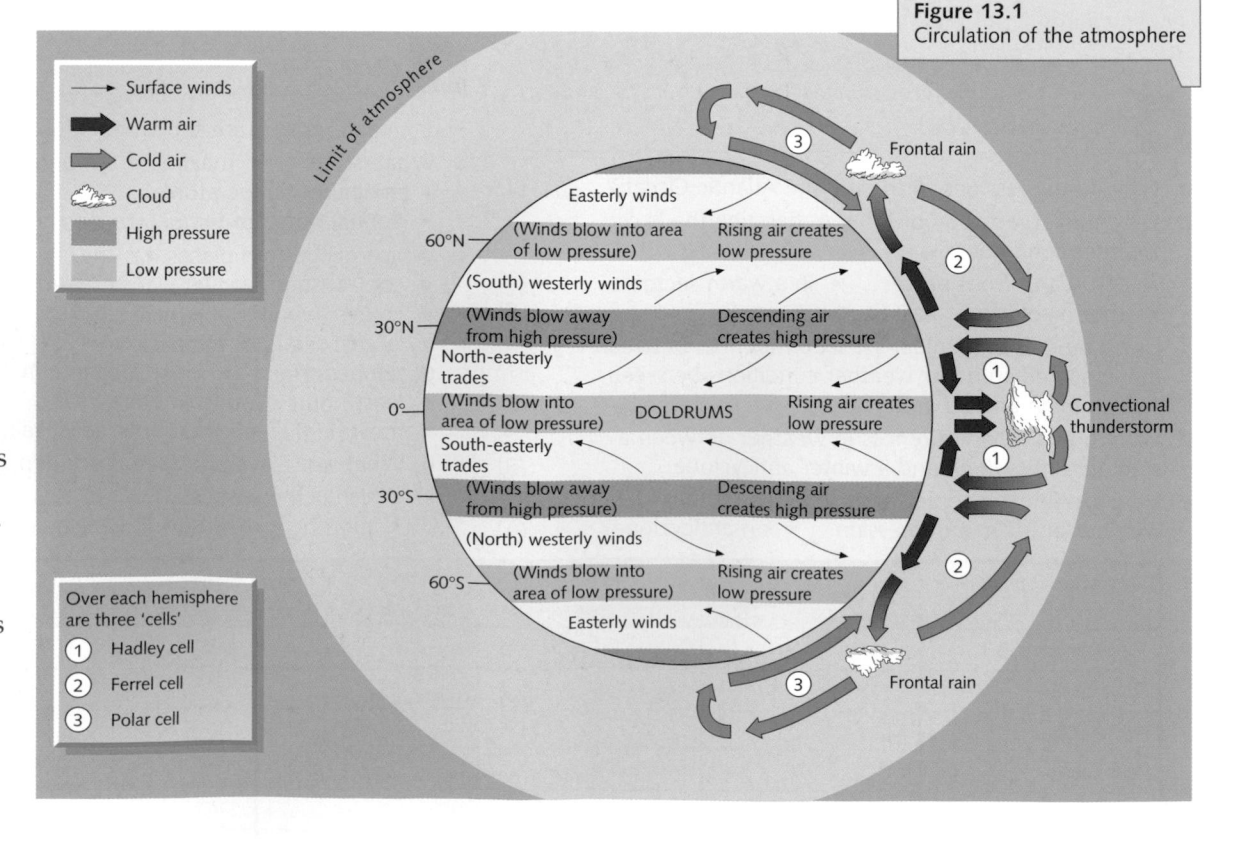

Figure 13.1
Circulation of the atmosphere

Factors affecting global temperatures

The five main factors that affect temperature have already been described on page 201 in relation to Britain's climate. Their effects, on a global scale, include:

Latitude Places in the tropics, being nearer the Equator, are much warmer than places towards the Poles (see Figure 12.2). The equatorial, tropical continental and monsoon climates are therefore much warmer than the British and cold climates.

Distance from the sea As land heats up more quickly during the summer and cools down more rapidly in winter than the sea, then places towards the centre of continents (e.g. tropical continental) will have warmer summers and colder winters than those with a coastal location (e.g. British climate).

Prevailing winds The seasonal difference in heating between land and sea also affects the temperature of the prevailing wind. Winds that blow from the warm sea in winter (e.g. British climate) or the warm land in summer (e.g. tropical continental) will raise temperatures whereas those that blow from the cold land in winter (e.g. cold climate) or the cooler sea in summer (British climate) will lower temperatures.

Ocean currents Ocean currents, which are classified as being either warm or cold, affect the climate of coastal areas. Warm currents tend to raise winter temperatures (e.g. British climate) while cold currents usually lower summer temperatures (e.g. where the hot deserts border the sea).

Altitude Temperature decreases, on average, 1°C for every 100 metres, which means that mountains are much colder than lowlands. The Andes in Ecuador are high enough to be snow-covered all year despite their location on the Equator.

Rainfall

Similarly the three main types of rainfall that affect the British Isles (pages 202 and 203) also have a major effect on world climate:

Relief (Figure 12.8) Rainfall is heavy where prevailing winds blow from the sea and are forced to rise over coastal mountains (e.g. the Indian monsoon in summer). Places protected by mountains from the prevailing rain-bearing winds are usually dry (e.g. the Indian monsoon in winter).

Frontal (Figure 12.9) Rainfall is also heavy where depressions, with their associated fronts, are frequent (page 204). Somewhat similar, in that they are also areas of low pressure that bring heavy rain, are **tropical storms** such as hurricanes (page 216).

Convectional (Figure 12.10) This occurs where temperatures are either continually high, as most afternoons in the equatorial climate, or seasonally, as in southern England in summer.

World climates

Figure 13.2 shows the location and extent of six types of climate described in this book. Maps that show the world climates are very generalised (simplified) as, due to their scale, they cannot show local variations. They also suggest that boundaries between climates are a thin line, whereas in reality any change is often gradual and extends across a broad transition zone.

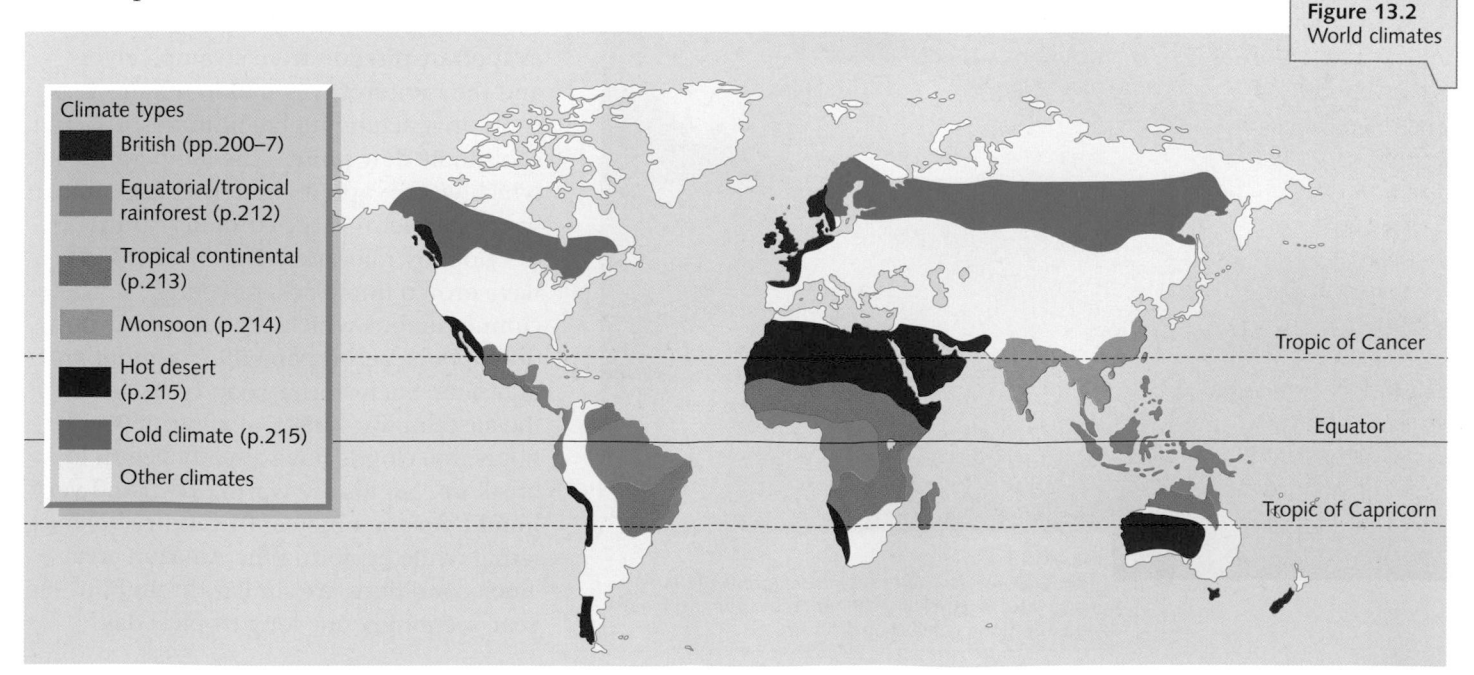

Figure 13.2
World climates

Climate types
- British (pp.200–7)
- Equatorial/tropical rainforest (p.212)
- Tropical continental (p.213)
- Monsoon (p.214)
- Hot desert (p.215)
- Cold climate (p.215)
- Other climates

Tropic of Cancer

Equator

Tropic of Capricorn

Equatorial and tropical continental climates

Equatorial

Places with an equatorial climate lie within 5° either side of the Equator. The main areas are the large drainage basins of the Amazon (South America) and Congo (Africa) rivers and the extreme south-east of Asia (Figure 13.2). Figure 13.3 is a climate graph for Manaus which is located 3° south of the Equator and in the centre of the Amazon basin in Brazil. It shows that temperatures are both high and constant throughout the year. The small annual range of 2°C is due to the sun always being at a high angle in the sky, even if not always directly overhead (Figure 12.2). Equatorial areas have annual rainfall totals in excess of 2000 mm, mainly due to convectional thunderstorms which occur during most afternoons throughout the year (Figure 12.10). These storms are due to the convergence of warm air (the trade winds). The warm air is forced to rise, creating an area of low pressure and giving heavy rain (Figure 13.1).

Some places, like Manaus, may have two or three drier, but not dry, months when the sun is overhead at the opposite tropic (which is Cancer in the case of Manaus), and most rain when the sun is closer to being overhead. Winds are generally light (the doldrums are areas of calm over equatorial oceans) and variable (there are no prevailing winds, Figure 13.1). The equatorial climate is characterised by its high humidity, a lack of seasonal change, and a daily weather pattern that remains remarkably uniform throughout the year.

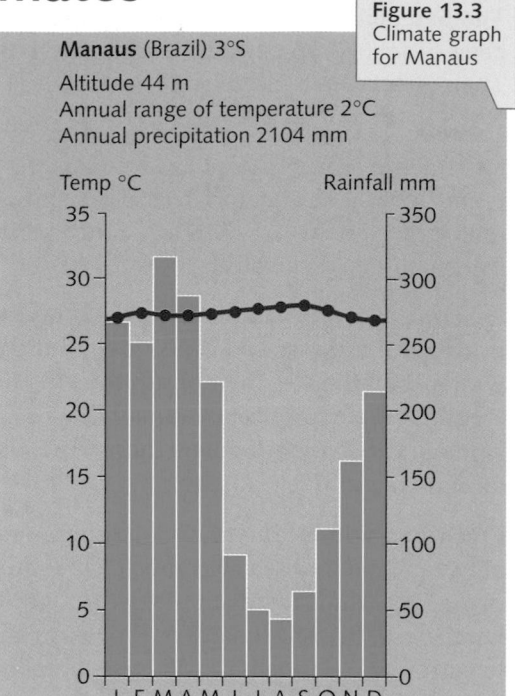

Figure 13.3
Climate graph for Manaus

Manaus (Brazil) 3°S
Altitude 44 m
Annual range of temperature 2°C
Annual precipitation 2104 mm

The daily pattern

One day is very much like another, with places receiving 12 hours of daylight and 12 hours of darkness. The sun rises at 0600 hours and its heat soon evaporates the morning mist, the heavy overnight dew, and any moisture remaining from the previous afternoon's storm. Even by 0800 hours temperatures are as high as 25°C and by noon, when the sun is near to a vertical position, they reach 33°C. The high temperatures cause air to rise in powerful convection currents. The rising air, which is very moist due to rapid evapotranspiration from swamps, rivers and the rainforest vegetation (Figure 13.4), cools on reaching higher altitudes. When it cools to its **dew point** – the temperature at which water vapour condenses back into water droplets – large cumulus clouds develop. By mid-afternoon these clouds have grown into black, towering cumulonimbus which produce torrential downpours, accompanied by thunder and lightning. Such storms soon cease, leaving the air calm. By sunset, at about 1800 hours, the clouds have already begun to break up. Nights are warm (23°C) and very humid. As one nineteenth-century botanist wrote while exploring the Amazon area, 'Such conditions are similar throughout the year, seemingly one long tropical day'.

Figure 13.4
The rainforest is a store for vast amounts of water. This water is recycled daily due to evapotranspiration and afternoon convectional storms

Tropical continental (interior)

This climate is found in central parts of continents, away from coasts, and mainly between latitudes 5° and 15° north and south of the Equator (Figure 13.2). This includes the Brazilian Highlands and parts of Venezuela (South America), northern Australia, and a large semi-circle surrounding the Congo basin (Africa). This last area includes Kenya which, although straddling the Equator, lies at an altitude too high to share the main features of the equatorial climate. The main characteristic of the tropical continental climate, as illustrated by the climate graph for Kano in northern Nigeria (Figure 13.5), is alternate wet and dry seasons, which are caused by the apparent movement of the overhead sun (Figure 13.6). The hot, wet season occurs at Kano when the sun is overhead in the northern hemisphere (21 June – Figure 13.6b). Temperatures rise as the sun takes a more vertical position and because places are too far inland to be affected by any moderating influence of the sea. However, temperatures fall slightly in July and August as cloud cover and rainfall increase. The frequent afternoon convectional thunderstorms mean that the climate resembles, at this time of year, that of equatorial areas. Unfortunately both the length of the rainy season and the total amounts of rain are unreliable and many areas, especially those with increasing distance from the Equator in Africa, have experienced severe drought in recent years (Case Study 13A).

The dry, slightly cooler season at Kano occurs when the sun is overhead in the opposite (i.e. southern) hemisphere (21 December – Figure 13.6d). The apparent movement of the sun southwards is accompanied by the movement of the equatorial low pressure and the global wind belts (Figure 13.6 and page 210) so that the area is now influenced by the prevailing trade winds. The trade winds, blowing from the east, will have shed any moisture long before they reach these inland areas. Indeed, as the sun is still at a sufficiently high angle to give quite high temperatures, the climate here at this time of year resembles that of the hot deserts. (Remember that for places south of the Equator, the seasonal pattern described here is reversed.)

Temp °C
Rainfall mm
35
30
25
20
15
10
5
0
350
300
250
200
150
100
50
0
J F M A M J J A S O N D
Kano (Nigeria) 12°N
Altitude 630 m
Annual range of temperature 8°C
Annual precipitation 920 mm

Figure 13.5
Climate graph for Kano

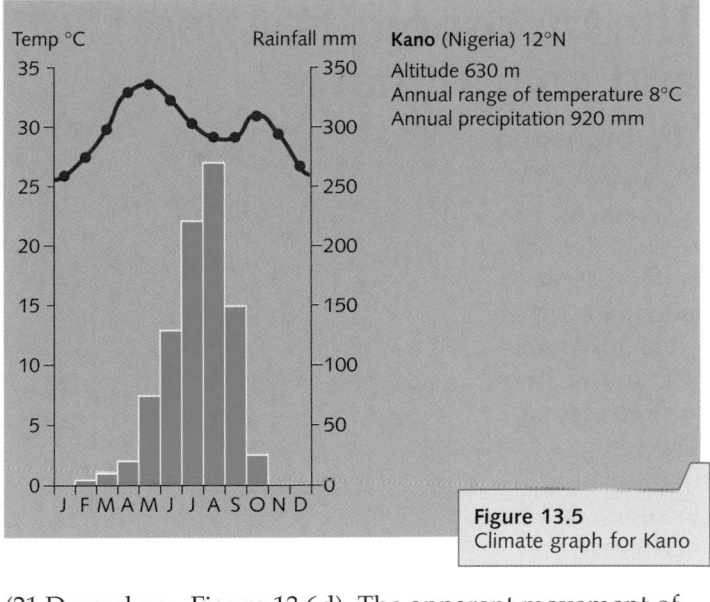

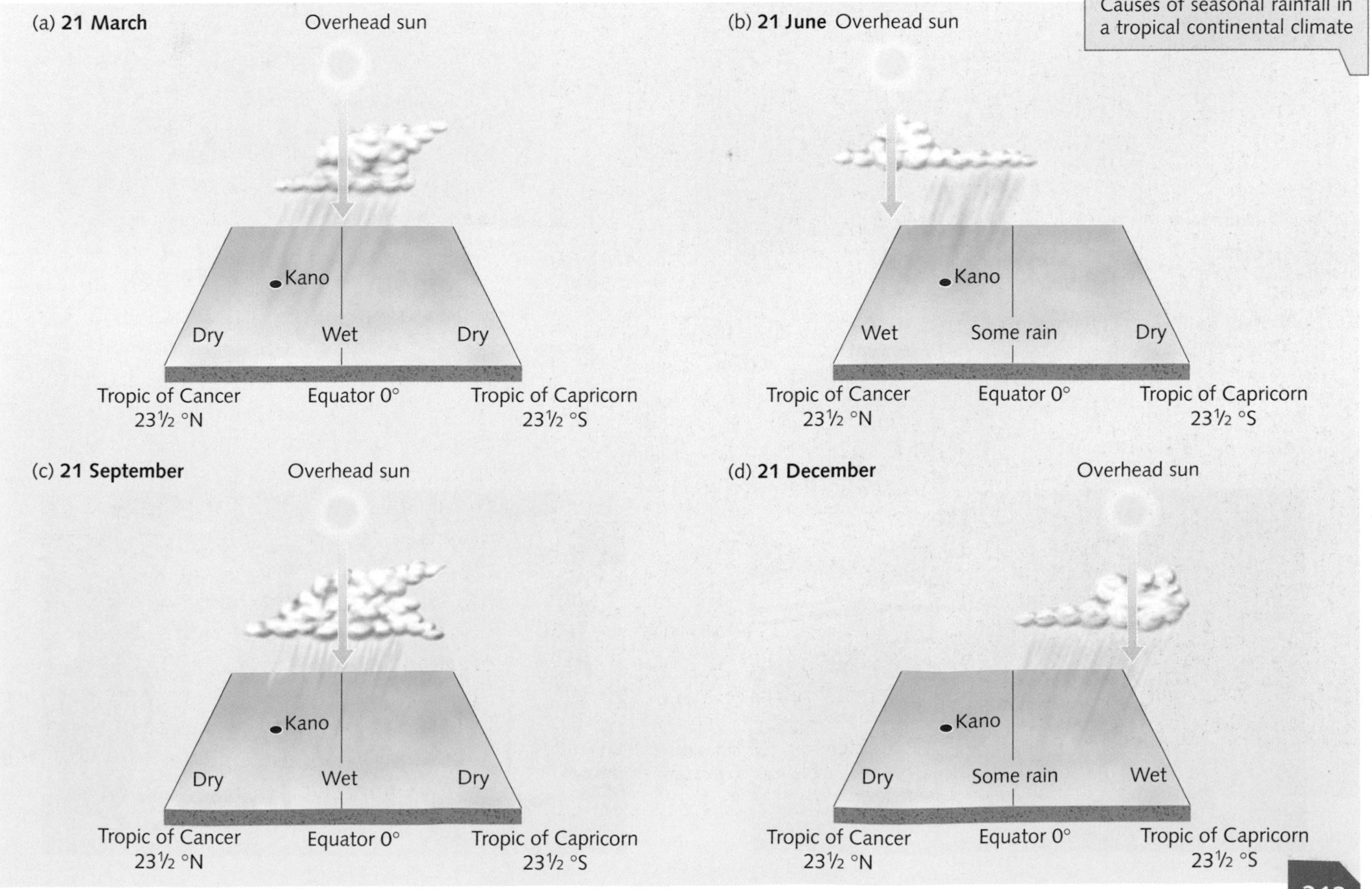

(a) 21 March
Overhead sun
Kano
Dry
Wet
Dry
Tropic of Cancer 23½ °N
Equator 0°
Tropic of Capricorn 23½ °S
(b) 21 June Overhead sun
Kano
Wet
Some rain
Dry
Tropic of Cancer 23½ °N
Equator 0°
Tropic of Capricorn 23½ °S
(c) 21 September
Overhead sun
Kano
Dry
Wet
Dry
Tropic of Cancer 23½ °N
Equator 0°
Tropic of Capricorn 23½ °S
(d) 21 December
Overhead sun
Kano
Dry
Some rain
Wet
Tropic of Cancer 23½ °N
Equator 0°
Tropic of Capricorn 23½ °S

Figure 13.6
Causes of seasonal rainfall in a tropical continental climate

The monsoon, hot desert and cold climates

The monsoon

The word **monsoon** means 'a season'. In much of South-east Asia, as seen in the climate graph for Mumbai (Figure 13.7), there are two seasons. These, the so-called south-west monsoon and the north-east monsoon, result from the reversal in the direction of the prevailing wind.

Figure 13.7
Climate graph for Mumbai

Mumbai (India) 19°N
Altitude 210 m
Annual range of temperature 6°C
Annual precipitation 2078 mm

The south-west monsoon (Figure 13.8) The sun appears to be overhead at the Tropic of Cancer in June. Places to the north of the Himalayas, which are a long way from any moderating influence of the sea, become extremely hot. As the hot air rises, an extensive area of low pressure is formed and warm, moist air is drawn northwards from the Indian Ocean. Where the air is forced to rise over mountains (e.g. the Western Ghats and the Himalayas), it gives large amounts of relief rainfall (Figure 12.8). Mumbai gets over 2000 mm in five months and Cherrapunji, reputed to be the wettest place on Earth, 14 000 mm. The rain, ideal for rice (Case Study 7), can cause extensive flooding in the Ganges basin, while places north of the Himalayas, being in the rainshadow, remain dry.

The north-east monsoon (Figure 13.9) During the northern winter, the overhead sun moves southwards. Places to the north of the Himalayas now become very cold and an extensive area of high pressure develops. Winds blow outwards from the high pressure area but, because they originate in a dry area, they give only small amounts of rain as they cross India – Mumbai, for example, only receives 45 mm in seven months. By the end of the 'dry' season, many places are at risk of drought.

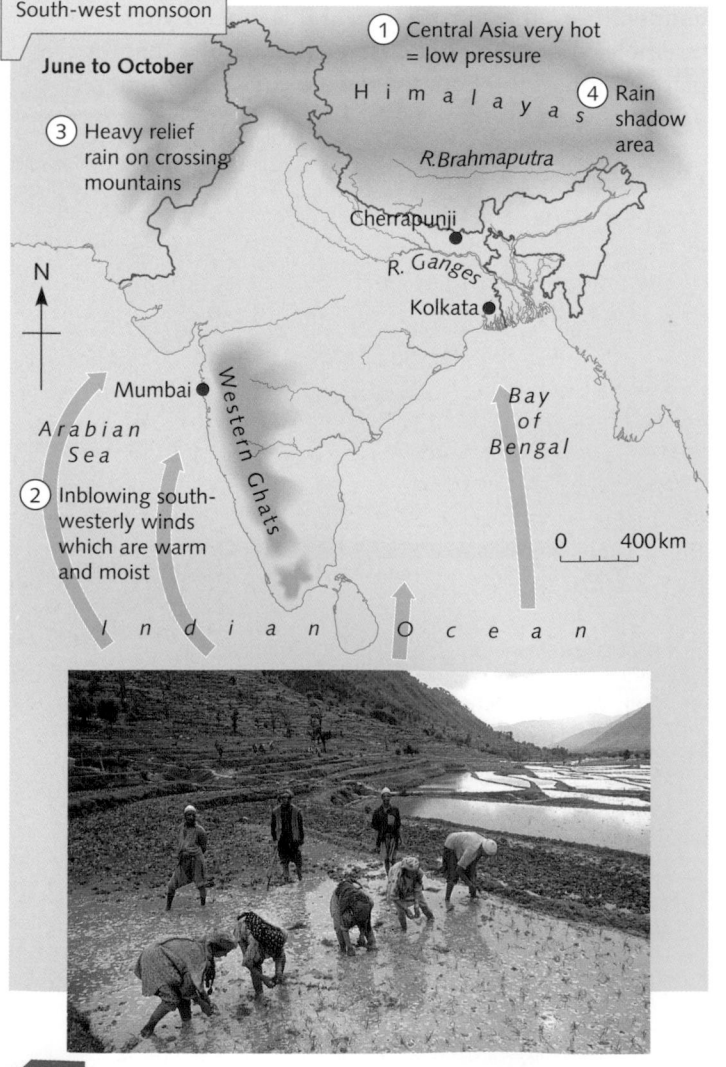

Figure 13.8
South-west monsoon

June to October

① Central Asia very hot = low pressure
③ Heavy relief rain on crossing mountains
④ Rain shadow area
② Inblowing south-westerly winds which are warm and moist

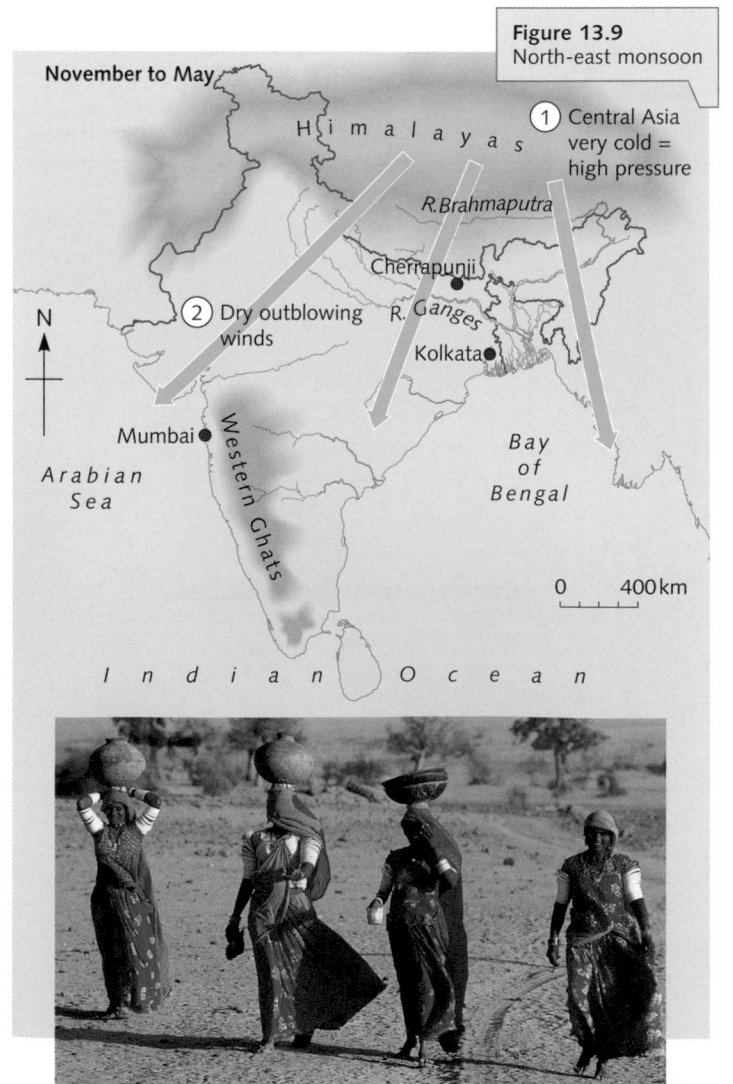

Figure 13.9
North-east monsoon

November to May

① Central Asia very cold = high pressure
② Dry outblowing winds

Hot desert

Figure 13.10 Climate graph for Ain Salah

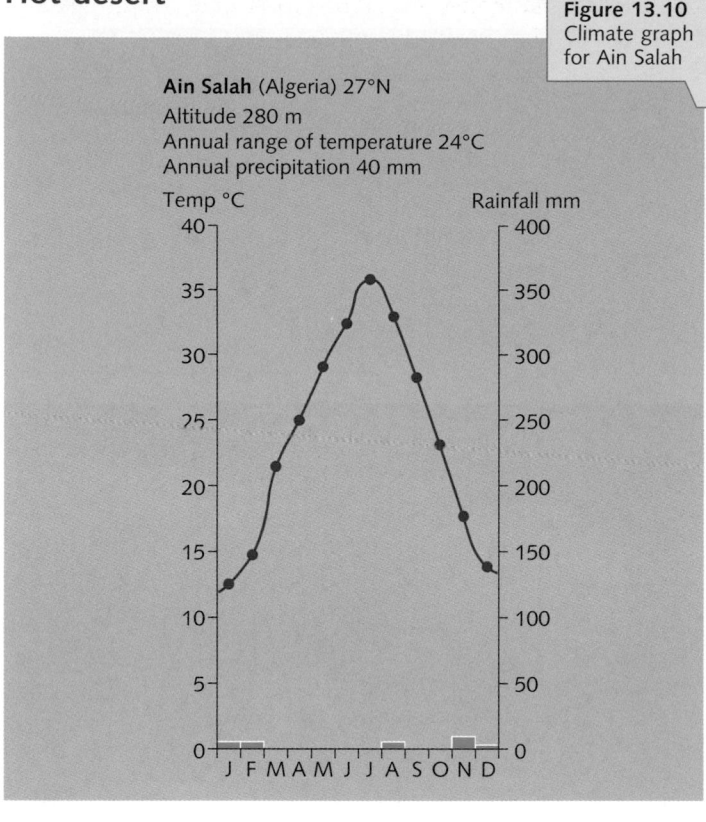

Ain Salah (Algeria) 27°N
Altitude 280 m
Annual range of temperature 24°C
Annual precipitation 40 mm

Cold climate

Figure 13.11 Climate graph for Fairbanks

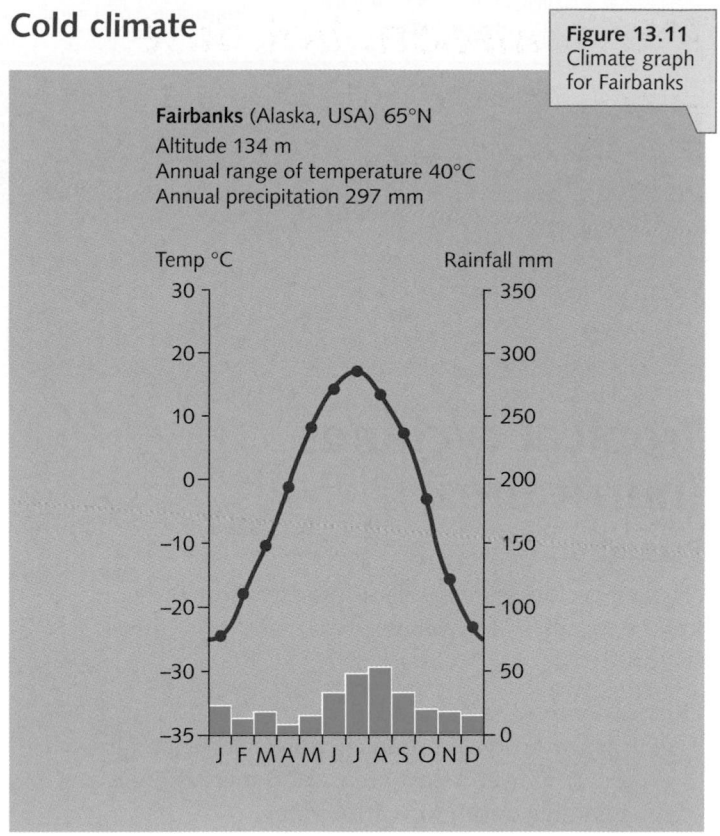

Fairbanks (Alaska, USA) 65°N
Altitude 134 m
Annual range of temperature 40°C
Annual precipitation 297 mm

Hot deserts are places with high annual temperatures, less than 250 mm of rain a year, and high evapotranspiration rates. Apart from the Sahara, which extends across Africa, most deserts are located on the west coast of continents, in the subtropical high pressure belt (Figure 13.1), and between 5° and 30° north and south of the Equator (Figure 13.2). They include the Atacama (South America), the Kalahari–Namib (southern Africa) and the Australian and Mexican deserts.

As the climate graph for Ain Salah in the Sahara Desert shows (Figure 13.10), temperatures are highest when the sun is overhead, but lower when it is in the opposite hemisphere. Coastal areas are much cooler, partly due to the moderating influence of the sea and partly because of cold offshore currents. Inland and away from the influence of the sea, cloudless skies allow daytime temperatures to rise to 50°C and night-time temperatures to fall to around freezing. Although deserts are very dry, all have some precipitation even if, as in the case of the Atacama and Namib, it comes mainly in the form of coastal fog. The lack of rain is due to a combination of factors:

- Prevailing winds blow from the dry land and so cannot pick up moisture.
- Prevailing winds have to cross mountain barriers which create rain shadows.
- Air that rose as convection currents at the Equator descends in these latitudes. As it descends it warms, can hold more moisture, and gives clear skies.
- When winds do blow from the sea, they are cool and unable to pick up much moisture.

The cold climate is found to the poleward side of 60°N in Eurasia, where it extends from Scandinavia across Russia to the Pacific, and across North America, central Alaska and northern Canada (Figure 13.2). The graph for Fairbanks (Figure 13.11) shows the large annual range in temperature that typifies this type of climate. Winters are very long, dark and extremely cold. This is partly due to the northerly latitude which, for places within the Arctic Circle, means a period of time when the sun fails to rise above the horizon, and partly due to the distance from any moderating influence of the sea. Strong winds can lower temperatures, increase the wind-chill factor and bring the risk of frost-bite. Summers are short but relatively warm for the latitude. This is partly due to the long hours of daylight (the sun does not set at all for a short time each year at places within the Arctic Circle, although its angle is always low in the sky) and partly due to its continental location far from the influence of the sea.

Precipitation is light throughout the year partly because the air is too cold to hold much moisture and partly because most places are a long way from any rain-bearing winds from the sea. It falls as snow in winter and may be blown about in blizzards. The summer maximum results from occasional convectional rainstorms occurring when temperatures are at their highest.

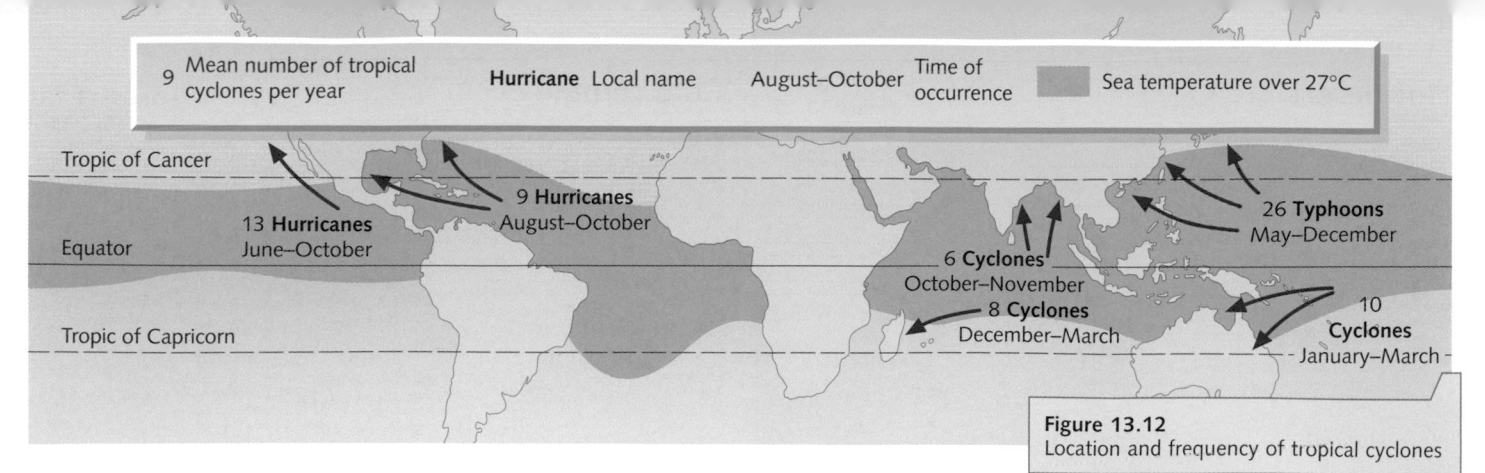

9 Mean number of tropical cyclones per year   **Hurricane** Local name   August–October Time of occurrence   Sea temperature over 27°C

Tropic of Cancer

13 Hurricanes
June–October

9 Hurricanes
August–October

Equator

26 Typhoons
May–December

6 Cyclones
October–November

8 Cyclones
December–March

10 Cyclones
January–March

Tropic of Capricorn

Figure 13.12
Location and frequency of tropical cyclones

Tropical cyclones (hurricanes)

Formation

Tropical cyclones are areas of intensive low pressure known locally as **hurricanes**, typhoons or cyclones (Figure 13.12).

Hurricanes tend to develop:
- over warm tropical oceans, where sea temperatures exceed 27°C over a vast area, and where there is a considerable depth of warm water
- in late summer and early autumn, when sea temperatures are at their highest
- in the trade wind belt between latitudes 5° and 20° north and south of the Equator.

Although their formation is not yet fully understood, they appear to originate when a strong vertical movement of air draws with it water vapour from the ocean below. As the air rises, in a spiral movement, it cools and condenses – a process that releases enormous amounts of heat energy. (It has been estimated that the heat energy released in a single day in a hurricane is equivalent to 500 000 atomic bombs the size of those dropped on Japan during the Second World War.) It is this heat energy that powers the storm and which must be maintained if the

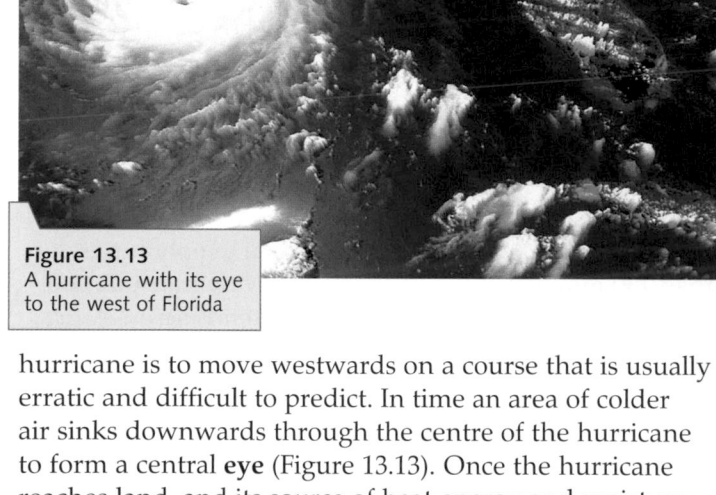

Figure 13.13
A hurricane with its eye to the west of Florida

hurricane is to move westwards on a course that is usually erratic and difficult to predict. In time an area of colder air sinks downwards through the centre of the hurricane to form a central **eye** (Figure 13.13). Once the hurricane reaches land, and its source of heat energy and moisture is removed, it rapidly decreases in strength. Its average lifespan is 7 to 14 days. The characteristic weather associated with the passing of a typical hurricane is described in Figure 13.14.

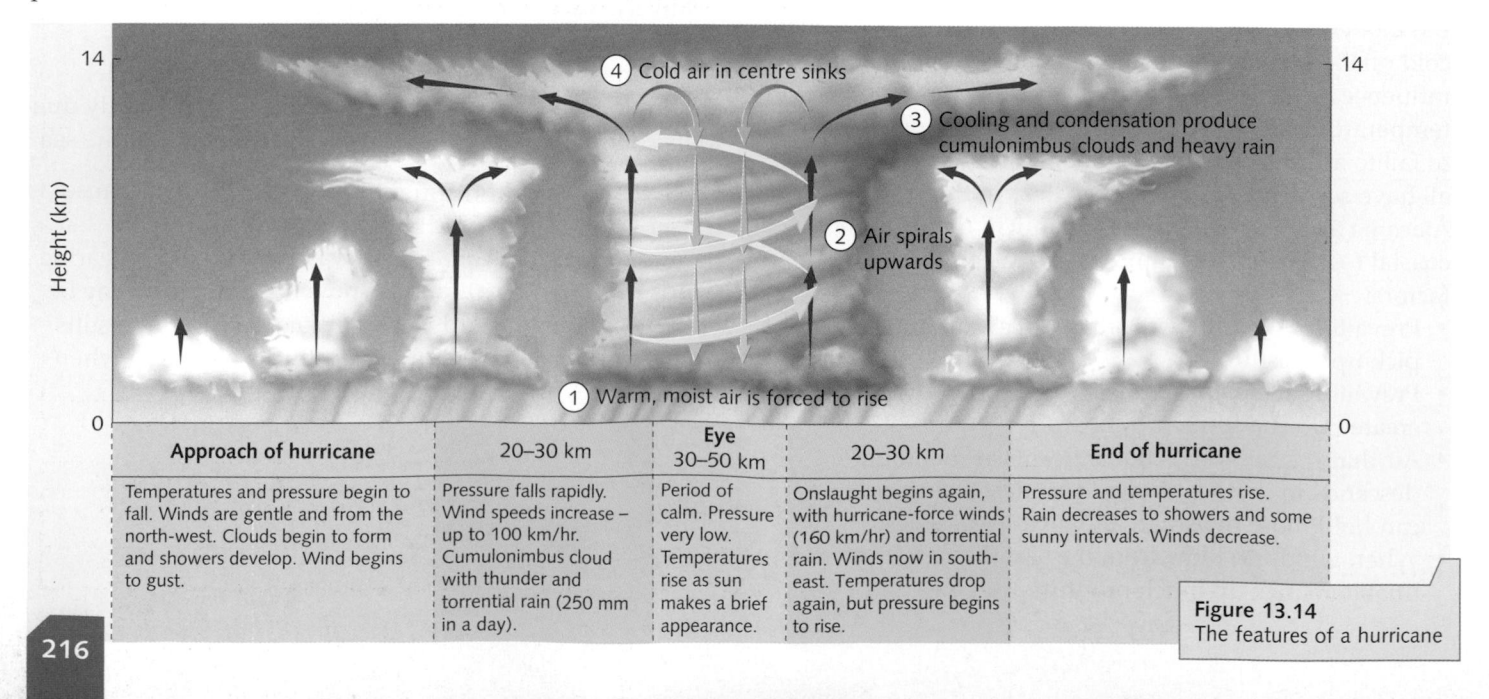

(4) Cold air in centre sinks

(3) Cooling and condensation produce cumulonimbus clouds and heavy rain

(2) Air spirals upwards

(1) Warm, moist air is forced to rise

Approach of hurricane	20–30 km	Eye 30–50 km	20–30 km	End of hurricane
Temperatures and pressure begin to fall. Winds are gentle and from the north-west. Clouds begin to form and showers develop. Wind begins to gust.	Pressure falls rapidly. Wind speeds increase – up to 100 km/hr. Cumulonimbus cloud with thunder and torrential rain (250 mm in a day).	Period of calm. Pressure very low. Temperatures rise as sun makes a brief appearance.	Onslaught begins again, with hurricane-force winds (160 km/hr) and torrential rain. Winds now in south-east. Temperatures drop again, but pressure begins to rise.	Pressure and temperatures rise. Rain decreases to showers and some sunny intervals. Winds decrease.

Figure 13.14
The features of a hurricane

Effects

Hurricanes, and other tropical storms, are a major hazard that can cause considerable loss of life and damage to property and to a country's economy.

- **Winds** often exceed 160 km/hr (and can reach 250 km/hr). In LEDCs, many of which lie in the tropical cyclone belt (Figure 13.12), whole villages may be destroyed and even in MEDCs, where people have money to reinforce buildings, houses and coastal developments can be severely damaged. High winds uproot trees and disrupt telephone and electricity power supplies. Worst hit are those LEDCs that have their only export crop destroyed (e.g Honduras, Figure 11.13).
- **Flooding**, caused by the torrential rain, is often the major cause of death. It can also pollute water supplies, increasing the risk of cholera.
- **Storm (tidal) surges**, up to 5 m and heightened by storm waves, flood low-lying coastal areas causing loss of life (Bangladesh, page 312) and blocking escape and relief roads.
- **Landslides** occur where heavy rainfull washes away buildings erected on steep, unstable slopes.

The Gulf of Mexico

Hurricanes are a major annual hazard to countries surrounding the Gulf of Mexico (Figure 13.12). However, their short-term and long-term effects are usually much greater in the LEDCs of Central America and the West Indies than in the MEDC of the USA (Figure 13.15). Two of the worst storms to hit each area in recent years are described in Figure 13.16.

Figure 13.15
The paths of two hurricanes

USA (MEDC)
Florida
Gulf of Mexico
CUBA
Caribbean Sea
JAMAICA
MEXICO
HONDURAS
Hurricane Wilma October 2005
Hurricane Mitch Oct/Nov 1998
GUATEMALA
NICARAGUA
EL SALVADOR
LEDCs
0   500 km

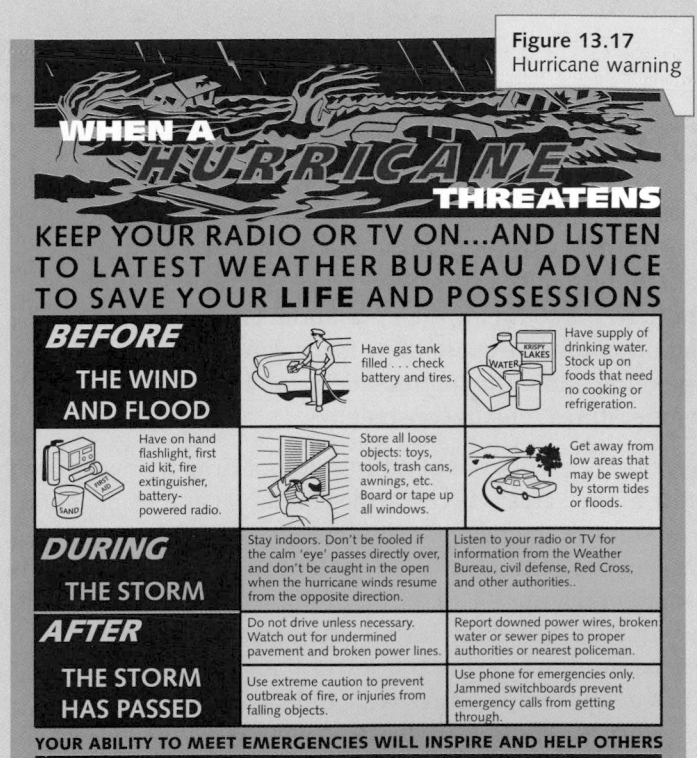

Figure 13.17
Hurricane warning

WHEN A HURRICANE THREATENS

KEEP YOUR RADIO OR TV ON...AND LISTEN TO LATEST WEATHER BUREAU ADVICE TO SAVE YOUR LIFE AND POSSESSIONS

BEFORE THE WIND AND FLOOD
- Have gas tank filled . . . check battery and tires.
- Have supply of drinking water. Stock up on foods that need no cooking or refrigeration.
- Have on hand flashlight, first aid kit, fire extinguisher, battery-powered radio.
- Store all loose objects: toys, tools, trash cans, awnings, etc. Board or tape up all windows.
- Get away from low areas that may be swept by storm tides or floods.

DURING THE STORM
- Stay indoors. Don't be fooled if the calm 'eye' passes directly over, and don't be caught in the open when the hurricane winds resume from the opposite direction.
- Listen to your radio or TV for information from the Weather Bureau, civil defense, Red Cross, and other authorities.

AFTER THE STORM HAS PASSED
- Do not drive unless necessary. Watch out for undermined pavement and broken power lines.
- Report downed power wires, broken water or sewer pipes to proper authorities or nearest policeman.
- Use extreme caution to prevent outbreak of fire, or injuries from falling objects.
- Use phone for emergencies only. Jammed switchboards prevent emergency calls from getting through.

YOUR ABILITY TO MEET EMERGENCIES WILL INSPIRE AND HELP OTHERS

US DEPARTMENT OF COMMERCE • WEATHER BUREAU

Figure 13.16
Hurricanes Mitch and Wilma

	LEDC – Hurricane Mitch, 1998	MEDC – Hurricane Wilma, 2005
Location	Nicaragua, Honduras, Guatemala, El Salvador	USA – Florida
Deaths	Over 12 000	12
Missing	Over 16 000	8
Homeless	700 000	60 000
Number affected	3 million	2 million
Time for help to arrive	Several days	Within hours
Effect on services	No telephones, electricity, fresh water or transport	No telephones, electricity or water
Effect on economy	Crops lost/damaged – often the only product for sale/export: **long-term**	Many businesses forced to close until electricity restored: **short-term**
Secondary effects	Shortages of food, clothing, medical supplies. Fear of cholera	Very few, apart from loss of crops
Prediction	Less effective early warning system. Some people without TV/radio to hear warning	Effective early warning system. People able to hear warnings on TV/radio (Figure 13.17)
Planning	Limited emergency services and equipment. Few available hospitals and doctors. Many houses flimsy and poorly built – easily blown down/washed away. Not insured	Emergency services well trained and equipped. Houses strongly built and reinforced. Many hospitals, doctors and ambulances. Houses insured (Figure 13.17)
Time to recover	Estimated 10 years	Within a few days

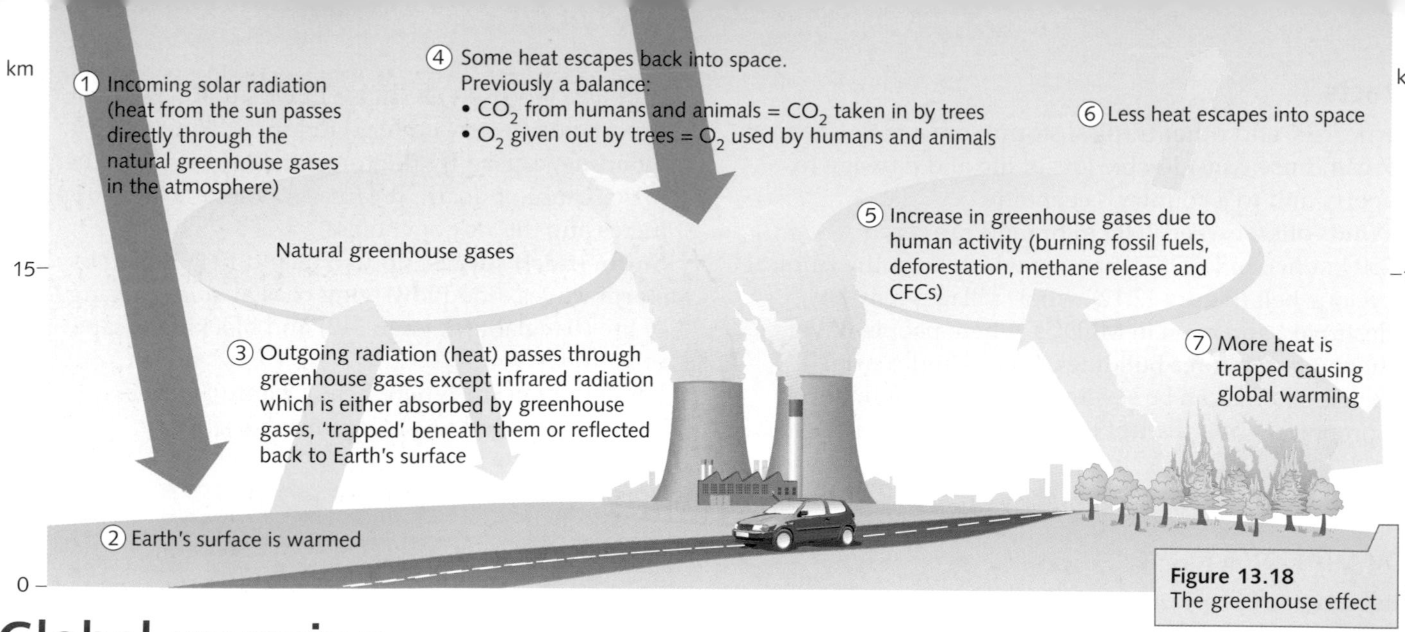

① Incoming solar radiation (heat from the sun passes directly through the natural greenhouse gases in the atmosphere)

④ Some heat escapes back into space. Previously a balance:
• CO_2 from humans and animals = CO_2 taken in by trees
• O_2 given out by trees = O_2 used by humans and animals

⑥ Less heat escapes into space

⑤ Increase in greenhouse gases due to human activity (burning fossil fuels, deforestation, methane release and CFCs)

Natural greenhouse gases

③ Outgoing radiation (heat) passes through greenhouse gases except infrared radiation which is either absorbed by greenhouse gases, 'trapped' beneath them or reflected back to Earth's surface

⑦ More heat is trapped causing global warming

② Earth's surface is warmed

Figure 13.18
The greenhouse effect

Global warming

The greenhouse effect

The Earth is warmed during the day by incoming radiation from the sun. The Earth loses heat at night through outgoing infrared radiation. Over a lengthy period of time, because there is a balance between incoming and outgoing radiation, the Earth's temperatures remain constant.

On cloudy nights, temperatures do not drop as low as on clear nights. This is because the clouds act as a blanket and trap some of the heat. Greenhouse gases in the atmosphere also act as a blanket, as they prevent the escape of infrared radiation (Figure 13.18). Without these greenhouse gases, which include carbon dioxide, the Earth's average temperature would be 33°C lower than it is today (during the Ice Age, temperatures were only 4°C lower than at present). Recent human activity has led to a significant increase in the amount, and type, of greenhouse gases in the atmosphere. This is preventing heat from escaping into space, and is believed to be responsible for a rise in world temperatures. World temperatures rose by 0.23°C in the 100 years between 1880 and 1979 compared with 0.62°C in the 30 years since then (Figure 13.19), with eight of the ten warmest ever recorded years falling in the last decade. The most recent UN report (produced by several hundred of the world's leading meteorologists) suggested, from the use of computer-generated models, that a further rise of 5.8°C could take place before the end of this century (replacing previous predictions of between 1.5°C and 4.5°C). The process by which world temperatures are rising is known as **global warming**.

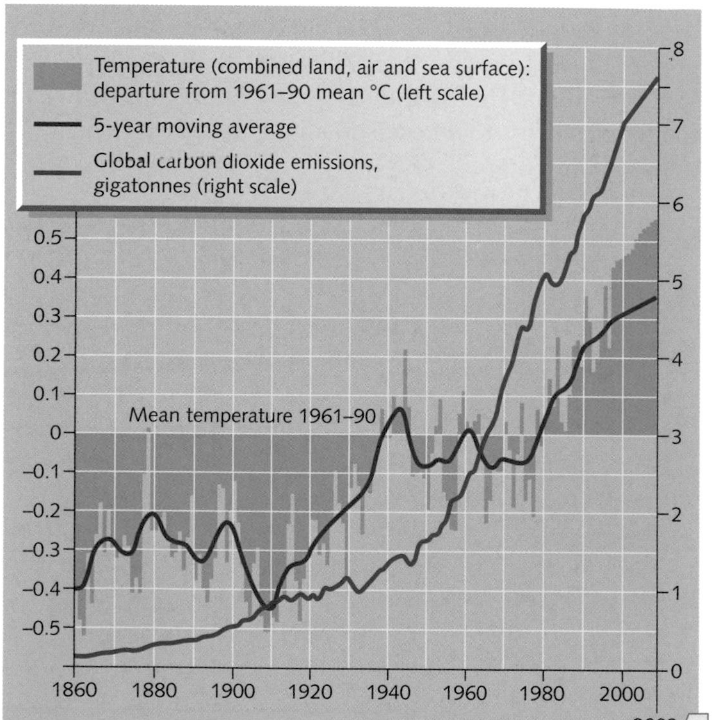

Causes of global warming

The major contributors to global warming are carbon dioxide and other pollutants released into the atmosphere (Figure 13.20).

• Carbon dioxide is the most important single factor in global warming. It is produced by road vehicles and by

Temperature (combined land, air and sea surface): departure from 1961–90 mean °C (left scale)

5-year moving average

Global carbon dioxide emissions, gigatonnes (right scale)

Mean temperature 1961–90

Figure 13.19
Increased carbon dioxide emissions and global warming

burning fossil fuels in power stations, in factories and in the home (Figures 13.21 and 13.22a). Since the more economically developed countries consume three-quarters of the world's energy, they are largely responsible for global warming. A secondary source of carbon dioxide is deforestation and the burning of the tropical rainforests (page 236).

• CFCs (chlorofluorocarbons) from aerosols, air conditioners, foam packaging and refrigerators are the most damaging of the greenhouse gases.

• Methane is released from decaying organic matter such as peat bogs, swamps, landfill sites, animal dung and farms (e.g. ricefields in South-east Asia).

• Nitrous oxide is emitted from car exhausts, power stations and agricultural fertiliser.

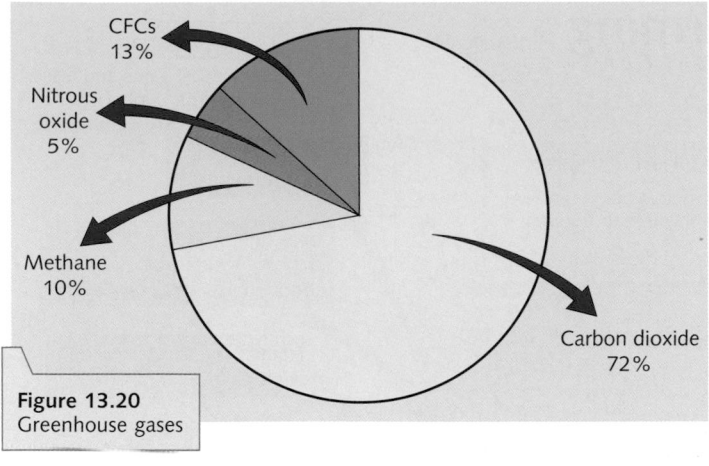

Figure 13.20
Greenhouse gases

Figure 13.21
Drax power station, Thorpe Marsh near Doncaster

Effects of global warming

The Intergovernmental Panel on Climatic Change (IPCC), reporting on behalf of the UN, described some effects that have already been observed, as well as predicting other effects likely to occur (Figures 13.23 and 13.24). These include the following:

- As sea temperatures rise, water in the oceans is expanding causing sea-levels to rise. IPCC scientists claim that sea-level rose by 0.2 metres during the last century, and give an average predicted rise of 0.5 metres for this century.
- Ice-caps and glaciers are already melting, and ice-shelves collapsing, at an unprecedented rate. Satellite data shows that sea-ice in the Arctic is retreating by 8 per cent a year and is 40 per cent less thick than it was in 1960.
- The release of water at present held in storage as ice and snow in the hydrological cycle (Figure 17.3) could raise the world's sea-level by a further 5 m. Even a rise of 1 m could flood 25 per cent of Bangladesh (page 312), 30 per cent of Egypt's arable land, and totally submerge several low-lying islands in the Indian and Pacific Oceans (Figure 13.24). At present, 40 per cent of the world's population live within 100 km of the coast – many people being vulnerable to rising sea-levels and storm surges (pages 310 and 312).
- The distribution of precipitation is predicted to change. The IPCC's super-computer models indicate that places with:
  - sufficient rainfall are likely to get more, resulting in increased flooding (e.g. northern Europe), a prediction made before the disastrous floods that affected Yorkshire and the East Midlands in 2007 (page 286).
  - insufficient rainfall are likely to get less in terms of both amount and reliability, giving increased drought (e.g. much of Africa) – a prediction which claims that the 1.7 billion people (one-third of the world's population) already short of water in 2000 could become 5 billion by 2025 (Case Study 13B).
- Ecosystems at all levels may be subject to such a rapid change that plants and wildlife may not have the time in which to adjust. This includes coral reefs, mangrove swamps, wetlands, coniferous and tropical forests, and tropical grassland. Already, trees are coming into leaf, insects are emerging and birds are laying their eggs much earlier in the year.
- Crop yields are expected to fall even further in Africa (page 111) as well as in parts of Asia and Latin America, increasing the risk of food shortages, although yields may increase in northern Europe and North America.
- A greater proportion of the world's population will be at risk from insect-borne (e.g. malaria and dengue fever) and water-borne (e.g. cholera) diseases.

Throughout the IPCC report, one prediction was constant: it will be the poor who will be hardest hit. The effects will be greatest in the LEDCs – in water and food shortages, diseases and natural disasters – places where people can least afford to adapt or are unable to migrate.

Figure 13.22
Gas emissions
(a) UK carbon dioxide emissions, 2007
(b) Main greenhouse gas emitters

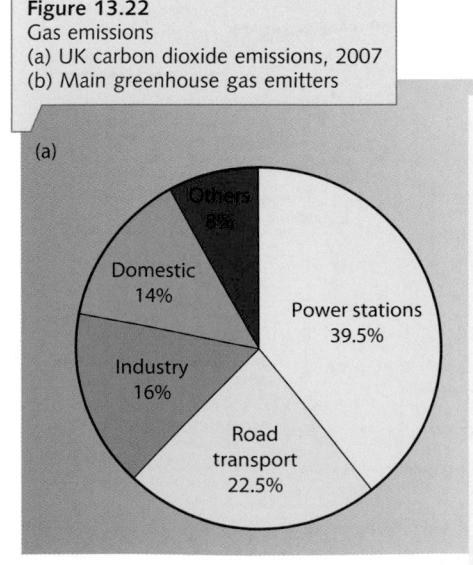

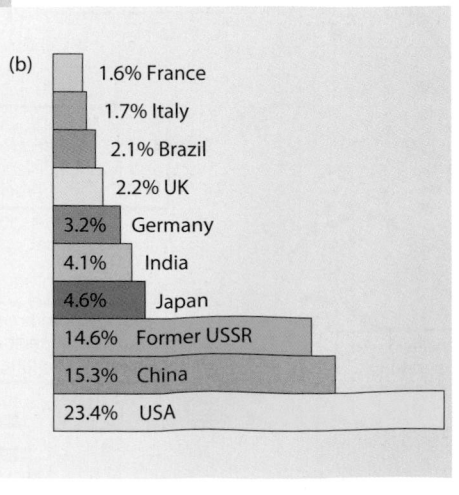

Predicted effects of global warming

In the UK

Computer models suggest that the UK's weather will change during this century. Winters will become milder and, especially in the north and west, both wetter and more windy, with an increase in the frequency of destructive storms and serious flooding. Summers in the south and east will become warmer, drier and sunnier – similar to the present-day Mediterranean climate with its heat-waves and drought. As the weather changes and sea-levels rise, coastal areas are likely to be affected by higher tides, storm surges (page 310) and flooding. Some of the possible gains and losses predicted to result from global warming are shown on Figure 13.23.

In the world

Some of the many computer-predicted changes are indicated on Figure 13.24.

Figure 13.23
Predicted UK gains and losses resulting from global warming

- • Cities likely to be subject to flooding
- ▬ Areas less than 5 metres above present-day mean sea-level, liable to flooding
- Gains   Losses

0   200 km

N

- Length of growing season will increase
- Trees such as Scots pine will grow higher up hillsides. More deciduous trees at lower levels
- Plants and shrubs will be able to grow further north
- Greater risk of forest fires
- Sea defences costing over £10 billion likely
- Valuable farmland in the Fens lost
- Loss of Norfolk Broads and salt marshes
- Houses of Parliament and Trafalgar Square flooded. The Thames Barrier will be ineffectual
- Arctic plants will face extinction
- Insufficient snow for winter sports in Aviemore to continue
- Area and yields of cereals will increase
- Higher yields of potatoes, sugarbeet and outdoor tomatoes
- Maize, vines, oranges and peaches will be grown in southern England
- More pests and diseases will exist due to mild winters
- Many coastal ports will be flooded
- Need for irrigation in summer

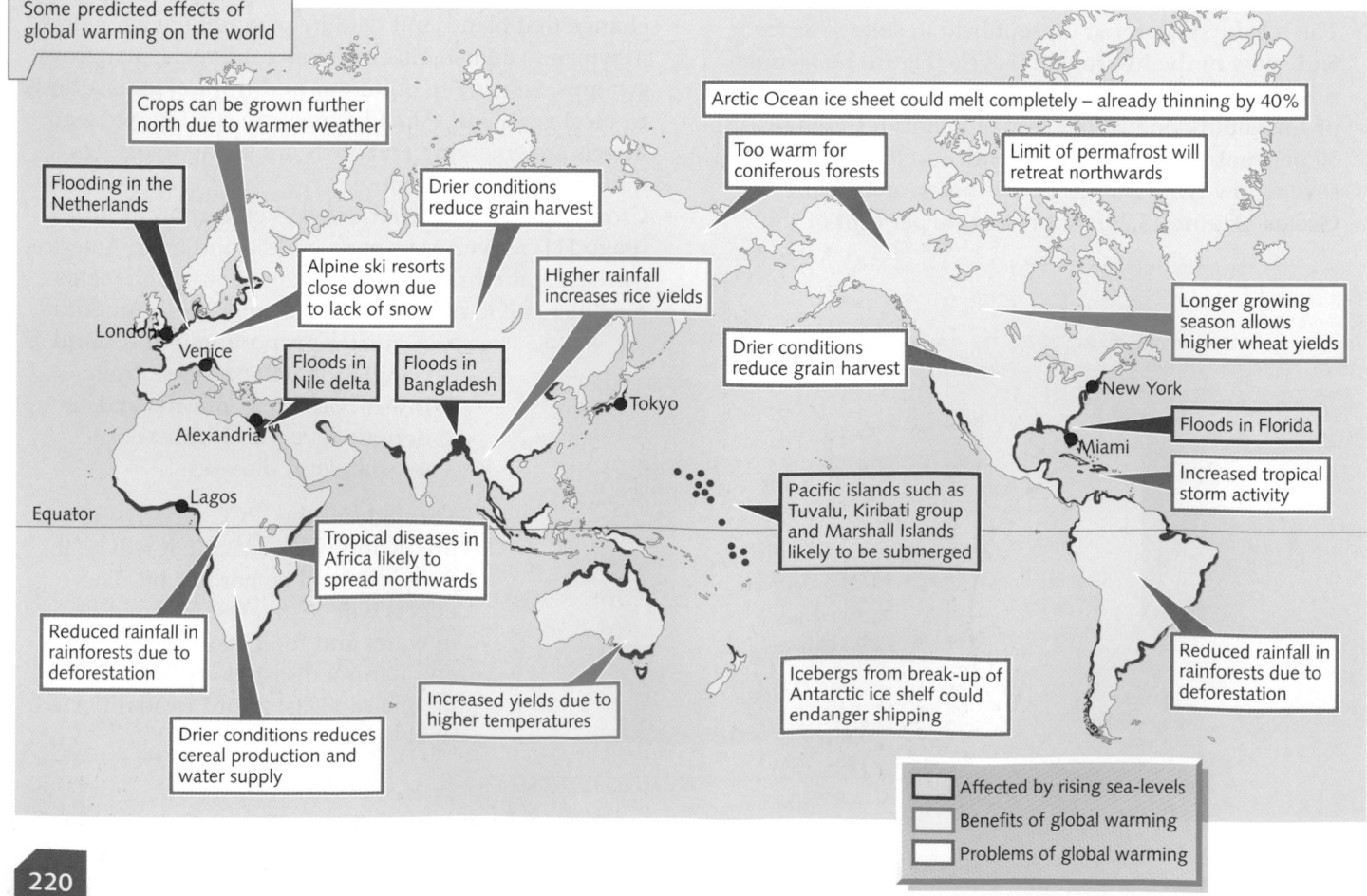

Figure 13.24
Some predicted effects of global warming on the world

- Crops can be grown further north due to warmer weather
- Flooding in the Netherlands
- Alpine ski resorts close down due to lack of snow
- Drier conditions reduce grain harvest
- Higher rainfall increases rice yields
- Too warm for coniferous forests
- Arctic Ocean ice sheet could melt completely – already thinning by 40%
- Limit of permafrost will retreat northwards
- Longer growing season allows higher wheat yields
- Drier conditions reduce grain harvest
- London
- Venice
- Floods in Nile delta
- Floods in Bangladesh
- Tokyo
- New York
- Floods in Florida
- Miami
- Increased tropical storm activity
- Alexandria
- Lagos
- Equator
- Tropical diseases in Africa likely to spread northwards
- Pacific islands such as Tuvalu, Kiribati group and Marshall Islands likely to be submerged
- Reduced rainfall in rainforests due to deforestation
- Drier conditions reduces cereal production and water supply
- Increased yields due to higher temperatures
- Icebergs from break-up of Antarctic ice shelf could endanger shipping
- Reduced rainfall in rainforests due to deforestation

- Affected by rising sea-levels
- Benefits of global warming
- Problems of global warming

Acid rain

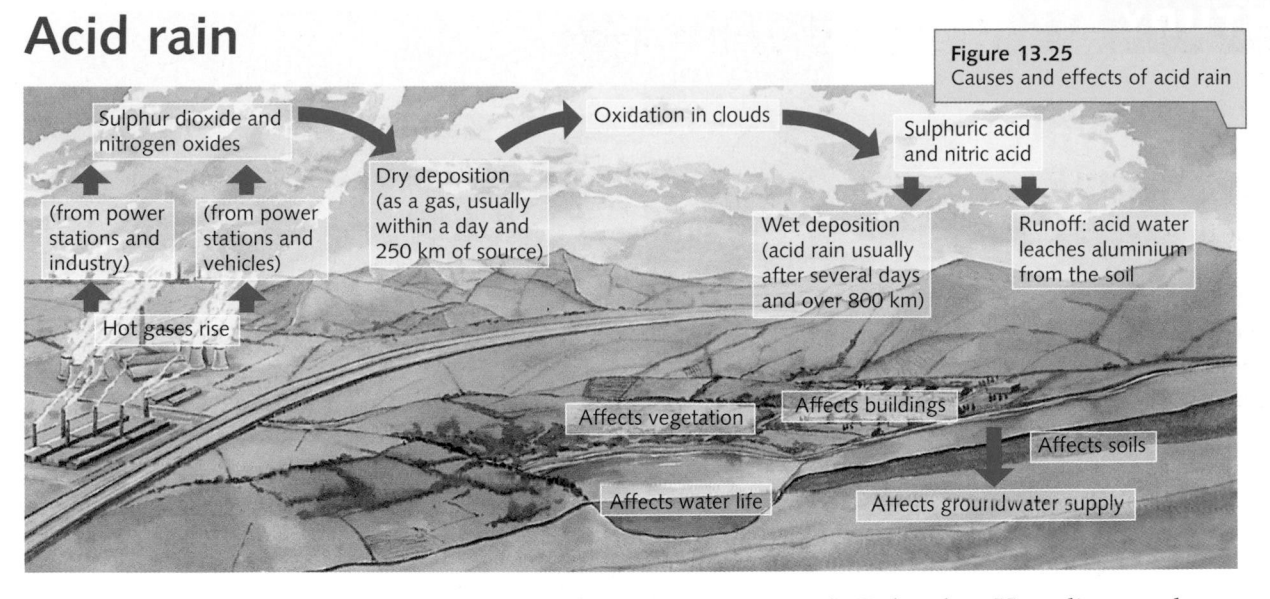

Figure 13.25
Causes and effects of acid rain

Sulphur dioxide and nitrogen oxides

(from power stations and industry) (from power stations and vehicles)

Hot gases rise

Oxidation in clouds

Dry deposition (as a gas, usually within a day and 250 km of source)

Sulphuric acid and nitric acid

Wet deposition (acid rain usually after several days and over 800 km)

Runoff: acid water leaches aluminium from the soil

Affects vegetation   Affects buildings

Affects soils

Affects water life   Affects groundwater supply

Acid rain was first noticed in Scandinavia in the 1950s when large numbers of freshwater fish died. Research showed that the water in which these fish had lived contained more than average amounts of acid. Later it was discovered that this extra acid had been carried by rain, hence the term **acid rain**. The acid is formed in the air from sulphur dioxide and nitrogen oxide which are emitted by thermal power stations, industry and motor vehicles (Figure 13.25). These gases are either carried by prevailing winds across seas and national frontiers to be deposited directly onto the Earth's surface (dry deposition), or are converted into acids (sulphuric and nitric acid) which then fall to the ground in the rain (wet deposition). Clean rainwater has a pH value of between 5.5 and 6 (pH7 is neutral). Today the pH readings are between 4 and 4.5 through much of north-west Europe, with the lowest ever recorded being 2.4 (Figure 13.26). A falling pH is the sign of increasing acidity. (Remember: when pH falls by one unit it means that the level of acid has increased ten times.) The lowest pH readings in European lakes were recorded in 1985, since when acidity has slowly decreased allowing both the lakes and rivers to recover.

Europe's pollution budget

Most European countries add acids to the air, with Britain being one of the major culprits. However, only about one-third of Britain's contribution actually lands back on British soil. Some falls into the North Sea but most is carried by the prevailing south-westerly winds towards Scandinavia. Despite being one of the least of the offenders, Scandinavia is one of the main sufferers from acid rain. It is because acid rain crosses national frontiers that it is an international problem which can only be solved through global management and co-operation.

The effects of acid rain

- The acidity of lakes has increased. Large concentrations kill fish and plant life.
- An increase in the acidity of soils reduces the number of crops that can be grown.
- Forests are being destroyed as important nutrients (calcium and potassium) are washed away (leached). These are replaced by manganese and aluminium which are harmful to root growth. In time, the trees become less resistant to drought, frost and disease, and shed their needles.
- Water supplies are more acidic and this could become a future health hazard. For example, the release of extra aluminium has been linked to Alzheimer's disease.
- Buildings are being eroded by chemical action caused by acid rain. The Acropolis in Athens and the Taj Mahal in India have both deteriorated rapidly in recent years.

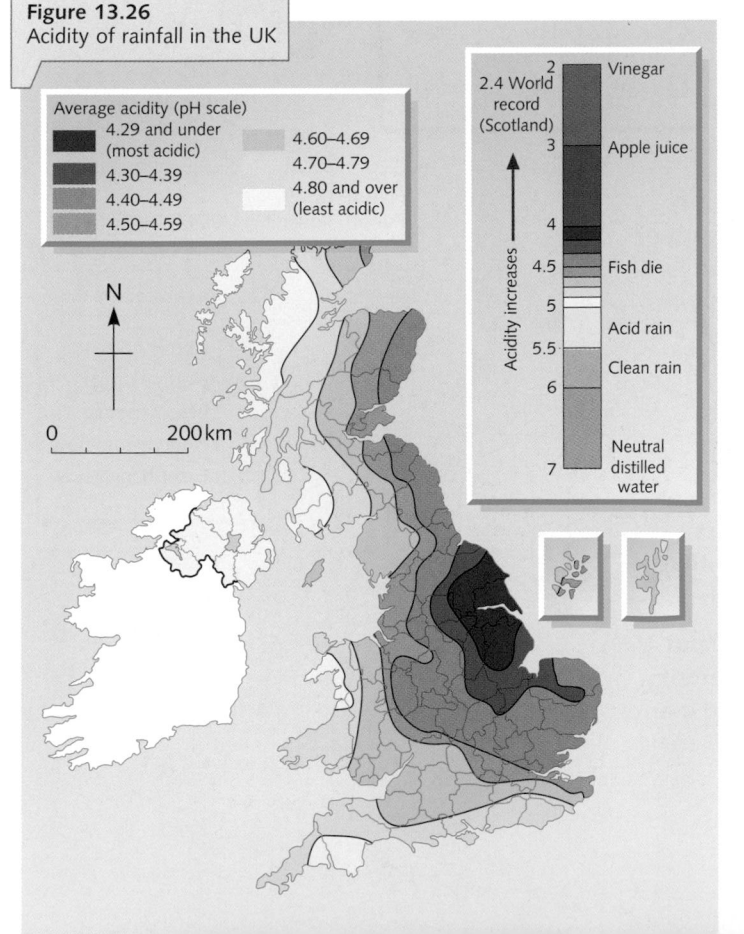

Figure 13.26
Acidity of rainfall in the UK

Average acidity (pH scale)
4.29 and under (most acidic)
4.30–4.39
4.40–4.49
4.50–4.59
4.60–4.69
4.70–4.79
4.80 and over (least acidic)

N

0   200 km

2   Vinegar
2.4 World record (Scotland)
3   Apple juice
4
4.5   Fish die
5
Acidity increases
Acid rain
5.5
6   Clean rain
7   Neutral distilled water

Drought and water supply in the UK

1995 and 1996

Perhaps it is because drought is less frequent in the UK than in some other parts of the world that Britain seems unable to cope with a water shortage. Although people still talked about the 1975–76 drought, water authorities seemed unprepared for the hot, dry summers of 1995 and 1996 (Figure 13.27). The National Rivers Authority (NRA), now the Environment Agency (EA), claimed in late August 1995 that river and reservoir levels across England and Wales were falling rapidly. By then, river flows in Cumbria were under 10 per cent and those in Wales and south-west England under 25 per cent of their normal level. The NRA claimed that this was a result of a combination of several months of below average rainfall during a time of exceptionally high temperatures. The low river levels meant that less water was available for extraction by the water companies, and to dilute pollutants. There was also less oxygen available for fish. By the autumn some reservoirs in Yorkshire were down to 10 per cent of their maximum capacity and Yorkshire Water was forced to buy water, coming from Kielder reservoir, from Northumbria Water. Water, already pumped from the Tyne to the Tees since that reservoir opened in 1980, was now also pumped to a reservoir in Cleveland before being transported by road to West Yorkshire by a fleet of 200 tankers working non-stop around the clock (Figure 13.28).

Figure 13.27
Haweswater Reservoir in the Lake District during the 1996 drought

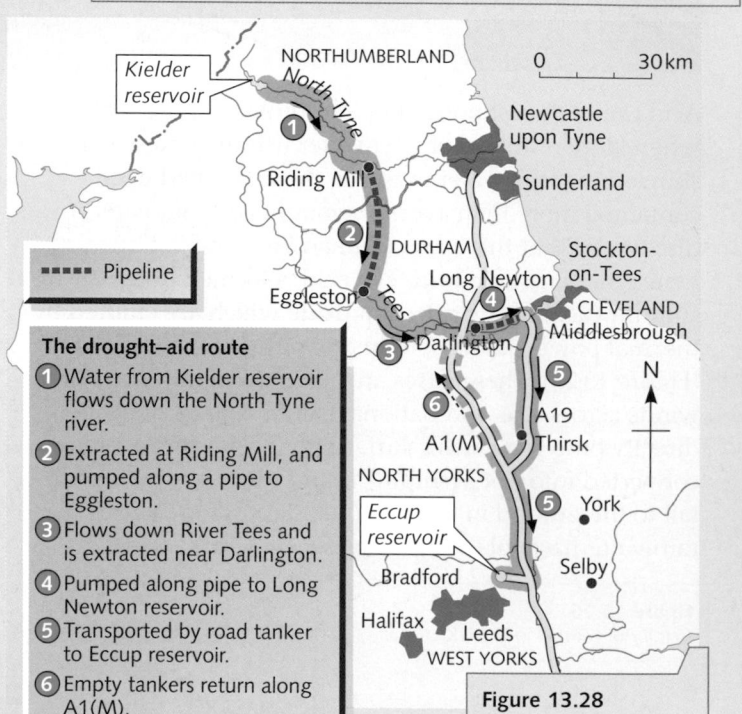

The drought–aid route
1. Water from Kielder reservoir flows down the North Tyne river.
2. Extracted at Riding Mill, and pumped along a pipe to Eggleston.
3. Flows down River Tees and is extracted near Darlington.
4. Pumped along pipe to Long Newton reservoir.
5. Transported by road tanker to Eccup reservoir.
6. Empty tankers return along A1(M).

Figure 13.28
The drought–aid route

2003

This year saw the driest nine months across much of the UK since 1921, and the south-east of England received less than 25 per cent of its normal rainfall. Despite a spectacular display of autumn leaves, livestock had to be fed their winter feed early, as there was no grass in the fields, and winter cereal crops failed to germinate in the parched ground.

2004 to 2006

A drought lasting from autumn 2004 to the end of 2006 was one of the worst experienced by south-east England for over 100 years (Figure 13.29). Eight water companies imposed hosepipe bans that affected 16 million people.

Although severe droughts, especially in the south-east of England, do appear to be on the increase – a predicted effect of global warming (Figure 13.23) – the summer of 2007 brought severe flooding to many parts of England and Wales. These events seem to point to an increasing scale in the unpredictability of the British weather.

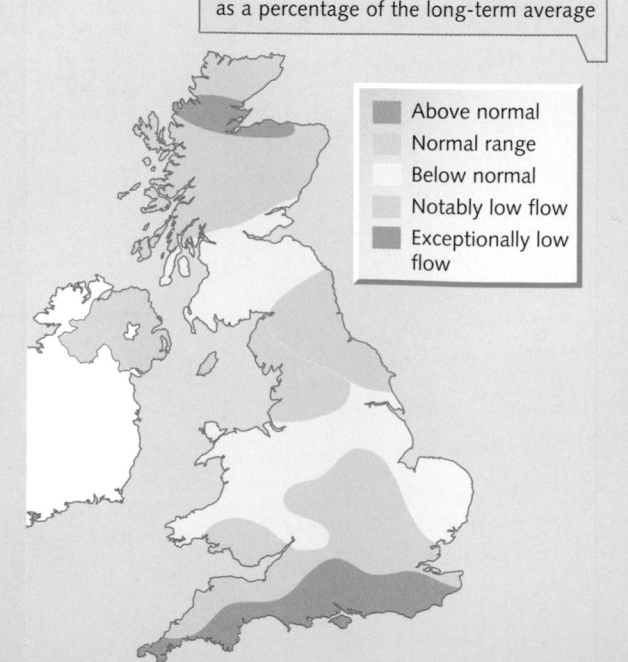

Figure 13.29
Runoff November 2004 – August 2006, as a percentage of the long-term average

- Above normal
- Normal range
- Below normal
- Notably low flow
- Exceptionally low flow

Water supplies and transfers

The UK receives more than enough rainfall in an average year to meet the present demand, and predicted future demand. Unfortunately this rainfall does not always fall where and when it is needed (Figure 13.30). Most rain falls in the higher north and west (where fewer people live and work) and in the winter (when demand is lowest). This means that surplus water has to be **stored** (on the surface or underground) until it is needed and then **transferred** (naturally or artificially) to areas which have a deficit (Figure 13.31).

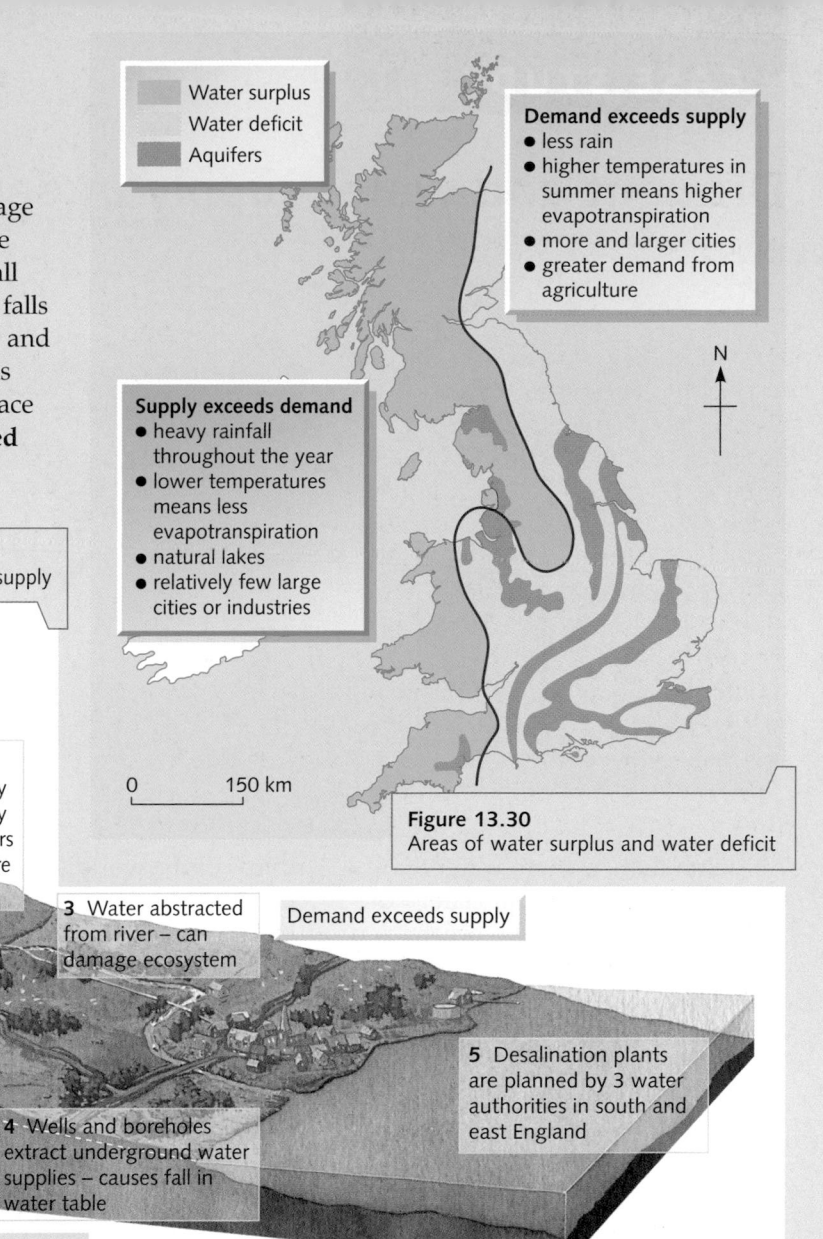

Water surplus
Water deficit
Aquifers

Demand exceeds supply
- less rain
- higher temperatures in summer means higher evapotranspiration
- more and larger cities
- greater demand from agriculture

N

Supply exceeds demand
- heavy rainfall throughout the year
- lower temperatures means less evapotranspiration
- natural lakes
- relatively few large cities or industries

0   150 km

Figure 13.30
Areas of water surplus and water deficit

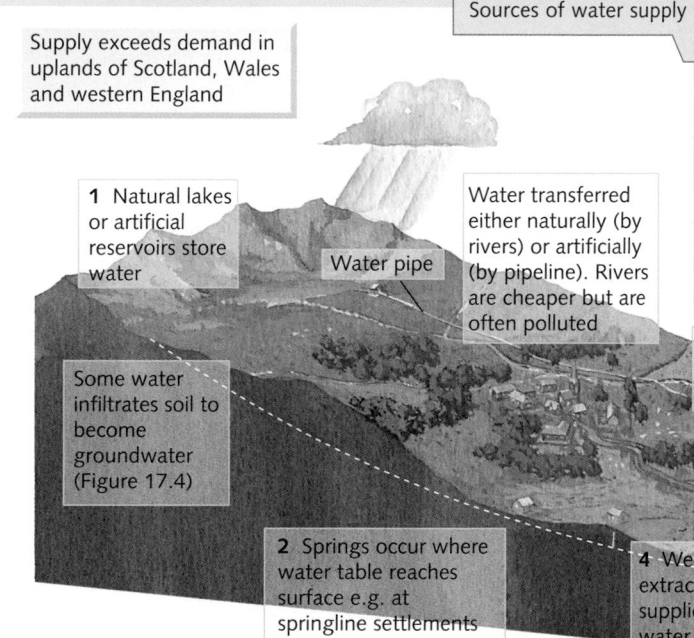

Figure 13.31
Sources of water supply

Supply exceeds demand in uplands of Scotland, Wales and western England

1 Natural lakes or artificial reservoirs store water

Water transferred either naturally (by rivers) or artificially (by pipeline). Rivers are cheaper but are often polluted

Water pipe

Some water infiltrates soil to become groundwater (Figure 17.4)

3 Water abstracted from river – can damage ecosystem

Demand exceeds supply

2 Springs occur where water table reaches surface e.g. at springline settlements

4 Wells and boreholes extract underground water supplies – causes fall in water table

5 Desalination plants are planned by 3 water authorities in south and east England

Figure 13.32
Estimated changes in demand for water, 1990–2021

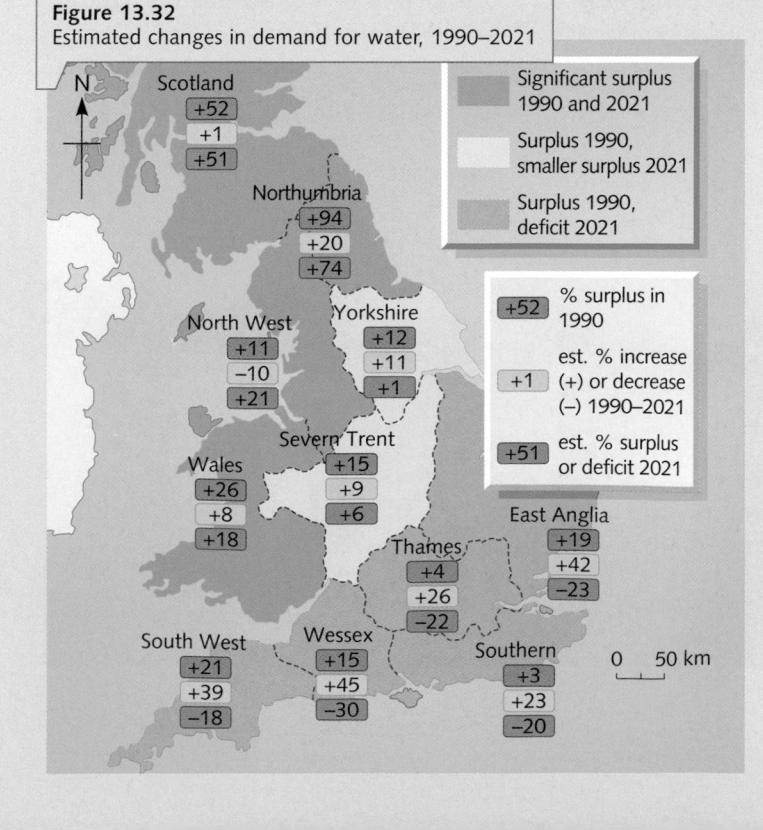

N

Scotland
+52
+1
+51

Significant surplus 1990 and 2021

Surplus 1990, smaller surplus 2021

Surplus 1990, deficit 2021

Northumbria
+94
+20
+74

North West
+11
−10
+21

Yorkshire
+12
+11
+1

+52 % surplus in 1990

+1 est. % increase (+) or decrease (−) 1990–2021

+51 est. % surplus or deficit 2021

Severn Trent
+15
+9
+6

Wales
+26
+8
+18

East Anglia
+19
+42
−23

Thames
+4
+26
−22

South West
+21
+39
−18

Wessex
+15
+45
−30

Southern
+3
+23
−20

0   50 km

Most of the UK's water supply is obtained from surface waters in rivers and lakes but an increasing amount is being taken from water-bearing rocks called **aquifers**, especially in the south-east of England (Figure 13.30). The dry years of 1995 to 1996 and 2003 to 2006 meant that far more water was extracted from aquifers than nature was able to replace. The resultant fall in the water table caused trees to die and some rivers, such as the Darent, to stop flowing altogether. Although excessive rainfall in the years following these events more than replenished underground supplies, and even led to the saturation of the ground and local flooding, the continual growth in demand for water, especially in the south-east of Britain, has led five of the water authorities in England and Wales to predict a significant water shortage by the year 2021 (Figure 13.32).

223

Drought and water supply in developing countries

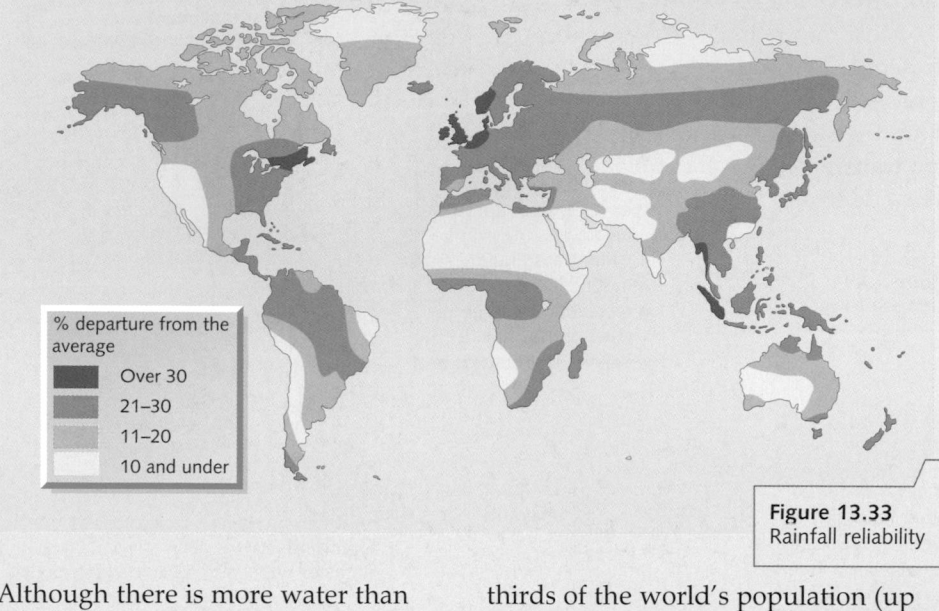

% departure from the average

- Over 30
- 21–30
- 11–20
- 10 and under

Figure 13.33
Rainfall reliability

Although there is more water than land on the Earth, 97 per cent of the total is found in seas and oceans and 2 per cent is stored as ice and snow. The remaining 1 per cent is constantly recycled in the hydrological (or water) cycle (Figure 17.3). As water is neither added to nor lost from the hydrological cycle, it is said to be a **closed system**, with the amount of fresh water available for life on Earth remaining constant, and it should be a renewable resource.

Over the last 300 years, when the world's population has increased sevenfold, the demand for water for domestic, agricultural and industrial use has increased by 40 times. This increased use of water has been accompanied by an increase in its pollution – so much so that the UN estimate that each day more than 5000 people die from using contaminated water – almost 2 million a year (page 328). This, coupled with the uneven distribution of fresh water, means that some 80 countries and 1.1 billion people are already experiencing 'severe water stress' either within certain regions or at certain times of the year. By 2050, according to UN estimates, two-thirds of the world's population (up to 5 billion), mainly living in LEDCs in Africa, Asia and Latin America, are likely to be short of reliable, clean water.

Reliability of rainfall

In Britain it is taken for granted that it is likely to rain every few days, and that rain will fall fairly evenly throughout the year. The situation in many developing countries is very different. Many countries experience a pronounced wet and dry season. If the rains fail during one year, the result for crops, animals and sometimes people, can be disastrous. The most vulnerable areas are desert margins and tropical interiors where average annual amounts are low and the rainy season is short (page 213). Here, where a small variation of 10 per cent below the average total can be critical, many countries experience a variation of over 30 per cent (Figure 13.33). The rain, when it does fall, often comes in torrential downpours which give insufficient time for it to infiltrate into the ground (Figure 17.4). Instead, surface runoff causes flash floods and water is lost to the local community. Few developing countries have the money or the technology to build dams to store water, and where they have been built, it has been with foreign financial 'aid' which has put the recipient country further into debt (page 190). Relatively few places, especially in rural areas, have piped water, and the daily walk to the local well or river can be a distance of several kilometres, taking several hours (Figure 13.34).

Figure 13.34
Carrying water, Burkina Faso

Clean water

According to the UN, an estimated 1.1 billion people lack a satisfactory or safe water supply (Figure 13.35). Rarely do people, even in developed countries, take sufficient care of this the most essential of resources. In developing countries, water supplies may be contaminated:

- in rural areas where rivers and streams are used for drinking, washing and the disposal of sewage
- in urban areas where, especially in shanty settlements, the absence of drains means that sewage may pollute water supplies (e.g. São Paulo page 82, Kolkata page 84 and Rio de Janeiro page 86)
- in many shanty settlements which either lack clean water or where large numbers of people are forced to share the same tap
- due to a lack of government legislation and implementation.

0–10%	
41–60%	
61–80%	
81–100%	
No data	

NORTH
SOUTH

Figure 13.35
Percentage of the population with access to safe water

Water that is dirty or polluted is no longer a renewable resource and often leads to outbreaks of diseases such as cholera, typhoid, dysentery and diarrhoea, which can affect up to half of the population of developing countries at the same time (page 328).

The ambitious UN 'International Drinking Water Supply and Sanitation Decade', launched in 1980 to provide 'water and sanitation for all by 1990', was never realised. Although there was some progress (Figure 13.36), especially with water supply – 50 per cent of people in rural areas now have an adequate supply compared with only 30 per cent in 1980 – the rapid growth in population and urbanisation means that 500 million more people lacked sanitation in 2008 than in 1980. The improvements in rural water supplies were mainly due to international charitable organisations like Oxfam, ActionAid, WaterAid and Practical Action which have helped local communities by introducing appropriate technology (page 328). These included self-help schemes such as:

- digging wells to reach permanent underground supplies
- lining the sides of the well with concrete (to prevent seepage) and adding a cover (to reduce evaporation)
- using modern pumps (Figures 11.7 and 13.37)
- teaching safe hygiene practices.

Figure 13.36
The world's clean water supplies

Figure 13.37
A water pump in Burkina Faso

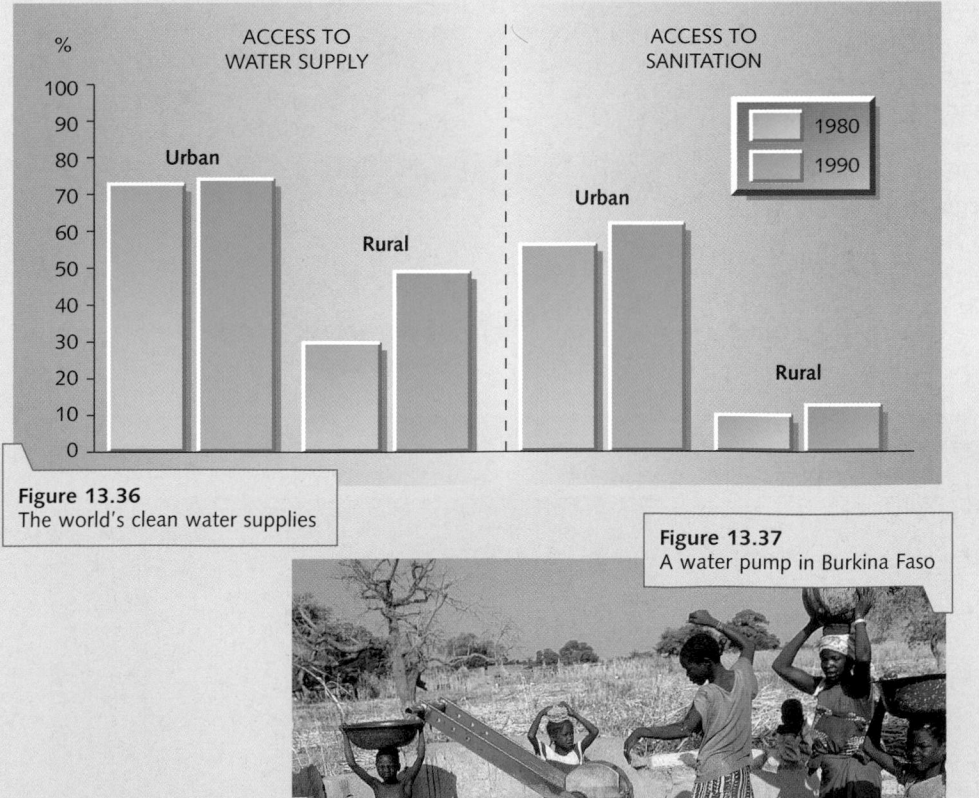

World climate

Key Words and Terms

a You should know the meaning of the following terms:
- atmospheric circulation • prevailing winds • trade winds • depressions
- relief rainfall • frontal rainfall • convectional rainfall
- annual range of temperature • dew point • the monsoon
- tropical storms/hurricanes • storm/tidal surge • global warming
- greenhouse effect/gases• acid rain • water supply • water shortage
- drought • water transfers • rainfall reliability.

b You should know the difference between:
- high pressure and low pressure
- depressions and tropical storms.

Key Ideas

You should know and understand the following:
- The global atmospheric circulation system.
- The effects of latitude, relief, land and sea, ocean currents and altitude on global temperatures.
- The effects of different amounts and reliability of rainfall on world climate.
- The global distribution of the major climate types.
- The differences between selected climate types: equatorial, tropical continental, monsoon, hot desert, north-west European/British and cold climates.
- The formation, distribution and effects of weather associated with tropical storms/hurricanes.
- Strategies to try to predict, plan for and manage tropical storms/hurricanes.
- Causes and effects of drought in MEDCs and LEDCs.
- Differences in the provision of an adequate water supply for MEDCs and LEDCs.
- Causes and effects of global warming.
- Causes and global effects of acid rain.

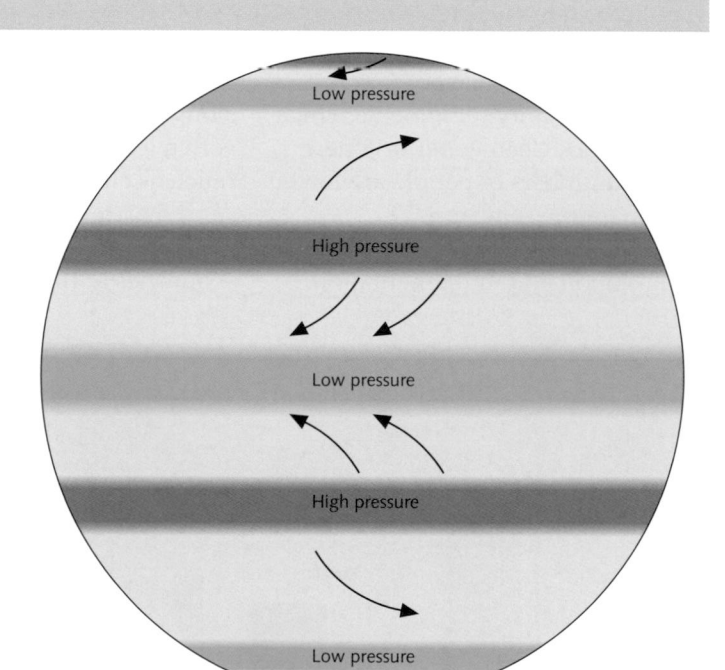

Skills, Theories and Models

- Interpret and use maps, diagrams and graphs
- Interpret and use climate graphs
- Identify features on a satellite image
- Understand the concept of global warming
- Theory of hurricane development
- Compare the effects of hurricanes in LEDCs and MEDCs
- Circulation of the atmosphere model

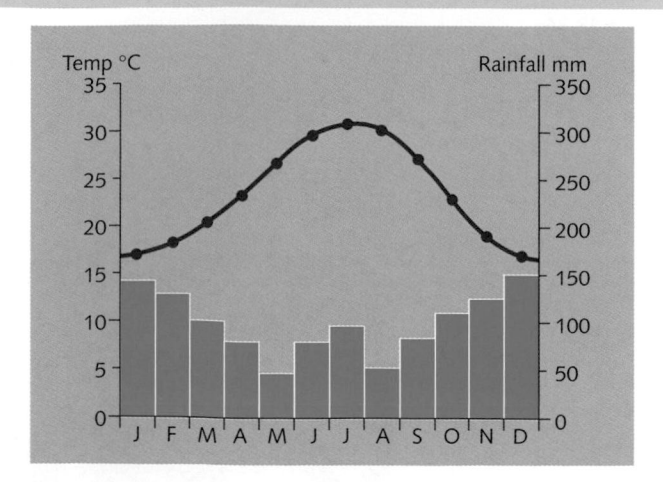

1 *(Pages 210 and 211)*

a Why is pressure:
- low at the Equator
- high about 30° from the Equator
- low at about 60° from the Equator? *(3)*

b How does this pattern of high and low pressure account for:
- the trade winds in the tropics
- the prevailing south-westerlies in Britain
- convectional rainfall at the Equator
- frontal rainfall in Britain? *(4)*

c i) Draw a simple diagram to show the general circulation of the atmosphere. Your diagram should show the three circulation cells. *(6)*

2 *(Pages 211 to 215)*

a Six climates are described in this book: British, equatorial, tropical continental (interior), monsoon, hot desert and cold climate. Their locations are shown on Figure 13.2. Match each of the six climates with the following descriptions:
- hot and dry all year
- hot, very wet summers; very warm, dry winters
- long cold winters, short warm summers, little rain
- hot, wet and humid all year
- cool summers; mild winters; rain all year
- hot, wet summers; warm, dry winters. *(6)*

b Make a copy of the table below.
- i) Fill in the left-hand column by choosing **three** of the climate types (**not** the British climate).
- ii) Complete the table by using information from the relevant climate graph and text. *(18)*

c For each of the **three** climate types which you selected in part (b):
- i) Name the part, or parts, of the world in which it is located. *(3)*
- ii) Explain how it is affected by:
  - latitude (angle of the sun)
  - distance from the sea
  - prevailing winds. *(3)*
- iii) Give reasons for any seasonal differences in climate. *(6)*

Climate type	Temperatures			Precipitation		
	Maximum	Minimum	Annual range	Annual total	Season with most rain	Type of rain
1						
2						
3						

ii) What relationships can you see between the circulation of the atmosphere and the pattern of world climate types? *(6)*

d i) Why do global temperatures increase towards the Equator? *(1)*
ii) Why do places in the centre of continents have warmer summers and colder winters than places that have a coastal location? *(2)*
iii) Why are prevailing winds from the sea:
- warmer in winter
- cooler in summer

than those blowing from the land? *(2)*
iv) How can ocean currents affect the temperatures of coastal areas? *(2)*
v) Why can snow be found in some places on the Equator? *(1)*

3 *(Page 216)*

a With reference to Figure 13.12:
- i) Name two continents that are severely affected by tropical storms (hurricanes) and for each name one likely source area. *(2)*
- ii) Describe the world distribution of hurricanes. *(2)*

b The map below is a synoptic chart for a hurricane centred over western Cuba at 0100 hours on 6 September.
- i) Describe the track of the storm:
  - before it reached Cuba
  - after it left Cuba. *(4)*
- ii) Describe the weather conditions over western Cuba at 0100 hours on 6 September. *(4)*

4 *(Pages 216 and 217)*

Refer to an atlas, the map and facts on Hurricane Gilbert.

a i) Name six islands that were affected by the hurricane. *(2)*

   ii) Name two countries on the mainland of Central America that were affected by the hurricane. *(2)*

   iii) Which country had the highest number of deaths? *(1)*

   iv) Seventy per cent of homes in Jamaica were damaged by the hurricane. Suggest two ways in which damage would have occurred. *(2)*

   v) Suggest why Galveston was getting ready for the hurricane from 15 September. *(1)*

   vi) What could the government and people of Galveston do to get ready for the hurricane? *(4)*

b i) Describe the route taken by the hurricane. *(2)*

   ii) Describe how people living in the area were affected by the hurricane. *(4)*

   iii) What measures might people living:
   - on an island in the West Indies
   - in Galveston (USA)

   have taken to reduce the impact of the storm? *(6)*

c Use the hurricane warning poster (Figure 13.17).

   i) List the precautions that people are advised to take when the warning is first given. *(3)*

   ii) Why might people in less economically developed countries either not hear, or choose to ignore, the warning? *(3)*

   iii) Why are people advised to take care immediately after the storm has passed? *(2)*

   iv) What damage may they find when they return home? *(3)*

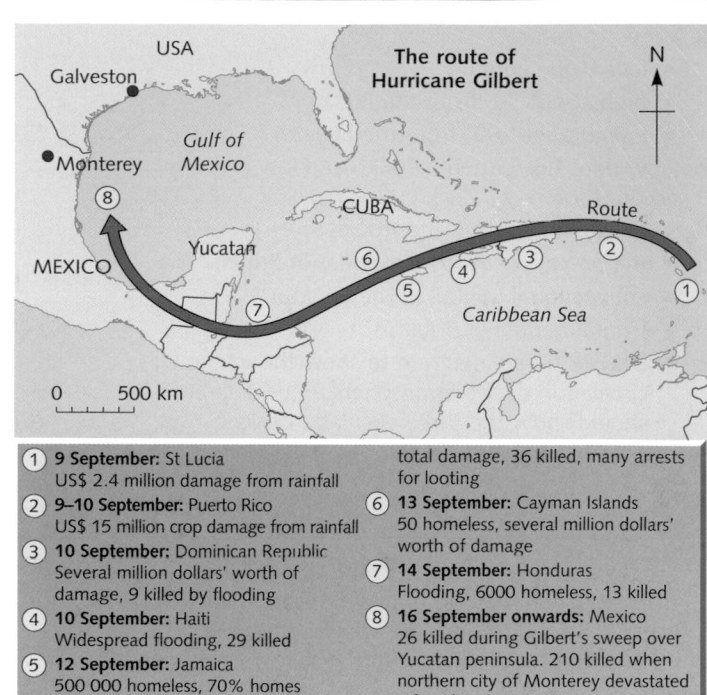

The route of Hurricane Gilbert

① **9 September:** St Lucia US\$ 2.4 million damage from rainfall		total damage, 36 killed, many arrests for looting
② **9–10 September:** Puerto Rico US\$ 15 million crop damage from rainfall	⑥ **13 September:** Cayman Islands 50 homeless, several million dollars' worth of damage	
③ **10 September:** Dominican Republic Several million dollars' worth of damage, 9 killed by flooding	⑦ **14 September:** Honduras Flooding, 6000 homeless, 13 killed	
④ **10 September:** Haiti Widespread flooding, 29 killed	⑧ **16 September onwards:** Mexico 26 killed during Gilbert's sweep over Yucatan peninsula. 210 killed when northern city of Monterey devastated a few days later	
⑤ **12 September:** Jamaica 500 000 homeless, 70% homes damaged, billions of dollars' worth		

d Describe the weather likely to be experienced with the passing of a typical hurricane. *(6)*

e i) How are hurricanes believed to form? *(3)*

   ii) Why do hurricanes rapidly decline on reaching land? *(2)*

---

5 *(Pages 218 and 220)*

a i) By how many degrees did the global temperature rise last century? *(1)*

   ii) By how many degrees is it predicted to rise by the end of this century? *(1)*

   iii) With the help of Figure 13.18, explain why global temperatures are rising. *(2)*

b Why are natural greenhouse gases in the atmosphere important to life on Earth? *(1)*

c i) Copy and complete the pie graph which shows how human activity has increased the amount of greenhouse gases in the atmosphere, by:
   - inserting the names of the four missing gases
   - naming one source for each of the four gases. *(4 × 2)*

   ii) How does each of the four gases contribute to global warming? *(4)*

   iii) Which country, at present, releases most greenhouse gases? *(1)*

d i) Copy and complete the table to show some of the effects, both good and bad, likely to result from global warming. *(6 × 3)*

   ii) How is global warming predicted to affect:
   - sea-level
   - precipitation
   - glaciers and icecaps
   - vegetation and wildlife? *(4)*

e i) List four advantages to Britain that are likely to result from global warming. *(4)*

   ii) List four problems likely to face Britain as a result of global warming. *(4)*

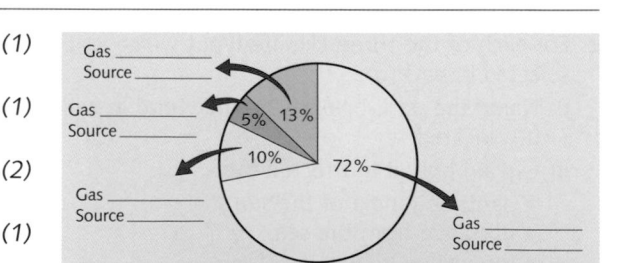

	Increase in temperature	Decrease in rainfall	Rise in sea-level
1 Area			
Effect			
2 Area			
Effect			
3 Area			
Effect			

6 (Page 221)

a i) What are the two main causes of acid rain? *(2)*

ii) Which two chemicals cause rain to be acidic? *(2)*

b i) Give three ways in which acid rain affects the natural environment. *(3)*

ii) How can acid rain affect buildings and people's health? *(2)*

c i) How is the level of acidity measured? *(1)*

ii) Which part of Europe suffers most from acid rain? *(1)*

iii) Why is this area so badly affected? *(2)*

d The map shows the origin of the sulphur content in acid rain that poisons lakes in Norway and Sweden.

i) Which two countries are the likely source of most of the sulphur? *(2)*

ii) Suggest reasons for the production of sulphur in those countries. *(2)*

e i) Why is international co-operation needed if the effects of acid rain are to be controlled? *(2)*

ii) How could the problem of acid rain be reduced? *(2)*

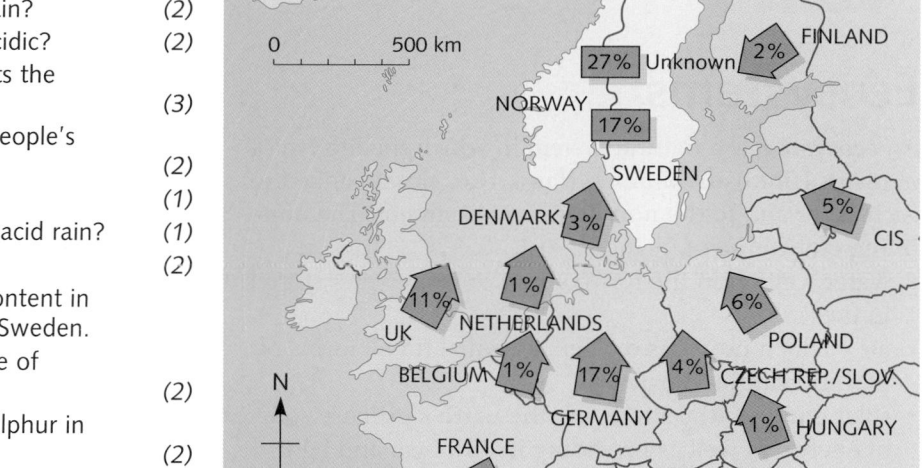

7 (Pages 222 and 223)

a i) Why was there a water shortage in much of England and Wales during 1995? *(1)*

ii) How did the drought affect rivers and reservoirs? *(1)*

iii) Which water authority had greatest difficulty in providing water? *(1)*

iv) How did this authority try to provide water? *(2)*

b i) Give three reasons why the north and west of Britain usually have a water surplus. *(3)*

ii) Give three reasons why the south and east of Britain usually have a water deficit. *(3)*

iii) Which three water authorities are predicted to have, in 2021:
- the largest water surplus
- the largest water deficit? *(6)*

c i) Using Figure 13.31, give four sources of fresh water supply. *(4)*

ii) How can an MEDC such as Britain manage to store water and to transfer it either during times of drought or to areas where demand is constantly high? *(4)*

8 (Pages 224 and 225)

a i) Which two climates on Figure 13.2 correspond most closely with the areas of seasonal or sporadic drought shown on this map? *(2)*

ii) Why do these two climates have seasonal or sporadic drought? *(2)*

iii) Give two reasons why, when rain does fall, it is often less effective than in Britain. *(2)*

b i) Using Figure 13.35, describe the location of places where fewest people have access to safe (clean) water. *(3)*

c i) Give three reasons why water supplies in many LEDCs may be contaminated in:
- urban areas
- rural areas. *(6)*

ii) What problems result from contaminated water? *(2)*

iii) Give three problems faced by LEDCs in trying to provide safe, clean water. *(3)*

Areas that usually have enough water

Areas with seasonal or sporadic drought

Areas that are always deficient in water

The world's uneven water supply

d i) Why is appropriate technology often more suitable in providing a safe, reliable supply of water to many people living in LEDCs than large schemes financed by overseas countries? *(4)*

ii) Describe how appropriate technology helps local communities to provide water for their people. *(3)*

14 Ecosystems

Ecosystems

An **ecosystem** is a natural system in which the life cycles of plants (**flora**) and animals (**fauna**) are closely linked to each other and to the non-living environment. The **non-living environment** includes:

- water – either in the form of rain or from water stored in the soil
- air – which provides oxygen, essential for all forms of life, and carbon dioxide
- solar energy – the sun being the Earth's primary source of energy as well as providing it with heat and light
- rocks – which provide nutrients and which may be permeable (allowing water to pass through them) or impermeable (page 245)
- soils – which vary in depth, acidity (pH), nutrients and fertility.

The **living environment** includes:
- plants, animals, insects and micro-organisms (most of which live in harmony with one another)
- people (who rarely seem to live in harmony with the environment).

Level	Examples	
Micro	Water droplets	Under a leaf or a stone
Meso (middle)	Freshwater pond	Woodland
	Sand dunes	Hedgerows
	Salt marsh	Wetland
Global (biome)	Tropical rainforest	Tropical grassland
	Coniferous forest	Tundra

Figure 14.1
Levels of ecosystems

Ecosystems vary in size, from extensive areas of rainforest or grassland (known as **biomes** – pages 232–5), to smaller areas of woodland (page 231) and wetland, down to under a stone or within a droplet of water (Figure 14.1). Any ecosystem depends on two basic processes: the flow of energy and the recycling of nutrients.

1 Energy flows Each ecosystem is sustained by the flow of energy through it. The main source of energy is sunlight which is absorbed by green plants and converted by the process of **photosynthesis**. Energy is then able to pass through the ecosystem in the **food chain** (Figure 14.2) in which plants are eaten by animals, and some animals consume each other. In other words, each link in the chain feeds on and obtains energy from the link preceding it. In turn it is consumed by and provides energy for the link that follows it. As energy only passes one way, the ecosystem is an **open system**, with inputs, flows, stores and outputs (page 279).

2 Recycling of nutrients Certain nutrients are continually circulated within the ecosystem and so are part of a **closed system**. Each cycle consists of plants taking up nutrients from the soil. The nutrients are then used by plants, or by animals which consume the plants. When the plants or animals die, they decompose and the nutrients are released and returned to the soil ready for future use (Figures 14.3 and 14.5).

Ecosystems can often be very fragile. Some, like the arctic tundra, tropical rainforest and areas of wetland, may take hundreds of years to develop fully. They can, however, be irretrievably damaged in a short time by human activity, e.g. the tundra in Alaska (Case Study 8) and the Amazon rainforest (page 236).

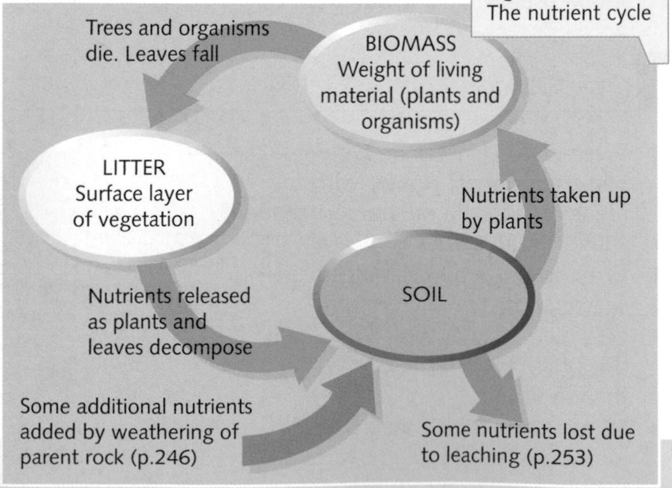

Figure 14.3
The nutrient cycle

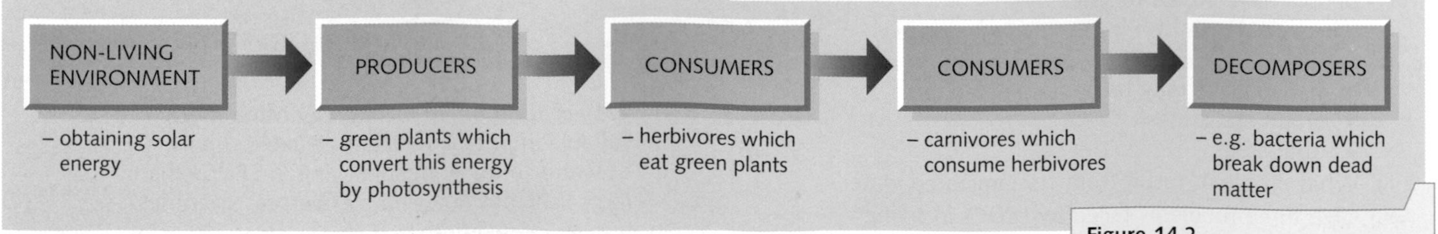

Trees and organisms die. Leaves fall

BIOMASS
Weight of living material (plants and organisms)

LITTER
Surface layer of vegetation

Nutrients taken up by plants

Nutrients released as plants and leaves decompose

SOIL

Some additional nutrients added by weathering of parent rock (p.246)

Some nutrients lost due to leaching (p.253)

| NON-LIVING ENVIRONMENT | → | PRODUCERS | → | CONSUMERS | → | CONSUMERS | → | DECOMPOSERS |

– obtaining solar energy

– green plants which convert this energy by photosynthesis

– herbivores which eat green plants

– carnivores which consume herbivores

– e.g. bacteria which break down dead matter

Figure 14.2
Transfers of energy in an ecosystem

A deciduous woodland ecosystem in Britain

A natural British deciduous woodland (Figure 14.4) consists of three layers of vegetation. The tallest layer, between 30 and 40 metres in height, consists of trees such as oak, ash, birch and beech. The middle layer, ranging from 5 to 15 metres, includes smaller trees and shrubs such as holly and hawthorn. The lowest, or ground, layer may include bracken, brambles, grass and ferns. Figures 14.5, 14.6 and 14.7 show a typical woodland ecosystem.

Figure 14.5 describes the open system of the ecosystem with its inputs (e.g. solar energy, carbon dioxide), flows (e.g. trees absorbing nutrients and water from the soil through their roots), processes (e.g. photosynthesis), stores (e.g. nutrients within the soil) and outputs (e.g. evapotranspiration, oxygen).

Figure 14.6 illustrates how nutrients are recycled in a closed system.

Figure 14.7 shows how energy is transferred through the ecosystem by means of the food chain.

Figure 14.4
A deciduous woodland

This woodland ecosystem, as shown in Figure 14.5, is said to be in balance with nature. Unfortunately, human interference often alters, and even destroys, ecosystems such as this through clearing land for development, e.g. farming and village settlement in earlier times, roads and recreation in more recent times. A vital issue for present generations is: 'How can we gain most benefit from an ecosystem without damaging or destroying it?' – the concept of **sustainable development**.

Figure 14.5
A deciduous woodland ecosystem in Britain

Evapotranspiration · Solar energy · Rainfall

CO_2 · O_2

Birds eat seeds

PHOTOSYNTHESIS

Leaf litter

Roots absorb nutrients from soil

Plants decay into humus

Animals eat plants and deposit faeces

GROUNDWATER

Nutrients leached out

Water drains into rivers

Underlying rocks weather into soil

- Inputs
- Processes and stores
- Outputs
- Recycling

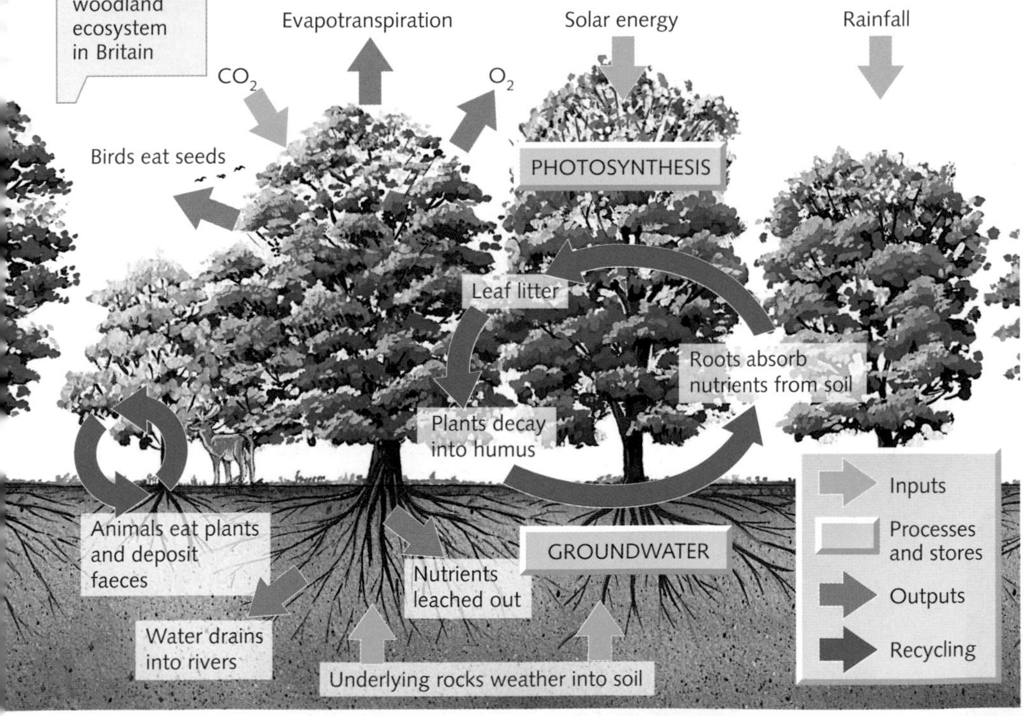

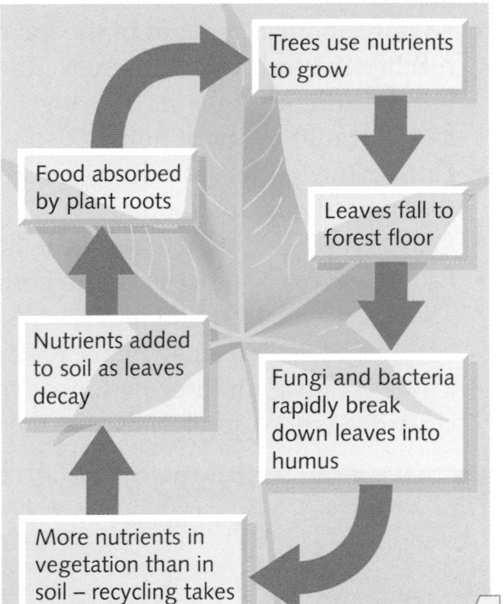

Figure 14.6
Recycling – the nutrient cycle in a British woodland

Trees use nutrients to grow

Food absorbed by plant roots

Leaves fall to forest floor

Nutrients added to soil as leaves decay

Fungi and bacteria rapidly break down leaves into humus

More nutrients in vegetation than in soil – recycling takes only 6 months

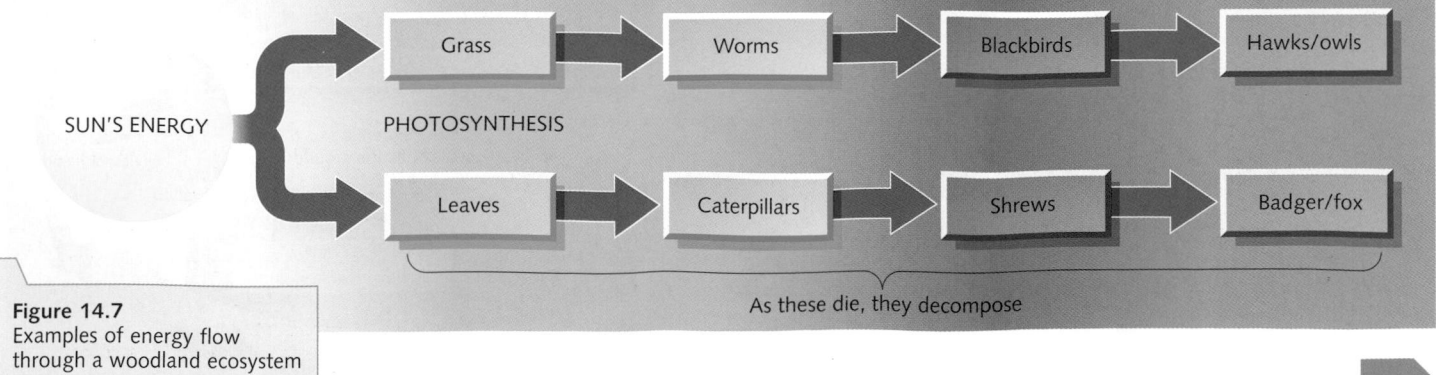

Figure 14.7
Examples of energy flow through a woodland ecosystem

SUN'S ENERGY

PHOTOSYNTHESIS

Grass → Worms → Blackbirds → Hawks/owls

Leaves → Caterpillars → Shrews → Badger/fox

As these die, they decompose

World biomes

Tropical rainforest

- Tropical rainforests grow in places that have an equatorial climate. The rainforest is the most luxuriant vegetation system in the world although its trees have had to adapt to the constant high temperatures, the heavy rainfall and the continuous growing season. Over one-third of the world's trees grow here.
- Although the trees are deciduous, the rainforest has an evergreen appearance as the continuous growing season allows trees to shed their leaves at any time.
- Vegetation grows in distinct layers (Figure 14.8). The lowest layer consists of shrubs. Above this is the under canopy, the main canopy and, rising above, the emergents, which can grow to 50 metres in height. Trees have to grow rapidly in order to reach the life-giving sunlight.
- Tree trunks are straight and, in their lower parts, branchless in their efforts to grow tall.
- Large buttress roots stand above the ground to give support to the trees (Figure 14.9).
- Lianas, which are vine-like plants, use the large trees as a support in their efforts to reach the canopy and sunlight.
- As only about 1 per cent of the incoming sunlight reaches the forest floor, there is little undergrowth. Shrubs and other plants which grow here have had to adapt to the lack of light.
- During the wetter months, large areas of land near to the main rivers are flooded.
- Leaves have drip-tips to shed the heavy rainfall.
- Fallen leaves soon decay in the hot, wet climate.
- There are over 1000 different species of tree, including such hardwoods as mahogany, rosewood and greenheart.
- There is dense undergrowth near rivers and in forest clearings where sunlight is able to penetrate the canopy.

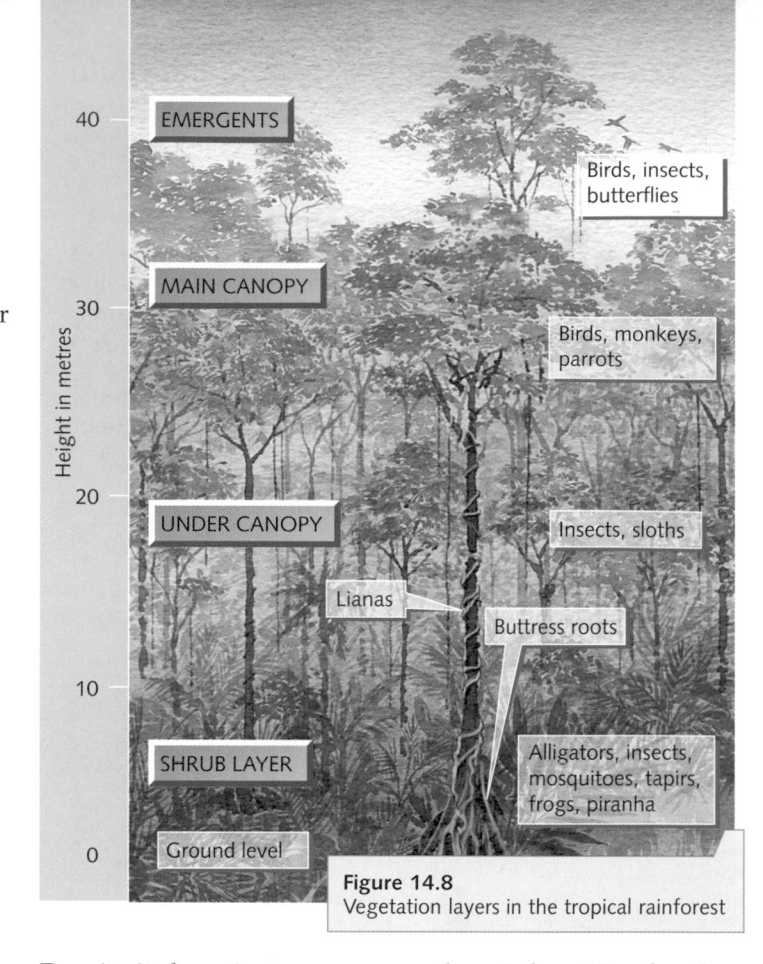

Figure 14.8
Vegetation layers in the tropical rainforest

Despite its luxuriant appearance, the rainforest is a fragile environment whose existence relies on the rapid and unbroken recycling of nutrients (Figure 14.10). Once the forest is cleared (deforestation page 236), then the nutrient cycle is broken. Humus is not replaced and the underlying soils soon become infertile and eroded. Not only is the rainforest unable to re-establish itself, but the land becomes too poor to be used for farming (page 237).

Figure 14.9
Inside the rainforest

Figure 14.10
The nutrient cycle before and after deforestation

Nutrient cycle developed in an area of natural tropical forest

Heavy daily convectional rainfall intercepted by tree canopy

Rich tree growth

Numerous fallen leaves

Rich soil

Ground is protected from the heavy rainfall

Leaves decay rapidly to form humus

Nutrients added to the soil

The cycle after an area of forest has been cleared

Clearance of forest – heavy rainfall hits the ground

Poorer-quality vegetation and soil erosion

Extra water washes away the soil (soil erosion)

No leaf fall to renew humus

Fewer leaves

Nutrients in soil washed downwards (leaching)

Less humus

Soil becomes less fertile

Few nutrients added to soil. Others are leached downwards and are lost to the plants

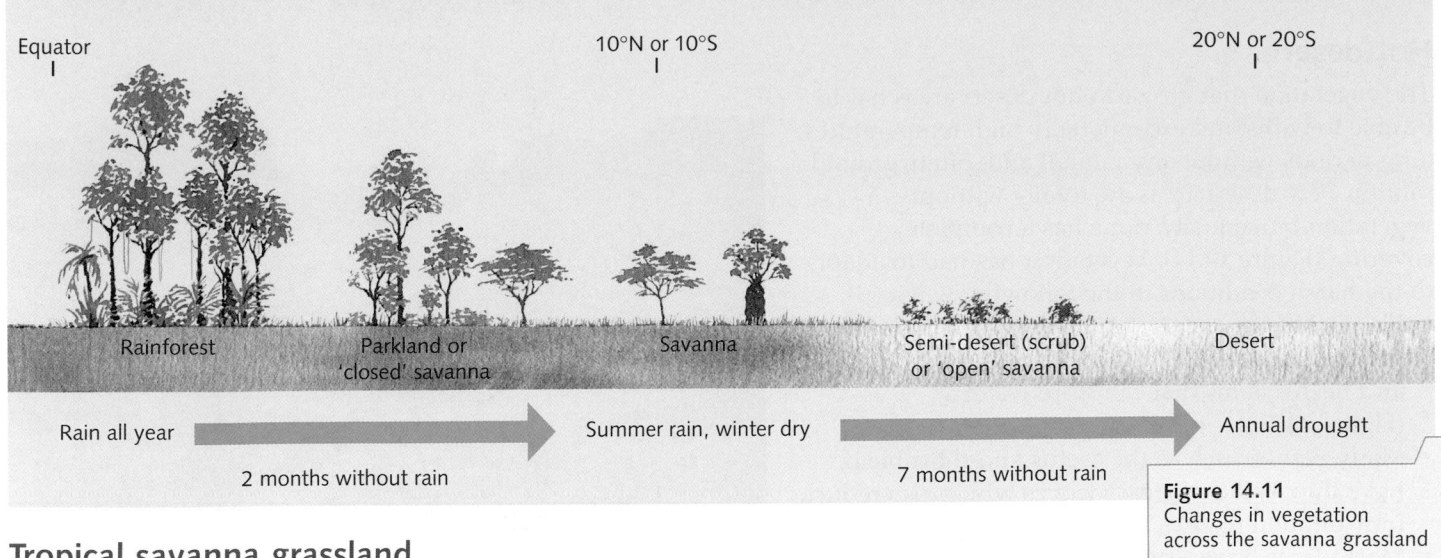

Equator | 10°N or 10°S | 20°N or 20°S

Rainforest | Parkland or 'closed' savanna | Savanna | Semi-desert (scrub) or 'open' savanna | Desert

Rain all year → Summer rain, winter dry → Annual drought

2 months without rain | 7 months without rain

Figure 14.11
Changes in vegetation across the savanna grassland

Tropical savanna grassland

A transect (section) across the savanna grasslands shows how the natural vegetation changes in response to the climate (Figure 14.11). Where the savanna merges with the tropical rainforest (rain all year), the vegetation is dense woodland with patches of tall grass. Moving away from these margins, the vegetation slowly changes to typical savanna grasslands with scattered trees (rain for half the year), and eventually to the drought-resistant bushes and odd clumps of grass on the desert margins (hardly any rain).

The dry season

The scattered deciduous trees lose their leaves, grasses turn yellow and dry up, and the ground assumes a dusty, reddish-brown colour. Some trees shed their leaves while others produce thin, waxy and even thornlike leaves to try to keep transpiration to a minimum. Most plants are **xerophytic** (drought-resistant) with very long roots to tap underground water supplies or with thick bark to store water in the trunk, like the baobab tree (Figure 14.12). Grasses grow in tufts, separated by patches of bare soil. As the dry season progresses, their stalks become stiff, yellow and straw-like and, in time, the plants wither.

The wet season

After the first rains, the grass seeds germinate and trees produce new leaves. Under the hot, wet conditions the grasses grow quickly and can reach a height of 3–4 metres before flowering and producing new seeds. The seemingly endless plains of the Serengeti (Tanzania) and Maasai Mara (Kenya) resemble a vast green sea occasionally interrupted by acacia trees (Figure 14.13). The acacias, with their crowns flattened by the trade winds, provide welcome shelter for wildlife.

The vegetation of these areas has been altered over a period of time by fire, either started deliberately or as a result of electrical storms. More recently, areas nearer the desert margins have experienced desertification (Case Study 15) mainly from pressures resulting from rapid population growth. Trees and shrubs have been removed for fuelwood. As settlements and cultivated areas increase, many nomadic herders, like the Fulani in West Africa and the Maasai in East Africa, find their traditional grazing grounds reduced in size. This leads to overgrazing and soil erosion in the areas to which they are restricted.

Figure 14.12
A baobab tree

Figure 14.13
Savanna grassland with acacia trees

Hot desert

The vegetation that grows in hot desert areas has to be able to withstand exceptionally high temperatures, long periods without any rainfall and, often, ground salinity. Few desert areas are totally without vegetation but, equally, none has a complete covering (Figure 14.14). Vegetation has had to adapt to the harsh conditions in the following ways:

- Plants, like the cactus, have thick, waxy skins to reduce the loss of moisture through transpiration, and fleshy stems that can store water (Figure 14.16).
- Many plants, such as the cactus and thornbush, have thin, spiky or glossy leaves which also reduce transpiration (Figure 14.16a).
- Most plants have either very long tap roots which enable them to reach deep down to underground supplies of water (acacias), or shallow roots that spread out over a wide area to allow them to make full use of any rain that falls (creosote bush).
- Seeds have the ability to lie dormant for several years and then to germinate quickly after a heavy shower and to complete their life cycle within two or three weeks. When these plants flower, the desert literally blooms, the bright flowers attracting insects to help with pollination (Figure 14.15).
- Many plants are also capable of growing in saline depressions where water collects following the short but heavy showers before rapidly evaporating to leave deposits of salt.
- Plants tend to be widely spaced so as to reduce competition for water (Figure 14.14).

Human impact on the desert biome has been fairly limited, partly due to the lack of commercial plants, other than the date palm, that grow there. Nevertheless, it is a fragile ecosystem which can be subject to an increased risk of desertification, especially in places where there is a growth in population (Case Study 15).

Figure 14.14
Desert vegetation, Arches National Park, Utah

Figure 14.15
The desert in bloom

b

a

Figure 14.16
Cacti
(a) prickly pear   (b) saguaro

Coniferous forest

The natural vegetation of the coniferous forest, or *taiga* as it is known in Russia, consists of vast uninterrupted stands of spruce, pine and fir (Figure 14.17). The most common trees are the Norwegian spruce and Scots pine in upland Britain and Scandinavia, and the Sitka spruce and Douglas fir in North America. Often, in contrast to deciduous woodlands, only one or two species of tree grow across a wide area. The trees have had to adapt to the harsh physical environment (Figure 14.18) where:

- winters are long and extremely cold, often with strong winds; precipitation falls as snow and, as groundwater is frozen, moisture is unavailable for plants
- summers are cool and the growing season is short, but there is some rain for plant growth
- soils are often thin and poor (see podsols, page 253).

Figure 14.17
Coniferous forest in the Canadian Rockies

Coniferous trees are softwoods. They are valuable for timber, as well as for pulp and paper. The coniferous forest, covering as it does vast tracts of mainly inhospitable land, has been less affected by human activity than have most other biomes.

Spruce
30 metres

Whorls

Distance between whorls indicates 1 year's growth

Evergreen – no need to renew leaves for the short growing season

Compact conical shape helps stability in the wind

Needles to reduce moisture loss

Trunk is usually straight and tall in attempt to reach the sunlight

Thick resinous bark acts as a protection against cold winds. Thin girth (trunk) due to rapid upward growth

Cones protect the seeds during very cold winters

Downward-sloping and springy branches allow snow to slide off

Very little undergrowth as trees are closely spaced and branches cut out sunlight

Covering of dead pine needles as the cold climate discourages decay

Shallow roots because either:
- soils are thin, or
- subsoil is frozen for much of the year, or
- cold boulder clay soil discourages deep root growth.

Long roots for anchorage against strong winds

Figure 14.18
Adaptation of conifers to cold climates

Deforestation – the destruction of an ecosystem

The Amazon Basin

Although one-third of the world's trees still grow in the Brazilian rainforest, their numbers are being rapidly reduced due to **deforestation**.

Why is Brazil's rainforest being cleared?

Brazil's rapid population growth since the 1960s has meant that more land was needed for people to live on, more farmland to produce food for the extra numbers, more jobs required for people to earn a living and more resources needed if people's standard of living was to improve. There was also the need to reduce the country's huge national debt (page 190). The Brazilian rainforest, at that time largely undeveloped, had the space and the resources.

Farming

Land is cleared for three types of farming.

1 'Slash and burn' is the traditional method used by the Amerindians of the rainforest. Although this is the most sustainable of the three types, it nevertheless causes considerable areas to be cleared, even if only temporarily, each year (Figure 14.19).

2 **Subsistence farming** has increased as a result of the government providing land to some of Brazil's 25 million landless people. In places, 10 km strips of land were cleared alongside highways, and settlers were brought in from places that were even poorer, like the drought areas in the north-east.

3 **Commercial cattle ranching** is run by large transnational companies which sell beef mainly to fast-food chains in developed countries. These companies burn the forest, replacing trees with grass (Figure 14.20).

Transport

Over 12 000 km of new roads have been built across the rainforest, the largest being the 5300 km Trans-Amazonian highway (Figure 14.21). These roads were built to develop the region and to transport timber,

Figure 14.19
Destruction of the rainforest

minerals, farm produce and people. A 900 km railway has been built from Carajas to the coast (Figure 14.24) and numerous small airstrips have been constructed.

Resources

The main types of resource are:

1 **Timber**, mainly hardwoods, is obtained by logging companies which fell trees for markets in developed countries. While timber is a valuable source of income for Brazil, little attempt has been made to replant deforested areas.

2 **Minerals** provide the region with a vast natural resource. They include iron ore (Figure 14.23), bauxite, manganese, diamonds, gold and silver.

3 **Hydro-electricity** is an important renewable source of energy but the building of dams and the creation of large lakes has led to large areas of forest being flooded (page 122).

Settlement

The development of Amazonia has led to an increase in population from 2 million (1960) to over 30 million (2000). Large tracts of forest have been cleared for the development of such new settlements as Maraba (150 000) and Carajas (Figure 14.24).

Figure 14.20
Commercial cattle ranching in the former rainforest

Figure 14.21
The Trans-Amazonian Highway

Figure 14.22
A member of the Kayapo tribe

Rates of forest clearance

Estimates vary as to how much of the Amazon rainforest has been deforested since clearances began in the early 1960s. The World Bank suggest 15 per cent. In contrast, some environmental groups claim the figure is up to 40 per cent which, if this is accurate, would mean that some 15 hectares (about 15 football pitches) have been cleared *every minute*. An American environmental organisation on one occasion claimed that there were 24 546 fires burning in the forest at the same time.

Effects of the clearances

- Of 30 million known species on the Earth, 28 million are found in the rainforest (99 per cent being insects). A typical patch of 10 km² could contain as many as 1500 species of flowering plant, 750 species of tree, 400 types of bird, 150 varieties of butterfly, 100 different reptiles and 60 types of amphibian. Many others have still to be identified and studied. Deforestation has destroyed the habitats of many of these species, some of which may have proven to be of considerable value. (We already get over half of our medicines from the rainforest – one of these, a recently discovered periwinkle, has reduced deaths from child leukaemia from 80 to 20 per cent.)

- There has been a huge reduction in the number of Amerindians (from 6 million when Europeans arrived to the present number of 200 000) and destruction of their traditional culture and way of life. Those remaining, such as the Kayapo, are often forced to live on reservations (Figure 14.22).

- The clearance of trees means that there is no canopy to protect the soil from the heavy afternoon rain, or roots to bind it together. The result is less interception and infiltration and more surface runoff and soil erosion (Figure 15.33). Deforestation also breaks the humus cycle (Figure 14.10) and existing nutrients are rapidly washed (leached) out of the soil leaving it infertile. This loss in fertility has already caused some of the new subsistence farms and the larger cattle ranches to be abandoned.

- Many rivers have been polluted due to mining operations.

- Deforestation is causing climatic change in two ways. With fewer trees there is less evapotranspiration and, therefore, less water vapour in the air (about one-quarter of the world's fresh water is, at present, stored in the Amazon Basin). With less moisture in the hydrological cycle (page 278), there is already evidence of reduced rainfall totals together with the threat of a possible increase in local droughts. At the same time, the burning of the forest is accelerating global warming by releasing huge amounts of carbon dioxide, the main greenhouse gas (page 218).

- It is possible that there are already changes in the composition of the atmosphere. Scientists claim that over one-third of the world's fresh oxygen supply comes from the tropical rainforest. This would be lost if the region is totally deforested.

Figure 14.23
Carajas iron ore mine

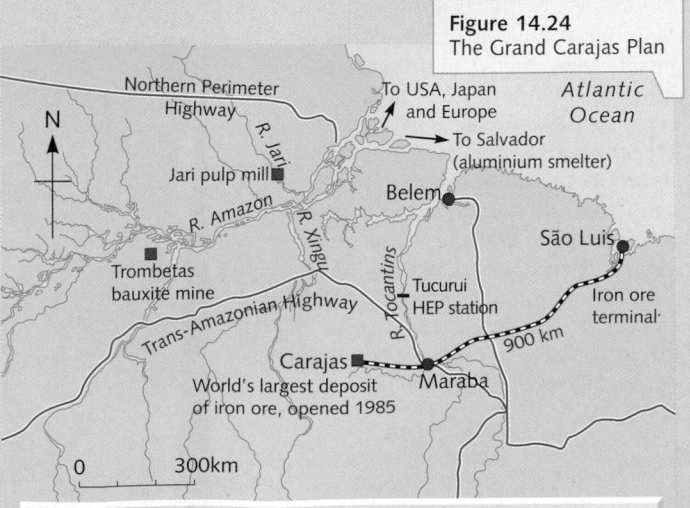

Figure 14.24
The Grand Carajas Plan

Chance discovery (1966) of one of world's largest iron ore deposits (high-quality 66% iron content). The Grand Plan covers $\frac{1}{10}$ of Brazil, but only 1.6% of this area has been cleared of forest. 7000 jobs created. Mine uses HEP from Tucurui. Local steelworks at Maraba. Rest of iron ore taken by rail (each train has 200 wagons) to port of São Luis for transport to steelworks in south-east Brazil, Japan, Korea and Germany.

CASE STUDY 14

Sustainable forestry in Malaysia

Rainforests in peninsular Malaysia

Non-tree-covered land use 29%
Secondary rainforest 49%
Tree crops 12%
Natural forest 10%

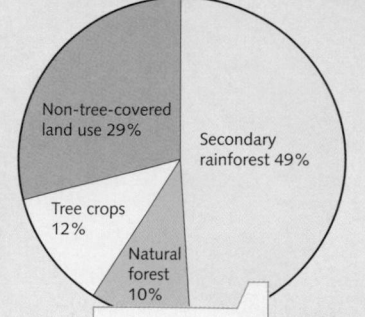

Figure 14.28
Forest land use

Appearance and extent

The natural vegetation of Malaysia is tropical rainforest, characteristically evergreen and rich in species (Figures 14.25 and 14.26). Over 70 per cent of the Malay peninsula is still covered in trees. This includes rainforest, both natural and secondary growth, and tree crops such as oil palm and rubber (Figure 14.28). However, estimates suggest that the rainforest area could decrease by more than half (down to 25 per cent) by 2020 at present rates of clearance.

Conversion to tree crops

During colonial times, extensive areas of natural forest were cleared and replanted with rubber trees (Figure 7.15). Rubber was grown on large plantations owned and operated by transnational corporations such as Dunlop and Guthries. Malaya, as it was then known, became the world's leading exporter of rubber. Since then there have been major changes:

The most luxuriant rainforest is found in the highlands that form the backbone to the peninsula (Figure 14.27). Here the forest has a closed canopy, some 35 to 45 m above ground level, consisting of the crowns of large trees closely fitted together and with buttress roots supporting stout, straight trunks (Figures 14.8 and 14.9). Larger trees occasionally rise above the canopy. Under the canopy are smaller trees, shrubs and herbs. Many smaller plants, especially orchids and ferns, grow on the trees (epiphytes). Others are parasitic. On the forest floor, fungi, aided by termites and other insects, rapidly break down dead trees and fallen leaves in nature's recycling process (Figure 14.10). The real opportunists of the rainforest are the lianas which climb up other plants until they can bask in the sunlight on the canopy. The most useful lianas belong to the palm family, their slender, flexible and tough stems being used to make rattan furniture and Melaka (Malacca) cane. Malaysia has 8000 species of flowering plants, 3000 species of tree, 500 species of fern and 60 species of bamboo.

Figure 14.25
Extract from the *Malaysian Official Yearbook*

THAILAND
Langkawi
Kota Bharu
N
Urban areas and other land uses
Forested areas, including virgin and commercial forest areas
Agriculture, mainly rubber and oil palm
National Park/wildlife sanctuaries
Penang
Ipoh
Cameron Highlands
TAMAN NEGARA
Straits of Malacca
Kuantan
South China Sea
Shah Alam new town
Kuala Lumpur
New international airport
Port Klang
Melaka
SINGAPORE
INDONESIA
0
100 km

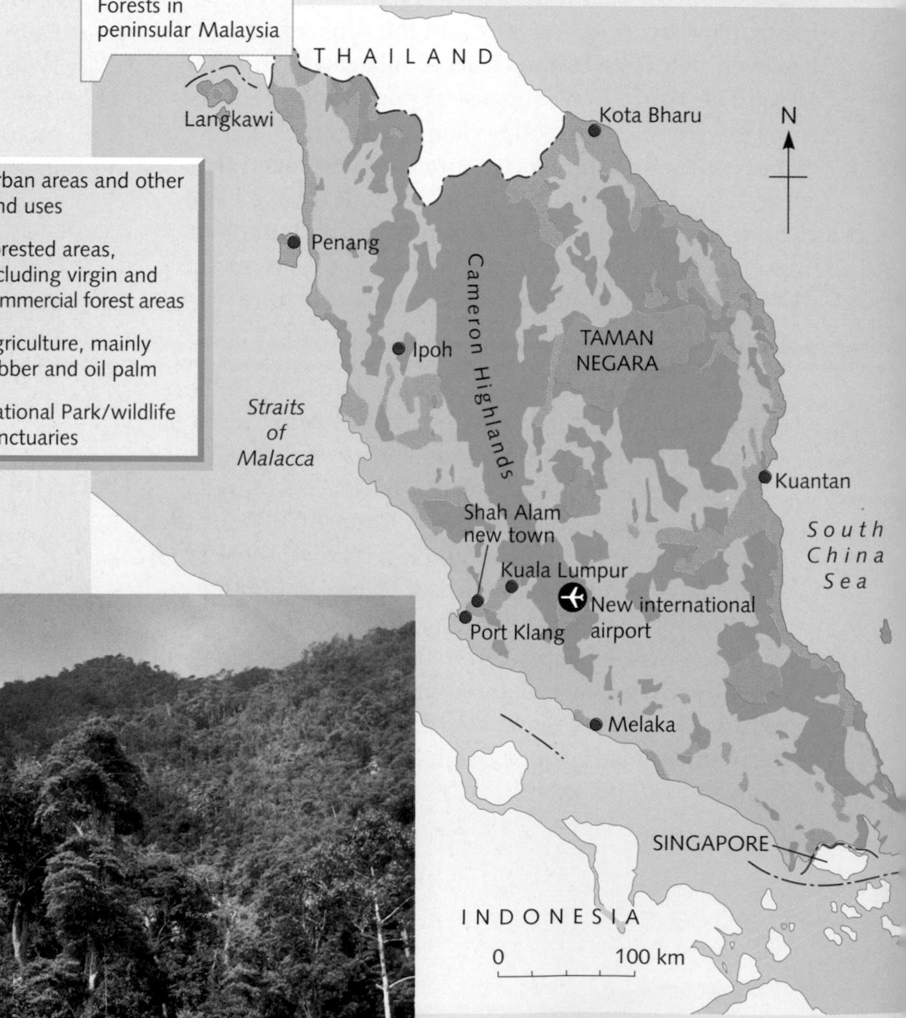

Figure 14.27
Forests in peninsular Malaysia

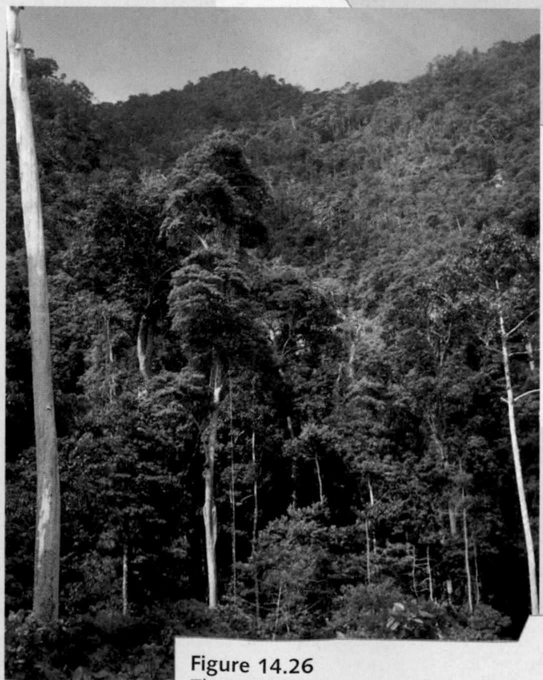

Figure 14.26
The rainforest in peninsular Malaysia

- There has been a sharp decline in the demand for natural rubber due to the use of synthetic rubber.
- The Malaysian government has taken over the largest plantations. Two-thirds of the former transnational plantations are now state-run, the remainder having been divided into 5 ha smallholdings for individual farmers.
- There has been a rapid increase in world demand for palm oil. Since the 1970s, large areas of lowland forest have been converted into oil palm plantations (Figure 7.16) under huge government 'rural development and resettlement schemes'. Malaysia is now the world's major exporter of palm oil.

Deforestation

Before the 1970s, there was little deforestation in peninsular Malaysia. Most of the population lived in rural *kampongs* (villages) which were small and virtually self-contained (Figure 14.29). There was little urban or industrial growth, and often the only forest clearance was for the extraction of minerals, notably tin.

Since then, Malaysia has become one of the newly industrialised countries (NICs) of the Pacific Rim (page 147), with a government working towards the country becoming 'fully developed' by the year 2020. This means that Malaysia needs more land for its growing population and more wealth for its continued development.

Deforestation therefore occurs where there is rapid urbanisation (page 78) and a resultant demand for housing, jobs and services. Most deforestation has taken place near the new town of Shah Alam (where most of Malaysia's new industry is located) and Kuala Lumpur (the capital city). Land has also been cleared for motorways and for the new international airport at Subuya (Figure 14.27).

Forest is also cut down when money is needed to repay the National Debt and to finance new housing, transport, services and industrial projects. Malaysia, like other developing countries, has had to export primary products (tin, timber, rubber and palm oil) to more economically developed countries (page 186). It is now faced with the option of either conserving its forests or using forest products to earn money (sawn wood is Malaysia's third most valuable export).

Methods of logging

Several different methods of logging have been used in Malaysia (Figure 14.31). Clearfelling was widely used before the 1980s, especially if the logging companies were non-Malaysian (Figure 14.30). This method of felling often resulted in the total destruction of the rainforest ecosystem, with the trees removed, wildlife habitats lost and the lifestyles of indigenous forest dwellers ruined – all for a rapid economic gain and to satisfy markets in developed countries. Although a clearfelled area can regenerate close to its former state in 30 years, it takes 60 years for a replacement giant Dipterocarp tree to reach the forest canopy.

Figure 14.29
A Malaysian *kampong*

Figure 14.30
Clearfelling

Figure 14.31
Methods of logging

To be cut in 5 years' time
Cut 15 years ago
Cut 10 years ago
Cut 5 years ago
Just cut

	(a) Selective cutting	(b) Integrated cutting	(c) Clearfelling	(d) Strip cutting
Description	Harvesting of selected single and groups of trees	Several different species harvested at the same time	Total removal of all trees	Variation of clearfelling, but often following the contours of the land
Advantages	Reduces crowding, promotes growth of younger trees, allows a more natural regeneration. More environmentally friendly	More economical than selective cutting, as trees for different uses (furniture, logs) are harvested at the same time.	Allows total afforestation, ideally with improved plants. Higher yields and profits	Clearing narrow strips avoids huge scars and allows easier regeneration. Constant income and replacement
Disadvantages	Expensive. Removal of some trees by machines usually damages others, especially if Dipterocarps are being harvested	Often poorer-quality trees left. Quality may degenerate over time	Leaves ugly scars. Destroys ecosystem and habitats. Encourages soil erosion and silting of rivers (see diagram on p.243)	Minimises erosion but still destroys sections of the ecosystem

Towards more sustainable forestry

Although Western environmental groups remain critical of Malaysia's logging policies, there is little doubt that forest management has improved dramatically since the pre-1980s. As the Malaysian Minister for Primary Products recently stated:

'The West has already become rich by destroying its forests. They have now adopted hypocritical principles which they did not apply to themselves but which they insist on imposing upon us. Yet we are doing a better job of preserving our forests than they ever did.'

Some of the ways by which the Malaysian government and the timber industry have tried to make forestry more sustainable include:

- Ensuring that logging companies only use selective cutting methods, do not fell trees under a minimum circumference and do not exceed the maximum number of trees per hectare that they are allowed to cut. Companies employing clearfelling methods do not get their licence renewed.

- Restricting the use of bulldozers and heavy destructive machinery (Figure 14.32) and keeping their entry tracks (known as skid trails) one kilometre apart.
- Spending money improving the Taman Negara National Park and other protected forest areas, together with increasing the number of recreational parks – 14 per cent of peninsular Malaysia's forest is protected from 'development' (Figure 14.27).
- Testing the use of helicopters – although 'helilogging' is a very expensive method of removing logs from the forest, it is far less destructive to the rainforest ecosystem.
- Implementing, through the Community Forestry Development scheme, three particular projects:
  - village forestry, which encourages the planting of traditional fruit trees to ensure a sustained food supply for local people
  - urban forestry, where trees are planted alongside main roads and in open spaces to stabilise temperatures, create shade and reduce soil erosion
  - forest recreation.

Figure 14.32
The destructive effects of heavy machinery in the rainforest

Ecosystems

Key Words and Terms

a You should know the meaning of the following terms:
- ecosystems • biomes • energy flows • photosynthesis • producers
- consumers • decomposers • nutrients • food chain
- biomass • litter • xerophytic • deforestation • tree crops
- logging • sustainable forestry.

b You should know the difference between:
- flora and fauna
- the living and non-living environment
- an open system and a closed system
- herbivores and carnivores
- prickly pear and saguaro cactus
- softwoods and hardwoods
- 'slash and burn', subsistence and commercial farming.

Key Ideas

You should know and understand the following:
- The concept of an ecosystem and how it works.
- The global distribution of the major ecosystems/biomes.
- The vegetation of selected world ecosystems/biomes – tropical rainforest, tropical savanna grassland, hot desert, temperate deciduous forest and the coniferous forest/taiga.
- The appearance of the major vegetation types and their adaptations to climate and soils.
- The fragile nature of ecosystems.
- Changes to, and exploitation of, the major vegetation types by human activity/mismanagement.
- The causes and effects of deforestation.
- The need for sustainable forestry and methods of stewardship/management of forests.

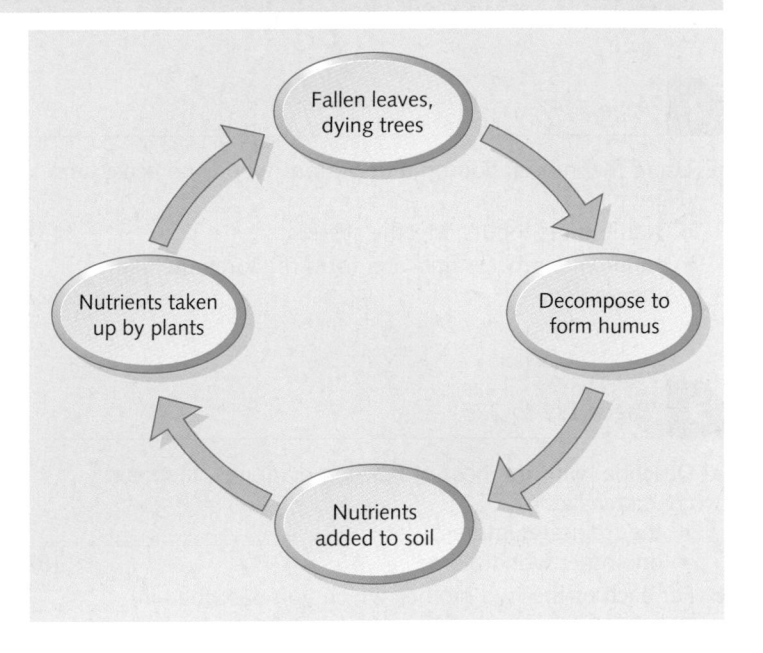

Skills, Theories and Models

- Annotate and interpret diagrams
- Recognise the distinctive character of places
- Understand the classification of world biomes
- Investigate the impact of resource development
- Appreciate the need to plan and manage resource development
- Understand ecosystems

QUESTIONS

1 (Page 230)

a i) Give three examples of ecosystems. (3)

   ii) Copy and complete the table by adding the following in the appropriate place.
   - fauna
   - solar energy
   - climate
   - flora
   - rocks and soil
   - people. (6)

Non-living environment	Living environment
e.g. 1	e.g. 1
2	2
3	3

b Explain the part played in an ecosystem by each of the following:
- energy flows
- photosynthesis
- food chain
- recycling of nutrients
- decomposition. (5)

c Make a copy of the diagram below. Complete it by adding two examples to each box from the following list:
- fungi
- leaves
- rabbit
- air
- green plants
- bacteria
- weasel
- caterpillar
- water
- owl. (10)

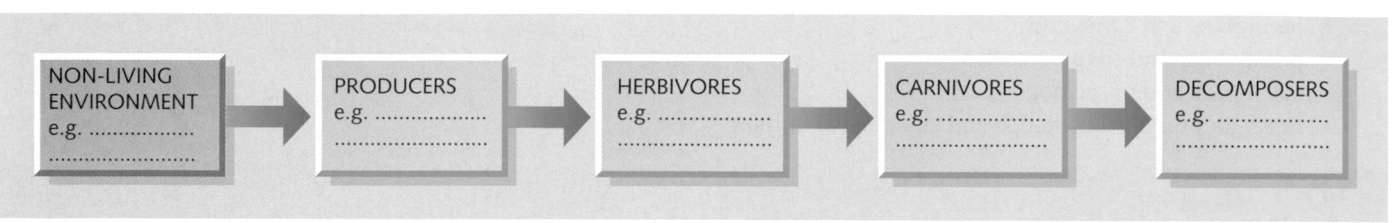

NON-LIVING ENVIRONMENT e.g. .... → PRODUCERS e.g. .... → HERBIVORES e.g. .... → CARNIVORES e.g. .... → DECOMPOSERS e.g. ....

2 (Page 231)

a Using the information on a deciduous woodland ecosystem (Figure 14.5):

   i) Name three inputs into the system. (3)

   ii) Name one process and one form of storage within the system. (2)

   iii) Name two outputs from the system. (2)

   iv) Give two examples of recycling within the system. (2)

b With the help of a labelled diagram, explain how one of the examples of recycling you named in part (**a**) iv works in a woodland ecosystem. (4)

3 (Pages 232 to 235)

a Describe, with the help of labelled diagrams, the main characteristics of:
- the tropical rainforest
- one other world biome. (8)

b For each of the two biomes which you described in part (**a**), explain how the natural vegetation has:
- adapted to the climatic conditions (6)
- been altered by human activity. (4)

Tropical rainforest   Tropical savanna grassland   Hot desert   Coniferous forest

4 (Pages 238 to 240)

a Describe the appearance of the Malaysian rainforest. (5)

b Why are tree crops an acceptable alternative to the rainforest? (2)

c Why is Malaysia's rainforest under threat? (3)

d Describe the advantages and disadvantages of:
- clearfelling
- selective felling. (4)

e Using the headings in the star diagram, describe how Malaysia is trying to develop a more sustainable form of forestry. (6)

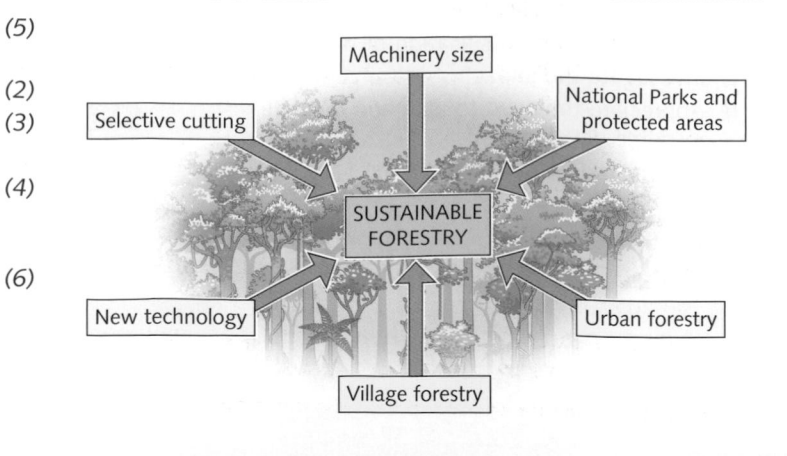

Machinery size

Selective cutting

National Parks and protected areas

SUSTAINABLE FORESTRY

New technology

Urban forestry

Village forestry

5 *(Pages 232, 236 and 237)*

a How does the tropical rainforest help to protect:
- the soil   • fresh water supplies   • wildlife   • the traditional Amerindian way of life? (8)

b Draw labelled diagrams to show the humus (nutrient) cycle:
- before deforestation   • after deforestation. (4)

c i) Give six reasons why parts of the Brazilian rainforest are being deforested. (6)

ii) Describe how deforestation is affecting:
- wildlife   • plant life   • the Amerindians   • the soil   • water supplies
- the local climate   • the world's climate   • oxygen supplies. (16)

d With the help of a labelled sketch map, describe the main features of the Grand Carajas Plan. (5)

e Brazilians often express different views on deforestation to those of people living in more economically developed countries. Read the extracts below, and study the diagram.

i) Why do some Brazilians feel the developed world has over-reacted? (2)

ii) Why do Brazilians resent being told by developed countries that they should protect the rainforest? (2)

iii) Can you suggest a compromise that might benefit Brazil and protect the rainforest? (2)

FORESTED AREA | DEFORESTED AREA

Evapotranspiration from trees adds moisture to the air

Heavy convectional storms most afternoons

Shifting cultivators live in harmony with nature. Only limited deforestation in Amazonia for fuel

Forested slopes, nature in balance

Habitat for wildlife (birds, animals and insects)

Fragile soils shielded from heavy rain by trees

Movement of rainwater through soil regulates river flow, prevents flooding and stores water for drier periods

Clean river usable for drinking

Tree roots control flow of water and stabilise the soil, preventing landslides

LITTLE ECONOMIC GAIN
CONSIDERABLE ENVIRONMENTAL GAIN

Fewer trees mean less evapotranspiration

Ranching mainly for poor-quality meat for hamburgers and frankfurters

Heavy rainfall (both in amount and intensity) washes away the unprotected surface soil

Plantation crops

Timber – loss of wildlife and many species of trees

Highways

Mining

Rapid surface runoff causes gully erosion and flooding

Lack of trees creates a fuel shortage

Muddy water undrinkable

No roots to hold soil together results in landslides

Heavy rainfall causes leaching, ruining the soil

Silt blocks rivers and fills reservoirs

REASONABLE SHORT-TERM ECONOMIC GAIN
CONSIDERABLE ENVIRONMENTAL LOSS

Views expressed by several Brazilians

A Brazilian tour guide: 'Europeans have exploited us for years and now they expect us not to develop or to use our own resources.'

An Indian guide: 'It is Western propaganda. What is the point in clearing the forest when the Amazon and its tributaries rise by 15 metres a year and flood 60% of the forest for several months?'

A geography lecturer at São Paulo University: 'It is an over-reaction by the Western world. At present we are making various studies of the forest to see how we can organise its development on a sustainable basis. Development cannot be stopped, but it should be made to harmonise with the environment.'

A Brazilian doctor: 'Conservation groups have to exaggerate if they are to raise funds.'

Views expressed by conservation groups

'Every minute, an area of rainforest the size of 15 football pitches is being destroyed. At the present rate of destruction, in 40 years' time there will be virtually no more rainforests. That means the Earth's climate will change. We will lose a precious source of medicine. We will be deprived of an important source of food and industrial products. And we will lose over half the world's animal and plant species.'

243

15 Rocks and soils

Rocks – types, formation and uses

Rock types

The Earth's crust consists of many different types of rock. It is usual to group these rocks into three main types. This simple classification is based on how each type of rock was formed (Figure 15.7).

1 **Igneous rocks** result from volcanic activity. They consist of crystals which formed as the volcanic rock cooled down, e.g. granite (Figure 15.1 and page 250) and basalt.

2 **Sedimentary rocks** have been laid down in layers. They usually consist either of small particles that have been eroded and transported, e.g. sandstone (Figure 15.2) and shale, or of the remains of plants and animals, e.g. limestone (Figure 15.3 and page 248), chalk (page 250) and coal.

3 **Metamorphic rocks** are those that have been altered either by extremes of pressure, e.g. shale is compressed into slate (Figure 15.4), or by extremes of heat, e.g. limestone is changed into marble.

Rock structure

The structure of a rock can, among other things, affect its **resistance** to erosion and its **permeability** to water.

Figure 15.1
Crystals in Shap granite, an igneous rock

Figure 15.3
Limestone fossils

Figure 15.2
Sandstone – a sedimentary rock

Figure 15.4
A slate quarry in North Wales – slate is a metamorphic rock

Resistance

Rocks have different strengths and so produce different landforms. For example:

- The harder a rock is, the more resistant it is likely to be to erosion. Harder rocks are therefore usually found as hills and mountains. The softer and less compact the rock, the more likely it is to be either broken up or worn away (see 'Weathering', page 246). Valleys are formed in soft rocks. In Figure 15.5, the hills consist of chalk while the valley has formed on clay (page 250).
- In a drainage basin (page 278), the more resistant the rock, the steeper the valley sides. Where resistant rock crosses a river's course, it is likely to create waterfalls and rapids (page 283).
- On coasts, resistant rocks form steep cliffs and stand out as headlands, whereas softer rocks form bays (page 300 and Figure 15.6).

Permeability

An **impermeable** rock is one that does not let water pass through it, in contrast to a **permeable** rock which does allow water to pass through it. Permeable rocks may either:

- consist of tiny pores through which water can pass – such rocks, which include chalk, are said to be **porous**; or
- contain areas of weakness, such as bedding planes, along which water can flow. Horizontal bedding planes, which separate individual layers of rock, can be seen in Figures 15.2 (sandstone) and 15.14 (Carboniferous limestone).

Rock structure, therefore, affects the landforms of an area and can produce distinctive types of scenery, e.g. limestone (page 248), chalk (Figure 15.5 and page 250) and granite (page 250).

Figure 15.5
A chalk escarpment and clay vale

Figure 15.6
Headlands and bays

Use of rocks

Many rocks have an important economic value. Such rocks are extracted, usually by mining or quarrying, for specific purposes. Some examples of the economic importance of rocks are given in Figure 15.7.

Figure 15.7
Type, formation and uses of rocks

	Rock	Formation	Uses
Igneous	Granite	Magma rises from the Earth's mantle (Figure 16.5) and slowly cools within the Earth's crust. The slow rate of cooling produces large crystals.	Building (Aberdeen is known as 'The Granite City'). Pottery (from kaolin, or china clay). Sites for reservoirs. Grouse moors (granite gives very poor soils). Some tourism (tors – Figure 15.23).
	Basalt	Magma reaches the Earth's surface as lava from a volcano, and cools very quickly. The rapid cooling creates small crystals.	Weathers into a fertile soil. Foundation material for roads. Some tourism (Giant's Causeway and Fingal's Cave).
Sedimentary	Coal	Fossilised remains of trees and plants that grew under very hot, wet conditions.	Thermal energy – power stations, industry, domestic use.
	Sandstone	Results from grains of sand being compressed and cemented together.	Building material – sites for many settlements.
	Limestone	Remains of shells and skeletons of small marine organisms, e.g. coral, which lived in warm, clear seas.	Sheep pasture (soils thin and porous), quarried for cement and lime. Stone walls. Tourism (caves, page 249).
	Chalk	A form of limestone.	Thin soils suited to cereals (wheat and barley). Also cement and lime. Spring line is ideal for settlement (Figure 15.5).
Metamorphic	Marble	Limestone changes by heat and pressure.	Monuments (e.g. Italy).
	Slate	Shales and clays changed by pressure.	Building material (roofs).

Weathering and mass movement

Weathering

Rocks that are exposed on the Earth's surface become vulnerable to weathering. Weathering is the disintegration (breaking up) and decomposition (decay) of rocks *in situ* – that is, in their place of origin. Weathering, unlike erosion, need not involve the movement of material.

There are two main types of weathering:

1 **Physical weathering** is the disintegration of rock into smaller pieces by physical processes without any change in the chemical composition of the rock. It is most likely to occur in areas of bare rock where there is no vegetation to protect the rock from extremes of weather. Two examples of physical weathering are **freeze–thaw** and **exfoliation**, to which a third, **biological weathering** by tree roots, can be added.

2 **Chemical weathering** is the decomposition of rocks caused by a chemical change within the rock. It is more likely to occur in warm, moist climates, as these encourage chemical reactions to take place. An example of chemical weathering is **limestone solution**.

Freeze–thaw, or **frost shattering** as it is sometimes called, occurs in cold climates when temperatures are around freezing point and where exposed rock contains many cracks. Water enters the cracks during the warmer day and freezes during the colder night. As the water turns into ice it expands and exerts pressure on the surrounding rock. When temperatures rise, the ice melts and pressure is released. Repeated freezing and thawing widens the cracks and causes pieces of rock to break off. Where broken-off rock collects at the foot of a cliff it is called **scree** (Figure 15.8).

Exfoliation, or **onion weathering**, occurs in very warm climates when exposed, non-vegetated rock is repeatedly heated and cooled. The surface layers heat up and expand more rapidly during the day and cool and contract more rapidly at night than do the inner layers. This sets up stresses within the rock which cause the surface layers to peel off, like the layers of an onion, to leave rounded rocks and hills (Figures 15.9 and 9.32).

Biological weathering is when tree roots penetrate and widen bedding planes and other weaknesses in the rock until blocks of rock become detached (Figure 15.10).

Carbonation, or **limestone solution**, is caused by carbonic acid (i.e. carbon dioxide in solution) which occurs naturally in rainwater. Although it is only a weak solution, it reacts chemically with rocks such as limestone which contain calcium carbonate. As the limestone slowly dissolves, it is removed in solution by running water. Solution widens bedding planes and joints to create the distinctive landforms described on pages 248 and 249.

Figure 15.8
Screes beside Moraine Lake in the Canadian Rockies

Figure 15.9
Exfoliation: Uluru (Ayers Rock), Australia

Figure 15.10
Biological weathering, Cumbria

Figure 15.11
Terracettes on a hillside

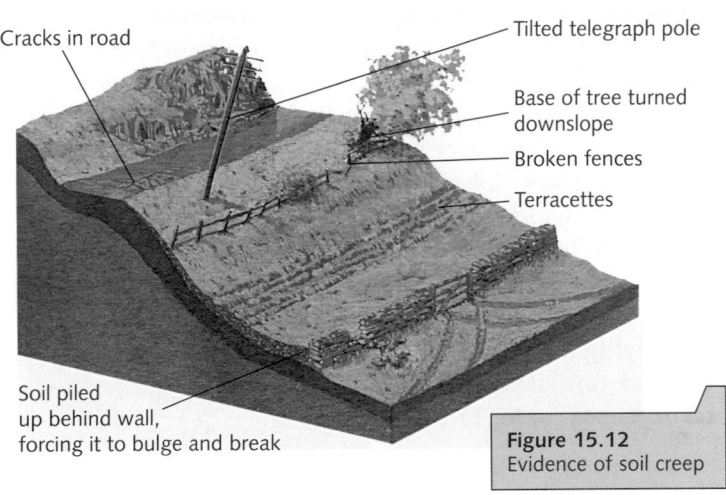

Cracks in road

Tilted telegraph pole

Base of tree turned downslope

Broken fences

Terracettes

Soil piled up behind wall, forcing it to bulge and break

Figure 15.12
Evidence of soil creep

Mass movement

Mass movement describes the downhill movement of weathered material under the force of gravity. The movement, which occurs on all slopes and is an almost continuous process, can cause the transport of soil, stones and rock. However, the speed of movement can vary considerably, from soil creep, where the movement is barely noticeable, to mudflows and landslides, where the movement becomes increasingly more rapid.

Soil creep is the slowest of downhill movements, occurring on very gentle and well-vegetated slopes. Although material may move by less than 1 cm a year, its results can be seen as step-like **terracettes** on banks and hillsides (Figure 15.11). Other effects of soil creep are shown on Figure 15.12.

Mudflows and **landslides** are rapid movements which often take place with little warning. They are most likely to occur on steep slopes.

- They may occur after periods of heavy rain when loose surface material becomes saturated and the extra weight causes material to move downhill. This can occur on steep valley sides as at Aberfan in South Wales (1966), on mountainsides as in Rio de Janeiro's *favelas* (Figure 5.24), or along coasts, especially where soil and softer rock overlie harder rock.
- They may follow earth movements such as earthquakes, as in Peru (1970), or after volcanic eruptions, as in the case of Nevado del Ruiz in Colombia (1985) and Pinatubo in the Philippines (1991).

Sarno, Italy

On 6 May 1998, following 48 hours of torrential rain, mudflows swept through several small towns and villages lying between Naples and Salerno. Rivers of mud, water and debris, arriving as mini-tidal waves up to 3 metres in height, burst into town centres, tearing down houses and bridges, swallowing cars, and causing residents in the densely populated rural areas near to Mount Vesuvius to flee for their lives. Early estimates suggested that over a thousand people had been made homeless, many having to spend the first night on rooftops or on the highest floors of apartment blocks until they could be rescued the next day by helicopters. Rescue teams, together with sniffer dogs trained to find avalanche victims, spent several days searching through the mud for up to 300 missing people. Worst-hit was Sarno, a small town of 2000 inhabitants (Figure 15.13). Of 143 bodies eventually recovered from the flood-affected area, over 100 came from this one town. Amongst the buildings affected was the local hospital where, the day after the event, the bodies of two doctors, a nurse and several patients were pulled from under the rubble and mud.

The authorities came to realise that up to 4000 similar settlements in Italy were equally ill-prepared for such an event. These included settlements on the slopes of Vesuvius, including suburbs of Naples which are extending up the mountainside – the lessons of ancient Pompeii and Herculaneum apparently long since forgotten.

Figure 15.13
Mudslide at Sarno, southern Italy

Limestone

Limestone consists mainly of calcium carbonate. There are several types of limestone including chalk, and Jurassic and Carboniferous limestone. Carboniferous limestone contains many fossils, including coral, indicating that it was formed on the bed of warm, clear seas.

Carboniferous limestone

Since its emergence from the sea, Carboniferous limestone has developed its own distinctive type of scenery, known as **karst**. The development of karst landforms is greatly influenced by three factors: the rock's structure, its permeability and its vulnerability to chemical weathering.

Structure Carboniferous limestone is a hard, grey sedimentary rock which was laid down in layers on the sea-bed. The horizontal junctions between the layers are called **bedding planes**. **Joints** are lines of weakness at right-angles to the bedding planes (Figure 15.14).

Permeability Permeability is the rate at which water can either be stored in a rock or is able to pass through it. Chalk, which consists of many pore spaces, can store water and is an example of a **porous rock**. Carboniferous limestone, which lacks pore spaces, allows water to flow along the bedding planes and down the joints, and is an example of **pervious rock**.

Vulnerability to chemical weathering Rainwater contains carbonic acid which is carbon dioxide in solution. Carbonic acid, although weak, reacts with calcium carbonate. The limestone is slowly dissolved, a process of chemical weathering known as **carbonation**, and is then removed in solution by running water. Chemical weathering, therefore, widens weaknesses in the rock such as bedding planes and joints.

Underground landforms
Carboniferous limestone areas are characterised by a lack of surface drainage. A river that has its source and headwaters on nearby impermeable rock will flow over the surface until it reaches an area of limestone (Figure 15.14). Various acids in the water, including carbonic acid derived from rainfall, begin to dissolve and widen surface joints to form **swallow holes**, or **sinks** (Figure 15.15). The river will, in time, disappear down one of these swallow holes. Once underground, the river will continue to widen joints and bedding planes through solution. Where solution is more active, underground caverns may form. The river will abandon these caverns as it tries to find a lower level. Should the river meet an underlying impermeable rock, it will have to flow over this rock until it reaches the surface as a **spring**, or **resurgence** (Figure 15.16).

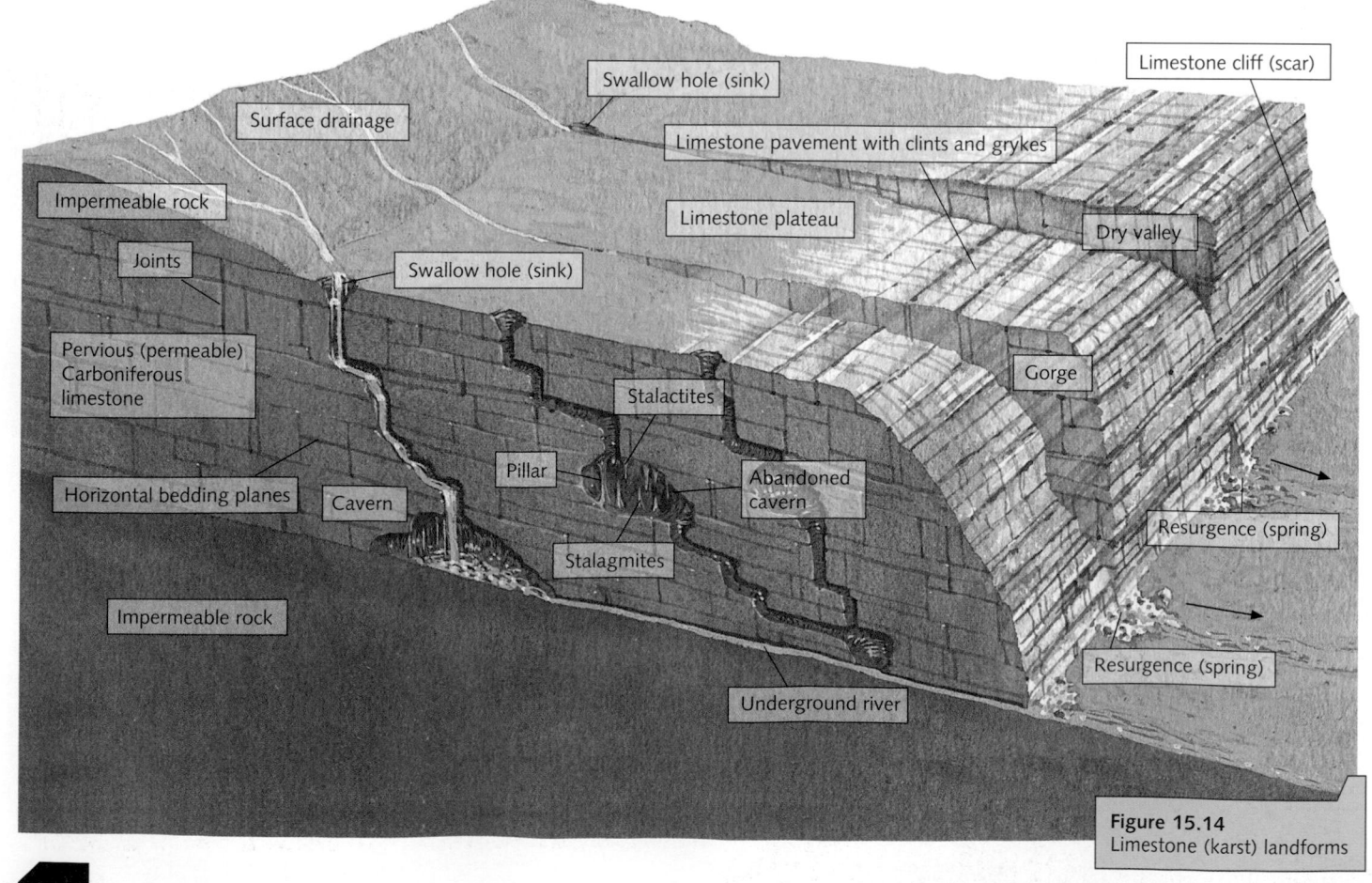

Figure 15.14
Limestone (karst) landforms

Surface landforms

Dry valleys are evidence that rivers once flowed on top of limestone (Figure 15.17). This might have occurred during the Ice Age when the ground was frozen and acted as an impermeable rock. The dry valleys are usually very steep-sided. Limestone areas often have a flat, plateau-like appearance. The flatness is due to the underlying horizontal bedding planes. Where there is no soil, the top bedding plane will be exposed as a limestone pavement (Figure 15.18). Many joints reach the surface along this pavement. They are widened and deepened by solution to form grooves known as **grykes**. The flat-topped blocks between grykes are called **clints**. Other surface landforms, often more developed in areas outside of Britain, result from limestone having collapsed. Where limestone collapses over an underground river it creates a gorge. If it collapses over a small cave then it forms a small depression called a **doline**; if over a series of caves, it produces a much larger depression known as a **polje**.

Deposition landforms

Water, containing calcium carbonate in solution, continually drips from the ceilings of underground caves. Although it is cold in these caves, some evaporation does take place allowing the formation of icicle-shaped **stalactites** (Figure 15.19). In caves in northern England stalactites only grow about 7.5 mm a year. As water drips onto the floor beneath the stalactite, further deposits of calcium carbonate produce the more rounded **stalagmites**. **Pillars** are the result of stalactites and stalagmites joining together.

Figure 15.15
Swallow hole, Malham

Figure 15.16
Resurgence, Malham Cove

Figure 15.19
Stalactites, stalagmites and pillars

Figure 15.17
Watlowes dry valley, Malham

Figure 15.18
Limestone pavement, Malham

Chalk and granite

Chalk

Chalk is a soft limestone that forms its own characteristic landforms in the south-east of England. As the rock is permeable, water is able to pass through it, making it relatively resistant to erosion. Where the rock is horizontal, the landscape is one of gently rounded hills, e.g. Salisbury Plain. Where the rock has been tilted, the porous and more resistant chalk alternates with impermeable and less resistant clays to form the **ridge and vale** scenery of the North and South Downs (Figure 15.5). The ridges, referred to as **escarpments**, consist of a steep scarp slope and a gentle dip slope (Figure 15.20).

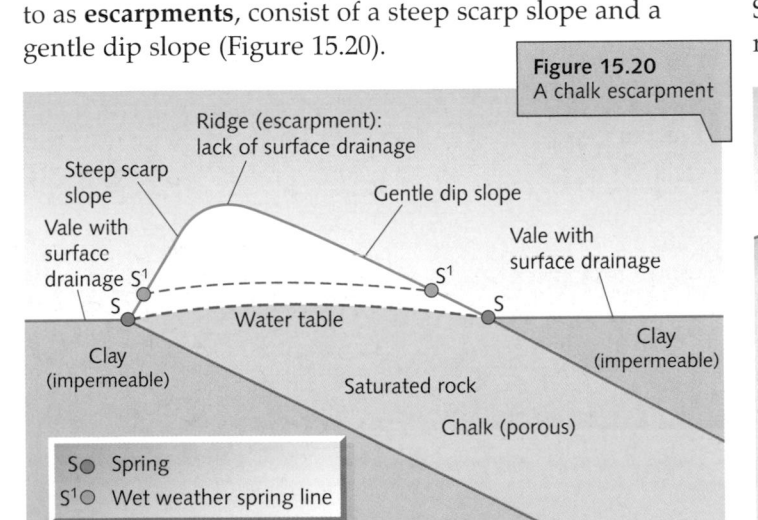

Figure 15.20
A chalk escarpment

Rivers and streams are usually absent on chalk, appearing as **springs** at the junction of the permeable and impermeable rocks. Springs occur where the **water table** reaches the surface (Figure 15.20). The water table, below which the ground is saturated, is likely to rise after wet weather and fall during times of drought. **Dry valleys** are another feature of chalk areas (Figure 15.21). They are believed to have formed during the Ice Age when the ground was frozen and so acted as an impermeable rock. The valley sides, although steep, are more gentle than those found on limestone (Figure 15.17).

Figure 15.21
Devil's Dyke, a dry valley in chalk, South Downs

Granite

Granite forms when magma (molten rock) from inside the Earth rises towards the surface and cools within the crust (Figure 15.22). The slow rate of cooling allows large crystals to form (Figure 15.1) and large cracks (joints) to develop (Figure 15.23). Over a lengthy period of time the less resistant rocks that overlie the granite may be removed by erosion. If this removal is by running water, rounded moorlands are formed, e.g. Dartmoor. If it is by ice, then jagged mountains are created, e.g. the Isle of Skye. Granite, being an impermeable rock, has numerous rivers and much surface drainage.

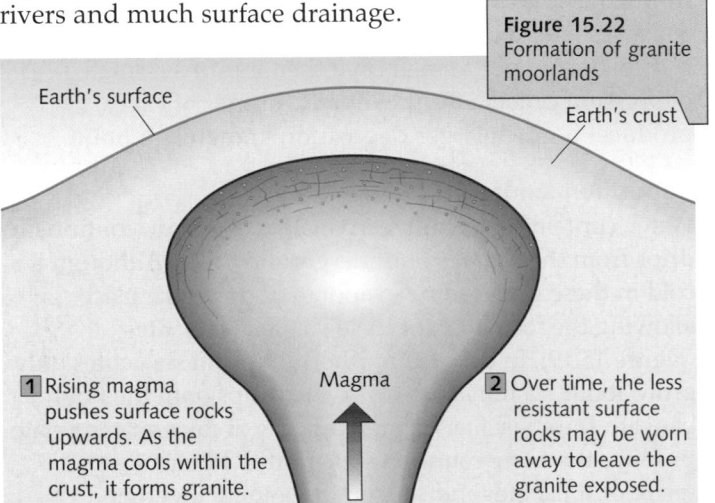

Figure 15.22
Formation of granite moorlands

The most distinctive granite landform is the **tor** (Figure 15.23). Tors are thought to result from the joints within the granite being widened by chemical weathering (page 246) when the granite is near to, but still underneath, the surface. The weathered rock is only later exposed on the surface.

Figure 15.23
Hound Tor, Dartmoor

Limitations and land use of different landscapes

Pages 248 to 250 have described how Carboniferous limestone, chalk and granite each produce their own distinctive landforms and types of landscape. These three landscapes will, in turn, have different types of land use, some of which may have economic value but equally may pose limitations to human activity.

Farming Carboniferous limestone has thin soil, lacks surface water and forms upland areas that are only suitable for sheep. Granite, being impermeable, tends to have a surfeit of surface water and poor soils that also makes it unsuitable for farming. In contrast, chalk, although also having thin soils and lacking surface drainage, is lower in altitude and found further south, enabling cereals such as wheat and barley to be grown successfully (Figure 15.24).

Water supply As Carboniferous limestone is permeable and chalk is porous, neither has surface water but both provide springs where either underground rivers (limestone – Figures 15.14 and 15.16) or the water table (chalk – Figure 15.20) reach the surface. Chalk is an **aquifer**, or water-bearing rock, from which water may be extracted during dry years and where, hopefully, it will be replenished when rainfall is heavier (page 223). Granite areas, such as Dartmoor, tend to be boggy in places and ideal for storing water in reservoirs (Figure 15.25).

Building material Granite provides an excellent building stone (Aberdeen is known as 'The Granite City') while limestone is used locally in parts of England (e.g. Cotswold stone).

Minerals and industry As Carboniferous limestone and chalk both consist of lime, they can be quarried for cement making (Figure 15.27) while weathered granite produces china clay which is used in the making of pottery.

Settlement The bleak upland moorland landscapes created on Carboniferous limestone and granite prove inhospitable for settlement, whereas the assured water supply found at the foot of chalk escarpments has led to a series of spring-line settlements (Figures 3.2 and 15.20).

Tourism The granite area of Dartmoor and the Carboniferous limestone regions of the Peak District and the Yorkshire Dales were among the first of Britain's National Parks, with the chalk South Downs about to become the latest addition (Figure 10.7). All these areas attract large numbers of visitors.

Figure 15.24
Cereal farming on the chalk South Downs

Figure 15.25
A reservoir on Dartmoor

Figure 15.26
Cheddar Gorge, a tourist attraction in Carboniferous limestone

Quarrying and mining in National Parks

National Parks, although created in areas of outstanding natural beauty, are places where people live and have to find work. Quarries are one source of employment. There are still some 30 working quarries in several National Parks in England and Wales including Snowdonia and the Lake District for slate, and the Peak District and Yorkshire Dales for Carboniferous limestone (Figures 15.27 and 10.10).

Working quarries pose a major conflict in National Parks because, although they can bring economic and some social benefits to people living there (Figure 15.28a), they, and more especially disused quarries and mines (e.g. copper and lead), can cause considerable environmental damage and some social loss (Figure 15.28b).

Figure 15.27
A working quarry in the Peak District National Park

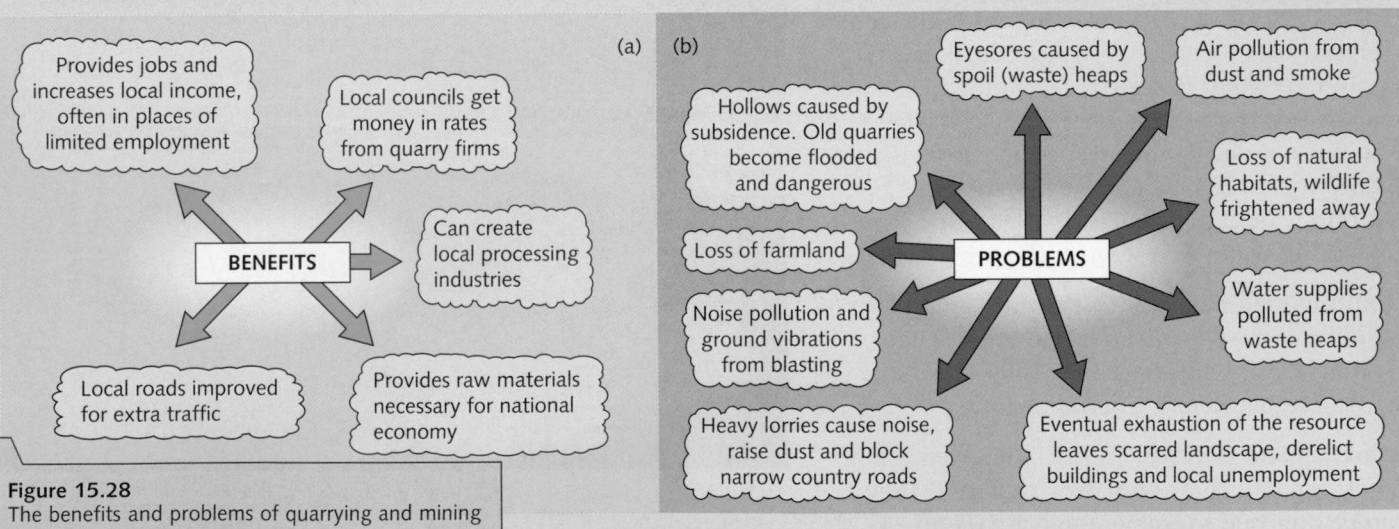

(a)

- Provides jobs and increases local income, often in places of limited employment
- Local councils get money in rates from quarry firms
- Can create local processing industries
- **BENEFITS**
- Local roads improved for extra traffic
- Provides raw materials necessary for national economy

(b)

- Eyesores caused by spoil (waste) heaps
- Air pollution from dust and smoke
- Hollows caused by subsidence. Old quarries become flooded and dangerous
- Loss of natural habitats, wildlife frightened away
- Loss of farmland
- **PROBLEMS**
- Water supplies polluted from waste heaps
- Noise pollution and ground vibrations from blasting
- Heavy lorries cause noise, raise dust and block narrow country roads
- Eventual exhaustion of the resource leaves scarred landscape, derelict buildings and local unemployment

Figure 15.28
The benefits and problems of quarrying and mining

National Park Authorities are responsible for all planning applications, whether it is for a new quarry development or the extension of an existing project. The authorities have to ensure that working quarries and mines are landscaped and screened and that disused quarries and mines are restored, wherever possible, to their pre-quarry appearance.

Figure 15.29 shows a quarry that, when it closed, left a hole 30 metres deep which flooded and became dangerous, along with piles of extracted waste and many disused buildings. Since then a wildlife habitat has been created, the lake has been cleaned and made safer, grass has been sown to create a picnic area and to stabilise steep slopes, old buildings have been turned into a toilet block and an information centre, and the whole site has been screened by trees.

Figure 15.29
A restored quarry in the Northumberland National Park

Soils

Soil forms a thin upper layer over most parts of the Earth's land surface. It provides the foundation for plant, and consequently animal, life on land.

Formation

There are two main stages in soil formation:

1 Parent, or underlying, rock is weathered to give a layer of loose, broken material known as **regolith**. Regolith may also result from the deposition of material by water, wind, ice and volcanic activity.

2 The addition of water, gases (air), living organisms (biota) and decayed organic matter (humus).

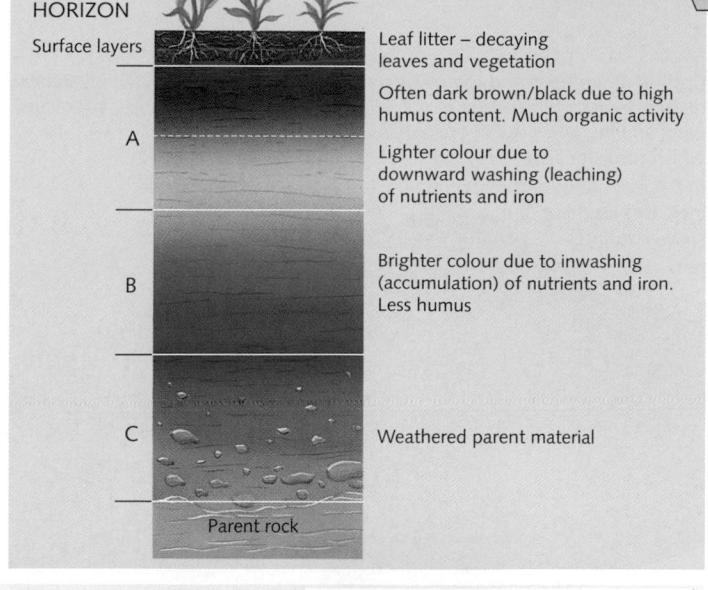

Figure 15.31
Soil profile model

HORIZON
Surface layers — Leaf litter – decaying leaves and vegetation

A — Often dark brown/black due to high humus content. Much organic activity

Lighter colour due to downward washing (leaching) of nutrients and iron

B — Brighter colour due to inwashing (accumulation) of nutrients and iron. Less humus

C — Weathered parent material

Parent rock

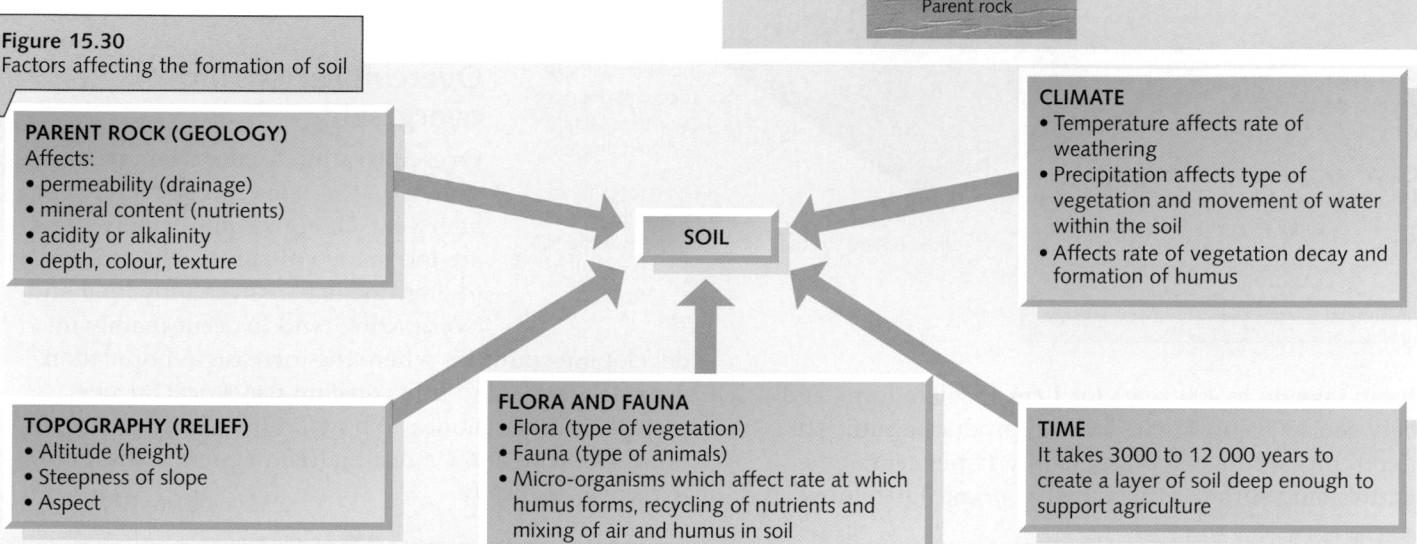

Figure 15.30
Factors affecting the formation of soil

PARENT ROCK (GEOLOGY)
Affects:
• permeability (drainage)
• mineral content (nutrients)
• acidity or alkalinity
• depth, colour, texture

CLIMATE
• Temperature affects rate of weathering
• Precipitation affects type of vegetation and movement of water within the soil
• Affects rate of vegetation decay and formation of humus

SOIL

TOPOGRAPHY (RELIEF)
• Altitude (height)
• Steepness of slope
• Aspect

FLORA AND FAUNA
• Flora (type of vegetation)
• Fauna (type of animals)
• Micro-organisms which affect rate at which humus forms, recycling of nutrients and mixing of air and humus in soil

TIME
It takes 3000 to 12 000 years to create a layer of soil deep enough to support agriculture

Soil formation is also dependent on five main factors, all of which are closely interconnected and interdependent. These factors are summarised in Figure 15.30. Increasingly, however, a sixth factor, that of human interference, is rapidly altering many of the world's natural soils, with the result that in many parts of the world soil, which is normally a sustainable resource (page 118), is being eroded and lost.

The soil profile

The soil profile is a vertical section that shows different layers, known as **horizons** (Figure 15.31). Apart from the surface layer, which consists mainly of fallen leaves and partly decayed vegetation, there are three main horizons, referred to as the A, B and C horizons. However, not all soils have the three horizons, nor the clear-cut boundaries between them that are shown on the model profile in Figure 15.31.

World soils

On a global scale, soils often develop a profile that reflects the climate and vegetation of the area in which they form. Figure 15.32 shows differences between three of these soils, two of which are found in the UK, the third in the tropical rainforest.

	Brown earths	Podsols	Tropical red earths
Location	Southern UK	Northern UK	Equatorial areas
Vegetation	Deciduous forest	Coniferous forest	Tropical rainforest
Weathering of parent rock	Quite fast	Extremely slow	Extremely fast
Soil depth	1–2 metres	under 1 metre	up to 30 metres
Leaf litter	Thick	Thin	Very thick
Movement of water/ leaching	Limited leaching	Some leaching	Rapid leaching
Recycling of nutrients	Rapid	Very slow	Very rapid
Earthworms/organisms	Many	Very few	Many
Fertility	Very fertile	Very infertile	Very fertile unless cycle is broken

Figure 15.32
Three soil types

Soil erosion

Cutting down trees and removing hedges leaves soil exposed to wind and increases water erosion, which forms gullies

In tropical areas, deforestation increases leaching and surface runoff

Mining ruins large areas

Wind blows away soil on exposed areas

Overcultivation impoverishes the soil

1 million hectares of arable land lost every year in USA to highways, urbanisation and industry (land is lost at an even greater rate near fast-growing cities in Latin America)

Ploughing up and down hillsides increases surface runoff

Overgrazing exposes land to erosion by wind and water

Heavy machinery compacts the ground

Agribusiness has little regard for the soil

Overcropping and monoculture impoverish the soil

Irrigation without adequate drainage can cause salinity and waterlogging

Figure 15.33
Some causes and effects of soil erosion

It can take up to 400 years for 1 cm of soil to form, and between 3000 and 12 000 years to produce a sufficient depth for farming. At present, only 11 per cent of the Earth's land surface is classified as prime agricultural land.

This land is needed to feed an ever growing world population. So far the increase in population has been matched by increases in food production but the capacity of the soil to produce enough food is being stretched to the limit (page 111). The problem is aggravated where human development is actually ruining this essential resource, through erosion or degradation (Figure 15.33).

Erosion is most rapid in areas where the land is mismanaged; where the protective vegetation cover is removed; where there is rapid population growth; where the land is steep; and where climatic conditions are extreme, especially if rainfall is seasonal, comes as downpours or is unreliable. The UN has estimated that an area the size of China and India combined has been lost in the last half-century, and that one-third of the remaining soil could be destroyed by the year 2030.

Removal of vegetation

Most damage to the soil results from the removal of vegetation, with deforestation being the major cause. Where vegetation is removed there will be no replacement of humus, no interception of rain by plants (page 279) and no roots to bind the soil together, so the surface will be left exposed to rain and wind. If the rainfall occurs as heavy thunderstorms and in areas of steep slopes, the soil will be washed downhill and removed by rivers to leave deep, unusable gullies (Figure 15.34). If heavy rain falls on more gentle slopes or if the climate is dry and windy, loose material will be removed to leave areas of bare rock. The removal of hedgerows (page 108), the ploughing of grassland and the collection of fuelwood (page 121) all increase the risk of erosion by water and wind.

Overcultivation and overgrazing

Overcultivation occurs when crops are grown on the same piece of land year after year. **Overgrazing** is where there are too many animals for the amount of grass available. Overcultivation and overgrazing tend to occur mainly in developing countries where the increase in population means that the land is in constant use. Local farmers neither have the money to buy fertiliser for their the land, nor the time to allow a resting (fallow) period for the soil to recover naturally.

Figure 15.34
Gully erosion

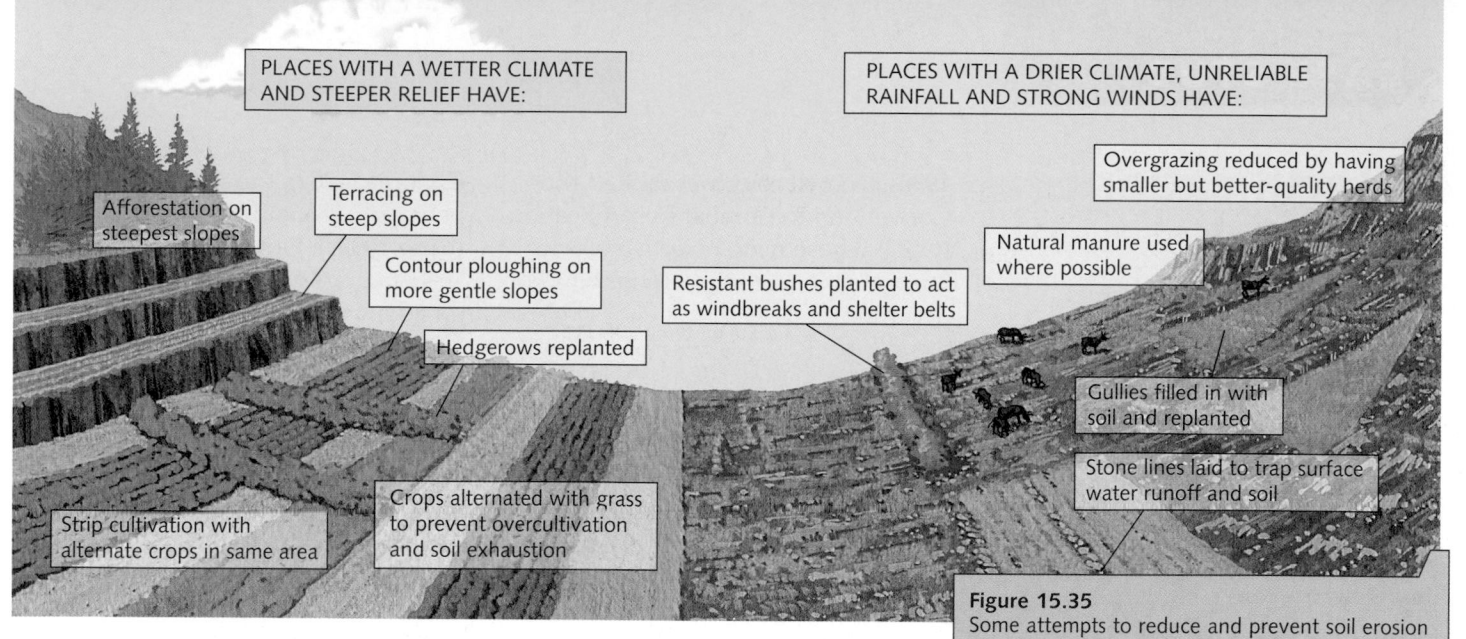

PLACES WITH A WETTER CLIMATE AND STEEPER RELIEF HAVE:

Afforestation on steepest slopes

Terracing on steep slopes

Contour ploughing on more gentle slopes

Hedgerows replanted

Strip cultivation with alternate crops in same area

Crops alternated with grass to prevent overcultivation and soil exhaustion

PLACES WITH A DRIER CLIMATE, UNRELIABLE RAINFALL AND STRONG WINDS HAVE:

Overgrazing reduced by having smaller but better-quality herds

Natural manure used where possible

Resistant bushes planted to act as windbreaks and shelter belts

Gullies filled in with soil and replanted

Stone lines laid to trap surface water runoff and soil

Figure 15.35
Some attempts to reduce and prevent soil erosion

Soil management

Soil is a sustainable resource, but only if it is carefully managed. Without careful management, soil, which may have taken centuries to form, can be lost within days and, under extreme conditions, sometimes even in minutes. Some suggested methods of preventing, or at least limiting, soil erosion while at the same time sustaining its productivity are shown on Figure 15.35.

Various techniques – such as terracing; the replanting of trees, grass and hedges; contour ploughing and strip cultivation – can reduce soil erosion by over 50 per cent. Also, as well as reducing erosion, the addition of mulch and manure increases soil fertility.

The best protection against soil erosion is to prevent it from being exposed to wind and rain. Trees, bushes and grass can act as windbreaks, improve water retention and bind the soil together. **Terracing** is when artificial 'steps' are cut into steep hillsides and the front of each flat terrace is edged with mud or stone walls known as **bunds** (Figure 15.36). The bunds trap water and soil. **Contour farming** is when crops are planted around the hillside rather than up and down the slope (Figure 15.37). **Strip farming** is when two or more crops are grown in the same field. Sometimes one crop may grow under the shelter of a taller crop – a technique that has long been used in Mediterranean areas. In India, vetier grass, with its deep roots to bind the soil, is being planted in contour strips across hillsides. In drier parts of Africa, stone lines (Figure 15.45) are used to trap surface water runoff and soil. Mulching, ideally using the waste from harvested crops, keeps the ground moist by protecting it from evaporation.

Figure 15.36
Rice terraces

Figure 15.37
Contour farming

Desertification in the Sahel

There are more than a hundred definitions of the term **desertification**. Taken literally it means 'turning the land into desert'. It is a process of land degradation, mainly in arid and semi-arid lands where the rainfall is unreliable, caused by human mismanagement of a fragile environment. The causes of desertification are complex, but it is now widely accepted that it results from a combination of physical processes and human activity.

In 2005, the UN claimed that there were 250 million people affected by desertification and a further 1 billion at risk. They live in countries that are amongst the world's poorest and have to rely upon the sustainable use of the land for their food, income and employment. While it is the Sahel region of Africa that has received most attention to date, the problem of desertification is far more widespread, as shown in Figure 15.38. In the late

1990s, over 60 countries ratified the Convention to Combat Desertification (CCD) whose main objective was to 'mitigate the effects of drought and/or desertification'.

The Sahel

The effects of desertification are greatest in the Sahel. The Sahel is a narrow belt of semi-arid land which lies immediately to the south of the Sahara Desert and which extends across most of Africa. (The map in Figure 15.38 was drawn at the Nairobi Conference on Desertification.) It has a climate that is a transition between the tropical continental (page 213) and the hot desert (page 215). Rainfall is confined to just one or two months of the year, but both the total amount of rain and the length of the 'rainy season' are unreliable. In some years the total rainfall may come in several downpours when the water is immediately lost through surface runoff. In other years the rains may fail altogether. Since the late 1960s, there have been several lengthy droughts including those in

Ethiopia (1983), the Sudan (four between 1984 and 2005) and Somalia (early 1990s). Each drought means less grazing land and fewer crops in an area where several countries have some of the highest birth and population growth rates in the world. In other words, there are fewer available resources to be shared by an increasing number of people (Figure 15.39).

Figure 15.39
The effects of drought on people

Figure 15.38
Areas at risk from desertification

Severe	
Moderate	
Slight	

SAHEL

Worst areas	% total population at risk
① Ethiopia	18
② Sudan	23
③ Chad	30
④ Niger	42
⑤ Somalia	26
⑥ Mali	32
⑦ Burkina Faso	21

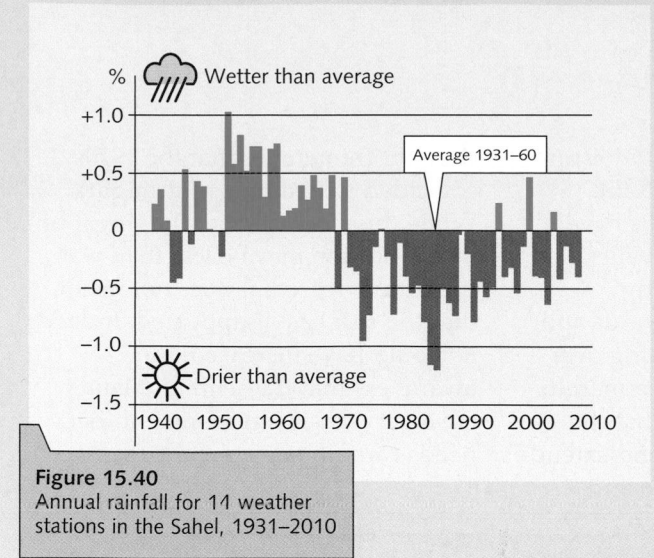

Figure 15.40
Annual rainfall for 11 weather stations in the Sahel, 1931–2010

Figure 15.41
Overgrazed land in Kenya

Causes of desertification

It is suggested that desertification results from a combination of:

- climatic change – a decrease in rainfall and, possibly, the effects of global warming
- population growth – an increase in numbers of animals and people.

Climatic change

Even at the best of times, water supply is always a problem in these semi-arid lands. More than a century of rainfall data for the Sahel shows an exceptionally wet period during the 1950s and 1960s followed by extremely dry years through the 1970s and 1980s (Figure 15.40). The resultant drought caused seasonal rivers and water holes to dry up, the water table to fall and non-drought-resistant vegetation, notably grass, to die. More recently, global warming has been suggested as a contributory cause of desertification. This is because higher temperatures mean increased evapotranspiration (the warmer the air the more vapour it can hold), reduced condensation and, therefore, less rainfall. Since 1990 until the present, rainfall has returned to levels slightly below the 1900–93 average, but year-to-year variability has been much higher.

Overgrazing

The basic economy of many Sahel communities is animal grazing – cattle, camels, goats and sheep. In some areas, livestock numbers increased by up to 40 per cent during the wetter years of the 1950s. When the drought began, it meant that there were too many animals for the amount of grass available and any protective grass covering was soon stripped away (Figure 15.41). Animals had to be taken to wells and water holes (the only source of available water) where the herds were too large for the amount of grass available. The result was the loss of vegetation at an increasing distance from the source of water, and the turning of the land into a desert (Figure 15.42).

Figure 15.42
The results of overgrazing

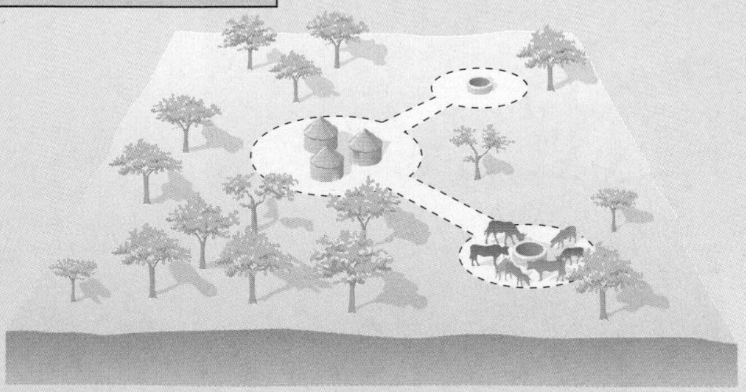

BEFORE
Traditional farming practices: bare ground created as a result of cutting wood for fuel, and grazing around wells.

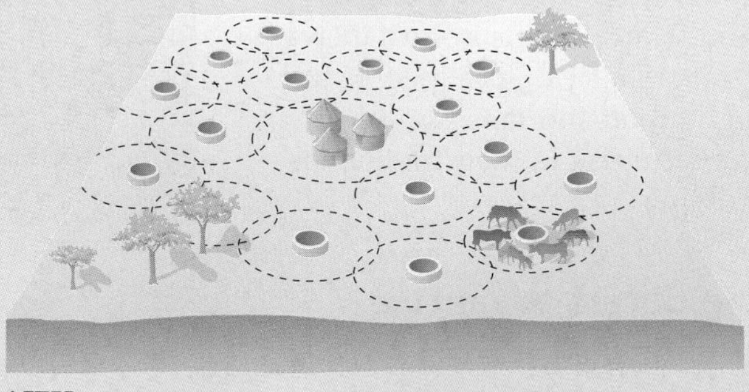

AFTER
As the population grew, more wells were bored, more trees cleared and the number of animals increased. There has been no change in the traditional herding practices, and more land has been left overgrazed and bare.

Population growth

The population of most Sahel countries, which has risen rapidly since 1950 (Figure 15.43), is increasing at an annual rate of between 3.0 and 3.3 per cent (2008). This growth has mainly been as a result of high birth rates within each country and, in several cases, immigration resulting from civil wars in neighbouring states. As the population has grown, so too has the pressure on the land. Farmers have been forced to grow crops on marginal land which is more vulnerable to erosion, and to grow the same crop year after year on the same piece of land which had previously been allowed 'rest', or fallow, periods. This has led to overcultivation, the exhaustion of the soil and, finally, to soil erosion. The increase in population has also led to a greater need for fuelwood for cooking. Villagers, who have to walk further each year in their search for wood, cause deforestation and accelerate desertification (Figure 8.8).

The results of desertification

Whether it is climatic change, overgrazing or population growth, in all cases the soil loses its protective vegetation cover, contains less humus and holds less moisture. As it becomes exposed to the wind and the occasional heavy downpour, it becomes increasingly at risk from erosion. These events in turn increase the vulnerability of local people to drought, food shortages and the longer-term risk to their traditional way of life.

Is the desert advancing?

In the 1970s there was concern that the Sahara Desert was advancing southwards by between 5 and 10 km a year. Attempts were made to try to halt its progress, such as planting trees and shrubs to create a 'green belt' along the southern fringe (Figure 15.44) or constructing stone lines to try to capture what moisture became available (Figure 15.45). Recent satellite photographs have shown, contrary to earlier belief, that the Sahara has not made any significant advance southwards and that sand dunes are retreating over the entire 6000 km southern margin. Instead it is now thought that the southern fringes contract and extend according to periods of rain and drought (remember that the 1970s coincided with a time of drought). While this suggests that the threat of desertification may be less than was previously feared, it does not mean that the risk has disappeared. Indeed, it should have increased our awareness that the semi-arid lands are a very fragile environment that needs careful management.

Map number (Figure 15.38)		Population (millions)			Growth for decade 1990–2000 (%)	UN predicted growth 2008–50 (%)
		1950	1995	2008		
1	Ethiopia	19.6	51.6	79.1	33	187
2	Sudan	9.2	30.0	39.4	32	85
3	Chad	2.7	6.3	10.1	40	102
4	Niger	2.4	9.1	14.7	32	261
5	Somalia	2.4	9.2	9.4	8	166
6	Mali	3.5	10.8	12.7	26	169
7	Burkina Faso	3.6	10.6	15.2	34	147
	UK	50.6	58.3	61.3	3	26

Figure 15.43
Comparative population change in the Sahel area and the UK

Figure 15.44
A green belt along the southern edge of the Sahara Desert

Figure 15.45
Stone lines in Burkina Faso

Rocks and soils

Key Words and Terms

a You should know the meaning of the following terms:
- rock structure • rock resistance • porous/porosity
- freeze–thaw/frost shattering • exfoliation • limestone solution/carbonation
- soil creep • mudflows/lahars • landslides • escarpment • soil profile
- soil horizons • parent rock • desertification.

b You should know the difference between:
- igneous, sedimentary and metamorphic rocks
- permeable and impermeable
- bedding planes and joints
- weathering, erosion and mass movement
- physical, chemical and biological weathering
- overgrazing and overcultivation.

Key Ideas

You should know and understand the following:
- The formation, characteristics, examples and uses of igneous, sedimentary and metamorphic rocks.
- The effect of structure/geology on landforms: resistance, permeability.
- The differences between physical, chemical and biological weathering: the processes of freeze–thaw/frost shattering, exfoliation and limestone solution/carbonation.
- The causes and effects of mass movement – soil creep, mudflows.
- The characteristic landforms, and their formation, associated with Carboniferous limestone/karst areas: swallow holes/sinks, springs/resurgences, dry valleys, limestone gorges, limestone pavements, clints and grykes, caverns, stalactites, stalagmites and pillars.
- The characteristic landforms, and their formation, associated with chalk areas: escarpments, scarp and dip slopes, springs, dry valleys and clay vales.
- The characteristic landforms, and their formation, associated with granite areas: tors.
- Different landscapes have different land uses, economic value and limitations for human activity.
- The advantages and problems of mining and quarrying, especially in areas of attractive scenery/National Parks.
- The formation of soils.
- Soils are a resource that can easily be eroded if they are not carefully managed.
- The causes and effects of desertification.

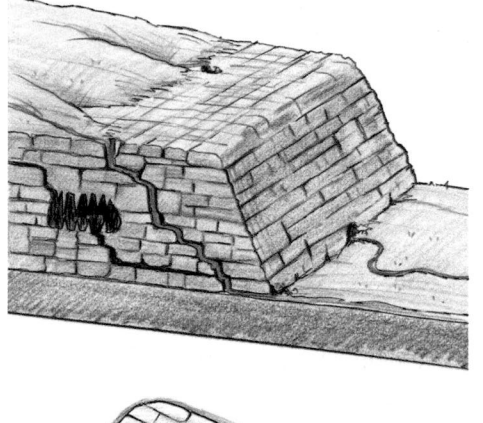

Skills, Theories and Models

- Interpret and use soil profiles
- Understand the classification of rocks
- Understand the classification of soils
- Recognise the distinctive character of places
- Investigate the impact of resource development
- Appreciate the need to plan and manage resource development

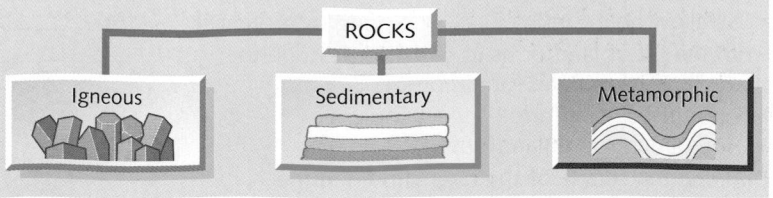

QUESTIONS

1 *(Pages 244 and 245)*

a Using the table, name:
  i) the oldest geological period. *(1)*
  ii) two igneous, three sedimentary and one
      metamorphic rock. *(6)*
b For one igneous, one sedimentary and one igneous rock:
  i) Describe how it was formed. *(3)*
  ii) Give two of its uses. *(3)*
c Of the rocks named in Figure 15.7, which one
   would you least expect to find in Britain? *(1)*

Geological period	UK rocks
Quaternary	(Ice Age)
Tertiary	Granite Basalt
Cretaceous	Chalk
Jurassic	Oolitic limestone
Triassic	New Red Sandstone
Permian	
Carboniferous	Coal Measures Limestone
Devonian	Old Red Sandstone
Silurian	Slate
Ordovician	Granite
Cambrian	

Younger
Older

2 *(Pages 246 and 247)*

a For each of the following types of weathering:
  • freeze–thaw (frost shattering)   • exfoliation
  • solution/carbonation   • biological.
  i) Describe the weathering process. *(4 × 2)*
  ii) Use the following list to link it to the conditions
     under which it is most likely to occur:
     • a hot, rocky desert
     • a limestone area
     • a cold, mountainous area
     • a woodland on a rocky slope. *(4)*
b Describe how:
  • mass movement occurs on slopes
  • soil creep can produce terracettes
  • mudflows and landslides can destroy property
    and cause loss of life. *(3)*

3 *(Pages 248 and 249)*

a Make a larger copy of the diagram showing landforms in an
   area of Carboniferous limestone. Complete it by adding the
   following labels to the correct boxes:
   • Bedding planes   • Cavern
   • Dry valley   • Clints and grykes
   • Limestone pavement   • Impermeable rock
   • Stalactites and stalagmites • Joints
   • Swallow hole (sink)   • Spring (resurgence). *(10)*
b From the list of landforms in part (**a**), describe the
   characteristics and the formation of:
   • one surface landform
   • one underground landform. *(6)*
c Describe the course of the river shown in the
   diagram as it crosses the area. *(4)*

4 *(Pages 250 and 251)*

a Choose either a chalk escarpment, a dry valley or a granite
   tor. For the feature that you have chosen, describe:
  i) Where it may be found in England. *(1)*
  ii) How, with the aid of labelled diagrams, it might have
      formed. *(3)*
b What are the economic uses and limitations of landscapes
   formed by each of Carboniferous limestone, chalk and
   granite for:
   • farming
   • water supply
   • tourism
   • settlement. *(2 × 4)*

5 *(Pages 252 and 253)*

a Describe how quarrying in a National Park can:
- benefit the local community
- harm the environment. *(4)*

b How can a National Park Authority:
- try to sustainably manage the quarry when it is still working
- restore the quarry after it closes? *(4)*

c Describe how the following affect the formation of soil:
- parent rock
- climate
- topography
- flora and fauna. *(8)*

d On a copy of the soil profile, label the following features:
- humus
- parent rock
- leaf litter
- direction of leaching
- A, B and C horizons
- weathered parent material. *(8)*

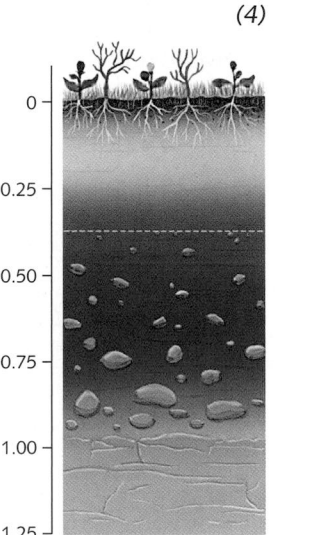

6 *(Pages 254 and 255)*

a Using Figure 15.33, describe five causes of soil erosion. *(5)*

b i) Why is soil management necessary? *(2)*
ii) Describe how each of the eight methods shown on this diagram may help to reduce soil erosion. *(8)*

Replanted trees — Terracing — Contour ploughing — Natural manure — Wind breaks — Varied crop types — Strip cultivation — Low-density grazing

7 *(Pages 256 to 258)*

a The map shows the Sahel, an area in Africa affected by severe desertification.
i) Which desert lies to the north of the Sahel? *(1)*
ii) Name four countries in the Sahel affected by desertification. *(4)*
iii) Name two other countries that would be affected if severe desertification were to spread 100 km further south. *(2)*

Mauritania · Algeria · Libya · Red Sea · Gambia · Mali · S A H A R A   D E S E R T · Eritrea · Senegal · Niger · Guinea Bissau · Burkina Faso · Chad · Sudan · Guinea · Benin · Sierra Leone · Togo · Nigeria · Ethiopia · Somali Rep. · Côte d'Ivoire · Ghana · Central African Rep. · Liberia · Cameroon · Atlantic Ocean · –10°

Area of severe desertification

N · 0 · 500 km

b Copy the diagram below, which shows the processes believed to be the cause of desertification. Complete it by putting the following words and phrases into the correct boxes:
- Deforestation
- Increased risk of soil erosion
- Less rainfall, increased drought
- Overcultivation
- Overgrazing
- Soil exposed to wind and rain
- Vegetation cannot re-establish itself
- Vegetation dies
- Vegetation cover removed. *(9)*

Climatic change → Population growth → More animals reared → More land needed for crops → More energy needed for cooking, etc. → Desertification

c Explain how each of the following can lead to desertification:
- climatic change
- overgrazing
- overcultivation
- deforestation. *(4 × 2)*

261

Tectonic activity

Earthquakes and volcanic eruptions are caused by movements within the Earth. While there are many thousand gentle earth movements each year, occasionally one is sufficiently violent to cause severe damage to property, to disrupt human activity, and to result in loss of life. Earth movements cannot, as yet, be predicted. Scientists do, however, know which parts of the world are most likely to be affected by these movements, even if they cannot say when, or with what severity, they will occur.

Earthquakes

Figure 16.1 identifies the location of places where earthquake activity is most frequent. It also locates some of the more recent major earthquakes. The map clearly shows that there is a well-defined distribution pattern, with most earthquakes occurring in long, narrow belts. These belts include those that:
• encircle the whole of the Pacific Ocean
• extend down the entire length of the mid-Atlantic Ocean
• stretch across southern Europe and Asia, linking the Atlantic and Pacific Oceans.

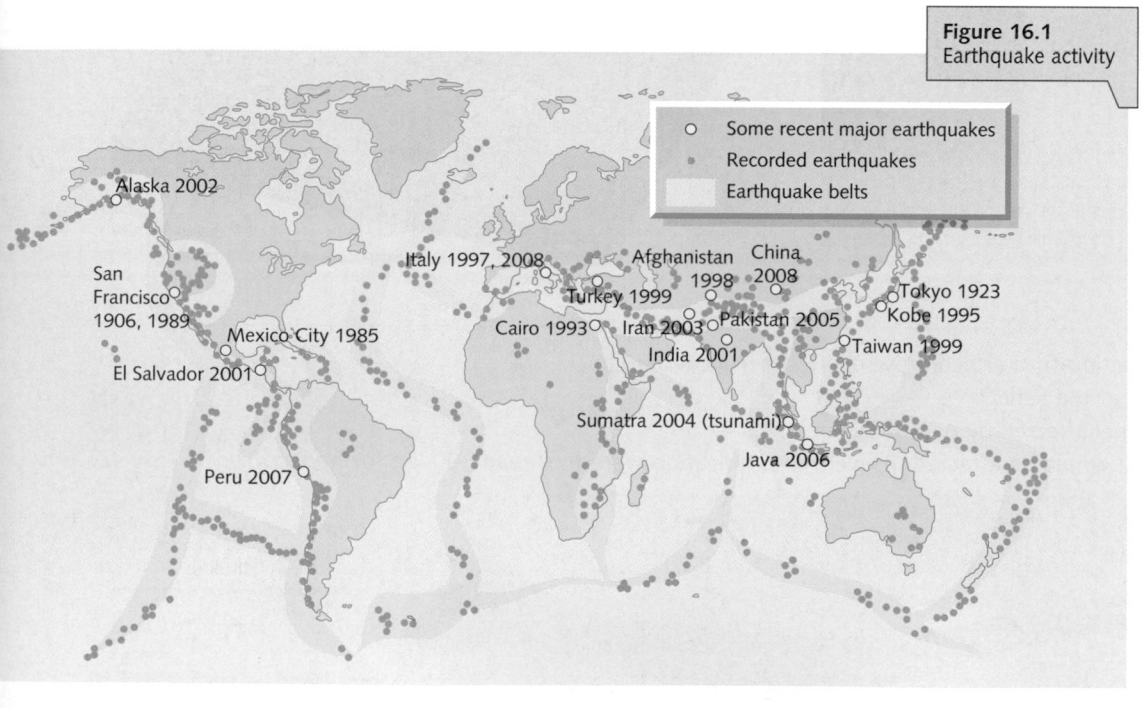

Figure 16.1
Earthquake activity

Volcanoes

The locations of the world's major centres of volcanic activity, including some of the most recent major volcanic eruptions, are shown on Figure 16.2. This map shows that volcanoes also occur in long, narrow belts. These belts include:
• the so-called 'Pacific Ring of Fire' which encircles the whole of the Pacific Ocean
• the one that extends down the entire length of the mid-Atlantic Ocean
• smaller areas in southern Europe, the Caribbean, east Africa and the mid-Pacific Ocean.

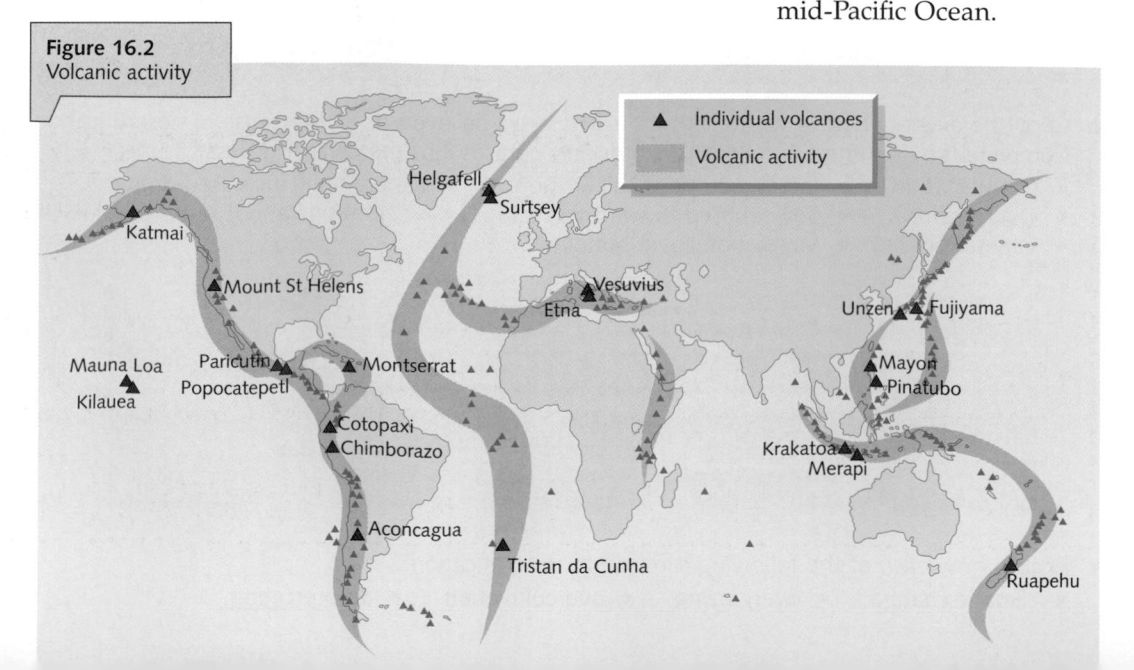

Figure 16.2
Volcanic activity

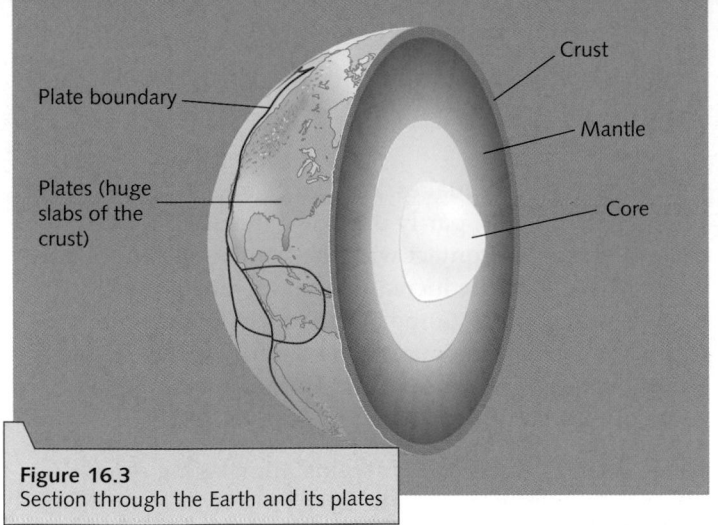

Figure 16.3
Section through the Earth and its plates

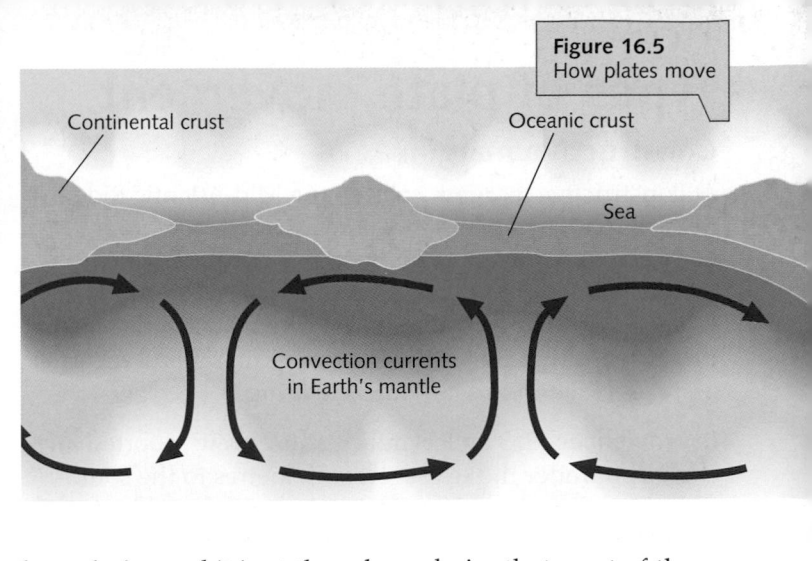

Figure 16.5
How plates move

Figures 16.1 and 16.2 show that earthquakes and volcanic activity often occur at similar places on the Earth's crust, and in narrow **zones of activity**. If the Earth were the size of an apple, its **crust** would be no thicker than the apple's skin. Underneath the crust is the **mantle** (Figure 16.3) where temperatures are so high that rock exists in a semi-molten state. The crust is broken into several large, and other smaller, segments known as **plates** which float, like rafts, on the mantle (Figure 16.4). Heat from within the Earth creates convection currents which cause the plates to move, perhaps by a few centimetres a year (Figure 16.5). Plates may either move away from, towards, or sideways past, neighbouring plates. Plates meet at **plate boundaries** and it is at these boundaries that most of the world's earthquakes and volcanic eruptions occur, and where high mountain ranges are located (compare Figure 16.4 with Figures 16.1 and 16.2). Very little activity takes place in the rigid centre of plates.

Plates consist of two types of crust: continental and oceanic. Continental crust is older, lighter, cannot sink and is permanent. Oceanic crust is younger, heavier, can sink and is constantly being destroyed and replaced. It is these differences in crust that account for the variation in processes and landforms and the level of activity at plate boundaries (Figure 16.6).

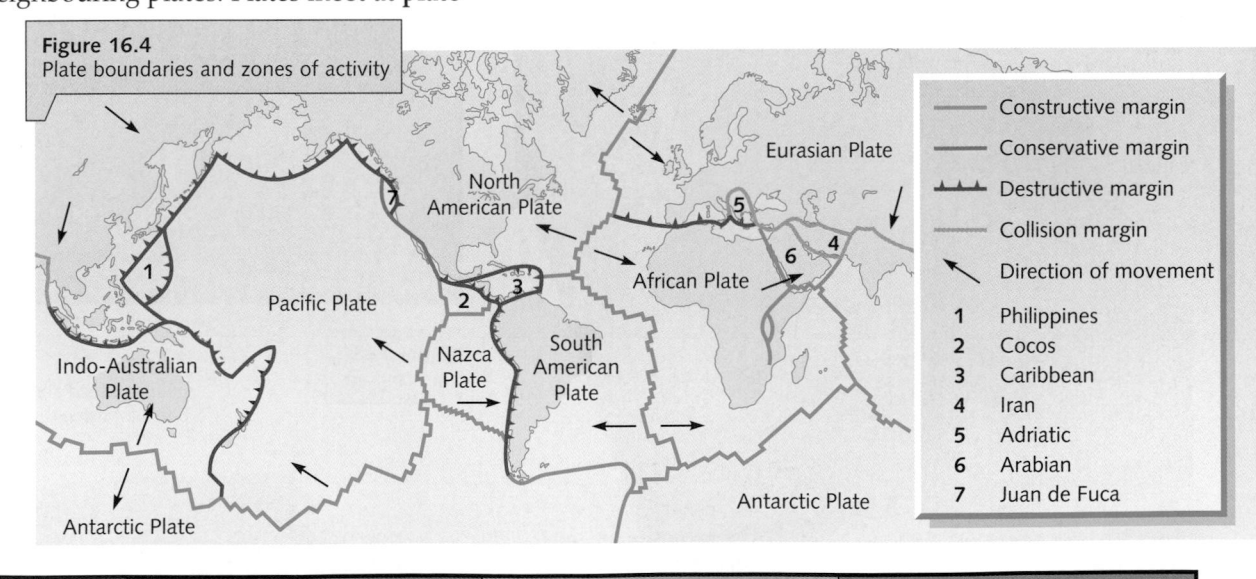

Figure 16.4
Plate boundaries and zones of activity

——	Constructive margin	
——	Conservative margin	
▲▲▲	Destructive margin	
——	Collision margin	
↖	Direction of movement	
1	Philippines	
2	Cocos	
3	Caribbean	
4	Iran	
5	Adriatic	
6	Arabian	
7	Juan de Fuca	

Figure 16.6
Activity at plate boundaries

Type of plate boundary		Description of changes	Earthquake/volcanic activity	Examples
A	Constructive margins	Two plates move away from each other. New oceanic crust appears, forming mid-ocean ridges with volcanoes	Gentle volcanic and earthquake activity	Mid-Atlantic Ridge, e.g. Iceland
B	i) Destructive margins	Oceanic crust moves towards continental crust but being heavier sinks and is destroyed, forming deep-sea trenches and island arcs with volcanoes	Violent volcanic and earthquake activity	Nazca and South American Plates, Cocos and North American Plates, Juan de Fuca and North American Plates
	ii) Collision zones	Two continental crusts collide and as neither can sink, are forced up into fold mountains	Earthquake activity (no volcanic activity)	Indo-Australian and Eurasian Plates, e.g. Himalayas
C	Conservative margins	Two plates move sideways past each other – land is neither formed nor destroyed	Can be violent earthquake activity (no volcanic activity)	Pacific and North American Plates, e.g. San Andreas, California

Types of plate movement

Constructive margins

At constructive margins, such as the Mid-Atlantic Ridge (Figure 16.7), two plates move away from each other. Molten rock, or magma, immediately rises to fill any possible gap and forms new oceanic crust. The Atlantic Ocean is widening by about 3 cm a year which means that the Americas are moving away from Eurasia and Africa – a process known as **sea-floor spreading**.

In November 1963 an Icelandic fishing crew reported an explosion under the sea several kilometres to the south-west of the Westman Islands. Further submarine explosions, accompanied by ejections of steam and ash, gave birth to the island of Surtsey. The permanence of the new island was guaranteed when, in April 1964, lava began to flow from the central vent. When the eruptions ended, in 1967, the island measured 2.8 km² and reached 178 metres above sea-level. Within a month plants and insects had begun colonisation.

Later eruptions elsewhere in Iceland caused the temporary evacuation of Heimaey in 1973 and the melting of part of the Vatnajökull ice cap in 1996.

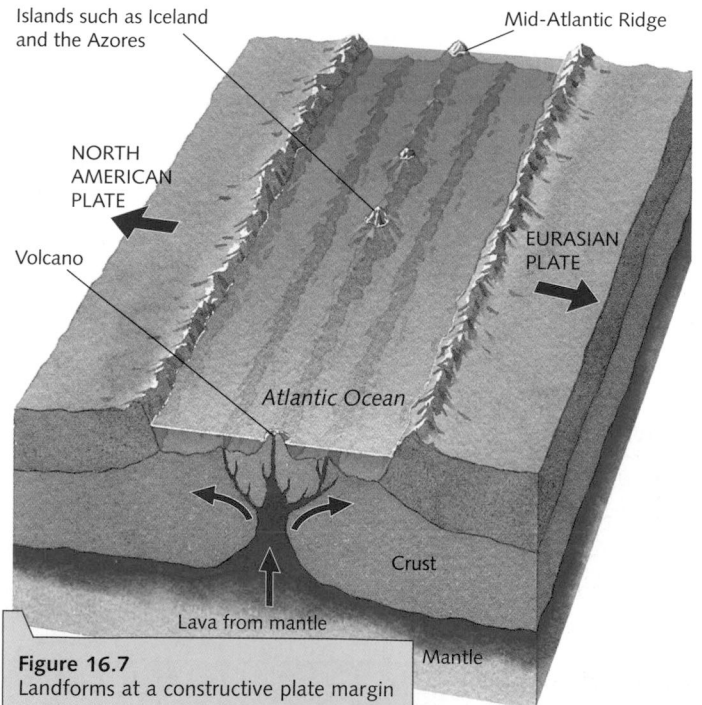

Figure 16.7
Landforms at a constructive plate margin

Destructive margins

Destructive margins occur where plates consisting of oceanic crust move towards plates of continental crust. In the south Pacific Ocean (Figure 16.8), the Nazca Plate (oceanic crust) is moving towards the South American Plate (continental crust). Where they meet, the Nazca Plate is forced downwards to form a **subduction zone** and an associated **deep-sea trench** (the Peru–Chile trench). The increase in pressure, as the plate is forced downwards, can

trigger severe earthquakes. As the oceanic crust continues to descend, it melts, partly due to heat resulting from friction caused by contact with the South American Plate and partly due to the increase in temperature as it re-enters the mantle. Some of the newly formed magma, being lighter than the mantle, rises to the surface to form volcanoes (e.g. Chimborazo and Cotopaxi) and a long chain of fold mountains (the Andes).

Sometimes, at destructive margins, the magma rises offshore to form **island arcs** such as the West Indies, Japan (page 270) and, off the south coast of Alaska, the Aleutian Islands (Figure 16.9).

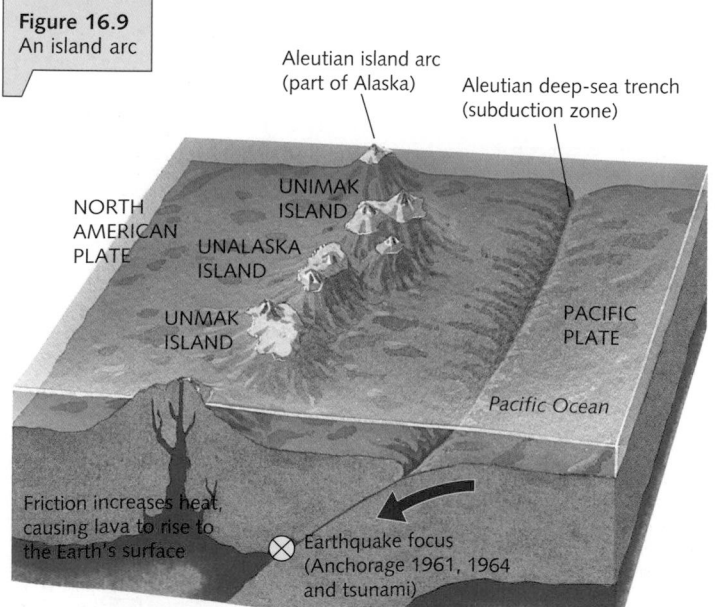

Figure 16.8
Landforms at a destructive margin

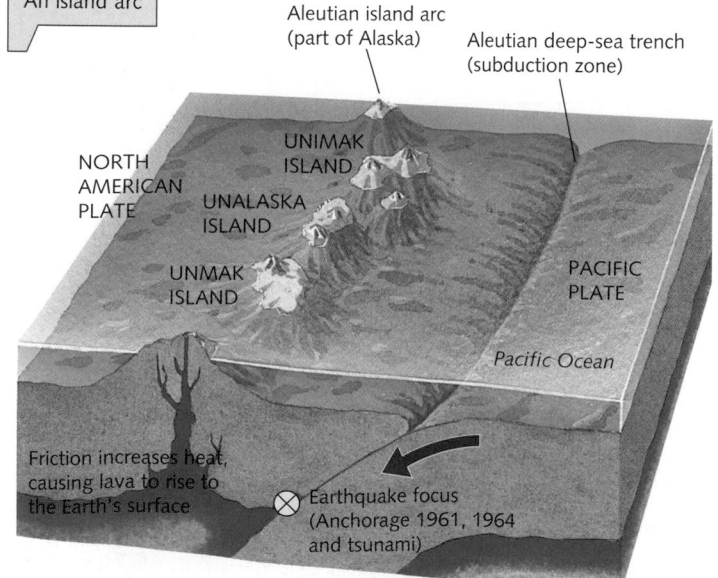

Figure 16.9
An island arc

Tsunamis are giant waves that are created by either a volcanic eruption (Krakatoa 1883) or a shallow submarine earthquake (Indian Ocean 2004) that occur at destructive plate margins. These waves can travel at great speeds across oceans, causing damage and loss of life far from their source of origin (300 000 people died in the Indian Ocean event – Figure 16.10).

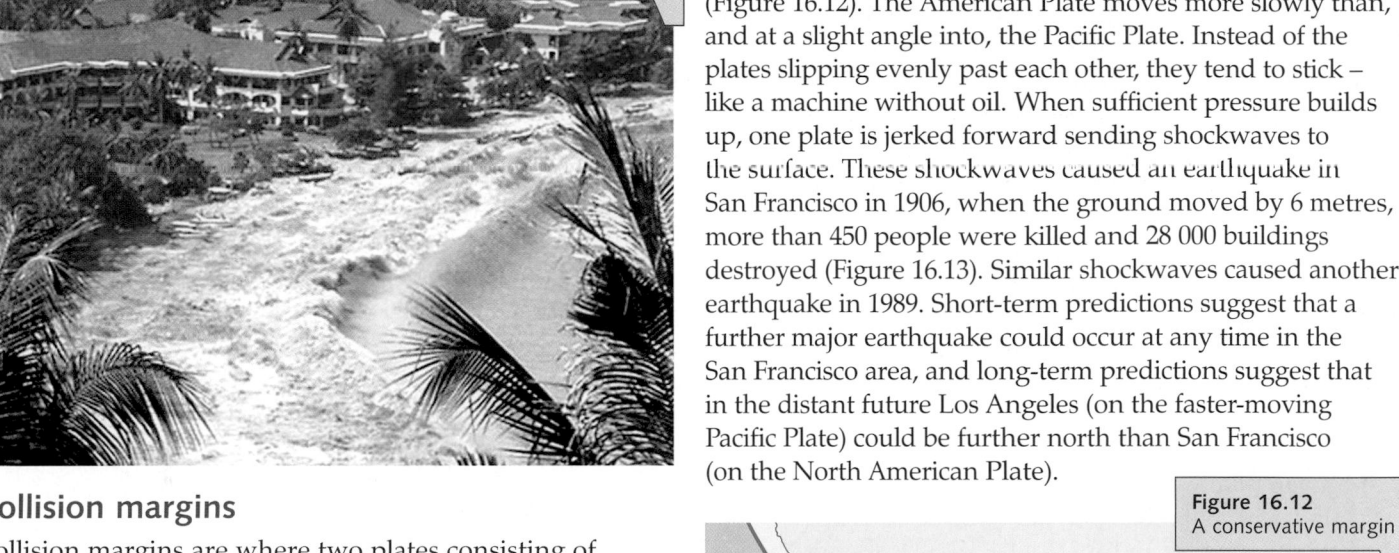

Figure 16.10
Indian Ocean tsunami

Collision margins

Collision margins are where two plates consisting of continental crust, move together. As continental crust can neither sink nor be destroyed, the rocks between them are forced upwards to form **fold mountains**. The Indian Plate is moving into the Eurasian Plate at a rate of 5 cm a year (Figure 16.11). The land between them, which was once the bed of the now non-existent Tethys Sea, has been buckled and pushed upwards to form the Himalayas – indeed evidence suggests that Mount Everest is increasing in height. This movement, which is still taking place, accounts for major earthquakes such as that which caused the deaths of 10 000 people and left another 150 000 homeless in central India in 1993, and killed 78 000 in northern Pakistan in 2005 and 80 000 in south-west China in 2008 (page 272).

Conservative margins

Conservative margins are found where two plates slide past one another. As crust is neither being formed nor destroyed at this plate boundary, new landforms are not created and there is no volcanic activity. However, earthquakes can occur if the two plates 'stick'. This is the situation in California, where the San Andreas Fault marks the junction of the Pacific and North American Plates (Figure 16.12). The American Plate moves more slowly than, and at a slight angle into, the Pacific Plate. Instead of the plates slipping evenly past each other, they tend to stick – like a machine without oil. When sufficient pressure builds up, one plate is jerked forward sending shockwaves to the surface. These shockwaves caused an earthquake in San Francisco in 1906, when the ground moved by 6 metres, more than 450 people were killed and 28 000 buildings destroyed (Figure 16.13). Similar shockwaves caused another earthquake in 1989. Short-term predictions suggest that a further major earthquake could occur at any time in the San Francisco area, and long-term predictions suggest that in the distant future Los Angeles (on the faster-moving Pacific Plate) could be further north than San Francisco (on the North American Plate).

Figure 16.12
A conservative margin

San Francisco

North American Plate moving north-westwards by 1 cm per year (but appears to be moving south-eastwards in relation to the faster-moving Pacific Plate)

San Andreas Fault

Pacific Plate moving north-westwards by 6 cm per year

USA

Los Angeles

N

San Diego

MEXICO

Figure 16.11
A collision margin

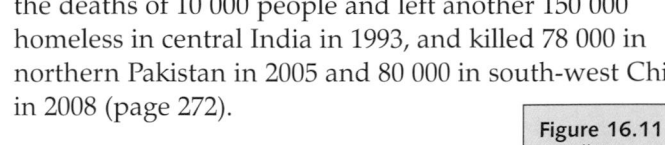

INDO-AUSTRALIAN PLATE (continental crust) moving north and east

Himalayas (including Mount Everest)

EURASIAN PLATE (continental crust) stationary

Collision zone

Mountain roots

Mantle

Figure 16.13
Destruction of San Francisco in 1906

Causes and effects of a volcanic eruption in an MEDC

Mount Etna, Italy

Causes of the eruptions

Between the continental crust of the African and Eurasian Plates is a smaller area of oceanic crust known as the Messina Plate. As the African Plate moves northwards, it pushes the oceanic crust into, and under, the Eurasian Plate forming a subduction zone with its associated deep-sea trench (Figure 16.14). The increase in pressure, caused by the Messina Plate being forced downwards, can trigger severe earthquakes in Sicily (60 000 deaths in Catania in 1693 and 100 000 in Messina in 1908) and southern Italy (Assisi in 1997 and L'Aquila in 2009). As the crust continues to descend it melts, partly from friction caused by contact with the Eurasian Plate and partly due to the increase in heat as it re-enters the mantle, forming magma. Periodically, some of this magma, being lighter than the surrounding mantle, rises to the surface to form volcanoes. These occur in Sicily where Mount Etna frequently erupts, as the small island arc of Vulcano and Stromboli and, on the Italian mainland, Vesuvius which erupts less frequently (Figure 16.14).

Eruptions 1970 to 2000

Mount Etna, which erupted thirteen times between 1970 and 2000, has already erupted four times this century. During the 1971 eruption most of the ski slopes and cable-car stations, together with the vulcanological observatory near the summit, were destroyed by lava. In the 1983 eruption, millions of tonnes of lava gushed out of a crater on the side of the mountain, eventually engulfing a hotel, 3 restaurants, 25 houses and numerous orange groves and vineyards. In time, the lava, flowing at an average speed of 15 km per hour along a path bounded by ridges of solidified rock, began to threaten several villages in its tracks (Figure 16.15). These villages were only saved by a series of controlled explosions that diverted a certain amount of the lava into a diversion channel.

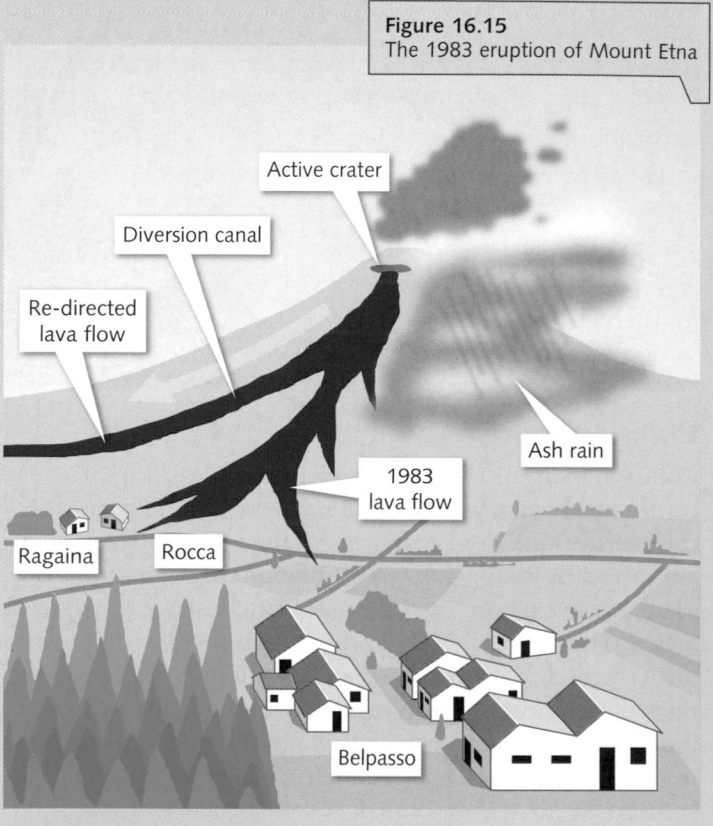

Figure 16.15
The 1983 eruption of Mount Etna

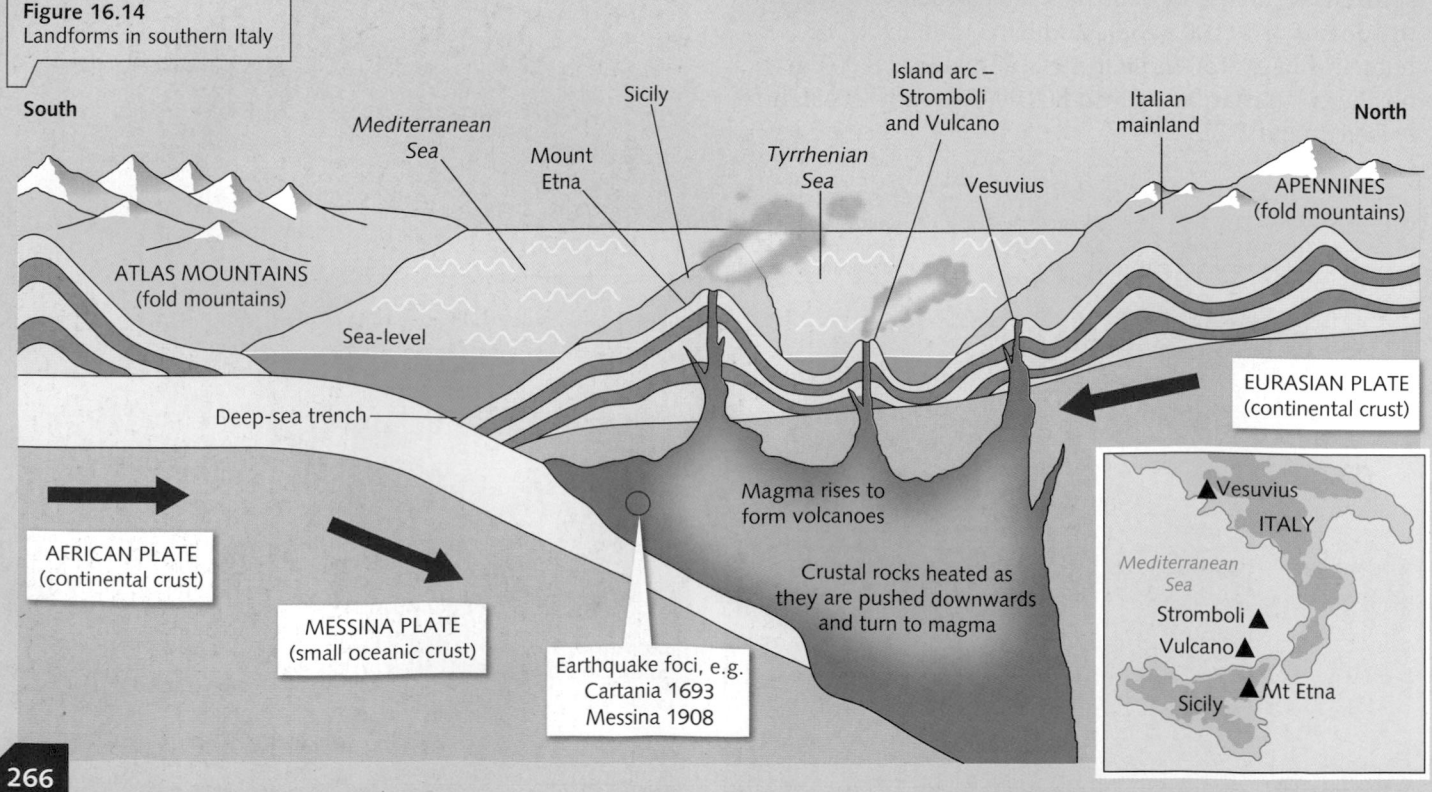

Figure 16.14
Landforms in southern Italy

27 October 2002

Despite attempts to improve the prediction of earthquakes and volcanic activity, there was no warning of this eruption (the author was standing with a group of people on the spot where the eruption was to occur less than nine hours later).

After a short period of seismic activity, a violent and devastating eruption occurred with lava being ejected along two **fissures** (i.e. along a line rather than from one central crater – Figure 16.17). One fissure was at a height of 2750 metres on the southern flank, the other between 2600 and 2500 metres on the north-eastern flank (Figure 16.16). Figure 16.18, taken on 30 October from the International Space Station, is looking south-east. Eruptive activity at two fissure segments on the northern flank shows as two small ash plumes in the lower central part of the image. Smoke from forest fires, started by the lava, is seen extending downhill to the left. A greyish plume is shown rising several kilometres into the atmosphere from the summit crater and, as it drifts at a high altitude south-eastwards towards Africa, it partly hides the eruptions on the southern flank. The dark streaks on the sides of the volcano denote lava solidified from earlier eruptions.

After a week of vigorous activity, when ash and rocks were thrown high into the sky, the eruptions became less intense on the northern flank but remained highly explosive to the south where lava continued to be ejected at irregular intervals. In early 2003, just over three months after the first eruption, volcanic activity ceased and Mount Etna was quiet again – until the next eruption in 2005.

Effects

Human life No deaths nor serious injuries were reported.

Settlements Most of the smaller hillside villages on the eastern slopes of Mount Etna, such as Santa Venerina and Linguaglossa, were destroyed by the initial earth tremors while, on the first day, the few buildings left standing in the tourist complex of Piano Provenzana were destroyed by a lava flow. Hundreds of families were evacuated from the area.

Communications Catania airport was closed for a number of days, and roads were covered by several centimetres of ash, paralysing local traffic – only four-wheel drives were allowed onto the mountain roads.

Farming Crops, especially oranges, vines and fruit, were affected by the heavy fall of ash.

Forests Many trees on the upper slopes, including pine forests near Linguaglossa, were destroyed by fires started by the lava and hot ash. It took fire-crews, aided by water dropped from helicopters, several days to regain control.

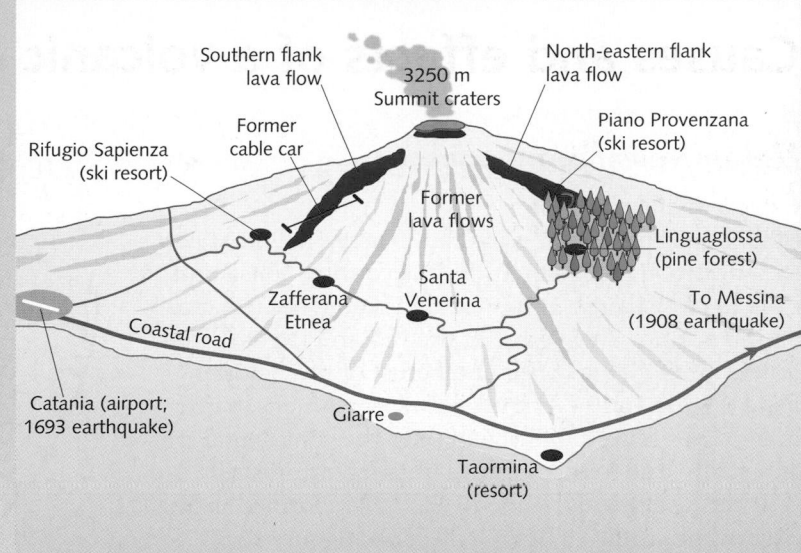

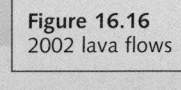

Figure 16.16
2002 lava flows

Figure 16.17
2002 fissure eruption

Figure 16.18
The 2002 eruption from space

Causes and effects of a volcanic eruption in an LEDC

Merapi volcano, Indonesia

Causes of the eruption

Indonesia is part of an island arc (Figure 16.9) that lies on a destructive plate margin (the country consists of over 13 000 islands). It has formed where oceanic crust on the north-eastern edge of the Indian Plate is forced downwards by the continental crust of the Eurasian Plate (Figure 16.19). As the oceanic crust melts, partly due to heat caused by friction with the continental crust and partly due to the increase in temperature as it re-enters the mantle, some of the newly formed magma rises to the surface to form volcanoes (Figure 16.8). Indonesia, with 130, has one-third of the world's active volcanoes. Of these, 34 are located on the island of Java, including Merapi ('Mountain of Fire') which is the most active (Figure 16.20). The volcano, which is considered sacred, has killed people on ten occasions since 1920 and has caused villages on its flanks to be evacuated even more frequently (Figure 16.21).

Effects of the 1994 eruption

When Merapi erupted in November 1994, it destroyed several villages and killed over 60 people – a large number for a volcanic event. Worst-hit was Turgo, a village 6 km away. The eruption was both sudden and violent and, before everyone could escape, the area was covered in a thick cloud of red-hot ash that made it too dark for people to see where they were going (Figure 16.22). The ash caused houses to collapse and burnt any exposed flesh of villagers working in the fields. As the cloud dispersed, survivors, with their few remaining possessions, fled to temporary shelters that are provided in anticipation of such an event.

The ejection of materials such as ash, rock and gases at the time of an eruption is said to be a **primary hazard** or **event**. However, this is often followed by later, or **secondary hazards/events**. Around Merapi, the main secondary hazard is mudflows, known as **lahars**. Lahars are caused when the heavy monsoon rain picks up ash and other loose volcanic material. The resultant mudflow rushes downhill covering settlements and farmland (Figure 16.23).

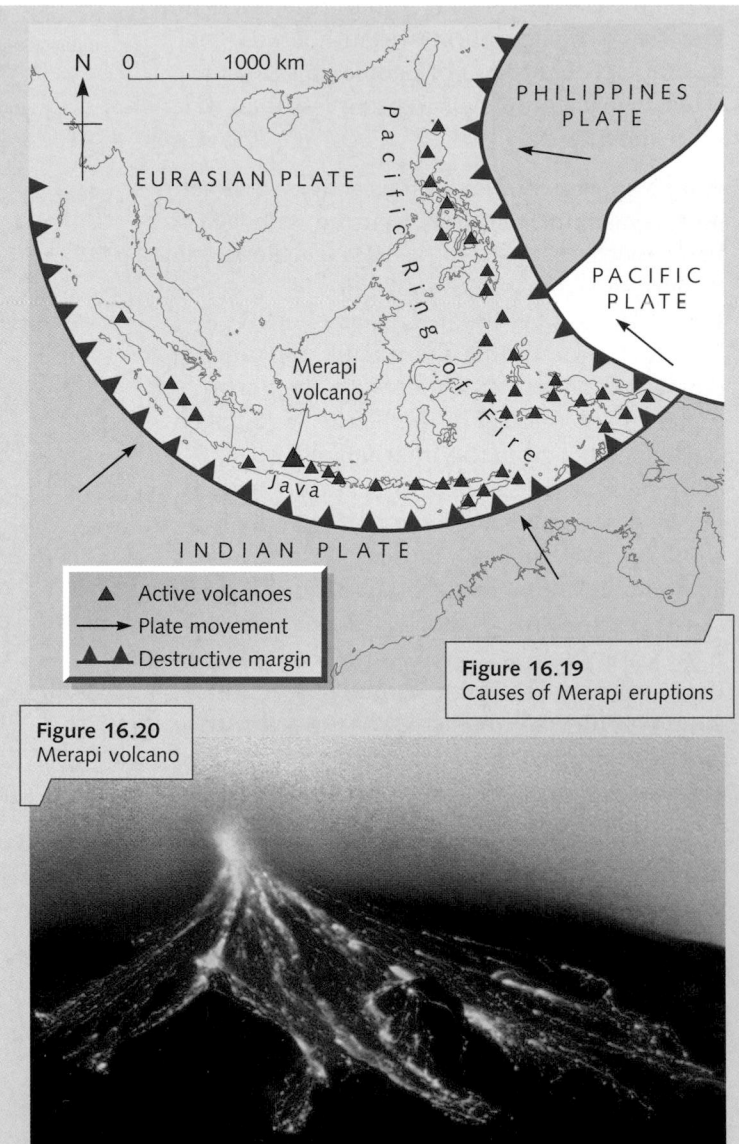

Figure 16.19
Causes of Merapi eruptions

Legend:
- ▲ Active volcanoes
- → Plate movement
- ▲▲ Destructive margin

Figure 16.20
Merapi volcano

Year	No. of people killed	Houses destroyed
1930 (E)	1369	1109
1954 (E)	64	144
1961 (E)	5	109
1969 (E)	3	91
1969 (M)	3	231
1974 (M)	9	29
1975 (M)	–	214
1976 (M)	27	385
1986 (M)	1	–
1994 (E)	61	25
2000 (E)	–	56
2000 (E)	–	–
2006 (M)	6	42
2008 (E)	–	–
(E) = Eruption	(M) = Mudflow	

Figure 16.21
Effects of recent eruptions

Figure 16.22
Cloud of red-hot ash

Figure 16.23
A lahar (mudflow)

Why do people live around Merapi?

Although people know that living near to the Merapi volcano is dangerous, they perceive that the risk is worth taking. This is because material ejected from the volcano rapidly breaks down (weathers) into a fertile soil that is rich in minerals needed for plant growth. The productive soil, which gives high crop yields, has encouraged intensive farming (page 97). Added to this, the hot, wet climate allows three crops a year to be grown on the same piece of land. However, high yields have led to a high population density (page 5) and a shortage of available farmland. This means that, following an eruption, farmers in the Merapi region have little choice but to return to their villages and rebuild their homes.

Responses to the Merapi hazard

These include trying to predict and giving advance warning of future eruptions, protecting people and their property in high-risk areas, educating people in evacuation procedures and, when requested, resettling people in lower-risk areas (Figure 16.24).

Figure 16.24
Responses to the Merapi hazard

Prediction – monitoring future eruptions

Volcanic activity in the Merapi volcano is monitored using seismometers. The equipment at individual sites is fairly simple. In one method, smoke created by the burning of kerosene produces a layer of carbon on a roll of paper. The roll of paper, which is turned slowly by a car battery, passes under a recording needle. Should there be any earth movement, the needle jumps and the tremor is recorded. Data from individual stations is collected and recorded centrally. Also, each month a team visits the summit to note any changes in pressure and temperature. Hopefully, this should enable scientists to give villagers sufficient time to seek safety before an eruption occurs.

Protection – mudflow dams

Dams have been built down each valley leading from Merapi to hold back lahars (mudflows). The dams trap the mud, which can later be spread over fields, and slowly release the water. Over 70 dams have been built to protect villages, roads, a major canal system and tourist sites.

Evacuation

A taskforce of local people has been specifically trained to give emergency treatment to the injured and to organise transport to evacuation camps and temporary accommodation.

Resettlement

Settlement is not allowed in a Prohibited Area near to the summit of the volcano, although farming is allowed there. Villagers, evacuated either from the Prohibited Area or from their village, are given two options:
1 They can live in relocation settlements where concrete houses are safer and have electricity, but where there is no land to farm.
2 They can move (a process of transmigration introduced by the government) to less crowded islands such as Sumatra.

In reality very few people elect to move, and most return to their villages however high the volcanic risk.

Causes and effects of an earthquake in an MEDC

Kobe, Japan

Earthquakes are measured on the **Richter scale** (Figure 16.25). Each level of magnitude on the scale is 10 times greater than the level below it. This means that the San Francisco earthquake of 1906 (measuring 8.2) was 10 times stronger than the one that affected Kobe in 1995 (which measured 7.2) and just over 100 times stronger than the one that hit Afghanistan in 1998.

Why did it happen?

Kobe is on a minor fault, the Nojima Fault, which lies above a destructive plate margin (Figure 16.26). It is here that the Philippines Plate (oceanic crust) is forced downwards on contact with the Eurasian Plate (continental crust). Over the centuries, volcanic activity at this plate boundary has created the Japanese island arc (page 264). However, associated earth movements mean that Japan is under constant threat from severe earthquakes (e.g. the 1923 Kanto earthquake which killed 140 000 people in Tokyo and Yokohama).

When did it happen?

The earthquake occurred at 5.46 on the morning of 17 January 1995. It recorded 7.2 on the Richter scale and lasted for 20 seconds. During this time the ground moved 18 cm horizontally and 12 cm vertically. The widespread devastation was due to the earthquake's focus being so near to the surface and its epicentre being so close to Kobe (Figure 16.27).

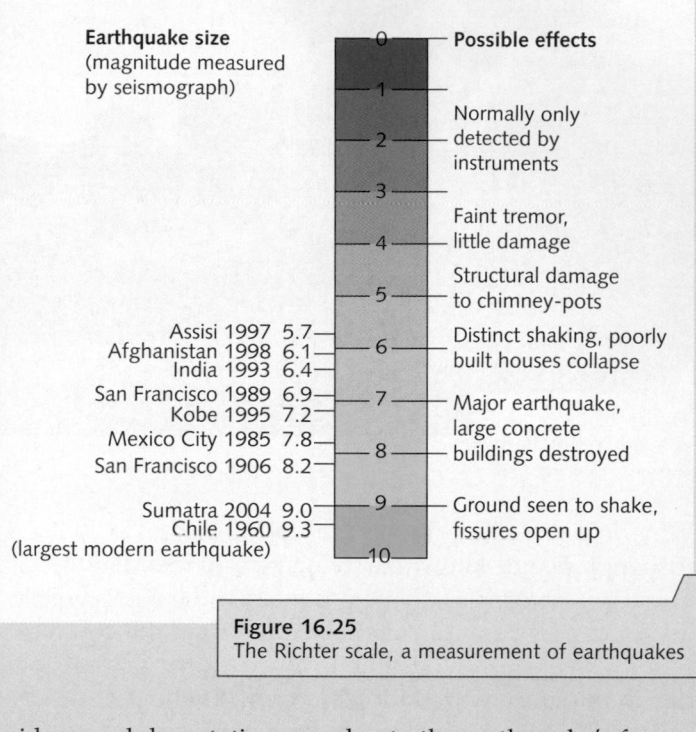

The Richter scale measures the size of the seismic waves during an earthquake.

Earthquake size (magnitude measured by seismograph)

Possible effects

Scale	Effect
0–2	Normally only detected by instruments
3–4	Faint tremor, little damage
5	Structural damage to chimney-pots
6	Distinct shaking, poorly built houses collapse
7–8	Major earthquake, large concrete buildings destroyed
9	Ground seen to shake, fissures open up

Assisi 1997 5.7
Afghanistan 1998 6.1
India 1993 6.4
San Francisco 1989 6.9
Kobe 1995 7.2
Mexico City 1985 7.8
San Francisco 1906 8.2
Sumatra 2004 9.0
Chile 1960 9.3
(largest modern earthquake)

Figure 16.25
The Richter scale, a measurement of earthquakes

Figure 16.26
Location of Kobe

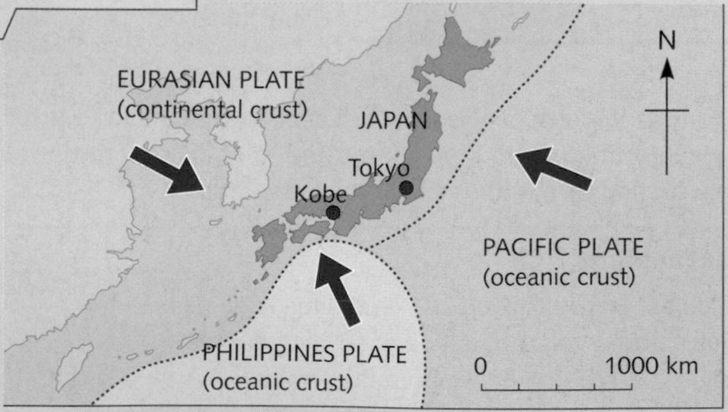

EURASIAN PLATE (continental crust)
JAPAN
Tokyo
Kobe
PACIFIC PLATE (oceanic crust)
PHILIPPINES PLATE (oceanic crust)
N
0 1000 km

Figure 16.27
The 1995 Kobe earthquake

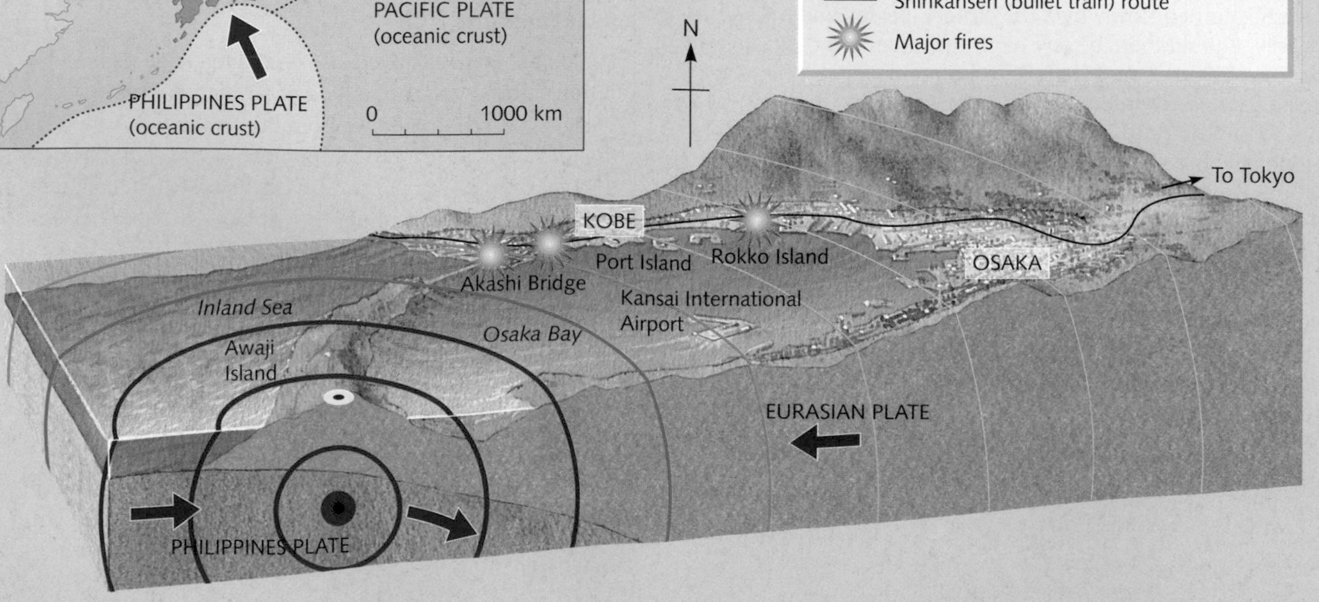

- Seismic shockwave travelling outwards
- Focus = point of earthquake
- Epicentre = point on ground above focus
- Shinkansen (bullet train) route
- Major fires

N

KOBE
Port Island
Akashi Bridge
Rokko Island
Kansai International Airport
OSAKA
To Tokyo
Inland Sea
Awaji Island
Osaka Bay
EURASIAN PLATE
PHILIPPINES PLATE

What were its effects?

The effects can be divided into primary and secondary types.

Primary effects

- Nearly 200 000 buildings collapsed, and also a 1 km stretch of the elevated Hanshin Expressway (Figure 16.28) and numerous bridges along a 130 km section of the bullet train route. Several trains on minor lines were derailed.
- 120 of the 150 quays in the port of Kobe were destroyed.

Secondary effects

- Electricity, gas and water supplies were disrupted.
- Fires (Figure 16.29), caused by broken gas pipes and ruptured electricity mains, raged for several days, destroying a further 7500 houses (many of which were made of wood). At one time, the wall of fire extended for over 400 metres.
- Roads were at gridlock, delaying ambulances and fire engines.
- An estimated 230 000 people were made homeless and had to live in temporary shelters (unheated school gyms or in open parks) at a time when night-time temperatures dropped to –2°C. There was a short-term shortage of blankets, clean water and food.
- People were afraid to return home as the 716 recorded aftershocks lasted several days after the main event (74 were strong enough to be felt by humans).
- Industries, including Mitsubishi and Panasonic (page 153), were forced to close.

Final figures put the death toll at 5500, injuries at 40 000 and houses totally destroyed at 180 000. However, the newly opened Kansai International Airport and Akashi Bridge were both undamaged – presumably due to their high-tech construction aimed at withstanding earthquakes (page 73).

What happened in the months after the earthquake?

- Kobe's infrastructure, including water, electricity, gas and telephone services, was fully operational by July.
- The area worst affected by fire had been cleared of rubble but little rebuilding had taken place. Most commercial buildings in central areas had been repaired.
- All rail services were back to normal by August.
- One year later the port of Kobe was 80 per cent functional but the Hanshin Expressway remained closed.
- Replacement buildings had to meet stronger earthquake-resistance standards. High-rise buildings had to have flexible steel frames, smaller buildings had to have concrete frames with reinforcing bars to absorb shockwaves, houses were not to be built just from brick (which shakes loose) or wood (which burns too easily) but with fire-resistant materials. New buildings had to be built on solid rock, not clay, as water rises to ground-level during an earthquake, causing clay to 'liquefy' into mud. This results in the collapse of buildings.
- There was an increase in the number of seismic instruments to record earth movements in the region.

Figure 16.28
The collapsed Hanshin Expressway

Figure 16.29
Fires following the earthquake

Causes and effects of an earthquake in an LEDC

Sichuan, China

Causes of the earthquake

The north-western part of the Chinese province of Sichuan lies along a major fault located at the border of the Indo-Australian and Eurasian Plates. As the Indo-Australian Plate moves north-eastwards by 5 cm per year, it pushes against the Eurasian Plate increasing the pressure at the fault line (Figure 16.30).

Effects of the earthquake

On **12 May 2008**, the pressure was suddenly released by an earthquake measuring 7.9 on the Richter scale and which lasted for 120 seconds. The epicentre was in a mountainous region 80 km north-west of Chengdu, the capital of Sichuan. The release of energy, which was 30 times greater than that during the Kobe earthquake (page 270), caused the land to move northwards by 9 metres and the shockwaves to be felt as far away as Beijing (1500 km) and Shanghai (1700 km). As the earthquake's focus was near to the surface and as this part of Sichuan is one of the most remote and poorest in China, the effects were devastating (Figure 16.31). Nearly all the buildings in the towns of Yingxiu, Beichuan and Dujiangyan were destroyed, including schools and hospitals (Figure 16.33). Yet, within hours of the event, 20 helicopters and 1300 soldiers were being sent to the scene although adverse weather was to limit flights and landslides blocked roads taken by rescue teams.

13 May First reports suggested a death toll of 8000 but with so many villages being remote and the roads to them destroyed, this figure was expected to rise rapidly. Huge mud and rock landslides knocked down telephone lines and mobile phone masts, preventing contact with the outside world. Reports were circulating over the collapse of many schools, trapping the students within them, as the quake occurred in mid-afternoon.

14 May The first soldiers eventually managed to reach the stricken area. Estimates suggested that up to 60 000 people were still missing, including a coach-load of Britons travelling to the nearby giant panda sanctuary at Wolong.

15 May An extra 90 helicopters brought in food, water and medical supplies and took out the worst-injured survivors. Over 100 000 troops were now searching the rubble, but only two people were found alive. Fears grew of flooding as cracks in a dam above Dujiangyan widened.

16 May The death toll rose to 20 000; a woman and an 11-year-old girl were pulled alive from the rubble in Dujiangyan; and the 'lost' British tourists were found safe.

16–25 May Unlike other earthquakes when aftershocks decrease rapidly in magnitude and frequency, here shocks in excess of 6.0 on the Richter scale continued for several weeks. These caused damaged buildings to collapse and added to the death toll, including 158 relief workers killed by a landslide. The latest threat was from the 34 'quake lakes' that had been formed by landslides blocking rivers and were now threatening to overflow.

Figure 16.30
Location of Sichuan, China

Magnitude (Richter scale)	7.9
Aftershocks over 6.0	8
Deaths	69 227
Injured	375 000
Missing	18 000
Homeless	4.8 million
People affected	Over 46 million
Classrooms collapsed	7000
Schoolchildren died	Over 19 000
Estimated damage (US$)	86 billion

Figure 16.31
Sichuan earthquake fact file

Figure 16.32
A man buried by the Sichuan earthquake

Why was the death toll so high?

The economic advances made by China's coastal cities and provinces had yet to reach outlying areas so that, in 2007, Sichuan was ranked 25th out of 31 provinces in terms of GDP per capita. This, together with adverse physical factors, meant that:

- Health care, education and communications were less well developed and able to cope with such a major event. Emergency services were also less well prepared.
- Many of the houses were poorly built and none was likely to have been designed for an earthquake of even a much lower magnitude than the one of 2008.
- Schools in particular seem to have been badly designed and built. This caused the collapse of 7000 classrooms and the deaths of over 19 000 schoolchildren.
- The mountainous terrain, together with landslides caused by both the initial earth movements and the subsequent heavy rainfall, disrupted transport and hindered the arrival of emergency aid.
- Many of the more remote areas were without telephones and mobile reception which increased their isolation from the rest of China and the world.

Why do people live in this earthquake-prone area?

Apart from this north-west corner, most of Sichuan is a relatively flat, low-lying river basin where the fertile soils enable high yields of rice to be grown – hence it is known as the 'rice bowl of China'. While this has created a high population density, it also means that poorer members of the community have been forced to live in less well-off, less developed places. Many of these people are either too poor to move or lack the education and skills needed to migrate to the better-paid factory jobs of eastern China.

Figure 16.33
Destruction in Yingxiu caused by the earthquake

Volcanoes and earthquakes in MEDCs and LEDCs

Volcanic eruptions

Prediction It is easier to predict volcanic eruptions than to predict earthquakes. This is because:

- tremors increase within the volcano
- as magma begins to rise, so do ground temperatures. These can be detected by satellites using heat-seeking cameras (new technology)
- the rising magma causes the volcano to swell and bulge (Figure 16.15). This can be measured by tiltmeters (which measure slope angles) and satellite global positioning systems
- immediately before an eruption, the volcano emits an increasing amount of gas and steam

Preparation This includes:

- setting up monitoring and warning systems, using new technology as described above
- preparing an evacuation plan and organising transport, accommodation and food for those who need to be moved
- training emergency services such as police, fire and ambulance crews
- organising emergency services such as water, food and power (Figure 16.34)
- organising post-eruption plans to include repairing communications and buildings.

MEDCs and LEDCs The effects of volcanic eruptions are usually much greater in LEDCs, i.e. the poorer countries of Indonesia (page 268), Montserrat and the Philippines (Mount Pinatubo). Human response to volcanic eruptions is usually much quicker and more efficient in MEDCs, i.e the richer countries of the USA (Mount St Helens) and Japan (with 22 observatories). LEDCs often have to rely on, and wait for, international aid.

Earthquakes

Prediction It is far harder to predict the time and location of earthquakes than of volcanic eruptions. However, it is possible to:

- install sensitive instruments that can measure an increase in earth tremors (seismometers), pressure, and any release of radon gas
- map epicentres and frequencies of previous earthquakes to see if there is a location and/or time pattern (this can only give possible timings of an event and not a precise location)
- observe unusual animal behaviour – dogs howling, fish jumping, mice fleeing houses (less scientific but has proved useful).

Preparation This includes:

- constructing buildings and roads to withstand earthquakes (e.g. San Francisco and Figure 16.35)
- training emergency services and having them available, e.g. helicopters, ambulances and fire engines

Figure 16.34
Relief following an earthquake

- organising emergency services to provide water, food and power
- setting up a warning and information system for use on TV and radio (earthquake preparation is part of the Japanese school curriculum).

MEDCs and LEDCs The effects of earthquakes are usually much greater in LEDCs, e.g. the poorer countries of China (page 272) and Iran (1997). Human response to earthquakes is much quicker and more efficient in MEDCs, e.g. the richer countries of Japan (page 270) and the USA (San Francisco 1989). LEDCs often have to rely upon, and wait for, international aid.

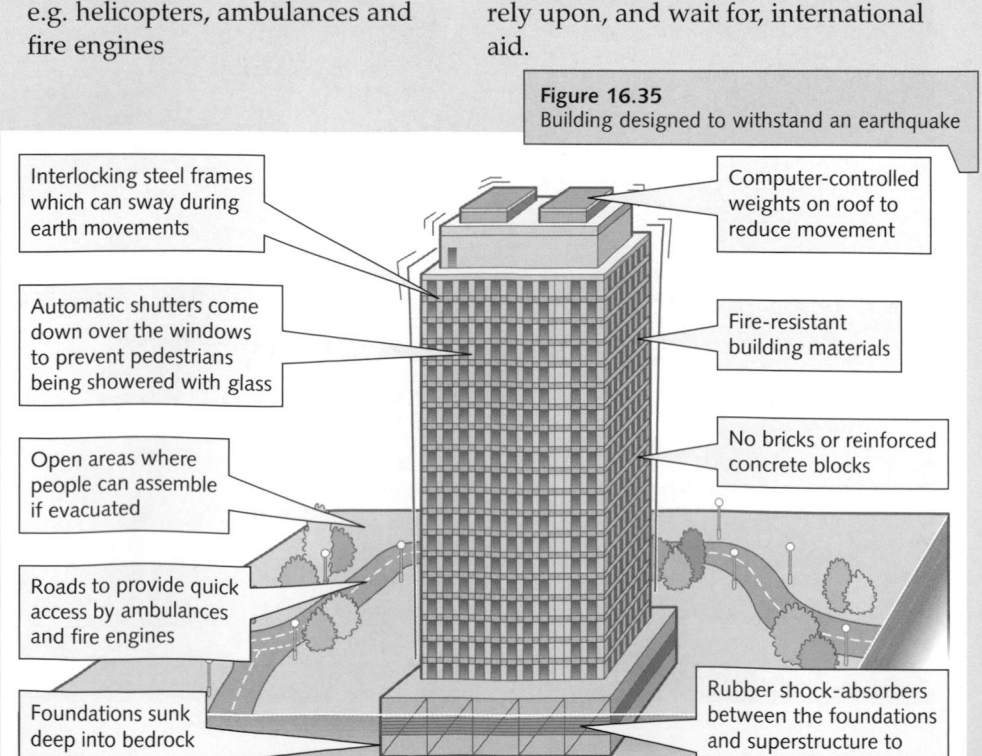

Figure 16.35
Building designed to withstand an earthquake

Interlocking steel frames which can sway during earth movements

Computer-controlled weights on roof to reduce movement

Automatic shutters come down over the windows to prevent pedestrians being showered with glass

Fire-resistant building materials

No bricks or reinforced concrete blocks

Open areas where people can assemble if evacuated

Roads to provide quick access by ambulances and fire engines

Foundations sunk deep into bedrock avoiding clay

Rubber shock-absorbers between the foundations and superstructure to absorb earth tremors

Plate tectonics

Key Words and Terms

a You should know the meaning of the following terms:
- tectonic activity • crust • mantle • zones of activity • plates
- plate boundaries/margins • constructive plate margin
- destructive plate margin • collision plate margin
- conservative plate margin • sea-floor spreading • subduction zone • deep-sea trench
- island arc • tsunamis • fold mountains • Richter scale • mudflows/lahars.

b You should know the difference between:
- continental and oceanic crust
- magma and lava
- the epicentre and focus of an earthquake
- predicting and preparing for a volcanic eruption and an earthquake.

Key Ideas

You should know and understand the following:

- The global distribution of earthquakes, volcanoes and fold mountains.
- The Earth's crust is unstable and divided into a series of tectonic plates.
- The global distribution of (tectonic) plates.
- That there is a relationship between earthquakes, volcanoes and plate boundaries/margins.
- The processes involved in plate activity.
- The causes and effects of crustal movement at three (four) plate margins – constructive, destructive, collision (and conservative).
- The processes involved and the landforms associated with each plate margin – fold mountains, island arcs, deep-sea ocean trenches and volcanoes.
- Causes, effects and responses to a volcanic eruption in an MEDC and in an LEDC – primary/short-term and secondary/long-term.
- Causes, effects and responses to an earthquake in an MEDC and an LEDC – primary/short-term and secondary/long-term.
- Differences in the effect and responses to volcanic eruptions/earthquakes between an MEDC and an LEDC.
- Strategies to plan for and manage in order to reduce the risk and effects of earthquakes and volcanic eruptions.
- Why people live in areas of volcanic activity/earthquakes.

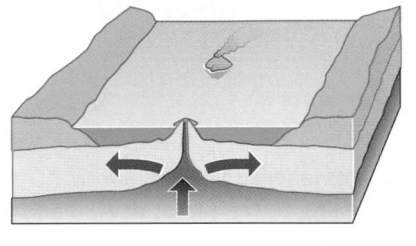

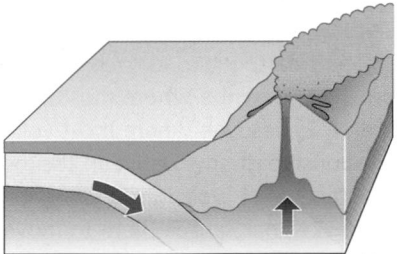

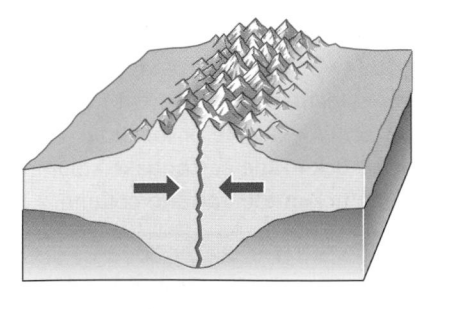

Skills, Theories and Models

- Interpret and use a variety of maps and annotated diagrams
- Classify types of plate boundary
- Compare the effects of volcanoes and earthquakes in MEDCs and LEDCs
- Understand the theory of plate tectonics

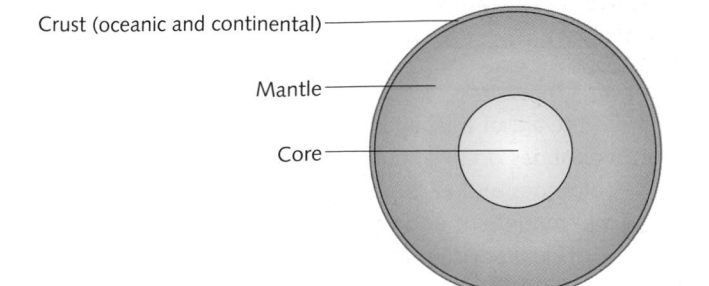

Crust (oceanic and continental)

Mantle

Core

1 *(Pages 262 to 265)*

a The map shows some of the major plate boundaries and the location of fold mountains.

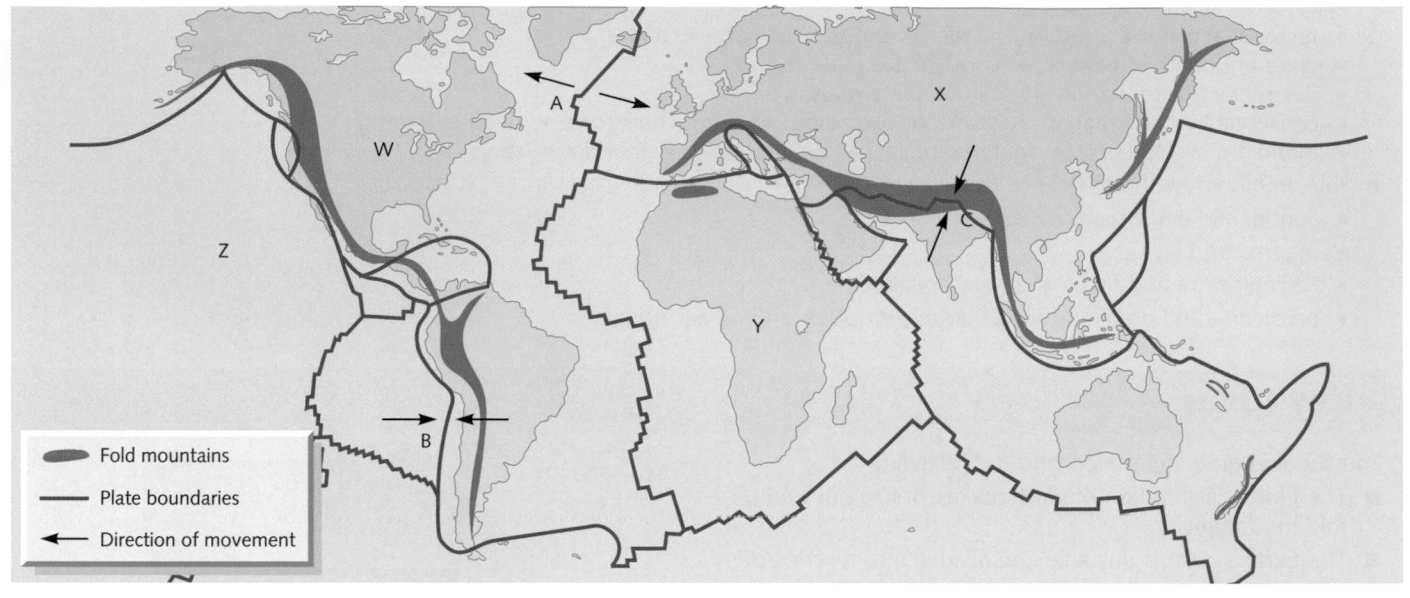

i)   Name the plates labelled W, X, Y and Z. (4)
ii) What is happening to the plates at A, B and C? (3)
iii) Name the type of plate boundary at A, B and C. (3)
b i)   Name one volcano in each of North America, South America, Europe and Asia. (4)
ii) Name one recent earthquake in each of North America, Europe and Asia. (3)
iii) Using Figure 16.1, describe where most earthquakes are likely to occur. (3)
iv) Using Figure 16.2, describe where most volcanic eruptions are likely to occur. (3)
c What is the relationship, if any, between the location of fold mountains and:
   • earthquake activity   • volcanic activity   • plate boundaries? (6)
d i)   What are plates? What causes them to move? (2)
ii) Give three differences between continental and oceanic crust. (3)
iii) Copy and complete the following table by:
   • naming the two types of crust involved (continental or oceanic)
   • giving the direction of movement (away from, towards or sideways past)
   • giving an actual example or location
   • putting one tick in the earthquake column
   • putting one tick in the volcano column. *(5 × 4)*

Plate margin	Two types of crust involved	Direction of plate movement	Example/ location	Earthquake			Volcano		
				Violent	Less violent	Rare	Violent	Less violent	Rare
Constructive									
Destructive									
Collision									
Conservative									

e With the aid of a labelled diagram, explain why:
   i)   volcanoes occur at constructive plate margins (3)
   ii) volcanoes and earthquakes occur at destructive plate margins (4)
   iii) fold mountains and earthquakes occur at collision plate margins (4)
   iv) earthquakes occur at conservative plate margins. (3)

2 *(Pages 266 to 269 and 274)*

a Redraw these two diagrams.
   i) Name the plates. (2)
   ii) On each diagram, draw an arrow to show in which direction plate A is moving. (2)
b i) Why does Mount Etna erupt so frequently? (2)
   ii) Describe the 2002 eruption of Mount Etna. (2)
   iii) How did the eruption affect:
   • settlement   • communications   • human activities? (3)

Mount Etna	Merapi
Plate A Plate B Plate C	Plate A Plate B

c i) Why does Merapi erupt so frequently? (2)
   ii) Describe the 1994 eruption of Merapi. (2)
   iii) How did the eruption affect:
   • settlement   • communications   • human activity? (3)
d i) Why is it hard to predict volcanic activity? (2)
   ii) Why is it harder to predict this activity in LEDCs like Indonesia (Merapi) than in MEDCs
       like Italy (Mount Etna)? (2)
   iii) Why is rescue work so difficult immediately after a volcanic eruption? (2)
   iv) Why is rescue work harder in LEDCs like Indonesia than in MEDCs like Italy? (2)
   v) Why does volcanic activity usually lead to a greater loss of life in LEDCs than in MEDCs? (2)

3 *(Pages 270 to 274)*

a Make a larger copy of the table.
   Complete it to show differences in:
   • the causes
   • the effects
   of the Kobe (Japan) and Sichuan
   (China) earthquakes. *(22)*
b Why might an earthquake
   have been expected at:
   • Kobe
   • Sichuan? *(6)*

		Kobe (Japan)	Sichuan (China)	
i)	Type of plate margin			(2)
ii)	Two plates bordering this margin			(4)
iii)	Is the area extremely urban or extremely rural?			(2)
iv)	Date of earthquake			(2)
v)	Richter scale reading			(2)
vi)	Primary effects			(4)
vii)	Secondary effects			(6)

c What precautions had the authorities
   taken in preparation for a possible earthquake in:
   • Kobe   • Sichuan? (4)
d Why might it have been expected that damage and loss of life would be much greater in Kobe
   than in Sichuan (population density, strength of the earthquake)? (2)
e i) Why were the Kobe authorities able to deal quickly with the effects of the earthquake? (4)
   ii) Why did it take longer for help to reach Sichuan? (3)
f Why do people continue to live in places like:
   • Kobe   • Sichuan
   when they know they are at risk from future earthquakes? (4)
g What can be done to predict earthquakes? (2)
h i) Why are the effects of earthquakes usually much greater in LEDCs than in MEDCs? (4)
   ii) Why is human response to earthquakes usually much quicker in MEDCs than in LEDCs? (4)

Drainage basins and rivers

Drainage basins

A **drainage basin** or **river basin** is an area of land drained by a main river and its tributaries (Figure 17.1). Its boundary, marked by a ridge of higher land, is called a watershed. A **watershed**, therefore, separates one drainage basin from neighbouring drainage basins. Some basins, like the Mississippi which drains over one-third of the USA, are enormous. Others, possibly that of your local river, can be small. However, size is less important than the drainage density. The drainage density is:

the total length of all the streams in the drainage basin
―――――――――――――――――――――――――――――――
the total area of the drainage basin

The drainage basin of the River Exe (Figure 17.2) has a much higher density than the basin in Figure 17.1. The density is highest on impermeable rocks and clays, and lowest on permeable rocks and sands. The higher the density the greater the risk of flooding, especially as a result of a flash flood (page 286).

Drainage basins can store rainwater, either within the river channel itself, or in lakes and in the ground. Excess water is carried back to the sea by rivers. Rivers form part of the hydrological (water) cycle (Figure 17.3). The seas and oceans contain 97 per cent of the world's water, but being salty it is not suitable for use by terrestrial plants, animals and people.

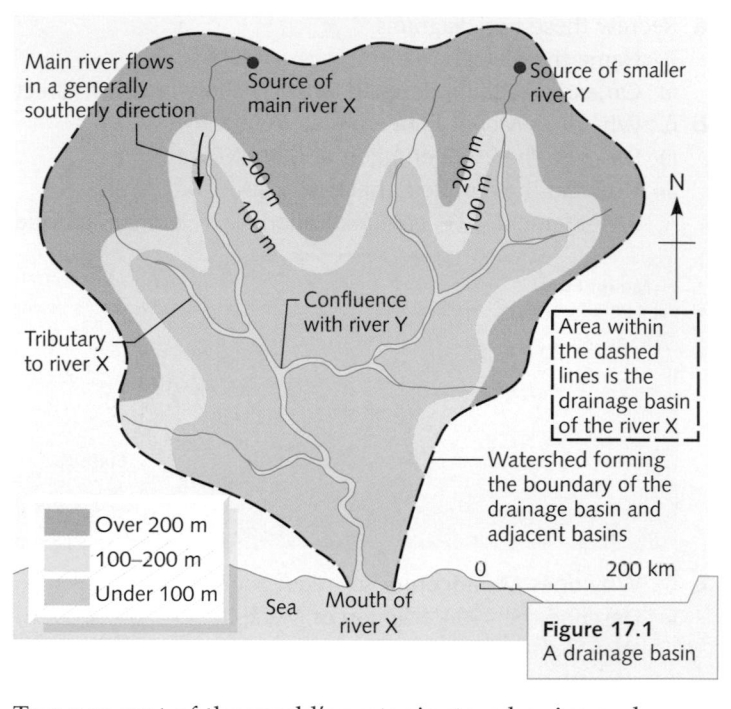

Figure 17.1
A drainage basin

Two per cent of the world's water is stored as ice and snow in arctic and alpine areas. That leaves 1 per cent which is either fresh water on land or water vapour in the atmosphere. As the amount of fresh water and vapour is limited, it has to be recycled over and over again. It is this constant recycling of water between the sea, air and land which is known as the **hydrological cycle**. As no water is added to or lost from the hydrological cycle, it is said to be a **closed system**.

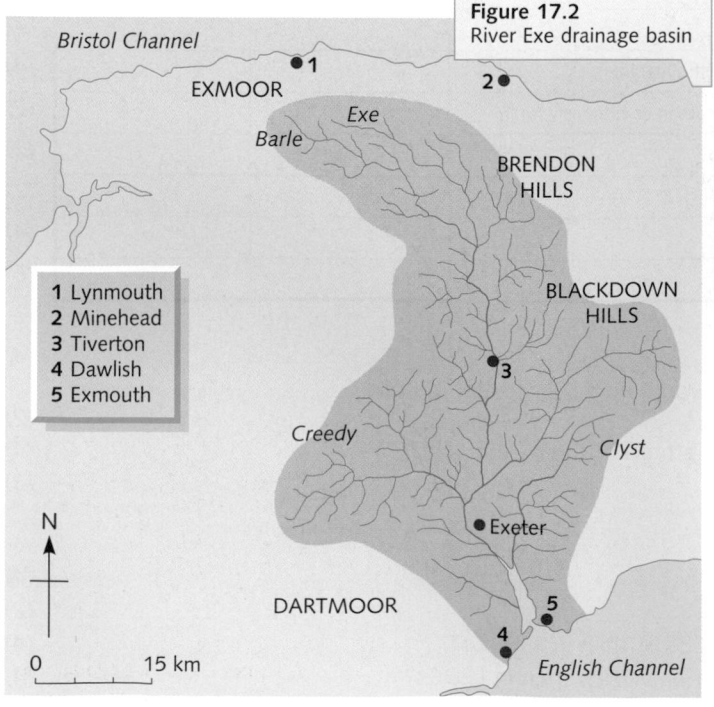

Figure 17.2
River Exe drainage basin

1 Lynmouth
2 Minehead
3 Tiverton
4 Dawlish
5 Exmouth

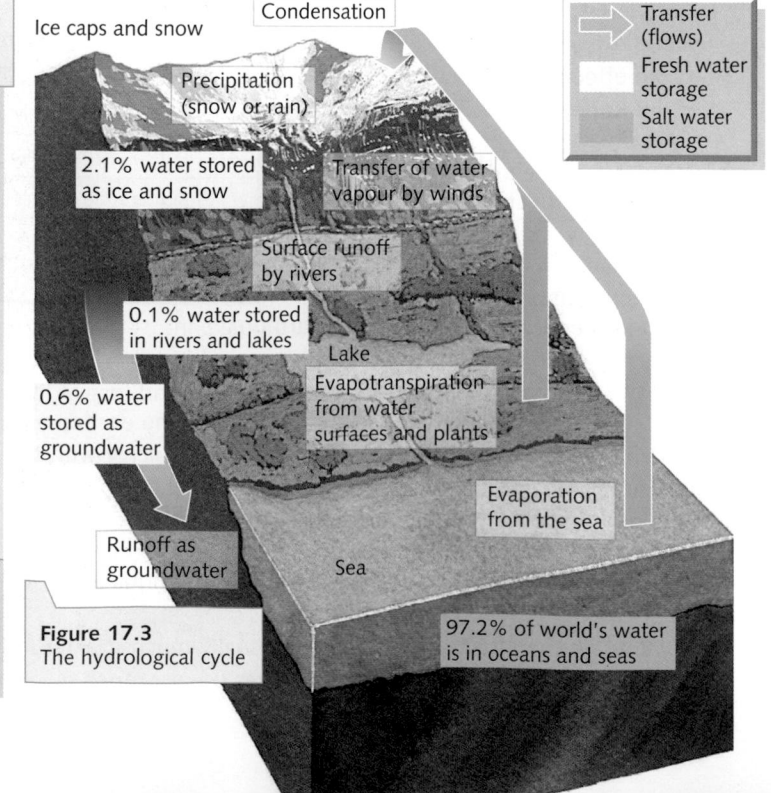

Figure 17.3
The hydrological cycle

The drainage basin system

A drainage basin forms part of the hydrological cycle but, unlike the hydrological cycle, it is an **open system**.

INPUTS ➡ FLOWS ➡ STORES ➡ FLOWS ➡ STORES ➡ FLOWS ➡ OUTPUTS

It is an open system because it has:
- **Inputs** where water enters the system through precipitation (rain and snow).
- **Outputs** where water is lost to the system either by rivers carrying it to the sea or through evapotranspiration. **Evapotranspiration** is the loss of moisture directly from rivers or lakes (evaporation) or from vegetation (transpiration).

Within the system are **stores** and **transfers** (flows).
- **Stores** are places where water is held, e.g. in pools and lakes on the surface or in soil and rocks underground.
- **Transfers** are processes by which water flows, or moves, through the system, e.g. infiltration, surface runoff, throughflow.

A typical drainage basin system is shown in Figure 17.4.

When it rains, most water droplets are intercepted by trees and plants. Interception is greatest in summer. If the rain falls as a short, light shower then little water will reach the ground. It will be stored on leaves and then lost to the system through evaporation.

When the rain is heavier and lasts longer, water will drip from the vegetation onto the ground. At first it may form pools (**surface storage**) but as the ground becomes increasingly wet and soft, it will begin to infiltrate. **Infiltration** is the downward movement of water through tiny pores in the soil. This downward transfer will be greatest in porous rock or soil such as chalk or sand, and least in impermeable rock or soil like granite or clay. The water will then either be stored in the soil or slowly transferred sideways or downwards. The movement of water sideways is called **throughflow** and it is likely to form, eventually, a spring on a valley side. When the movement of water is downwards it is called **percolation**. Percolation forms **groundwater**, which is water stored at a depth in rocks. Groundwater flow is the slowest form of water transfer. The fastest process of water movement is surface runoff. **Surface runoff**, sometimes referred to as **overland flow**, occurs when either the storm is too heavy for water to infiltrate into the soil, where the soil is impermeable, or when the soil has become saturated. The level of **saturation**, i.e. when all the pores have been filled with water, is known as the **water table**. Although some rain may fall directly into the river, most water reaches it by a combination of surface runoff, throughflow and groundwater flow. Rivers carry water to the sea where it is lost to the system.

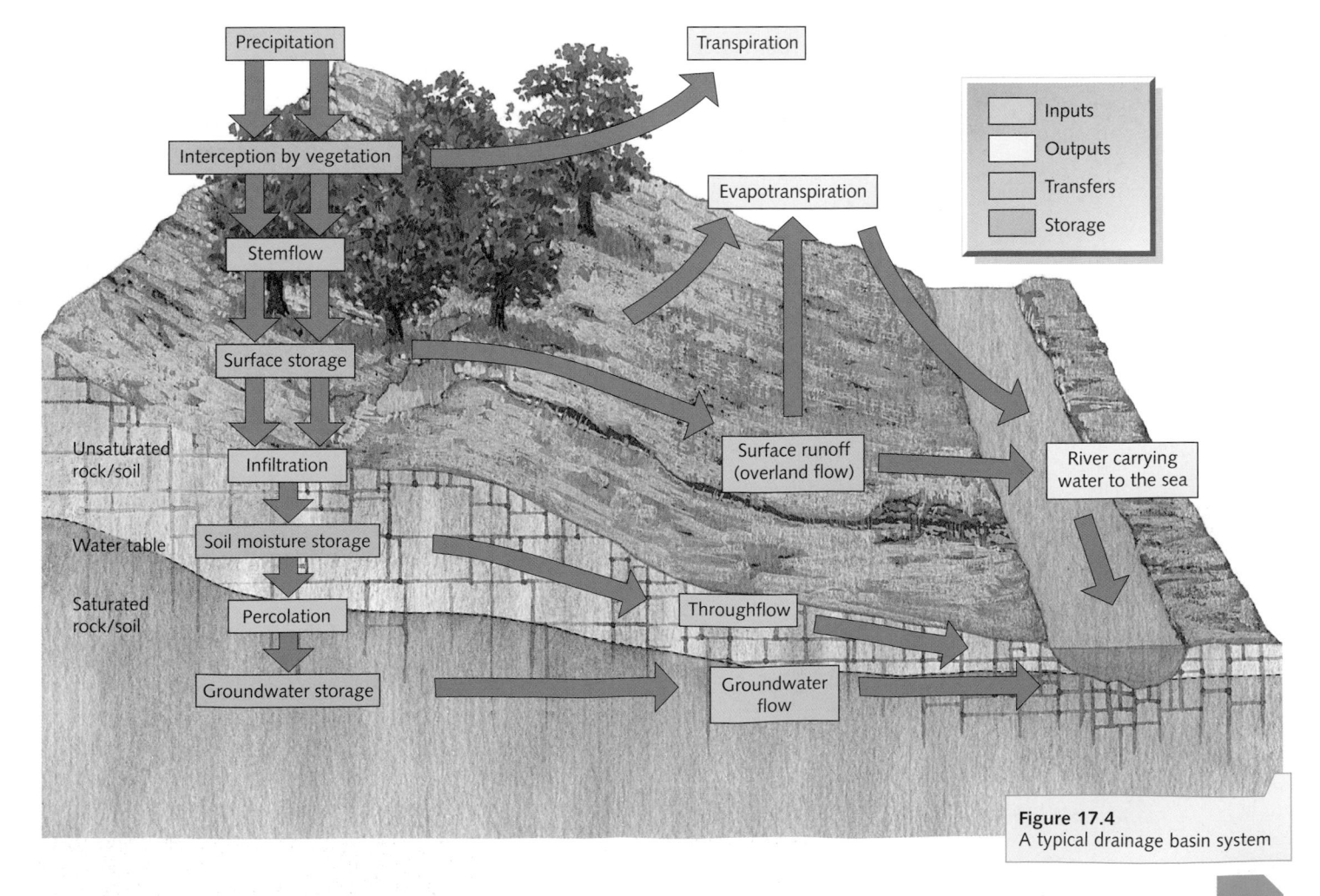

Figure 17.4
A typical drainage basin system

River discharge and flood hydrographs

Discharge depends on the river's velocity and volume.

Velocity is the speed of the river. It is measured in metres per second.

Volume is the amount of water in the river system. It is the cross-sectional area of the river's channel measured in square metres.

Discharge is the velocity of the river times its volume. It is the amount of water in the river passing a given point at a given time, measured in cumecs (cubic metres per second).

In some drainage basins, discharge and river levels rise very quickly after a storm. This can cause frequent, and occasionally serious, flooding. Following a storm in these basins, both discharge and river levels fall almost as rapidly, and after dry spells become very low. In other drainage basins, rivers seem to maintain a more even flow.

The flood (storm) hydrograph

A **hydrograph** is a graph showing the discharge of a river at a given point (a gauging station) over a period of time. A flood or storm hydrograph shows how a river responds to one particular storm (Figure 17.5). When a storm begins, discharge does not increase immediately as only a little of the rain will fall directly into the channel. The first water to reach the river will come from surface runoff, and this will later be supplemented by water from throughflow. The increase in discharge is shown by the rising limb.

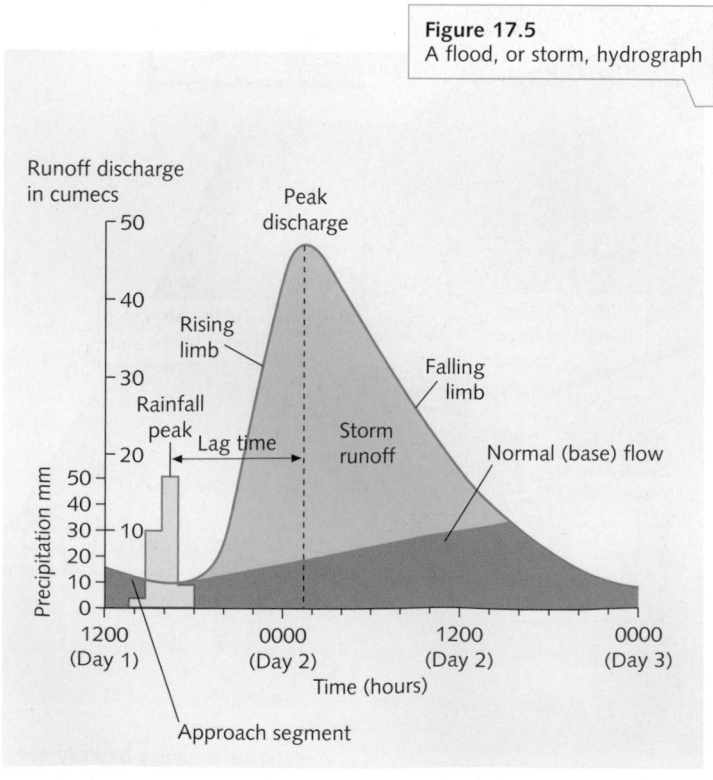

Figure 17.5
A flood, or storm, hydrograph

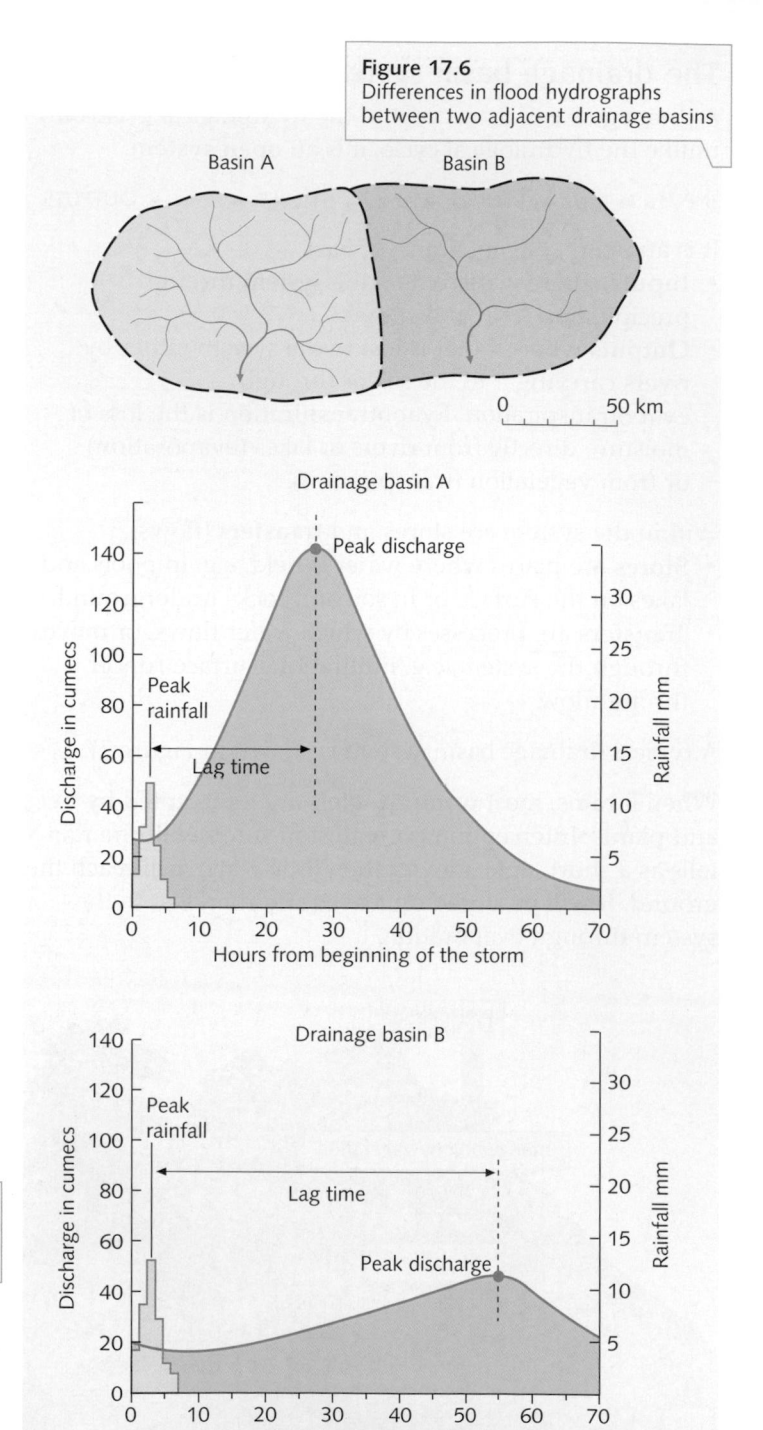

Figure 17.6
Differences in flood hydrographs between two adjacent drainage basins

The gap between the time of peak (maximum) rainfall and peak discharge (highest river level) is called **lag time**. A river with a short lag time and a high discharge is more likely to flood than a river with a lengthy lag time and a low discharge.

Factors affecting the shape of the flood hydrograph

It is possible that two drainage basins, located side by side, can receive approximately the same amount of rainfall and yet have very different hydrograph shapes (Figure 17.6). The river in basin A is likely to flood regularly, the river in basin B probably never. The reasons for this difference may be due to one factor, or a combination of factors (Figure 17.7).

Factor	Drainage basin A	Drainage basin B
Relief	Faster runoff on steep slopes	Slower runoff on more gentle slopes
Rock type	Surface runoff on impermeable rock	Throughflow and groundwater flow as rainfull inflitrates into permeable (porous) rock
Soil	Very thin soil; less infiltration	Deeper soil; more infiltration
Natural vegetation	Thin grass and moorland; less interception	Forest; most interception. Roots delay thoroughflow and absorb moisture. Evapotranspiration reduces chances of water reaching river
Land use	Urbanisation; increased tarmac (impermeable layer) and drains (increased runoff). Arable land exposes more soil	Rural area with little tarmac, concrete or drains. Tree crops and arable farming increase interception
Use of river	Limited use	Water extracted for industry, domestic use and irrigation. Dam built to store water
Drainage density	Higher density means more streams to collect water quickly	Lower density; fewer streams to collect water

Figure 17.7
Reasons for differences in flood hydrographs for two drainage basins

Extreme weather conditions are often the major cause of a river flooding. A torrential thunderstorm, continuous rainfall for several days, or a heavy snowfall melting while it rains all increase the discharge of a river. Although high summer temperatures increase evapotranspiration and reduce the amount of water available to reach a river, they can make the ground hard, reducing infiltration when it does rain. Freezing conditions in winter can make the ground impermeable.

The River Torridge, Devon

Hydrographs can also be drawn for longer periods of time. The Environment Agency produces hydrographs to show, among other things, if there is a relationship between the two variables of rainfall and discharge. Figure 17.8, which covers a very wet month in late 1992, was provided by the then National Rivers Authority (NRA) – now the Environment Agency – for the River Torridge in North Devon. Notice that:

- discharge is dependent on rainfall, but rainfall does not depend on discharge
- most of the discharge peaks are a half or full day after the rainfall peak (lag time)

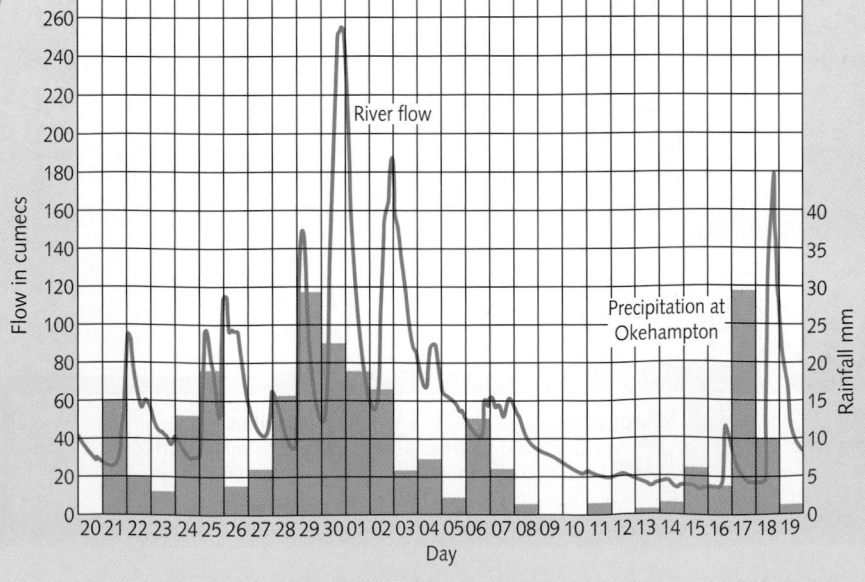

Figure 17.9
A gauging station

- the highest discharge peak came after several very wet days during which river levels had no time to drop, rather than after the wettest day which happened to follow a relatively dry spell when the river level had had time to fall
- the Torridge responds very quickly to heavy rainfall (it flows over impermeable rock) and so would appear to pose a flood risk.

Figure 17.8
Hydrograph for the River Torridge at Torrington

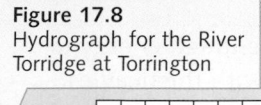

River flow

Precipitation at Okehampton

Flow in cumecs

Rainfall mm

Day

River processes

Energy is needed in any system, not just the drainage basin, for transfers to take place (page 230).

In the case of a river most of this energy, an estimated 95 per cent under normal conditions, is needed to overcome friction. Most friction occurs at the **wetted perimeter**, i.e. where the water comes into contact with the river's banks and bed. The channel of a mountain stream, often filled with boulders, creates much friction (Figure 17.10). As a result, water flows less quickly here than in the lowlands where the channel becomes wider and deeper (Figure 17.11).

Following a period of heavy rain, or after the confluence with a major tributary, the volume of the river will increase. As less water will be in contact with the wetted perimeter, friction will be reduced and the river will increase its velocity. The surplus energy, resulting from the decrease in friction, can now be used to pick up and transport material. The greater the velocity of a river the greater the amount of material, both in quantity and size, that can be carried. The material that is transported by a river is called its **load**.

Transportation

A river can transport its load by one of four processes: **traction** and **saltation**, along its bed; and **suspension** and **solution**, within the river itself (Figure 17.12).

Erosion

A river uses the transported material to erode its banks and bed. As the velocity of a river increases, so too does the load it can carry and the rate at which it can erode. A river may erode by one of four processes:

- **Attrition** is when boulders and other material, which are being transported along the bed of the river, collide and break up into smaller pieces. This is more likely to occur when rivers are still flowing in highland areas.
- **Hydraulic action** is when the sheer force of the river dislodges particles from the river's banks and bed.

Figure 17.10
Rapids at Aberglaslyn, North Wales

Figure 17.11
Velocity, discharge and cross-section profile in the upper and lower course of a river

(a) An upland stream
5 metres
Uneven river banks
Angular boulders on river bed

Despite waterfalls where the velocity is locally high, the large number of angular boulders and uneven banks give a large wetted perimeter which increases friction and reduces velocity.

(b) A lowland river
50 metres
Smooth banks
Small rounded pebbles

Here there is a relatively smaller wetted perimeter in comparison with the volume. Due to this, and the smooth banks and bedload, friction is reduced, allowing velocity to increase.

- **Corrasion** occurs when smaller material, carried in suspension, rubs against the banks of the river. This process is more likely in lowland areas by which time material will have been broken up small enough to be carried in suspension. River banks are worn away by a sand-papering action which is also called **abrasion**.
- **Corrosion** is when acids in the river dissolve rocks, such as limestone, which form the banks and bed. This can occur at any point of the river's course.

Deposition

Deposition occurs when a river lacks enough energy to carry its load. Deposition, beginning with the heaviest material first, can occur following a dry spell when the discharge and velocity of the river drop, or where the current slows down (the inside of a meander bend or where the river enters the sea).

Figure 17.12
Processes of transportation

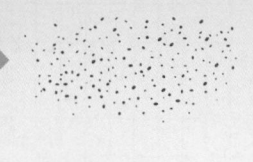

River flow

| **Traction** Rolling stones along the bed. (This needs the most energy.) | **Saltation** Sand-sized particles bounce along the bed in a 'leap-frog' movement. | **Suspension** Silt and clay-sized particles are carried within the water flow. | **Solution** Some minerals dissolve in the water. (This needs the least energy.) |

River landforms in a highland area

Figure 17.13
V-shaped valley and interlocking spurs: the Peak District

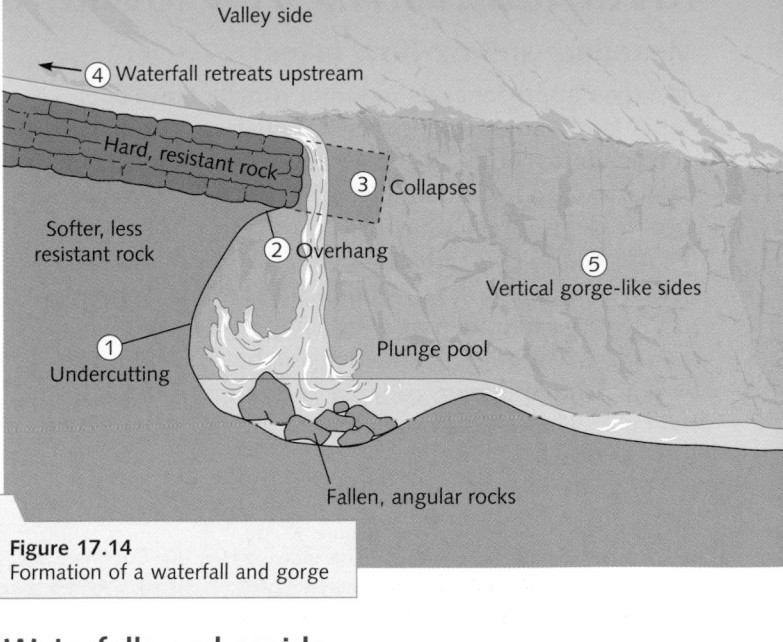

Valley side

④ Waterfall retreats upstream

Hard, resistant rock

③ Collapses

Softer, less
resistant rock

② Overhang

⑤ Vertical gorge-like sides

① Undercutting

Plunge pool

Fallen, angular rocks

Figure 17.14
Formation of a waterfall and gorge

V-shaped valleys and interlocking spurs

Any spare energy possessed by a river near to its source will be used to transport large boulders along its bed. This results in the river cutting rapidly downwards, a process called **vertical erosion**. Vertical erosion leads to the development of steep-sided, narrow valleys shaped like the letter V (Figure 17.13). The valley sides are steep due to soil and loose rock being washed downhill following periods of heavy rainfall. The material is then added to the load of the river. The river itself is forced to wind its way around protruding hillsides. These hillsides, known as **interlocking spurs**, restrict the view up or down the valley.

Waterfalls and rapids

Waterfalls form when there is a sudden interruption in the course of a river. They may result from erosion by ice (Figure 19.9), changes in sea-level (Figure 18.12), and earth movements. However, many waterfalls form when rivers meet a band of softer, less resistant rock after flowing over a relatively hard, resistant rock (Figure 17.14). The underlying softer rock is worn away more quickly, and the harder rock is undercut. In time the overlying harder rock will become unsupported and will collapse. After its collapse, some of the rock will be swirled around by the river, especially during times of high discharge, to form a deep plunge pool. This process is likely to be repeated many times, causing the waterfall to retreat upstream and leaving a **steep-sided gorge** (Figure 17.15). The Niagara Falls are retreating by one metre a year. **Rapids** occur where the layers of hard and soft rock are very thin, and so no obvious break of slope develops as in a waterfall.

Figure 17.15
Niagara Falls

1 American Falls
2 Horseshoe Falls
3 Niagara River and gorge
4 Goat Island
5 Canada
6 USA
7 Plunge pool
8 Resistant rock
9 Less resistant rock

River landforms in a lowland area

Meanders and ox-bow lakes

As a river approaches its mouth it usually flows over flatter land and develops increasingly large bends known as **meanders** (Figure 17.16). Meanders constantly change their shape and position. When a river reaches a meander most water is directed towards the outside of the bend (Figure 17.17). This reduces friction and increases the velocity of the river at this point. The river therefore has more energy to transport material in suspension. This material will erode the outside bank by corrasion. The bank will be undercut, collapse and retreat to leave a small river cliff. The river is now eroding through **lateral**, not vertical, **erosion**.

Meanwhile, as there is less water on the inside of the bend, there is also an increase in friction and a decrease in velocity. As the river loses energy it begins to deposit some of its load. The deposited material builds up to form a gently sloping slip-off slope (Figure 17.18).

Continual erosion on the outside bends results in the neck of the meander getting narrower until, usually at a time of flood, the river cuts through the neck and shortens its course. The fastest current will now be flowing in the centre of the channel and deposition is more likely next to the banks. The original meander will be blocked off to leave a crescent-shaped **ox-bow lake**. This lake will slowly dry up, except during periods of heavy rain.

Figure 17.16
A meandering river: the Cuckmere in East Sussex

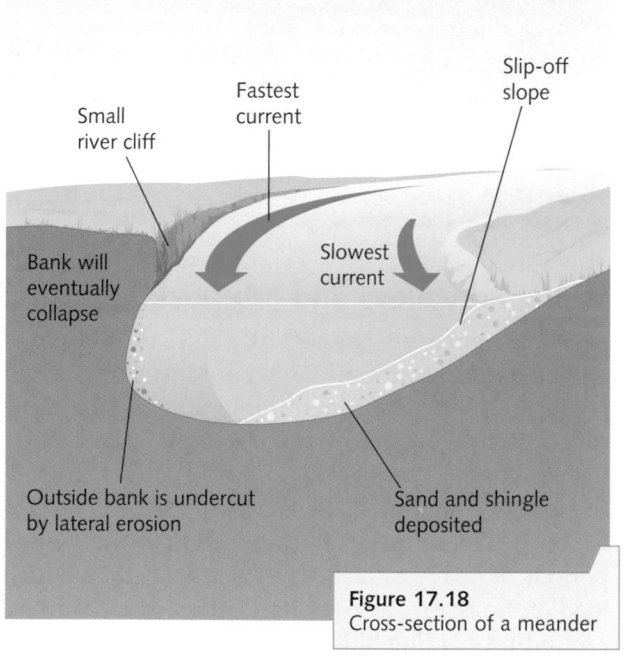

Figure 17.18
Cross-section of a meander

Small river cliff

Fastest current

Slip-off slope

Bank will eventually collapse

Slowest current

Outside bank is undercut by lateral erosion

Sand and shingle deposited

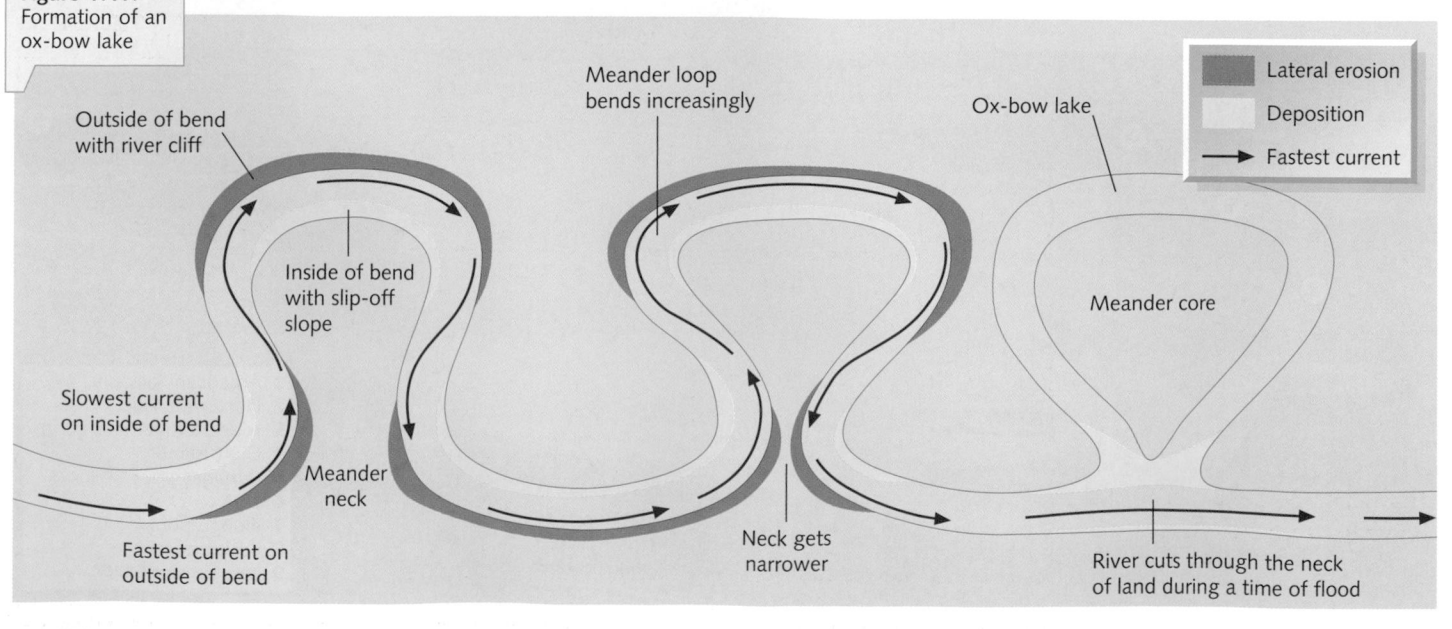

Figure 17.17
Formation of an ox-bow lake

Outside of bend with river cliff

Inside of bend with slip-off slope

Slowest current on inside of bend

Fastest current on outside of bend

Meander neck

Meander loop bends increasingly

Neck gets narrower

River cuts through the neck of land during a time of flood

Ox-bow lake

Meander core

Lateral erosion

Deposition

Fastest current

Floodplain and levées

The river widens its valley by lateral erosion. At times of high discharge, the river has considerable amounts of energy which it uses to transport large amounts of material in suspension. When the river overflows its banks it will spread out across any surrounding flat land. The sudden increase in friction will reduce the water's velocity and fine silt will be deposited. Each time the river floods another layer of silt is added and a flat **floodplain** is formed (Figure 17.19). The coarsest material will be dropped first and this can form a natural embankment, called a **levée**, next to the river. Sometimes levées are artificially strengthened to act as flood banks. Some rivers, like the Mississippi, flow between levées at a height well above their floodplain. If, during a later flood, the river breaks through its levées, then widespread flooding may occur.

Figure 17.19
Cross-section of a river floodplain

Labels: Gentle valley side; River flowing above height of floodplain; Gentle valley side; Flat floodplain; Levée artificially heightened and strengthened; Coarse material forms natural levées; Layers of silt deposited during several floods

Deltas

As large rivers approach the sea, they have the energy to carry huge amounts of fine material in suspension. On reaching the sea, the river current may suddenly be reduced, allowing the material to be deposited. Sometimes deposition occurs in the main channel, and blocks it. The river has then to divide into a series of smaller channels, called **distributaries**, in order to reach the sea. Over a period of time the deposited material of sand and silt may build upwards and outwards to form a **delta** (Figure 17.20). Deltas are only likely to form where the amount of material brought down by a river is too great for sea currents to remove it (e.g. Mississippi and Ganges) or in seas that are virtually tideless (e.g. Nile and Rhône in the Mediterranean). Deltas can also form when a river flows into the gentle waters of a lake.

Figure 17.20
The Mississippi delta

Labels: Material deposited; Main channel of Mississippi River; Earlier deposits now built up to sea-level; very marshy; Sand and silt; River divides into numerous distributaries; Edge of land; Gulf of Mexico

Changes in the channel and valley of a river from source to mouth

No two rivers are alike but most show similar changes from upland areas where they probably have their source, to lowland areas where they approach the sea. Figure 17.21 gives the likely changes in the **cross-section** (cross-profile) and **long-section** (long-profile) of a river, in the river's channel and discharge, and the valley shape.

Figure 17.21
Changes in a river from source to mouth

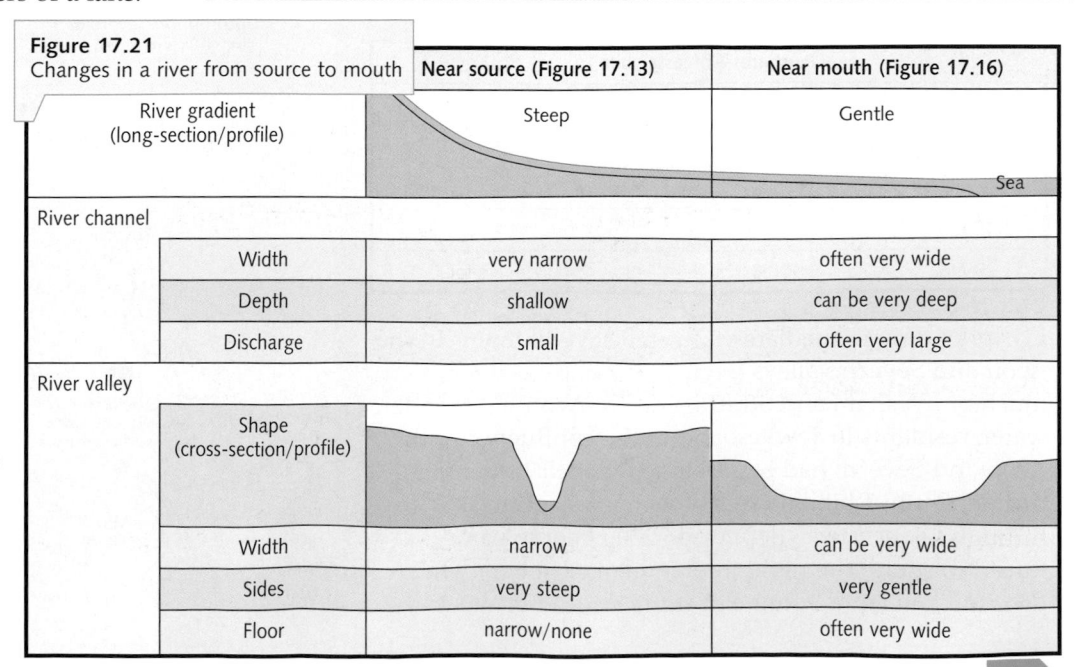

		Near source (Figure 17.13)	Near mouth (Figure 17.16)
River gradient (long-section/profile)		Steep	Gentle / Sea
River channel	Width	very narrow	often very wide
	Depth	shallow	can be very deep
	Discharge	small	often very large
River valley	Shape (cross-section/profile)		
	Width	narrow	can be very wide
	Sides	very steep	very gentle
	Floor	narrow/none	often very wide

River flooding in an MEDC

Rivers throughout the world provide an attraction for human settlement and economic development. They provide a water supply for domestic, industrial and agricultural use; a source of power, food and recreation; and a means of transport. However, under extreme climatic conditions, and increasingly due to human mismanagement, rivers can flood and cause death and widespread damage. The effects of flooding can be equally severe in MEDCs and LEDCs although the former are often better equipped to respond to specific events.

PLACES

English Midlands

For many parts of England and Wales, 2007 was the wettest year, and certainly the wettest summer, ever recorded. The main reason was that depressions, which usually travel to the north of the British Isles in summer, on this occasion frequently took the more southerly route normally associated with winter. These depressions brought with them westerly winds (page 204) which, as they passed over the Atlantic Ocean which was much warmer at this time of year, were able to collect more moisture and, on reaching the British Isles, to give heavier rainfall. Torrential rain during June had caused severe flooding in Hull, Doncaster and Sheffield before a very slow-moving depression crossed the English Midlands during 19–20 July (Figure 17.22) where many weather stations in this region were to record their highest ever daily rainfall.

Although forecasters had warned of heavy rain for up to a week beforehand and the Met Office had issued a severe weather warning two days in advance, no one quite expected the downpour of 20 July. Two months of rain fell in 2 hours, and three times July's normal total in 24 hours, on soil that was already saturated and where many rivers were already close to overflowing their banks. Pershore, in Worcestershire, recorded over 120 mm in the one day (Figure 17.23).

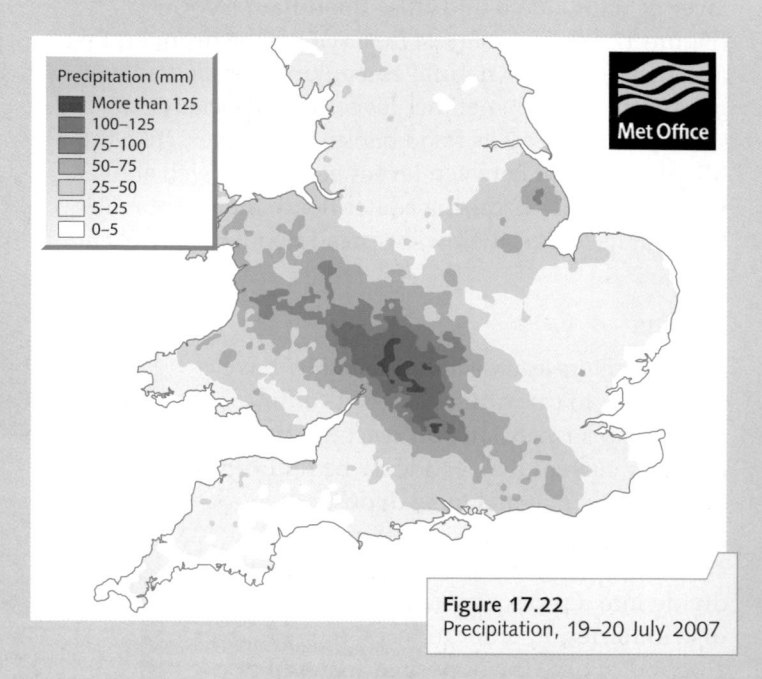

Figure 17.22
Precipitation, 19–20 July 2007

Figure 17.23
Data for Pershore, Worcestershire

Pershore, Worcestershire		
July average	July 2007	% of average
46.6 mm	252.4 mm	541
Daily rainfall		
19 July	36.6 mm	
20 July	120.8 mm	(157.4 mm total)

Flash flooding immediately affected several towns in the Avon and Severn Valleys (Figure 17.24). By evening much of Evesham and Stratford-upon-Avon were under water, residents in Tewkesbury, at the confluence of the Avon and Severn, had begun to leave their homes and, further south, 1 billion litres of water was pouring through Gloucester where up to 2000 people were forced to spend the night in emergency shelters. The flooding and extra volume of traffic caused gridlock on

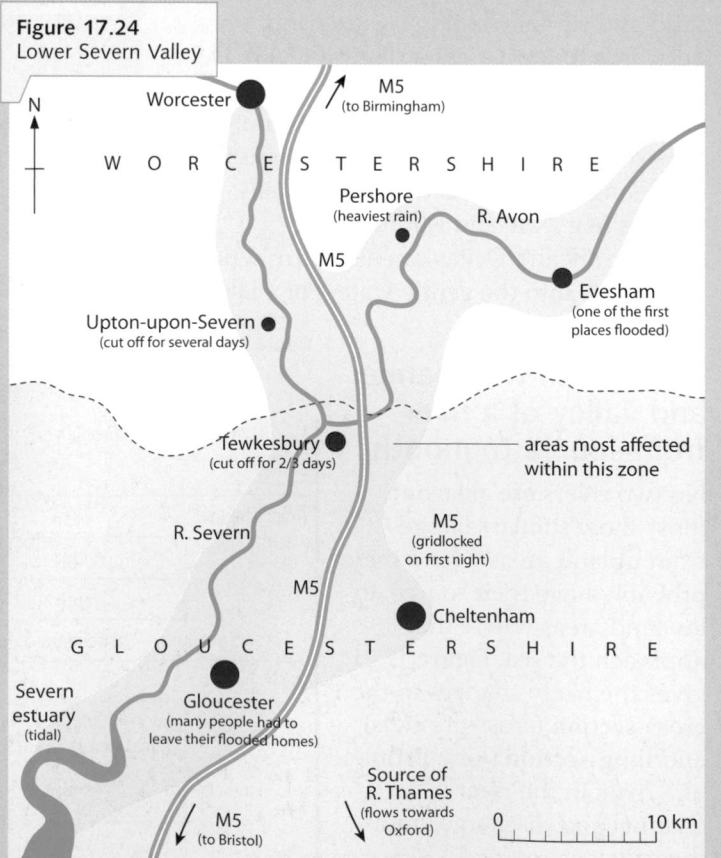

Figure 17.24
Lower Severn Valley

Figure 17.25
Traffic on a flooded road near Gloucester

Figure 17.27
Flooding in Gloucester

major roads (Figure 17.25), with an estimated 10 000 motorists left stranded for up to 10 hours on the M5 between Worcester and Gloucester. This gridlock prevented emergency services moving equipment, such as portable steel flood barriers, to places like Upton-on-Severn threatened by flooding, and hampered their attempts to rescue people already trapped. The result was the largest deployment of helicopters and the biggest ever peacetime emergency in the UK.

Further rain and increased runoff made the situation even worse. All 75 000 residents of Tewkesbury were cut off (Figure 17.26) and a nearby water treatment works was forced to close, leaving 350 000 people without water for washing, cooking or sewerage. Further south, floodwater began seeping into an electricity sub-power station, threatening to cut off supplies to 600 000 homes –

a disaster narrowly averted by the military being called in to help construct a 1 km embankment around the station (Figure 17.27). This was achieved hours before a high tide at nearby Gloucester caused the level of the Severn to peak at almost 8 metres above its usual level. Meanwhile, further rain was causing floodwater to affect parts of the Upper Thames Basin.

Three days later, much of Gloucestershire was still without water and many residents were receiving supplies from freshwater tankers. While those without water were told they might remain in this situation for two weeks, residents who had been forced to evacuate their flooded homes were warned it might be over a year before they could return. Meanwhile it was now people living close to the Thames in Oxfordshire who were facing the real threat of flooding (Figure 17.24). Although floodwater was now receding from most places in the Severn Valley and mopping-up operations were beginning, the clean-up was predicted to take several months and to cost over £2 billion.

Figure 17.26
Tewkesbury at the confluence of the flooded Avon and Severn

River flooding in an LEDC

The effects of a major river flood may be as severe in an MEDC as in an LEDC. However, relief operations in an LEDC, where available capital and technology are limited, is more likely to be slower and less effective. It also takes an LEDC much longer to recover from the event.

Bangladesh

Flooding is an annual event in Bangladesh. The monsoon rains cause rivers such as the Jamuna (Brahmaputra) and Padma (Ganges) to overflow their banks between July and mid-August (Figure 17.28). Most of Bangladesh's 153 million inhabitants live on the floodplains of these rivers (Figure 17.29). For most of them, the seasonal flood is essential for their survival as it brings water in which to grow the main crops of rice and jute, as well as silt to fertilise their fields. Flooding of 20 per cent of the country is considered beneficial for crops and the ecological balance (Figure 17.30). However, a figure much less than that can result in food shortages, while an inundation much in excess can cause considerable loss of life, ruin crops and seriously damage property. In 1998, 68 per cent of the country was flooded, some places for more than 70 days. This flood was unprecedented in terms of both its magnitude and duration. Apart from destroying crops and basic infrastructural features such as roads and bridges, it caused the deaths of over 1000 people, destroyed 7 million homes and left more than 25 million people homeless (Figures 17.29 and 17.31).

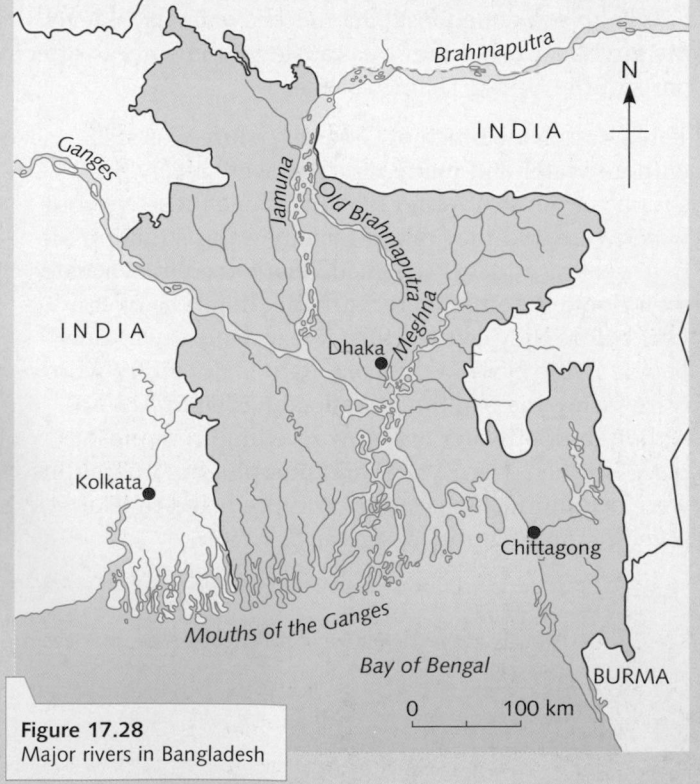

Figure 17.28
Major rivers in Bangladesh

Figure 17.29
Floodwaters covering low-lying areas of the country

May and June There was very little rainfall in the drainage basin area, 92 per cent of which lies beyond the national border of Bangladesh.

July Heavy rainfall at the beginning of the month caused the Brahmaputra, Meghna and Ganges rivers to rise rapidly. When the rivers reached a peak on the 28th (Figure 17.32), 30 per cent of the country was under water.

August The flood situation slowly improved until the 20th when most rivers again exceeded their danger levels. A second flood peak, reached on the 30th, saw floodwaters cover over 40 per cent of the country (Figure 17.32).

September Further heavy rain, unusually late in the year, caused the rivers to reach a third, and the highest, peak on the 7th. By that time almost 70 per cent of the country was, or had been, under water. Although the situation improved after that date, it was the end of the month before the floodwaters drained away from many areas (Figures 17.32).

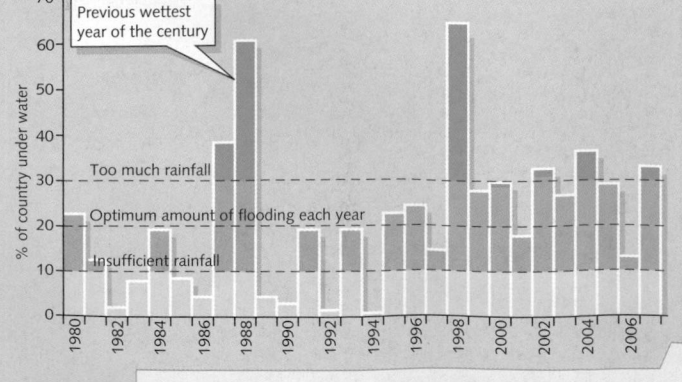

Figure 17.30
Proportion of Bangladesh inundated by the annual flood

Figure 17.31
Effect of the 1998 flood

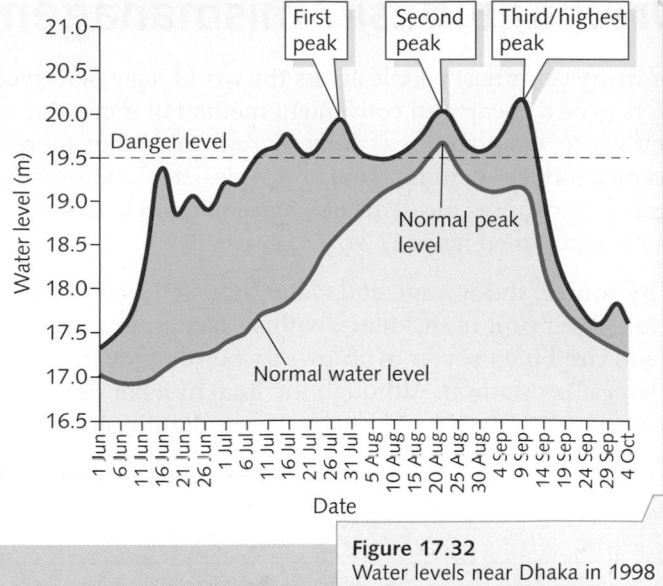

Figure 17.32
Water levels near Dhaka in 1998

Effects

At its peak, the flood was so deep that in some places only the tops of trees and buildings could be seen. Railways, roads and bridges had been swept away. Most parts of the country were without electricity for several weeks and, because floodwater polluted wells, there was no safe drinking water. Hospitals were already full of people suffering from dysentery and diarrhoea, and the threat of disease, especially cholera, was increasing. Parts of Dhaka, including the international airport, were under 2 metres of water. Delivering overseas aid, including food and medical supplies, became almost impossible. On 26 August, two weeks before the third peak and a month before the floodwaters finally receded, urgent supplies were requested.

Causes

- Much of Bangladesh consists of the floodplain of the Ganges and Brahmaputra rivers. Over half the country lies below 6 metres above sea-level. Once the rivers overflow their banks or breach the protective floodbanks (levées), the water can spread over vast distances, inundating roads, railways, farmland and settlements.
- Bangladesh has a monsoon climate (pages 112 and 214), with most places receiving between 1800 and 2600 mm of rain a year (the average for London is about 600 mm). However, 80 per cent of that total is concentrated in just four or five months (June to September).
- The period of heavy rain coincides with that of the highest temperatures. These temperatures melt ice and snow in the Himalayas where the Ganges, Brahmaputra and their tributaries have their headwaters, adding to the discharge of the rivers.
- Global warming (page 219) is causing glaciers in the Himalayas to melt (increasing runoff) and sea-level in the Bay of Bengal to rise.

- The effect of tropical storms (page 216) and associated storm surges in the Bay of Bengal (Case Study 18B) can prevent floodwater from rivers escaping into the sea.
- Silt is brought down by the rivers and deposited within their channels. This, together with further silt that blocks irrigation channels, results in the heightening of the river beds, increasing the risk of flooding.
- Human mismanagement has increased the magnitude and frequency of flooding by:
  - building on the floodplains to house the country's growing population (the process of urbanisation, Figure 17.7)
  - cutting down trees in the upper drainage basin (e.g. Nepal) – deforestation increases the rate and amount of runoff (Figure 17.7).

Flooding since 1998

Since the droughts of the 1970s and 1980s, the years with less than sufficient rainfall have decreased dramatically in number, whereas those with above average rainfall since 1998 have increased significantly (Figure 17.30). However, although in:

- 2002, 7 people died and 1000 were left homeless
- 2004, 70 people died, 3.8 million were affected and half of Dhaka was flooded
- 2007, over half a million people were affected

none of these events proved to be as extreme or as disastrous as the one that occurred in 1998.

Figure 17.33
Recent floods in Bangladesh

Drainage basin mismanagement

For many centuries, people across the world have perceived rivers to be a cheap and convenient method of removing their waste, whether it was domestic, agricultural or, more recently, industrial. In England and Wales, the Environment Agency categorises incidents of water pollution by both source and type (Figure 17.34).

- **By source**, the sewage and water industry accounted for 30 per cent of incidents, with surface water outfall and combined sewer overflows accounting for the largest proportion. Although the total number of substantiated incidents has increased slightly since the 1990s, mainly in agriculture (e.g. fertiliser and farm slurry – page 108), the number of major incidents has decreased considerably. This decrease in major incidents has been credited partly to the efforts of the Environment Agency and partly to increased awareness by the general public in reporting incidents. Of concern, however, has been the increase in major incidents relating to road accidents.

- **By type**, sewage and oil accounted for nearly half of the substantiated cases (Figure 17.34b), with oil and chemicals being the cause of most major incidents.

(a) Sources		Number of major incidents	(b) Type	Number of major incidents
Agriculture	• Fertiliser and pesticides • Farm slurry, silage	32	Organic wastes	28
Industry	• 'Warm' water discharge after cooling • Toxic wastes	62	Chemicals	48
Mining	• Minerals/chemicals from disused mines	8		
Transport	• Oil/petrol spillages	16	Oil	45
People	• Untreated sewage • Rubbish – plastic, polythene, tins	48	Sewage	37
Others	• Poisonous substances, acid rain	23	Others	24

Figure 17.34
Recent water pollution incidents in England and Wales by source and by type

Figure 17.35 shows that in the USA the main causes were factories and cities releasing untreated waste into lakes and rivers, and runoff from farms, construction sites, cities and mines.

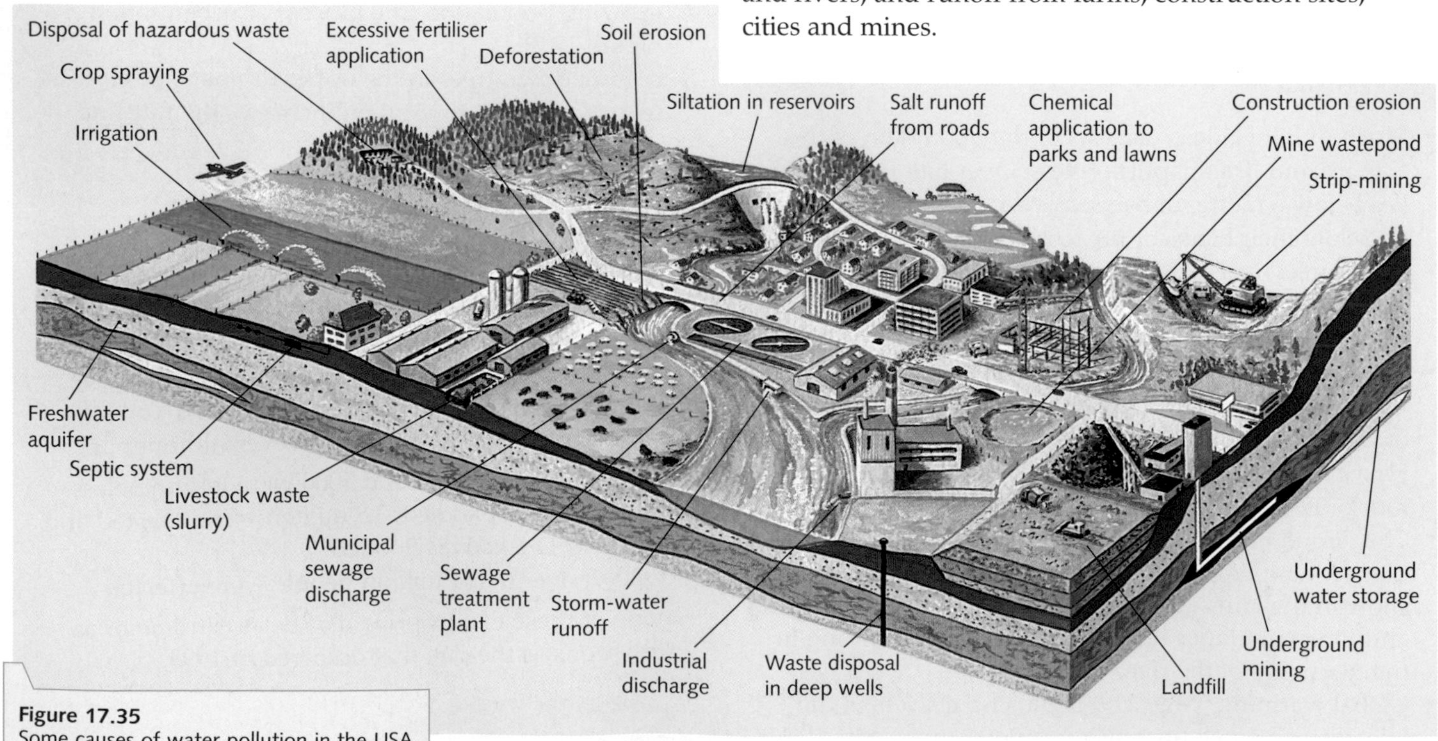

Figure 17.35
Some causes of water pollution in the USA

Drainage basin management

Should rivers be allowed to flood?

Some rivers, like the Mississippi in the USA, the Po in Italy and the Huang He in China, have built up their beds so that they flow above the level of their floodplains. Such rivers can only be prevented from flooding by building dams, diversion channels and artificial embankments (i.e. levées – Figure 17.19). The problem arises when a serious flood risk occurs. If the levées break, or if they have not been built high enough, then large areas of land are flooded and numerous lives are put at risk. Following the Mississippi floods of 1993, scientists and engineers began asking whether the dams, diversion channels and levées had actually aggravated that and other floods. There are two schools of thought. One accepts that as floods are part of a river's natural cycle and the drainage basin's ecosystem, then flooding should be allowed as a natural event (Figure 17.36). The other argues for better flood defences and a more effective control of rivers (Figure 17.37).

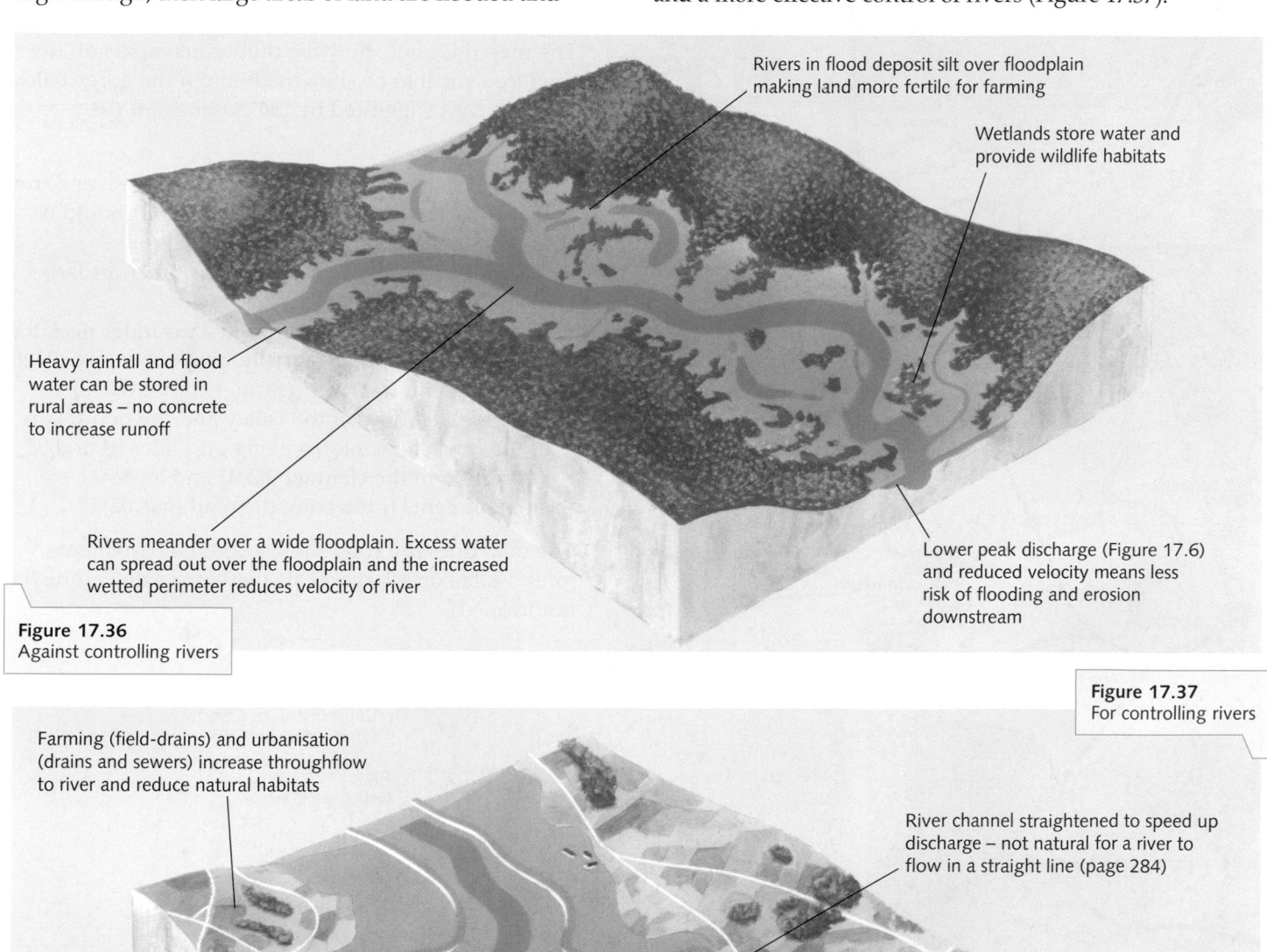

Rivers in flood deposit silt over floodplain making land more fertile for farming

Wetlands store water and provide wildlife habitats

Heavy rainfall and flood water can be stored in rural areas – no concrete to increase runoff

Rivers meander over a wide floodplain. Excess water can spread out over the floodplain and the increased wetted perimeter reduces velocity of river

Lower peak discharge (Figure 17.6) and reduced velocity means less risk of flooding and erosion downstream

Figure 17.36
Against controlling rivers

Figure 17.37
For controlling rivers

Farming (field-drains) and urbanisation (drains and sewers) increase throughflow to river and reduce natural habitats

River channel straightened to speed up discharge – not natural for a river to flow in a straight line (page 284)

If levées built too close to the river, there is no room for excess water

River channel between levées builds up above the level of the floodplain

Much expense and environmental degradation involved in forcing a river to flow in a particular course

Breached levée

Higher peak discharge (Figure 17.6) and increased velocity means greater risk of flooding and erosion downstream

River landforms on OS maps

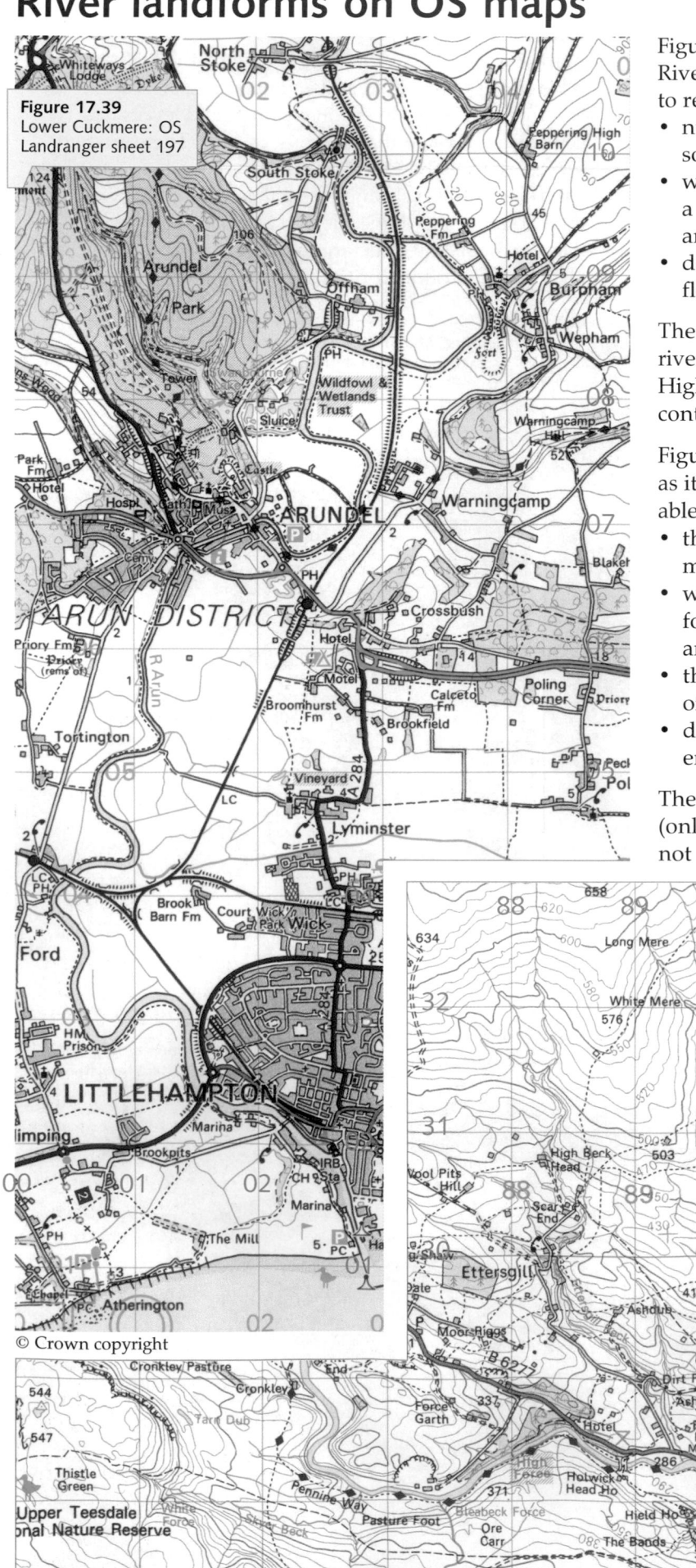

Figure 17.39
Lower Cuckmere: OS Landranger sheet 197

© Crown copyright

Figure 17.38 shows part of the upper course of the River Tees in the northern Pennines. You should be able to recognise:

- numerous small tributary streams, each with its source and narrow channel (8932)
- waterfalls large enough to be named (9028) – *force* is a Scandinavian word for a waterfall (High Force 8828 and Low Force 9027)
- deep, steep-sided V-shaped valleys with little or no floodplain (8831 and 8628).

The map does not show the numerous rapids on the river (too small to be shown) although the gorge below High Force is suggested by the closeness of the contours.

Figure 17.39 shows the lower course of the River Arun as it approaches the English Channel. You should be able to recognise:

- the wide channel of the river and numerous large meanders (0004)
- where the river has cut through a meander neck to form a cut-off and, eventually, an ox-bow lake (0308 and 0210) and the river's former course (0210)
- the wide, flat floor of the valley and the floodplain of the river (no contours along grid lines 04 and 05
- deposition in the channel (0201) and levées/ embankments (protecting the road in 0102).

The map does not fully show levées/embankments (only visible on a larger-scale map) nor a delta (one has not formed).

Figure 17.38
Upper Teesdale: OS Landranger sheet 92

© Crown copyright

Flooding on the Yangtze – management and mismanagement

Natural flooding

The Yangtze rises on the Tibetan Plateau and flows 6380 km before entering the East China Sea just north of Shanghai. Although it has always been prone to flooding, both the frequency and intensity of floods have increased in recent years. During the same time there has been a rapid growth in both population size and in human activity on the river's floodplain, which has led to an increase in the adverse effects of flooding. Some recent floods are described in Figure 17.40.

1931 – 140 000 deaths and 40 million made homeless.
1954, 1981 and 1991 – large areas inundated.
1995 – 1300 deaths and 1000 million homeless.
1998 – worst flood for over half a century with the river, in places, reaching its highest ever level – a level that was maintained for over two months. There were an estimated 3000 deaths, 5.6 million homes washed away, 30 million people made homeless and huge areas of crops destroyed. The river rose steadily during June and by early July, 1000 soldiers and police were working around the clock for seven days a week, strengthening floodbanks (levées) between Yichang and Wuhan (Figure 17.42). In just *one* day, Wuhan received about half of London's *annual* total, and the next day the city witnessed the first of several flood peaks that it was to experience over the next two months. Later – by which time several smaller cities had been submerged for over a month – the Chinese authorities admitted that they were allowing smaller towns to flood in order to protect larger cities. As one official said, 'By diverting the Yangtze some villages were destroyed but the very real threat to Wuhan's 7 million residents was averted'.
1999 – 240 deaths and 2 million evacuated.
2002 – Wuhan again under threat.

Figure 17.40
The flood hazard

Causes of flooding

- Large areas of the Yangtze drainage basin receive high annual amounts of rainfall. Most of this is seasonal, falling in intensive storms during the summer monsoon (page 214).
- The rains arrive as temperatures rise. One result of this is the rapid melting of snow and glaciers where the Yangtze has its source.
- Large areas of the upper basin were deforested in the 1950s and 1960s during Chairman Mao's so-called 'Great Leap Forward'. Estimates suggest that one-third of the area was cleared between then and 1988. Deforestation (page 236) reduces interception and increases runoff and soil erosion – the latter causing the bed of the river to rise even higher above its floodplain near Wuhan (Figure 17.19).
- The building of houses, factories, roads and railways on the floodplain (urbanisation) has reduced the area of land that once could be safely allowed to flood.

Figure 17.41
Scenery within the Three Gorges

Flood control – the Three Gorges Dam

The 50 000 km² area of land between Yichang and Wuhan was only protected by a 180 km long, 12–16 metre high levée. This land, because it was flat and consisted of fertile silt deposited by Yangtze floods, had become settled by over 15 million people. Should the levée have broken, or been overtopped, the result would have been a huge economic disaster and an unimaginable loss of human life. For several decades, many Chinese had been convinced that the only way to reduce the flood risk was to build a dam across the river. This dam has now been built – against much international and some internal advice – at Sandouping where the Yangtze leaves its scenic Three Gorges section (Figures 17.41 and 17.42). Completed in 2003 as part of the Three Gorges Project (TGP), the dam is 2309 metres long and 185 metres high.

Figure 17.42
Location of the Three Gorges Project

	Area to be flooded
●	Large town or city
▬	Three Gorges Project
◯	Gorge
A	Upstream to Yunnan and source of river
B	Downriver to Wuhan and Shanghai – flood-risk area

0   1000 km

Beijing
Yangtze
Yichang Shanghai
Wuhan
Chongqing

N

Wanxian

0   50 km

Yangtze River

Chongqing

Fengdu
(Figure 17.47)

Qutang Gorge 13 km
Wu Gorge 43 km

Xiling Gorge 75 km
Dam (Figure 17.43)

Yichang
Sandouping

A

B

The Three Gorges Project (TGP)

Deliberate flooding

One benefit of the dam will be its ability to store water from a flood whose size is not expected to occur more than once in every 100 years (the previous biggest flood before 1998 occurred in 1881). By controlling the summer's flood and releasing the water gradually during the dry season, millions of people living downstream in places like Wuhan now have the risk of **natural flooding** greatly reduced. However, it has also meant that, for the thousands of people who lived upriver, their homes and land have become permanently submerged under the newly created lake – **deliberate flooding**.

The TGP, however, does more than control flooding. It is a multipurpose project which includes the world's largest hydro-electricity scheme, generating vast amounts of a clean, renewable form of energy. Also, by creating a 600 km lake, navigation is improved upriver as far as Chongqing. The lake is held back by a dam (Figure 17.46), built at a point where a resistant band of granite crosses the valley just after the Yangtze has passed through three spectacular limestone gorges. At the dam are 34 electricity generators, a pair of ship-locks and a ship-lift (Figure 17.43).

However, the TGP, the scale of which has never previously been attempted, generated massive controversy, mainly due to the known effects on local people and the unknown effects of interference with the natural functioning of a major river and the surrounding environment. Serious doubts remain as to the TGP's long-term economic, social and ecological advantages and costs (Figures 17.44–17.48).

Figure 17.44
Fengdu in 1999 showing usual low water and flood levels and the 1998 record level, together with the projected lake level in 2009

180
175 — Lake level by 2009
Height of multi-storey flats
164
1998 record flood level
152
Usual annual flood level
149
Usual annual low-water level
140
Metres

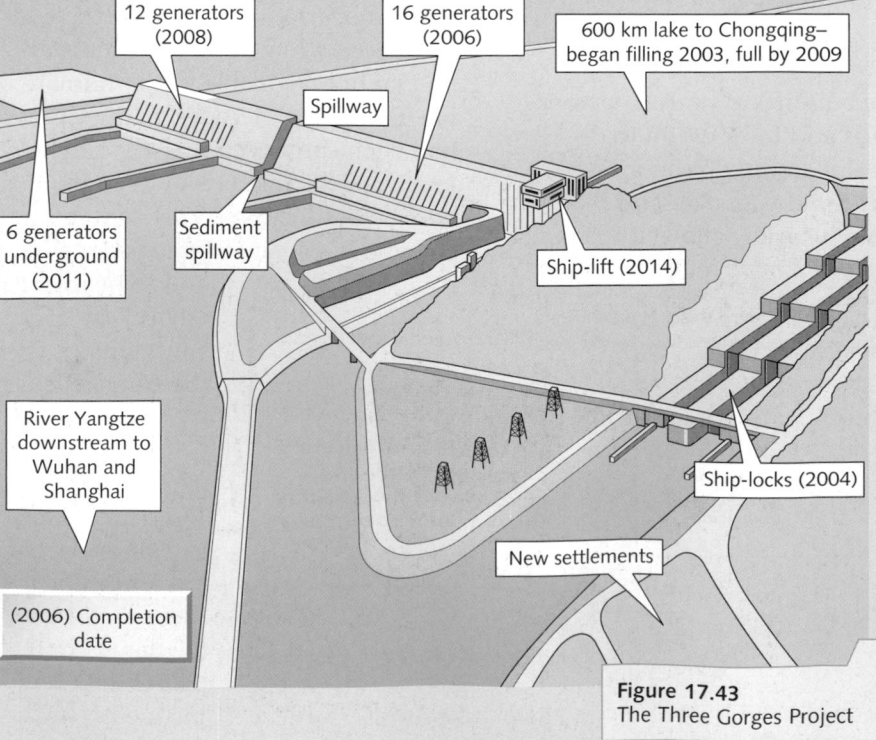

12 generators (2008)
16 generators (2006)
600 km lake to Chongqing– began filling 2003, full by 2009
Spillway
6 generators underground (2011)
Sediment spillway
Ship-lift (2014)
River Yangtze downstream to Wuhan and Shanghai
Ship-locks (2004)
New settlements
(2006) Completion date

Figure 17.43
The Three Gorges Project

Figure 17.45
A limestone gorge in 1999 showing low water and flood levels and existing housing, together with new settlements built at the 2009 predicted lake level

Figure 17.46
The Three Gorges Dam

Advantages of the TGP

- The main object was to protect up to 100 million people living downriver who had experienced considerable loss of life and damage to property in an area that was frequently flooded during the 1990s and 2002 (Figure 17.40). The dam, by creating a large lake behind, now prevents the seasonal variation in the level of flow (up to 8 metres between summer and winter) and controls discharge by releasing floodwater when river levels are low.
- The second advantage is the generation of hydro-electricity from the 34 generators at the dam. Apart from the need for electricity, if China is to develop its industries and improve its people's standard of living, hydro-electricity is a clean fuel and a renewable resource. Before 2000 most of China's energy came from coal (it uses 40 per cent of the world's total) – a major cause of both global warming and acid rain. By reducing coal consumption, the TGP should mean less air pollution and risk to people's health.
- The Yangtze is the main transport route inland from the coast. Before the dam was built, ships of up to only 1500 tonnes could reach Chongqing. This was because the river flowed too fast during the summer flood and was too low, exposing rocks, in the dry season. The 600 km lake now allows safe navigation throughout the year for vessels up to 10 000 tonnes. Since 2004 these have been able to pass through the dam by ship-locks and, after 2014, will be able to use the ship-lift. Tourism should benefit, as cruise ships will be able to sail at all times.
- The many new settlements that have been created (Figure 17.45) have improved housing (with electricity, running water and, in time, sewerage), transport links and services.
- Influenced by the lake, the winter climate should become milder and wetter, increasing the length of the growing season.

Figure 17.47
Fengdu as it was in 1999 – the city is now totally submerged

Figure 17.48
Rubbish and sewage at Yichang

Disadvantages of the TGP

- The TGP has meant the relocation of 1.3 million people whose homes were drowned as the lake water rose, submerging, in total, 4 cities, 8 towns and 356 villages (Figure 17.47). People forced to leave their homes were given little choice as to where they could resettle. Although each family was given some compensation towards moving and rebuilding costs, it was often insufficient to pay for the modern, but more expensive, housing. As the best sites for settlement had already been used or were to be flooded, many people (despite promises to the contrary) had to move to more distant provinces. The lake also flooded many sacred temples and historic sites as well as the best farmland, and thousands of small factories located alongside the river, giving rise to fears that these, once submerged, might release toxic waste and poison the lake's water.
- The vast amount of silt which is transported each year by the Yangtze will now be trapped behind the dam. Although the dam has been designed with special sluices to allow sediment to pass downriver, there are fears that:
  - a delta will form at the head of the lake, eventually blocking the port of Chongqing
  - as the lake fills with silt, its storage capacity will be reduced
  - reduced sediment downriver will harm agriculture as farmers rely on fertile silt to improve their fields.
- Before the dam, sewage from settlements alongside the Yangtze was allowed to flow downriver and out to sea. Presumably, until treatment works are built, this and other rubbish will be trapped behind the dam (Figure 17.48).
- There is concern that the dam could break due either to the sheer weight of water, to earthquakes, bad construction or terrorist activity.
- Tourism might be affected if the scenic quality of the Three Gorges is lost (Figure 17.41).
- Endangered species such as the Yangtze sturgeon and river dolphin may become extinct.
- Landslides are reported to be occurring along the lakeside.

Drainage basins and rivers

Key Words and Terms

a You should know the meaning of the following terms:
- drainage basin • watershed • drainage density • hydrological (water) cycle
- storm/flood hydrograph • wetted perimeter • lag time • load
- V-shaped valley • interlocking spurs • waterfall with gorge • meander
- ox-bow lake • floodplain • levées • delta.

b You should know the difference between:
- inputs, stores, transfers and outputs in a system
- permeable and impermeable rocks
- velocity, volume and discharge of a river
- rainfall peak and peak discharge
- vertical and lateral erosion.

Key Ideas

You should know and understand the following:
- The hydrological cycle and the drainage basin system.
- That the Earth's surface is modified by fluvial/river processes that result in distinctive landforms.
- The processes of river erosion – attrition, hydraulic action, corrasion/abrasion and corrosion/solution.
- The processes by which a river transports material – traction, saltation, suspension and solution.
- The conditions under which a river deposits material.
- The processes that lead to the formation of distinctive river landforms – V-shaped valley, interlocking spurs, waterfall with gorge, meanders, ox-bow lake, floodplain, levées and deltas.
- Changes in the long-profile and the cross-profile of a river.
- Changes downriver/downvalley in channel/valley width, discharge, gradient, cross-section.
- Causes and effects (primary and secondary) of a river flood in an MEDC and an LEDC.
- That river flooding illustrates the interaction between natural/physical processes and human activity.
- Attempts to control/manage (short-term/long-term) flooding in an MEDC and an LEDC.
- The need for sustainable management of river flooding.
- Recognition of river landforms on an OS map.

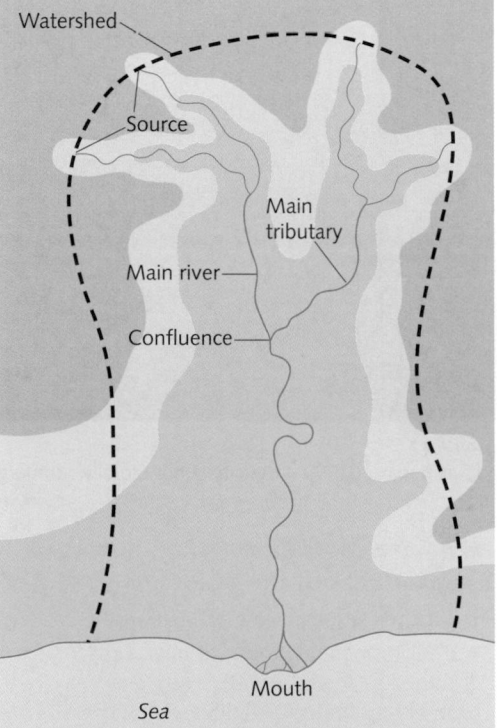

Skills, Theories and Models

- Interpret and use diagrams and maps
- Analyse and use photographs
- Interpret and use OS maps
- Interpret and use flood hydrographs
- Compare the effects of flooding in MEDCs and LEDCs
- Investigate the impact of resource development
- Understand the hydrological (water) cycle
- Understand the drainage basin system
- Appreciate the need to plan and manage resource development

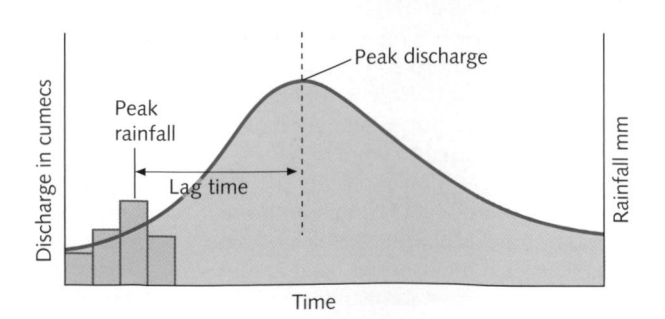

1 (Page 278)

a i) Write these words in the correct boxes on a copy of the diagram:
  - condensation
  - evaporation
  - transpiration
  - throughflow
  - surface runoff
  - evaporation. *(6)*

ii) Name two ways, shown by arrows, by which rainwater finds its way back to the sea. *(2)*

b With reference to Figure 17.3:
  i) List two ways in which surface water may be stored. *(2)*
  ii) Where else may surplus water be stored? *(1)*

2 (Page 279)

a The diagram below shows a partly completed drainage basin system. Complete it by matching the following terms with the appropriate numbered boxes:
  - evaporation
  - groundwater
  - infiltration
  - interception
  - surface runoff
  - throughflow
  - transpiration. *(7)*

b Draw a table with four columns. Head these columns:
  Inputs   Stores   Flows   Outputs.
  Fill in your table by adding the 14 labels from the drainage basin system which you completed in part (**a**). *(14)*

c What is the difference between:
  - evaporation and transpiration
  - infiltration and interception
  - surface runoff and throughflow? *(6)*

3 (Pages 280 to 281)

a On the hydrograph below:
  i) What time was peak rainfall? *(1)*
  ii) How many hours was lag time? *(1)*

b i) Why was there only a slight rise in discharge at point X? *(1)*
  ii) What was the cause of the steep rising limb at Y? *(1)*
  iii) Why was the falling limb at Z less steep than the rising limb? *(1)*

c The diagram below shows two hydrographs.
  i) Complete a table by saying which hydrograph, A or B, is the more likely to correspond to each of the following pairs.
  - a long period of gentle rain and a short, heavy thunderstorm
  - a basin with steep valley sides and a basin with gentle sides
  - an area of impermeable rock and an area of permeable (porous) rock
  - an area of forest and an area of bare rock
  - a mainly urbanised basin and a mainly rural basin
  - a river with a dam built across it and a river with no dam
  - a basin with a high drainage density and a basin with a low drainage density. *(7)*
  ii) Give one reason for each of your answers in part (i). *(7)*

4 (Page 282)

a Describe four processes by which a river can transport material. *(4)*

b Describe four processes by which a river can erode its banks and bed. *(4)*

c Under what conditions will a river deposit material? *(2)*

5 *(Pages 283 and 284)*

a Fieldsketch A shows a waterfall and gorge.
  i) On a copy of the sketch, mark on and label:
     an area of hard rock, an area of soft rock, a plunge
     pool, the position of the gorge, and a place where
     rock is likely to collapse. *(5)*
  ii) Describe and explain the main features of the
     waterfall. *(3)*
b Explain how a gorge forms. *(3)*

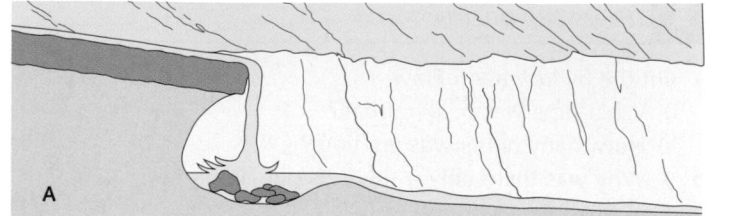

A

c Fieldsketch B shows a meander.
  i) On a copy of the sketch, mark on and label:
     areas of erosion, areas of deposition, and the
     position of the fastest current. *(3)*
  ii) Why is erosion and deposition taking place where
     you have shown? *(6)*
d On a cross-section of a meander, label a river cliff
   and slip-off slope. *(3)*
e Explain how an ox-bow lake forms. *(3)*

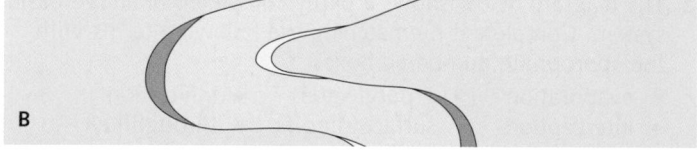

B

6 *(Pages 282 to 285)*

a Fifteen river features have been labelled **a** to **o** on the
   fieldsketch. Match up the correct letter with the feature
   named in the following list:
   delta, floodplain, gorge, interlocking spur, levée, meander,
   mouth, ox-bow, plunge pool, rapids, river cliff, slip-off
   slope, source, V-shaped valley, waterfall. *(15)*
b The sketch has also divided the river valley into two sections
   labelled **A** and **B**. For each pair of terms below, say which is
   likely to occur in zone A and which in zone B. In each case
   give a reason for your answer.
   • vertical erosion and lateral erosion
   • attrition and corrasion
   • traction and suspension. *(9)*

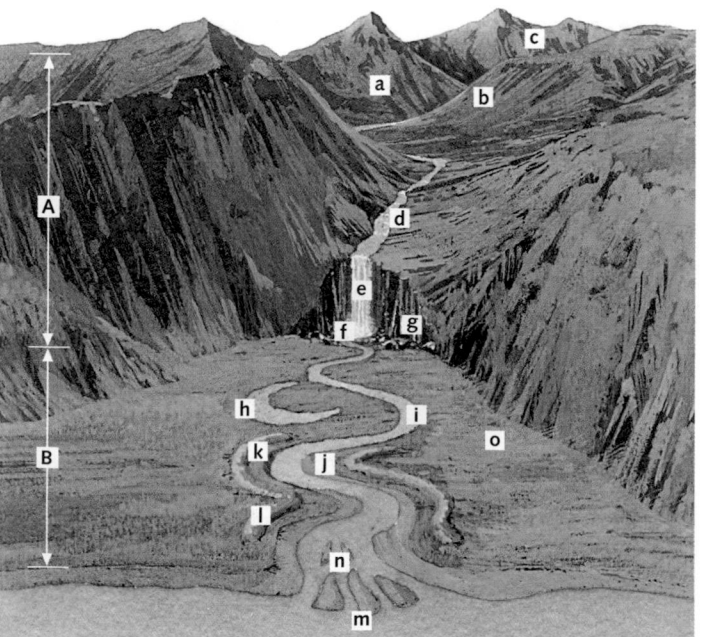

7 *(Pages 286 to 289)*

The English Midlands experienced severe flooding in 2007.
a i) Why was the summer of 2007 much wetter
     than usual? *(2)*
  ii) Why did so much rain fall on 19–20 July? *(2)*
b How did the 2007 flood affect:
   • transport
   • public services
   • residential property? *(3)*
c How did the emergency services respond? *(2)*
d Give one advantage and one disadvantage of the site
   of Tewkesbury. *(2)*
e What natural (physical) features of the Ganges–
   Brahmaputra drainage basin make Bangladesh
   vulnerable to river flooding each year? *(6)*
f How has human activity in the Ganges–Brahmaputra
   basin increased the risk of damage from flooding? *(3)*
g How did extreme weather conditions in 1998 cause
   the worst flooding ever known in Bangladesh? *(3)*
h What were the effects of the 1998 flood on:
   • human life
   • property
   • the economy of Bangladesh? *(6)*
i Why did Bangladesh need overseas aid after the flood?
   What types of aid were needed most? *(6)*
j Why are the effects of river flooding usually much worse
   in an LEDC than in an MEDC? You should consider:
   • precautions taken before the flood
   • responses to the flood, both immediate and
     longer term. *(6)*

8 *(Pages 287 and 291)*

a Why was a flood prevention scheme thought necessary for Exeter? (4)
b Describe the main points of the flood prevention scheme. (5)
c How successful do you think the scheme has been to date? (2)

Map: River Exe flood prevention scheme
- A377, A396 roads
- **1a** New flood relief channel can take 250 cumecs
- **1b** River Exe can take 450 cumecs at this point
- River Creedy, River Exe
- Cowley Weir
- **2** Radial gates close automatically as river levels rise; limits discharge passing through Exeter
- Exwick Weir
- **3** Exwick flood relief channel
- Length 12 km
  Max. discharge capacity
  1960 = 450 cumecs
  1990 = 700 cumecs
  1960 flood (peak) = 600 cumecs
- St David's railway station
- **4** Stilling basin to reduce speed of flood discharge from relief channel
- Cathedral
- **7** River widened and deepened between the two relief channels
- **5** Two new road bridges. New arch is 55 m wide compared with older narrow bridge of 46 m
- Museum
- Trew's Weir
- St James's Weir
- **8** Land cleared to leave floodplain as an area of open space
- **6** Trew's Weir flood relief channel
- Exeter Canal
- To Exmouth
- Country park

The River Exe has its source on Exmoor from where heavy rain can reach Exeter in 24 hours. Over the centuries the city has been flooded on many occasions, the worst being in 1960 when 1000 homes were flooded in October and 1200 homes and businesses in December. Once costs had been balanced against risks, the Exeter Flood Protection Scheme was implemented. Although since its completion in 1977 the overflow channel has been needed, on average, three or four times a year, no property has been flooded.

9 *(Pages 290 to 291)*

For any river (or lake) that you have studied which suffers from water pollution:
a Name the river (or lake).
b i) On a copy of the star diagram, list six ways by which it may have become polluted. (6)
  ii) Describe any four of these ways. (4)
c Describe three attempts made by the authorities to clean up the river (or lake). (6)

Star diagram: Farming (1, 2), Industry (1, 2), Water pollution, Domestic (1), Power stations (1)

d With the help of Figure 17.37, describe five ways by which water authorities have tried to prevent rivers from flooding. (5)
e With the help of Figure 17.36, give five reasons why some groups of people now feel that it would be better to allow rivers to flood. (5)
f Do you think that rivers should be allowed to 'flow free' and flood, or do you think every effort should be made to control them? Give what you consider to be the two most important points to support your decision. (2)

10 *(Pages 293 to 295)*

a i) Give two reasons why flooding is a regular occurrence in the Yangtze basin. (2)
  ii) Give two reasons why it has become more frequent and severe. (2)
b i) Why does the Three Gorges attract tourists? (1)
  ii) Why was Sandouping a good site for a dam? (1)
c Using the table, describe:
  i) four predicted benefits of the Three Gorges Project (4)
  ii) four predicted problems created by the project. (4)
d Match each of the letters A to I on the sketch (based on Figure 17.43) with an item in the following list:
- power station with 16 generators
- power station with 12 generators
- underground power station with 6 generators
- lake • Yangtze River
- ship-locks • ship-lift
- spillway • dam. (9)

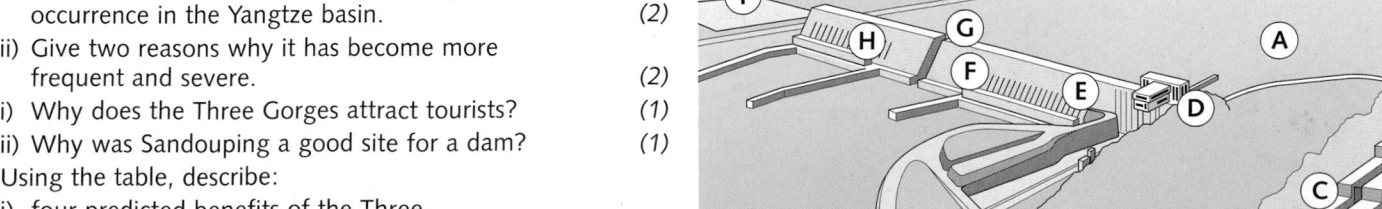

Benefits	Problems
Flood control	Settlements
Hydro-electricity	Silt
Navigation	Sewage
Housing	Dam failure

Coastal processes and erosion landforms

The **coast** is a narrow contact zone between land and sea. It is constantly changing due to the effects of land, air and marine processes. On many coastlines the dominant process results from the action of waves. Waves are usually created by the transfer of energy from the wind blowing over the surface of the sea (the exceptions are tsunamis which result from submarine earth movements). The larger the wave, the more energy it contains. The largest waves are formed when winds are very strong, blow for lengthy periods and cross large expanses of water. The maximum distance of water over which winds can blow is called the **fetch**. In the case of south-west England the fetch is from the south-west. This also coincides with the direction of the prevailing, or most frequent, wind. In eastern England the fetch is generally from the east.

Water particles within a wave move in a circular orbit (Figure 18.1). Each particle, or a floating object, tends to move vertically up and down. It is only the shape of the wave and its energy that is transferred horizontally towards the coast. However, when a wave reaches shallow water, the velocity at its base will be slowed due to friction with the sea-bed, and the circular orbit is changed to one that is more elliptical (Figure 18.1). The top of the wave, unaffected by friction, becomes higher and steeper until it breaks. Only at this point does the remnant of the wave, called the **swash**, actually move forwards. The swash transfers energy up the beach, the **backwash** returns energy down the beach.

There are two types of wave (Figure 18.2).

1 **Constructive waves** have limited energy. Most of this is used by the swash to transport material up the beach.
2 **Destructive waves** have much more energy. Most of this is used by the backwash to transport material back down the beach.

Erosion

Waves, like rivers (page 282), can erode the land by one of four processes:

- **Corrasion** (abrasion) is caused by large waves hurling beach material against a cliff.
- **Attrition** is when waves cause rocks and boulders on the beach to bump into each other and to break up into small particles.
- **Corrosion** (solution) is when salts and other acids in seawater slowly dissolve a cliff.
- **Hydraulic pressure** is the force of waves compressing air in cracks in a cliff.

Headlands and bays

Headlands and **bays** form along coastlines where there are alternating outcrops of resistant (harder) and less resistant (softer) rock (Figure 18.3). Destructive waves erode the areas of softer rock more rapidly to form bays. The waves cannot, however, wear away the resistant rock as quickly and so headlands are left protruding out into the sea. The headlands are now exposed to the full force of the waves, and become more vulnerable to erosion. At the same time they protect the adjacent bays from destructive waves (Figure 15.6).

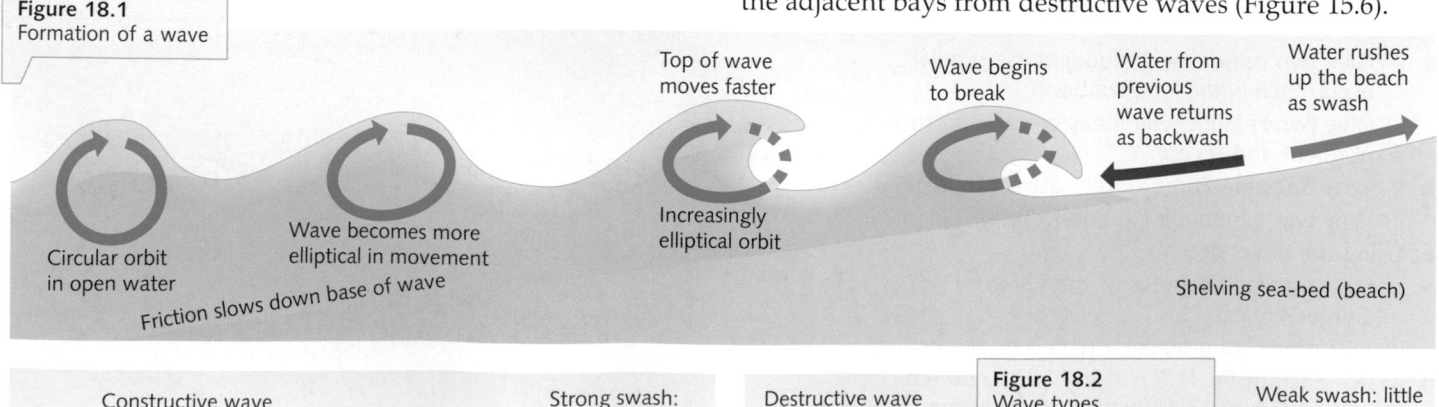

Figure 18.1
Formation of a wave

Top of wave moves faster
Wave begins to break
Water from previous wave returns as backwash
Water rushes up the beach as swash

Circular orbit in open water
Wave becomes more elliptical in movement
Friction slows down base of wave
Increasingly elliptical orbit
Shelving sea-bed (beach)

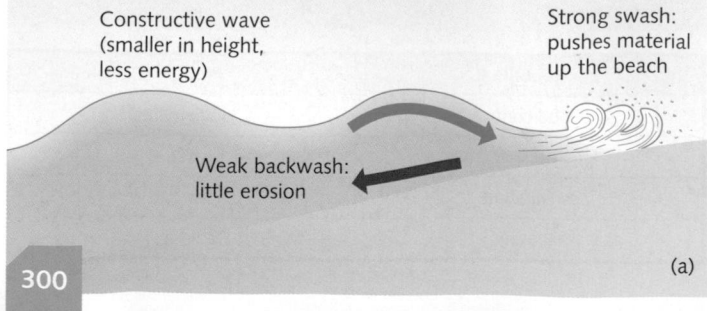

Constructive wave (smaller in height, less energy)
Strong swash: pushes material up the beach
Weak backwash: little erosion

(a)

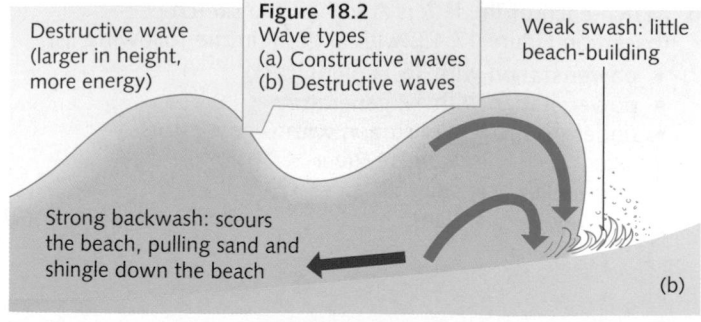

Destructive wave (larger in height, more energy)

Figure 18.2
Wave types
(a) Constructive waves
(b) Destructive waves

Weak swash: little beach-building
Strong backwash: scours the beach, pulling sand and shingle down the beach

(b)

Cliffs, wave-cut notches and wave-cut platforms

Wave erosion is greatest when large waves break against the foot of the cliff. With wave energy at its maximum, the waves undercut the foot of the cliff to form a **wave-cut notch** (Figure 18.4). Over a period of time the notch enlarges until the cliff above it, left unsupported, collapses. As this process is repeated, the cliff retreats and, often, increases in height. The gently sloping expanse of rock marking the foot of the retreating cliff is called a **wave-cut platform** (Figure 18.5). Wave-cut platforms are exposed at low tide but covered at high tide.

Caves, arches and stacks

Cliffs are more likely to form where the coastline consists of resistant rock. However, within resistant rocks there are usually places of weakness, such as a joint or a fault (Figure 18.6). Corrasion, corrosion and hydraulic action by the waves will widen any weakness to form, initially, a **cave**. If a cave forms at a headland, the cave might be widened and deepened until the sea cuts through to form a **natural arch**. Waves will continue to erode the foot of the arch until its roof becomes too heavy to be supported. When the roof collapses it will leave part of the former cliff isolated as a **stack** (Figure 18.7). In time, further wave action will result in the stack collapsing to leave a **stump**.

Figure 18.3
Formation of headlands and bays

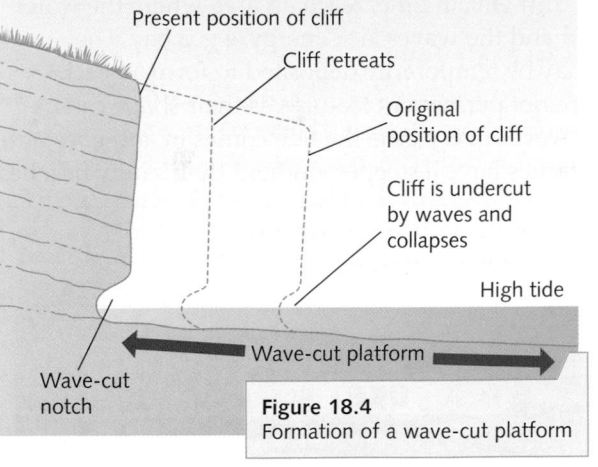

Figure 18.4
Formation of a wave-cut platform

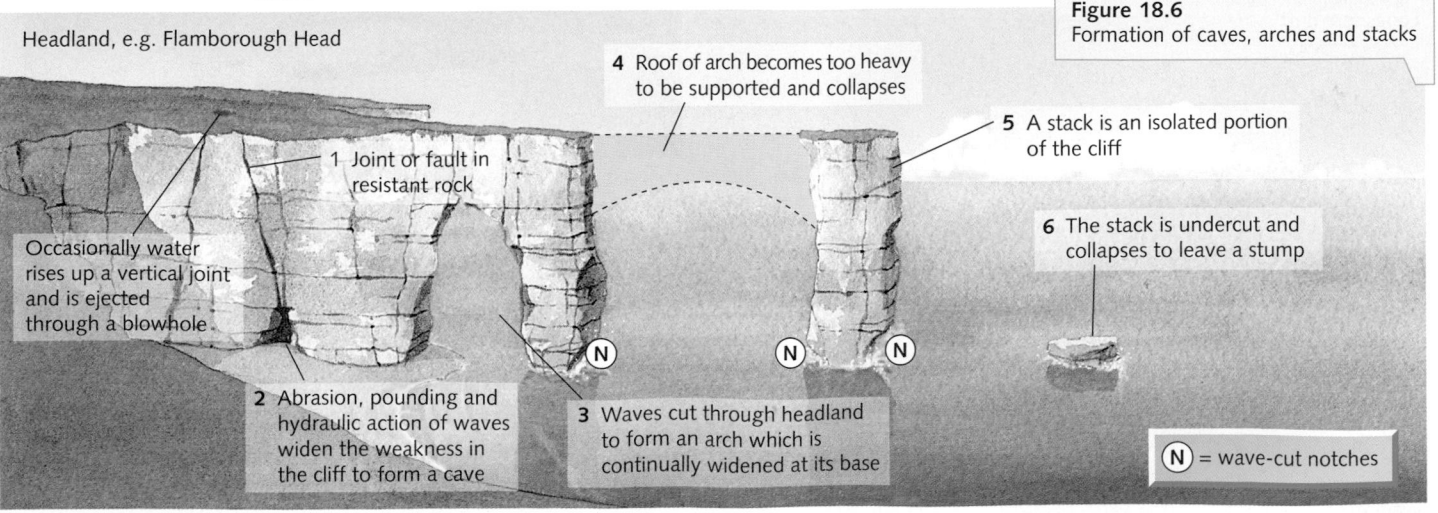

Figure 18.5
Wave-cut notch and platform: Flamborough Head, Yorkshire

Figure 18.6
Formation of caves, arches and stacks

Coastal processes and deposition landforms

Figure 18.7
The Foreland, Isle of Purbeck (see Figure 18.17)

A – headland   E – caves
B – wave-cut notch   F – arch
C – wave-cut platform   G – stacks
D – fault

Transportation

Although waves do carry material up and down a beach, the major movement is along the coast by a process called **longshore drift** (Figure 18.8). Waves rarely approach a beach at right-angles, but rather from a direction similar to that from which the wind is blowing. When a wave breaks, the swash carries material up the beach at the same angle at which the wave approached the shore. As the swash dies away, the backwash returns material straight down the beach, at right-angles to the water, under the influence of gravity. Material is slowly moved along the coast in a zig-zag course. The effect of longshore drift can best be seen when wooden groynes have been built to prevent material from being moved along the beach (Figure 18.9).

Deposition

Sand and shingle being transported along the coast by longshore drift will, in time, reach an area where the water is sheltered and the waves lack energy, e.g. a bay. The material may be temporarily deposited to form a beach. Beaches are not permanent features as their shape can be altered by waves every time the tide comes in and goes out. Shingle beaches have a steeper gradient than sandy beaches.

Figure 18.9
Groynes at Worthing, West Sussex

Figure 18.8
Longshore drift

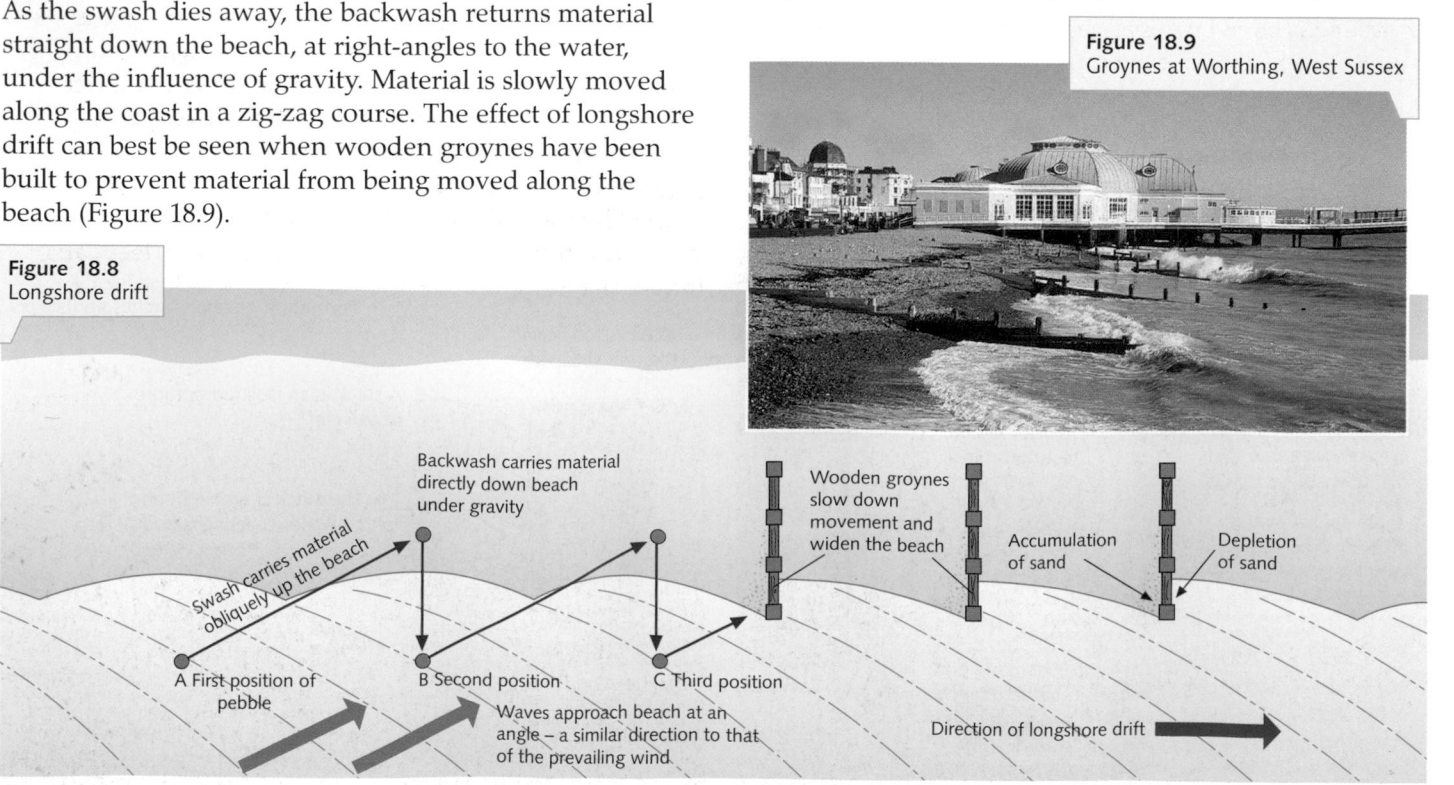

Backwash carries material directly down beach under gravity

Swash carries material obliquely up the beach

Wooden groynes slow down movement and widen the beach

Accumulation of sand

Depletion of sand

A First position of pebble

B Second position

C Third position

Waves approach beach at an angle – a similar direction to that of the prevailing wind.

Direction of longshore drift

Figure 18 10
Hurst Spit, Hampshire

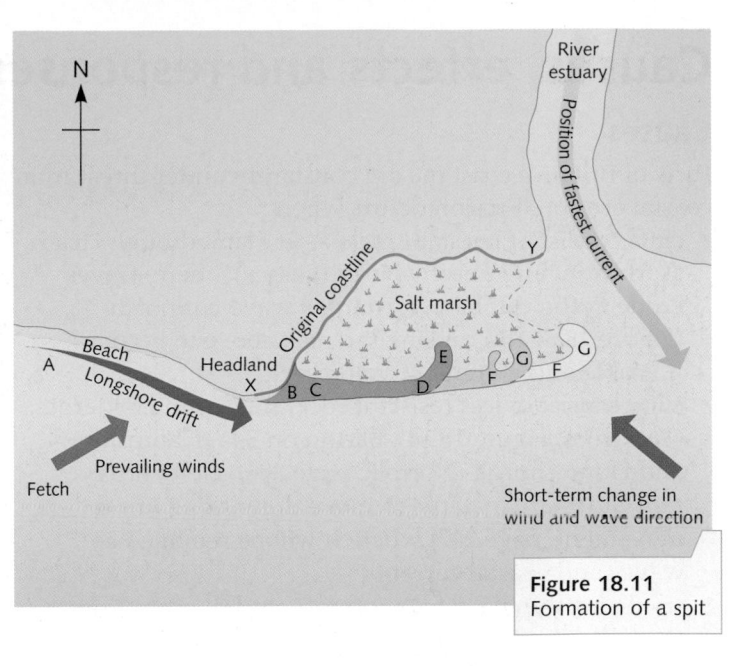

Figure 18.11
Formation of a spit

Spits

A **spit** is a permanent landform resulting from marine deposition. It is a long, narrow accumulation of sand or shingle, with one end attached to the land, and the other projecting at a narrow angle either into the sea or across a river estuary. Many spits have a hooked or curved end (Figure 18.10). They form where longshore drift moves large amounts of sand and shingle along the coast, and where the coastline suddenly changes direction to leave a shallow, sheltered area of water. In Figure 18.11 line X to Y marks the position of the original coastline. As the fetch and prevailing winds are, in this example, from the southwest, material is moved eastwards along the coast by longshore drift. After headland X the direction of the original coastline changes and larger material (shingle) is deposited in water sheltered by the headland (B). Further deposition of finer material (sand) enables the feature to build up slowly to sea-level (C) and to extend length (D). Occasionally the wind changes its direction (e.g. it comes from the east). This in turn causes the waves to alter their direction (e.g. approach from the south-east).

During this time some material at the end of the spit may be pushed inland to form a curved end (E). When the wind returns to its usual direction the spit resumes its growth eastwards (F). Spits become permanent when sand is blown up the beach, by the prevailing wind, to form **sand dunes**. **Salt marsh** is likely to develop in the sheltered water behind the spit. The spit is unable to grow across the estuary as the river current carries material out to sea. Should there be no river, the spit may grow across the bay to form a **bar**.

Sea-level changes

During the Ice Age large amounts of water were held in storage as ice and snow. This interruption in the hydrological cycle (page 278) caused the world's sea-level to fall. After the Ice Age the sea-level rose as the ice and snow melted. Many coastal areas were drowned, creating landforms such as fiords and rias. Both **fiords** and **rias** are drowned valleys. Fiords are found where glaciers overdeepened valleys until they were below sea-level (Figure 18.12) while rias occur in valleys which were formed by rivers (Figure 18.13). Fiords are long, narrow inlets with high, cliff-like sides. They are very deep, apart from a shallow entrance. Rias are more winding with relatively low, gentle sides. The depth of a ria increases towards the sea.

Figure 18.12
Geiranger Fiord, Norway

Figure 18.13
A ria: the Fal estuary

Causes, effects and responses to cliff erosion

Causes

Parts of Britain's coastline are constantly under threat from coastal erosion. Erosion occurs where:

- cliffs consist of resistant rock, as at Flamborough Head (Yorkshire) and Beachy Head (Sussex) – here, waves erode at the cliff base, forming a wave-cut notch (Figure 18.5) and causing the rock above to become unstable and, in time, to collapse
- cliffs consist of less resistant rock, such as in Holderness (Yorkshire, Figure 18.14), Barton on Sea (Hampshire) and Happisburgh (Norfolk, page 306). Here, heavy rain can cause material to move downslope (mass movement, page 247) where it will be removed by waves and coastal currents
- sand and shingle, which forms a natural protection at the foot of a cliff, is removed by human activity enabling erosion by waves to accelerate
- people have built on cliff tops, as this adds weight which can cause cliff collapse.

Figure 18.14
A collapsing cliff, Holderness

Effects

In 1999, a section of chalk 200 metres long and 15 metres deep, fell 180 metres into the sea at Beachy Head. A cliff collapse near Scarborough in 1993 destroyed a hotel, while further south, in Holderness, the coastline is retreating by an average of 2 metres a year (Figure 18.14). Here, it is now 3 km further west than it was in Roman times, and some 50 villages mentioned in the Domesday Book of 1086 have been lost to the sea (Question 1, page 315). Cliffs at Happisburgh have retreated 125 metres since 1992 (Figures 18.22 and 18.23).

Figure 18.16
Methods of coastal protection at Felixstowe, Suffolk
(a) Sea-wall, revetments and groynes
(b) Riprap

Responses

Attempts to protect the coastline involve either 'hard' or 'soft' engineering (Figures 18.15, 18.16 and 18.21). It is becoming increasingly expensive to provide new defences and maintain existing ones (a £1.3 million scheme at Barton on Sea was destroyed in four years), while the construction of sea defences in one place often seems to increase the rate of erosion elsewhere. It is now accepted that decisions have to be made between defending the coast at all costs or allowing nature to take its course.

Hard	Wooden groynes (Figure 18.9) help reduce the force of waves and can trap material being moved along the beach by longshore drift. They help widen beaches and protect cliffs.
	Concrete sea-walls may be curved at the top to divert the force of the waves back out to sea (Figure 18.16a).
	Gabions are wire cages stacked together and filled with rock.
	Slatted revetments help dissipate the force of the waves (Figure 18.16a and question 2, page 315).
	Concrete blocks, called riprap, absorb the power of the waves (Figure 18.16b).
	Artificial offshore breakwaters and reefs (Figure 18.24b).
Soft	Beach replenishment/nourishment, by which the Environment Agency replaces lost beach material.
	Cliff stabilisation, in which pipes are inserted to remove excess water, and vegetation is planted to stabilise exposed soil.
	Managed retreat/realignment encourages salt marsh to develop as a buffer zone (Figure 18.26).

Figure 18.15
Hard and soft sea-defence engineering

Coastal landforms on OS maps

Figure 18.17 shows part of the Isle of Purbeck on the south coast of England. Using this, together with the geology map in Figure 18.18, you should be able to recognise:

- headlands (The Foreland 0582 and Peveril Point 0478) and bays (Swanage 0379 and Studland 0383)
- cliffs (Durlston Bay 0377 and Ballard Point 0481)
- caves (Tilly Whim 0376) and stacks (Old Harry 0582 and Figure 18.7)
- groynes (0379) indicating longshore drift.
- sandy beaches (0379 and 0383), sand dunes (0384) and salt marsh (0385).

The map does not show wave-cut notches, wave-cut platforms or arches (features too small to be shown at this scale) nor a spit with a hooked end (none has formed at this location).

Figure 18.18
Geology of the Isle of Purbeck

| 0 | 3 km |

Softer, less resistant sands and clays   *Studland Bay*
(Studland)   The Foreland
Harder, more resistant chalk   Ballard Point
Softer, less resistant clays   (Swanage)   *Swanage Bay*
    Peveril Point
Harder, more resistant limestone   Durlston Head

Former coastline

N

Figure 18.19 shows part of the south coast of England some 25 km east of the Isle of Purbeck. Together with Figure 18.10 you should be able to recognise:

- cliffs (2692)
- groynes (2891) indicating longshore drift from these cliffs towards the east
- shingle (2791) and sandy (3090) beaches
- a spit with several hooked ends (Hurst Spit 3189 and 3190)
- sand dunes on the spit (3189), and salt marsh behind it (3090).

© Crown copyright

Figure 18.17
Part of the OS Landranger sheet 195

Figure 18.19
Part of the OS Landranger sheets 195 and 196

© Crown copyright

The need for coastal management

The coastline of Britain, like all other parts of the world, is dynamic and always changing. These changes may result from:

- either natural (physical) processes of climate, mass movement, waves and the tide
- or human intervention and activity.

Although Britain's coasts are rarely affected by extreme events such as the Indian Ocean tsunami (page 265), storm surges as in the Bay of Bengal (page 312) or tropical storms as in the Gulf of Mexico (page 217), large stretches are under threat from one or more sources (Figure 18.20).

These threats may result from one, or a combination of:

- natural events such as flooding and erosion
- human demands that include – often in competition – settlement, economic activities and recreational use.

To try to reconcile these threats, there is the need for a national sustainable management plan. Such a plan has to try, amongst other considerations, to balance on one hand the rapidly increasing costs of providing new sea defences and maintaining existing ones at a time of a rising global sea-level, and on the other the need to protect people and their property.

The Norfolk coast

Who is responsible for coastal management?

The Department for Environment, Food and Rural Affairs (Defra) has overall responsibility for coastal management in England and Wales, although the Environment Agency (EA) has powers to reduce flooding in tidal waters. To fulfil its role Defra has the task of producing a **shoreline management plan (SMP)**. To do this means understanding processes along any given stretch of coastline. To achieve this, England and Wales have been divided into 11 self-contained cells, with Norfolk being part of one that extends from the Wash to the Thames estuary. The cells were chosen so that coastal processes within each had a minimum effect on adjacent cells. Each SMP has two basic concepts:

- natural processes should not be interfered with unless it is necessary to protect life or property
- all schemes must be economically viable and undergo a cost-benefit analysis to ensure that they make good use of public money.

What are the options?

An SMP has four alternative options:

1. Do nothing other than monitor the situation.
2. Hold the existing defence line by maintaining sea-walls and riprap.
3. Advance the existing defence line, e.g. by adding groynes, or beach replenishment.
4. Realign the existing defence line through a managed retreat.

Threat
Increased risk of flooding
• Rising sea-level linked to global warming
• Higher high tides
• Risk of increased number of storm surges
Increased risk of erosion
• Larger waves (generating more energy)
• Human activity (use of footpaths, building on cliff-tops)
Overuse and/or misuse
• Settlements and economic development
• Leisure and tourism (caravan and car parks, golf courses)

Figure 18.20
Threats to Britain's coasts

What are the factors to consider?

The following must all be considered for a new scheme or SMP:

1. The cost of its implementation and later maintenance.
2. Its benefits, e.g. effects on lives, homes, farmland and transport routes.
3. Its likely effectiveness.
4. Its possible effects on adjacent coastlines.
5. Its appearance.

How do various defence schemes compare?

Figure 18.21 gives the advantages and disadvantages of several types of sea defence.

Figure 18.21
Advantages and disadvantages of different sea defences

Sea defence	Cost per metre	Advantages and disadvantages
Curved sea-wall	£6000	Very strong and very expensive. Built to reflect waves (reducing energy) but can cause turbulence. Unsightly.
Riprap	£4500	Absorb wave energy but can break up over time. Unsightly.
Wooden revetments	£3000	Also absorb wave energy but need to be replaced regularly. Not very attractive.
Offshore bar	£5000	Protects base of cliffs but rocks need to be large or will be removed in storms. Interferes with nature.
Groynes	£40 (£4000 each)	Trap sand to widen beach but stops longshore drift, causing sand loss and increased erosion down the coast.
Beach replenishment	£2500	Absorbs wave energy and protects cliffs. Visually the most natural/attractive but needs constant replacing.

Coastal management in Norfolk

Erosion has always been a major problem along much of the Norfolk coast, although in the extreme south of the county, near to Yarmouth, flooding is the major danger (as experienced in the North Sea floods of 1953, Figure 18.32). The present-day SMP must aim to strike the seemingly impossible balance between protecting the coastline at a viable cost and trying to minimise the disruption of natural processes. Indeed the coast of East Anglia, of which Norfolk is a part, is in the centre of the controversy as to whether people and their property should be protected at all costs or whether nature should be allowed to take its course. This controversy is likely to intensify as climate change causes sea-level to rise, tides become higher and storms more frequent and severe (Figure 18.20).

Coastal erosion at Happisburgh

Happisburgh is one example of a settlement at the centre of the controversy. At present it has a population of 1400 living in about 600 houses. The village has a fourteenth-century stone church, several listed buildings and a landmark red-and-white striped lighthouse (Figure 18.22). Although it is now a coastal village, it was once located inland, being separated from the sea by Whimpell, a settlement that has long since been lost to the sea. Records for Happisburgh show that between 1600 and 1850 erosion by the sea resulted in the coastline retreating by over 250 metres. Attempts were then made to prevent further loss and these seem to have at least slowed down the rate of retreat. However, a lack of maintenance has led to the defences falling into disrepair and erosion to accelerate. The soft cliffs, varying in height between 6 and 10 metres, have retreated by 100 metres in the ten years since 1998 (compare Figures 18.22 and 18.23). This retreat is resulting in the loss of at least one property per year and considerable amounts of good agricultural land. The inhabitants of Happisburgh, naturally, want new sea defences built to protect their property despite the enormous costs that this would involve.

Figure 18.22
Happisburgh 1998

Figure 18.23
Happisburgh February 2009

Coastal defences at Sea Palling

Much of the north Norfolk coastline from Cromer southwards is protected by expensive sea defences.

At Sea Palling the beach is backed by sand dunes which, in earlier times, helped to provide a natural defence.

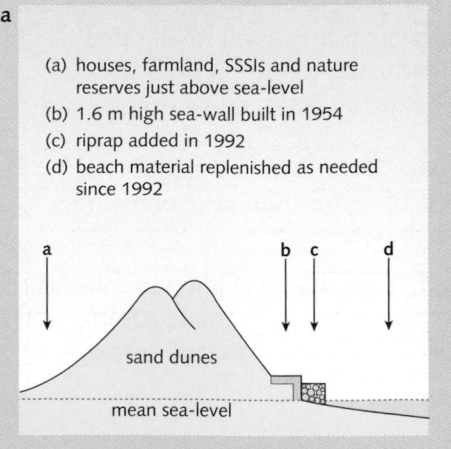

(a) houses, farmland, SSSIs and nature reserves just above sea-level
(b) 1.6 m high sea-wall built in 1954
(c) riprap added in 1992
(d) beach material replenished as needed since 1992

sand dunes

mean sea-level

Figure 18.24
Sea defences at Sea Palling:
(a) 1952–92
(b) Since 1995

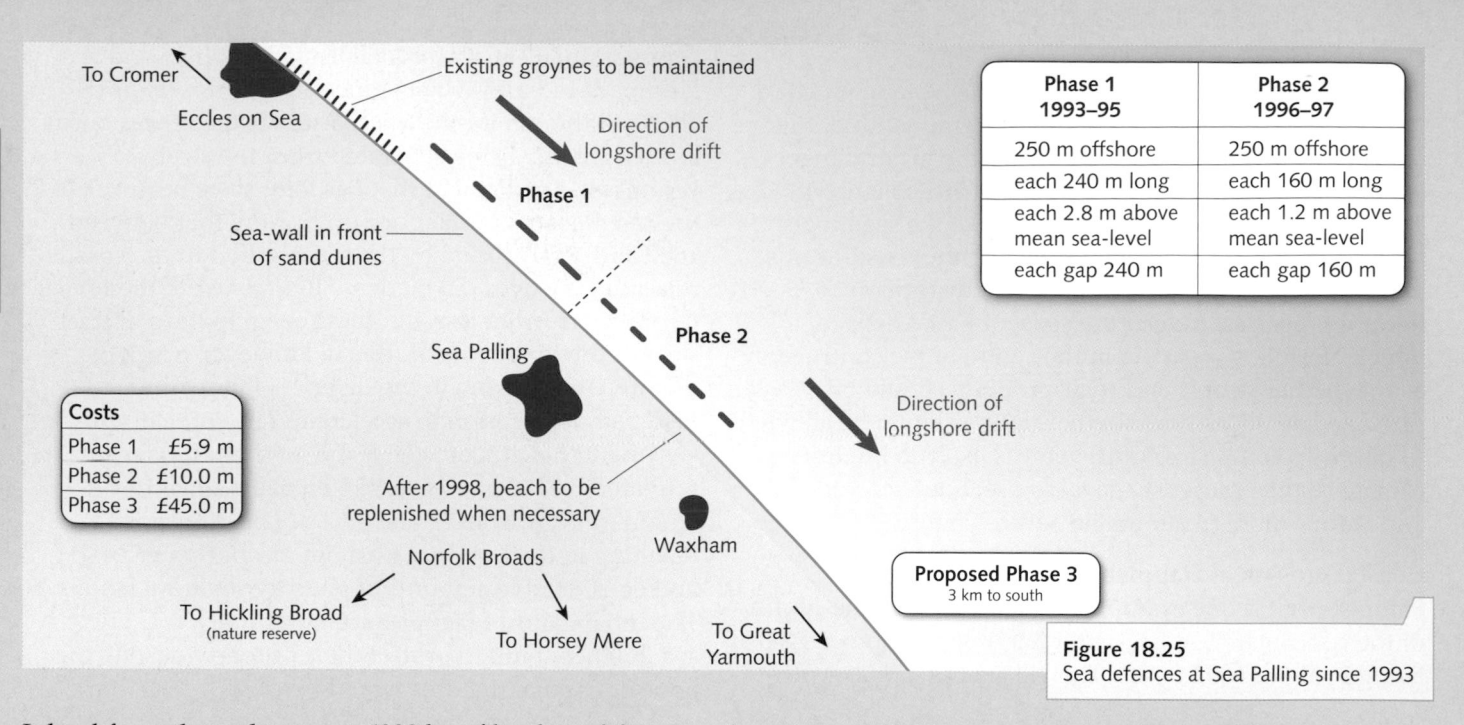

Phase 1 1993–95	Phase 2 1996–97
250 m offshore	250 m offshore
each 240 m long	each 160 m long
each 2.8 m above mean sea-level	each 1.2 m above mean sea-level
each gap 240 m	each gap 160 m

Costs

Phase 1	£5.9 m
Phase 2	£10.0 m
Phase 3	£45.0 m

Figure 18.25
Sea defences at Sea Palling since 1993

Inland from these dunes are 6000 ha of land used for settlement, farming and (this being part of the Norfolk Broads) tourism and wildlife. The 1953 storm surge (Case Study 18A) broke through the coastal defences, flooding large areas and, at Sea Palling itself, washing away houses and drowning seven people. Following the 1953 flood, a sea-wall was built in front of the sand dunes and there was some replenishment of beach material (Figure 18.24a). During the next 40 years the beach became increasingly narrow. This was due to material being transported southwards by longshore drift at times of northerly and easterly gales which, presumably due to the onset of climate change, seem to be increasing in both frequency and intensity. In 1991, following several severe winter storms, riprap was placed against the sea-wall as a temporary measure and in 1992 a beach management strategy was introduced. The aims of this strategy were that it should:

- have as little effect as possible on adjacent coastal areas
- have minimal environmental impact
- be cost-effective.

The strategy was implemented by placing 150 000 tonnes of rock in front of the sea-wall, replenishing the beach with nearly 1.5 million cubic metres of sand and – the major part – constructing four offshore reefs. The reefs were designed to protect the beach by reducing wave energy while at the same time allowing some longshore drift so as not to deplete the supply of sand to beaches further along the coast (Figure 18.25).

Almost immediately after their completion in 1995, a previously unpredicted problem arose: that of sand accumulating in the sheltered areas behind the reefs. This meant that the reefs became joined to the mainland (Figure 18.24b), which in turn interrupted longshore drift. To try to overcome this problem, the next five reefs were made:

- shorter – to reduce the area of shelter behind them
- lower – to allow more overtopping by waves
- closer together – to prevent erosion in the gaps between them.

A further five reefs are planned 3 km to the south but, as Figure 18.25 shows, the estimated cost has shot up in comparison with that of earlier reefs.

Proposed 'managed retreat'

Controversial plans to flood parts of Norfolk became known in 2008. If accepted, this proposal would mean that, for the first time, Britain is admitting defeat in the battle to maintain all of its coastal defences. Many experts believe that it would be far less expensive and a more practical option to 'realign the coast' in a 'managed retreat'. This would involve (Figure 18.26):

- building a new sea-wall further back from the present coastline, as this would cost only a fraction of that of trying to maintain the existing defences
- breaching the existing defences to allow the predicted higher tides and storm waves to flood an area between the existing defences and the proposed new sea-wall, which in time would develop into a salt marsh that would act as a buffer, reducing the strength of the waves (Figure 18.27 shows the first scheme of this kind in Britain).

In response to the proposal, Defra said that it was 'committed to the sustainable protection of people and property in Norfolk and elsewhere'.

If the scheme went ahead, it would mean allowing the sea, over a period of time, to breach 25 km of the Norfolk coastline between Eccles on Sea and Winterton-on-Sea (Figure 18.28).

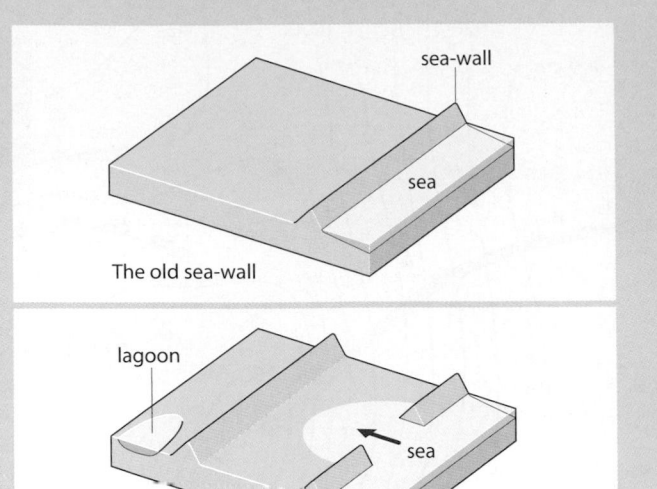

The old sea-wall

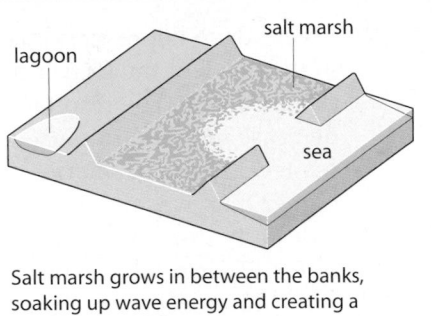

A new bank is built well back using soil dug out to create lagoons.
A hole is made in the old wall, allowing the sea in.

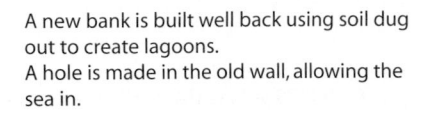

Salt marsh grows in between the banks, soaking up wave energy and creating a habitat for wildlife.

Figure 18.26
Breaching of an old sea-wall to create a salt marsh

This would lead to the creation of an area of saltwater lake and salt marsh covering 65 km². Within 50 years this lake could eliminate the four coastal villages of Eccles on Sea, Sea Palling, Waxham and Horsey and two further inland, at Hickling and Potter Heigham. The result would be the loss of some 600 houses, many hectares of good-quality farmland and five of the freshwater lakes that currently

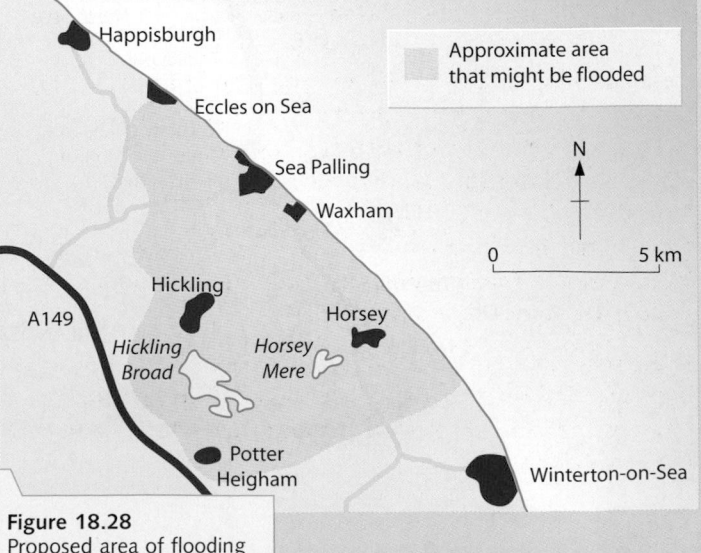

Figure 18.28
Proposed area of flooding on the North Norfolk coast

Figure 18.27
Tollesbury managed realignment in Essex

form part of the Norfolk Broads National Park. Of these lakes, Hickling Broad is a well-known tourist area and Horsey Mere has rare fauna and flora. Naturally a proposal such as this is likely to provoke strong opposition.

Opponents to the scheme claim that:
- it would mean in the short term making homes unsaleable and in the long term the relocation of hundreds of families, and having to pay them compensation
- a millennium of history would vanish under the waves and with it villages like Hickling, mentioned in the Domesday Book, and Sea Palling, which the sea failed to destroy in the 1953 flood
- the loss of valuable farmland and five freshwater wildlife habitats
- a loss of jobs in agriculture and tourism.

Proposers suggest that:
- the plan is economically more sustainable; to restore and maintain the existing sea-wall could cost £5000 a metre, whereas an inland retreat with a resulting buffer zone to dissipate the power of the waves might only be £500 a metre
- the salt marsh could prove ideal grazing for cattle and, at times of a storm surge (page 310), an area for storing excess seawater
- the salt marsh would provide a welcome haven for coastal wildlife when little of Britain's salt-marsh ecosystem remains, and more of what does is being lost each year.

No final decision has been made, but while the outcome might have a major effect on people living in this part of Norfolk, it is likely to have wider repercussions on many more living in similar environments who are at risk from the sea elsewhere in Britain.

Coastal flooding in MEDCs

South-east England and the Netherlands

Coastal flooding, when it occurs in densely populated areas, can cause considerable loss of life, damage to property and disruption to everyday life. Such an event affected south-east England and parts of the Netherlands in 1953.

Causes

- Many coastal areas have been drained for farming, some since the sixteenth century. One result of this drainage was a shrinking in the level of the land so that parts of the present-day English fenlands and the Dutch polders now lie below sea-level.
- Many of the flat, low-lying areas behind the coast are protected either naturally by sand dunes (Lincolnshire and the Dutch coast north of the Rhine) or artificially by sea-walls (East Anglia) and dykes (Netherlands). Although it needs extreme conditions to breach sand dunes, sea-walls and dykes are more vulnerable if they are not continually maintained.
- Increased building and economic activity in these areas adds to the number of people and amount of property at risk.
- **Storm surges** occasionally affect the southern North Sea. These storm surges result from depressions passing over the northern North Sea. As air pressure falls it allows the level of the sea to rise. At the same time, the associated northerly gale-force winds push water southwards into an area where the North Sea becomes both narrower and shallower. A surge of water can cause sea-level to rise by over a metre, topped by waves that themselves may be several metres in height.
- One consequence of global warming (page 219) is the predicted rise in sea-level together with higher tides, more storm surges and increased coastal flooding (Figure 13.23).

The 1953 storm surge

On the night of 31 January/1 February 1953, four major factors became linked in a fateful combination.

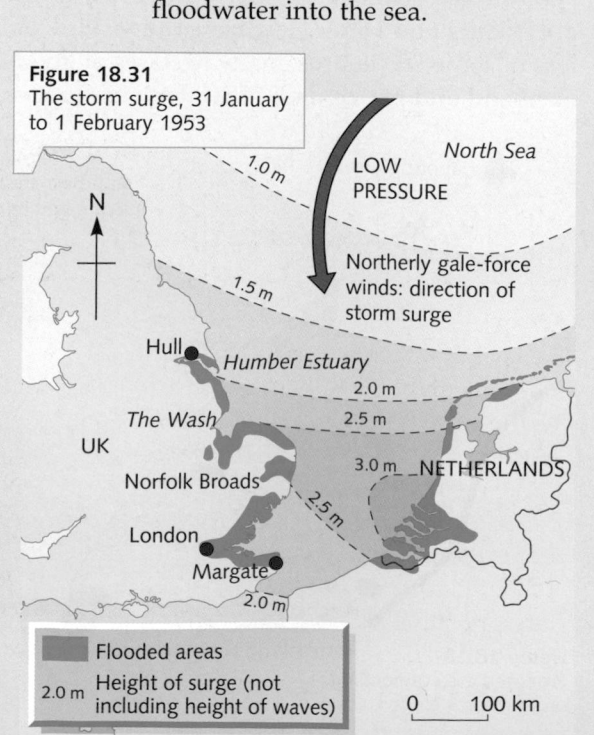

Figure 18.29
Weather chart for 00.00 GMT, 1 February 1953

1 A very deep depression passed to the north of the British Isles (Figure 18.29). The resultant decrease in pressure caused sea-level to rise 0.5 metres higher than was predicted.

2 Northerly storm-force winds (force 12) drove a wall of water southwards down the North Sea, creating waves 6 metres in height (Figure 18.30).

3 There was a full-moon spring tide (i.e. the highest high tide in a monthly cycle).

4 Rivers, especially in eastern England, were discharging floodwater into the sea.

Figure 18.30
Storm waves along the North Sea coast

Figure 18.31
The storm surge, 31 January to 1 February 1953

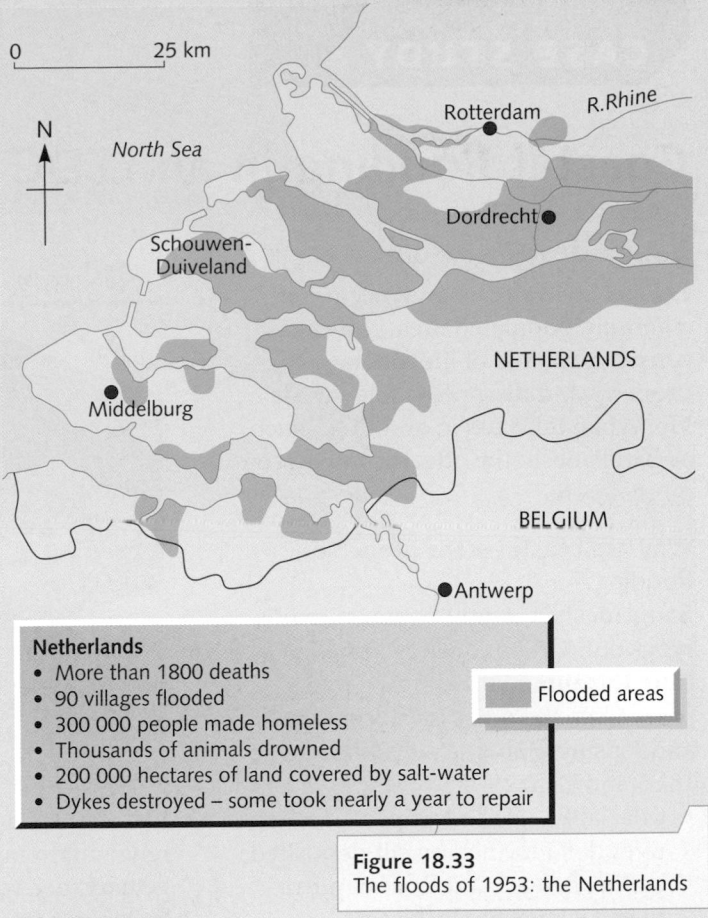

Humber Estuary
First reports of flooding not taken seriously

Hull

Grimsby

Lincolnshire coast
43 deaths

North Sea

Train collides with bungalow swept onto track
Flooding causes 65 deaths

Hunstanton

Wells

King's Lynn

Boat lifted by waves onto quay

Flooding in Norfolk Broads

1200 houses flooded
8 deaths

Felixstowe

Harwich

37 drowned, bungalows destroyed

London

R. Thames

Canvey Island

Canvey Island
Lies below sea-level – bungalows destroyed
58 drowned, 11 000 homeless

Isle of Sheppey
Much of island flooded

0   100 km

Eastern England
- 300 deaths
- 24 000 houses destroyed
- 30 000 people evacuated
- Thousands of animals drowned
- 1000 km² of farmland made infertile by salt-water
- Sea defences destroyed

Flooded areas

Figure 18.32
The floods of 1953: eastern England

0   25 km

N

North Sea

Rotterdam

R. Rhine

Dordrecht

Schouwen-Duiveland

NETHERLANDS

Middelburg

BELGIUM

Antwerp

Netherlands
- More than 1800 deaths
- 90 villages flooded
- 300 000 people made homeless
- Thousands of animals drowned
- 200 000 hectares of land covered by salt-water
- Dykes destroyed – some took nearly a year to repair

Flooded areas

Figure 18.33
The floods of 1953: the Netherlands

Figure 18.34
A breached dyke north of Dordrecht in the Netherlands

Effects

The resultant storm surge was up to 3 metres higher than the predicted spring tide (Figure 18.31). The surge, coupled with the huge storm waves, broke through sea defences in eastern England on the evening of 31 January, and in the Thames estuary and the delta region of the Netherlands in the early morning of 1 February. The event caused the deaths of more than 2100 people and made another 330 000 temporarily homeless, as well as drowning thousands of farm animals and ruining farmland (Figures 18.32, 18.33 and 18.34).

Responses

Although a storm surge of that magnitude is only predicted to occur once in every thousand years, nevertheless both Britain and the Netherlands learnt from the disaster and both took major precautions should the event repeat itself sooner than that prediction. These included:

- building tidal barriers across river estuaries (the Thames Barrier in England and the Delta Scheme in the Netherlands)
- building higher and stronger sea-walls and dykes
- improving weather forecasting and implementing early warning flood warning systems.

Although the event has, fortunately, not been repeated, neither country feels entirely safe, due to the predicted rise in sea-level expected as a result of global warming. More recently there has also been a growing number of people who feel that sea defences and dykes, apart from being unsightly, are too expensive to build and then maintain. These people feel that, as with rivers (page 291), nature should be allowed to take its course and that areas prone to flooding by the sea should be allowed to flood. Naturally, people who live and work in these areas disagree and want increased protection.

Coastal flooding in an LEDC

Bangladesh

Coastal flooding, when it occurs in a densely populated area, can cause considerable loss of life, damage to property and disruption to everyday life. When it occurs in an LEDC, such as Bangladesh, the effects can be even more severe.

Why are coastal areas prone to flooding?

Bangladesh is caught between two types of flooding: one resulting from the annual flooding by rivers (page 288), the other by storm surges and a rising global sea-level (Figures 18.35 and 18.37).

- Coastal Bangladesh is part of a huge river delta formed by silt deposited by the Ganges and Brahmaputra rivers as they enter the sea (page 285). As the silt accumulates upwards and outwards, it forms flat islands which divide the river into numerous distributaries. As the islands are marshy, have a rich soil and are ideal for rice growing, they have attracted a large population. The continual deposition of silt, however, increases the risk of flooding as it blocks the main river channels and raises the beds of the rivers.

- Tropical cyclones (page 216) affect Bangladesh in autumn. As they are funnelled up the Bay of Bengal, the force of the wind increases and sea-water is pushed northwards. Where the Bay of Bengal becomes narrower and shallower towards Bangladesh, the water builds up to form a **storm surge**. The surge may be 6 metres in height (some have been higher) and topped by waves reaching a further 4 metres (Figure 18.36). The wall of water, similar to a tidal wave, sweeps over the flat, defenceless islands of the delta, carrying away the often flimsy buildings and any life-form in its path.

- Local inhabitants, most of whom rely on farming or fishing, may not get advance warning as they may be too poor to own their own telephone or television or, even if they do, they cannot find land high enough upon which to escape the rising floodwaters.

- Although Bangladesh's contribution to global warming is minimal (page 218), the effects of this process on the country are expected to be considerable as the world's sea-level rises. Latest scientific measurements suggest that the world's sea-level is rising 2 metres per 100 years, a rate that is accelerating and is predicted to be 4 metres by the end of this century (page 219). A rise of only 1 metre would be sufficient to permanently flood one-quarter of

Bangladesh, affecting one-half of the present population of 153 million.

Effects

The effects of two recent storm surges are described in Figure 18.38. The two worst recorded surges were in 1876 (up to 250 000 deaths) and, more recently, in 1970 (500 000 deaths).

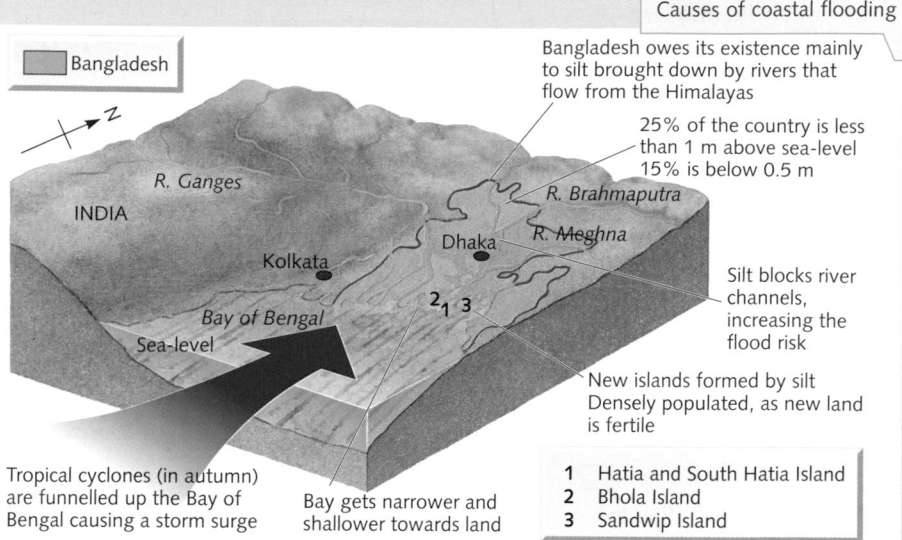

Figure 18.35
Causes of coastal flooding

Bangladesh owes its existence mainly to silt brought down by rivers that flow from the Himalayas

25% of the country is less than 1 m above sea-level 15% is below 0.5 m

Silt blocks river channels, increasing the flood risk

New islands formed by silt Densely populated, as new land is fertile

Tropical cyclones (in autumn) are funnelled up the Bay of Bengal causing a storm surge

Bay gets narrower and shallower towards land

1	Hatia and South Hatia Island
2	Bhola Island
3	Sandwip Island

Figure 18.37
Effects of storm surges, 1985–2008

Year	Height of storm surge (m)	Deaths (estimated)
1985	5.7	40 000
1988	4.8	25 000
1990	6.3	140 000
1991	6.1	135 000
1994	5.8	25 000
1999	5.3	10 000
2007	5.1	4000
2008*	6.3	146 000

* Myanmar

Figure 18.36
Development of a storm surge in the Bay of Bengal

High winds and tides combine to produce a storm surge topped by waves reaching 8 m in height

Low pressure (typhoon) over Bay of Bengal gives winds gusting up to 180 km/h

Top of surge

4 m

Normal high-tide level

4 m

Low-lying coastal area with little protection from flooding. Intensively farmed

Funnel-shaped bay getting shallower towards coast

May 1985

Three days after the tropical cyclone hit the coastal islands of Bangladesh, countless bodies are floating in the Bay of Bengal. Hundreds of survivors on bamboo rafts and floating roof-tops, stalked by sharks and crocodiles, are awaiting rescue. The Red Cross suggested that the tidal wave, 9 metres in height and extending 150 km inland, may have claimed the lives of 40 000 people. An official source in Hatia claimed that 6000 people, many in their sleep, were washed out to sea and the only survivors were those who climbed to the top of palm trees and clung on despite the 180 km/hour winds. No links have yet been made with the more remote islands. Already there is the threat of typhoid and cholera, as fresh water has been contaminated. Famine could result as the rice crop has been lost, and it will take next year's monsoon rains to wash the salt out of the soil. Thousands of animals and most of the coastal fishing fleet seem likely to have been lost. The people of Bangladesh, already amongst the world's poorest, will be even more destitute.

April 1991

The tropical cyclone which hit the coast of Bangladesh in late April brought with it winds of 225km/hour and waves 7 metres in height. it swept over unprotected off-shore islands and the flat delta of the Ganges–Brahmaputra rivers where the land is never more than one or two metres above sea-level. To escape the flood, people climbed trees and onto roof-tops – but the wind was too strong for them to cling on for long. Over 150 000 people and half a million cattle were drowned, entire villages swept away, thousands of hectares of crops lost, electricity supplies cut off, and roads and fishing boats destroyed. When the floods eventually subsided, people were faced with food shortages and disease caused by water supplies contaminated with sewage and dead bodies.

Figure 18.38
Two recent coastal floods

What can be done to reduce the flood risk?

Bangladesh, with its tropical storms, storm surges, rising sea-level and flat, low-lying land, is always going to be vulnerable to flooding by the sea. All that can be done is to try to reduce its impact. Unfortunately, Bangladesh is one of the world's poorer countries and lacks both the money and the technology needed to implement major protection schemes (compare the Dutch Delta Scheme and the Thames Barrier in Case Study 18A). Any flood protection scheme would, therefore, need large loans (which would mean Bangladesh falling further into debt) and technical help from richer organisations and/or countries. The following are two extreme suggestions:

- The Flood Action Plan (FAP), which is a massive proposal, supported by several wealthy countries and to be funded by the World Bank, to:
  – set up regional planning groups to study and monitor local river processes
  – construct a huge flood bank, several hundred kilometres long and up to 10 metres in height, to protect the land from coastal flooding.

The scheme has, however, provoked considerable opposition from groups that are concerned with its huge cost, its sustainability and its social and environmental implications. An alternative would be to allow flooding to continue, as it is essential for farming and fishing, but to try to control the extremes of flooding, ideally through the use of appropriate technology (page 150). For example:

- The building of flood shelters where people can seek refuge during times of flood. The shelters, built on stilts, can house up to 150 families, for several weeks if needed, until the floodwaters subside (Figure 18.39). There is space for fresh water, grain and other emergency food and medical supplies. Also available are toilets, boats (to rescue those who fail to make the shelters in time) and, sometimes, even a school for younger children (Figure 18.40).
- To improve early warning systems. This has, in part, been achieved by government and donor agencies printing and distributing information leaflets and enlisting 33 000 volunteers to educate people on preparing and coping with future storms and surges. The volunteers are also expected, when an event is forecast, to cycle around remoter areas giving residents warning by megaphone.

Figure 18.39
Flooding in the delta region

Figure 18.40
A coastal flood shelter

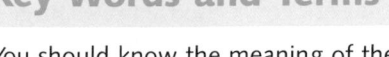

Coasts

Key Words and Terms

a You should know the meaning of the following terms:
- waves • corrasion/abrasion • attrition • corrosion/solution
- hydraulic pressure • longshore drift • groynes • mass movement
- storm/tidal surges • depressions • tropical storms • global warming
- tidal barrier • sea defences • managed coastal retreat/realignment.

b You should know the difference between:
- swash and backwash
- constructive and destructive waves
- hard and soft engineering (sea defences).

Key Ideas

You should know and understand the following:
- That the coast is modified by processes which result in distinctive landforms.
- The processes of coastal erosion – corrasion/abrasion, attrition, corrosion/solution, hydraulic pressure.
- The process by which the sea transports material – longshore drift.
- The conditions under which the sea deposits material.
- The processes that lead to the formation of distinctive coastal landforms – headlands and bays, cliffs, wave-cut notches and wave-cut platforms, caves, arches, stacks, beaches, spits and bars.
- Causes, effects and responses to cliff/coastal erosion/retreat.
- Coastal landforms are influenced by geology.
- Recognition of coastal landforms on an OS map.
- The need for and the strategies to manage and protect coastal areas.
- Causes (physical/human) and effects (primary/secondary) of a coastal flood in an MEDC and an LEDC.
- Attempts to predict/control/manage coastal flooding in an MEDC and an LEDC.

Skills, Theories and Models

- Interpret and use diagrams and maps
- Analyse and use photographs
- Interpret and use OS maps
- Appreciate the need for coastal management
- Identify ways to manage and protect coasts
- Appreciate other people's values and attitudes
- Compare the effects of coastal flooding in MEDCs and LEDCs

1 *(Pages 300 to 304)*

a How do waves form and why do they break? *(2)*

b Describe four processes by which waves can erode the land. *(4)*

c Explain why cliffs (headlands) and bays have formed in area A of the map. *(3)*

d With the aid of a labelled diagram, describe how a stack found at Flamborough Head might have formed. *(4)*

e i) What map evidence is there to show that the Holderness coast has retreated since Roman times? *(1)*

ii) Why is the coast retreating so rapidly at this point? *(1)*

iii) If you were a coastal engineer, what steps might you suggest to prevent further erosion along the Holderness coast? *(4)*

f i) With the aid of a labelled diagram, describe the process of longshore drift. *(3)*

ii) In which direction is longshore drift on the Holderness coast? *(1)*

iii) Why have groynes been built along parts of the Holderness coast? *(1)*

g With the aid of a labelled diagram, describe how the spit at Spurn Head might have formed. *(4)*

2 *(Pages 306 to 309)*

a i) Why is coastal management needed in Britain? *(2)*

ii) Who is responsible for coastal management in Britain? *(1)*

iii) What four options are available in a shoreline management plan (SMP)? *(4)*

b i) Which two sea-defence schemes are shown in the photograph? *(2)*

ii) How do they satisfy the 'factors to consider' in a sea-defence scheme (page 306)? *(2)*

iii) Describe two other sea-defence schemes that could have been used. Why do you think they were not accepted? *(2)*

c With reference to the Norfolk coast, describe:
- why the coast is eroding so rapidly at Happisburgh
- the attempts made to protect the coast at Sea Palling
- why a 'managed retreat' has been suggested. *(3×3)*

3 *(Pages 310 to 313)*

a South-east England and the Netherlands

i) Give four reasons why coastal areas in this region:
- are at constant risk of flooding by the sea
- were flooded during the winter of 1953. *(8)*

ii) What were the effects of the 1953 flood on:
- people • property • economic activities? *(6)*

iii) What were the responses of Britain and the Netherlands to try to lessen the effects of future flooding in this area? *(4)*

iv) Why do some groups of people want the coastline to be protected from flooding while other groups think nature should be allowed to take its course? *(6)*

b Bangladesh

i) Describe three physical and one human cause of flooding by the sea in Bangladesh. *(8)*

ii) With the aid of a labelled diagram, describe the causes and effects of a storm surge. *(5)*

iii) Why do so many Bangladeshis live in this flood-prone area? *(2)*

iv) What are the:
- primary (short-term) • secondary (longer-term)
problems that result from flooding by the sea? *(6)*

v) What can be done to try to reduce the effects of flooding by the sea? *(4)*

vi) Why is it difficult to implement flood prevention schemes in countries like Bangladesh? *(4)*

Glacial systems, processes and landforms

Although it is too warm for glaciers to be found in Britain today, much of the spectacular highland scenery of Scotland, Wales and northern England owes its attractiveness to the work of ice in earlier times. At the height of the Ice Age, northern Britain was covered by a large ice sheet. At other times glaciers extended down valleys leading from the higher mountains.

The glacier system

A glacier, like a river, behaves as a system with inputs, stores, flows and outputs (Figure 19.1). Inputs come from precipitation in the form of snow falling directly onto the glacier, or from avalanches along the glacier sides. Inputs mainly occur near to the head of a glacier in the **zone of accumulation** (Figure 19.1). During fresh falls of snow, air is trapped between the flakes. As more snow falls, the underlying layers are compressed, the air is squeezed out and the snow becomes firmer (like making a snowball). As more snow accumulates, the underlying layers are compressed into ice. Ice without any air (oxygen) left in it turns blue. The glacier itself is water held in storage and, as ice, flows (transfers) downhill under the force of gravity.

Outputs from the system are mainly meltwater with a limited amount of evaporation. When a glacier melts it is called **ablation** (Figure 19.1). Inputs (accumulation) are likely to exceed outputs (ablation) near to the head of a glacier and in winter. Ablation will exceed accumulation in summer and at lower altitudes where temperatures are higher. If, over a period of time, the annual rate of accumulation exceeds ablation, then the glacier will advance. If ablation exceeds accumulation, the glacier will retreat. At present most, but not all, of the world's remaining glaciers are retreating (Figure 19.2).

Erosion

A glacier can erode much faster than a river, but, like a river, it can only erode if it has a continuous supply of material. The main source of material for a glacier results from the process of freeze–thaw weathering (or frost shattering) (page 246). Freeze–thaw occurs in rocks that have many joints and cracks in them, and where temperatures are frequently around freezing point. Water, which gets into the cracks during the day, freezes at night. As it freezes it expands and puts pressure on the surrounding rock. When the ice melts, pressure is released. Repeated freezing and thawing widens the cracks and causes jagged pieces of rock to break off. The glacier uses this material, called **moraine**, to widen and deepen its valley.

There are two main processes of glacial erosion:
1 **Abrasion** is when the material carried by a glacier rubs against and, like sandpaper, wears away the sides and floor of the valley. It is similar to corrasion by a river, but on a much larger scale.
2 **Plucking** results from glacial ice freezing onto solid rock. As the glacier moves away it pulls with it large pieces of rock.

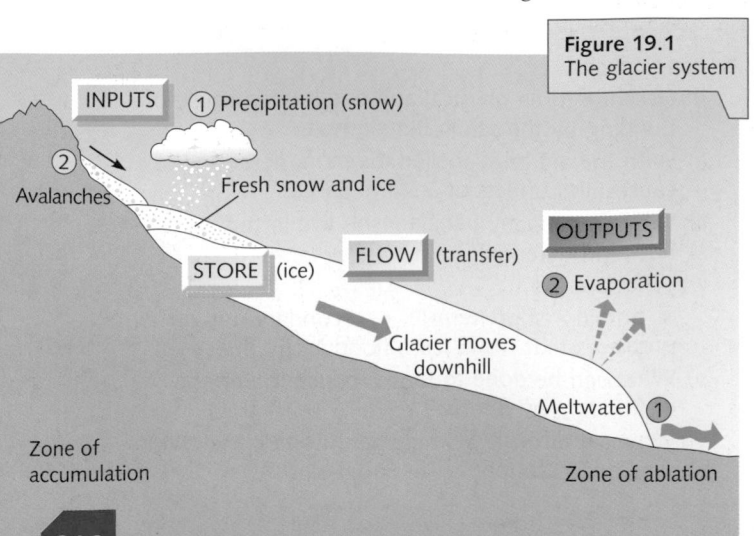

Figure 19.1
The glacier system

INPUTS ① Precipitation (snow)
②
Avalanches
Fresh snow and ice
STORE (ice) FLOW (transfer)
OUTPUTS
② Evaporation
Glacier moves downhill
Meltwater ①
Zone of accumulation
Zone of ablation

Figure 19.2
Yerupaja Glacier, Peru

Glacial landforms

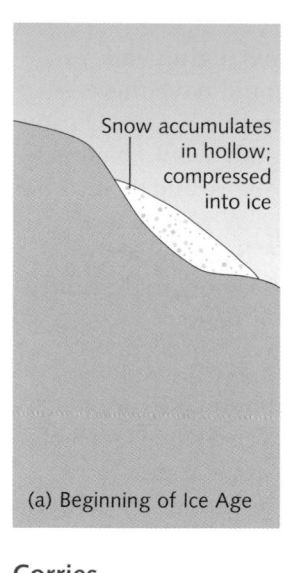

(a) Beginning of Ice Age

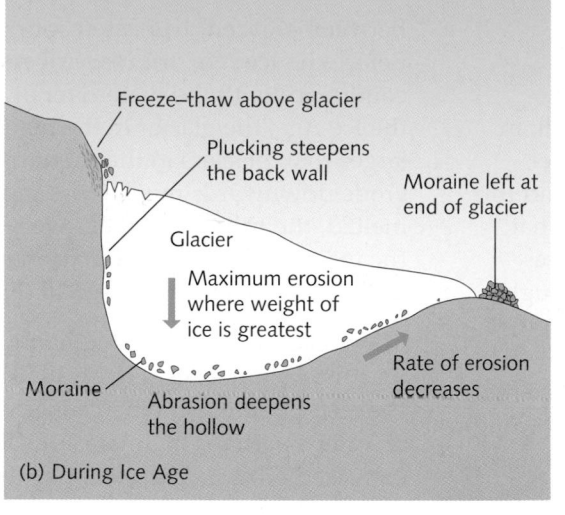

(b) During Ice Age

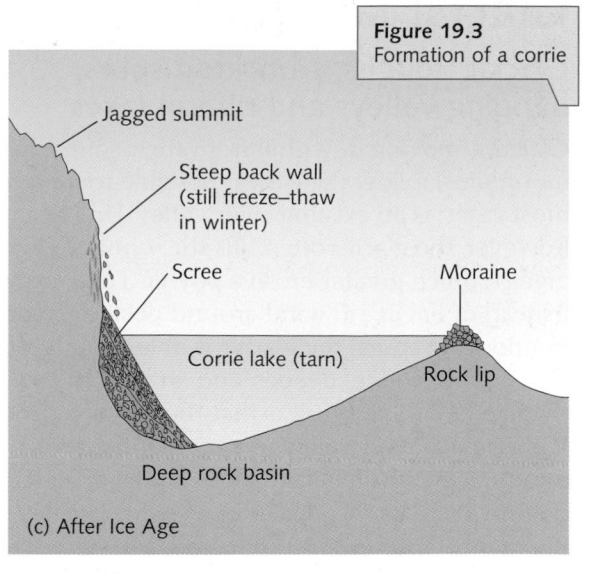

Figure 19.3
Formation of a corrie

(c) After Ice Age

Corries

Corries, which are also known as **cirques** and **cwms**, are deep, rounded hollows with a steep back wall and a rock basin. They began to form at the beginning of the Ice Age when snow accumulated in hollows on hillsides, especially in hollows with a less sunny north- and east-facing aspect (Figure 19.3a). Snow turned into ice, and the ice moved downhill. Freeze–thaw and plucking loosened and removed material from the back of the hollow, creating a steep back wall (Figure 19.3b). Moraine, dragged along the base of the glacier, deepened the floor of the hollow by abrasion, and formed a rock basin. A **rock lip** was left where the rate of erosion decreased. This lip was often heightened by the deposition of moraine. After the Ice Age the rock lip and moraine acted as a natural dam to meltwater, and many rock basins are now occupied by a deep, round **corrie lake** or **tarn** (Figures 19.3c and 19.4).

Figure 19.4
A corrie in Snowdonia, North Wales

Arêtes and pyramidal peaks

When two or more **corries** develop back to back (or side by side), they erode backwards (or sideways) towards each other. The land between them gets narrower until a knife-edged ridge, called an **arête**, is formed (Figure 19.5). Where three or more corries cut backwards into the same mountain, a **pyramidal peak**, or **horn**, develops (Figure 19.6). Arêtes radiate from the central peak.

Figure 19.5
A pyramidal peak: Machhappuchare, Nepal

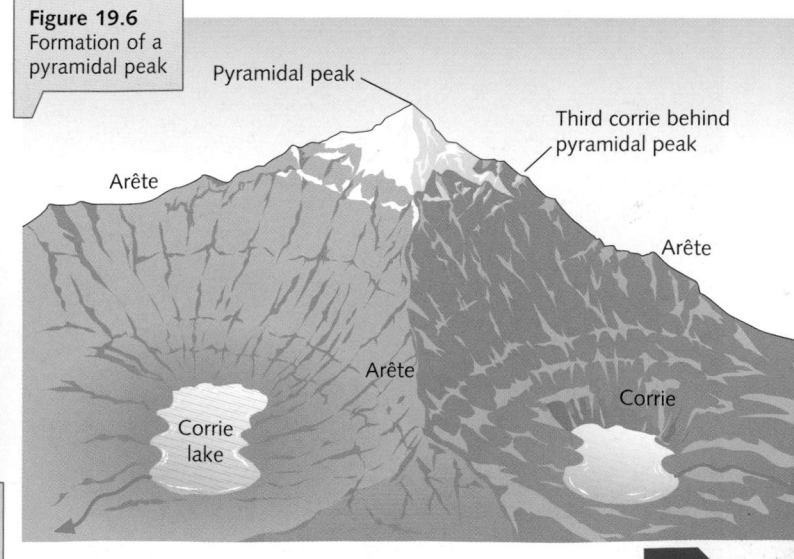

Figure 19.6
Formation of a pyramidal peak

Glacial landforms

Glacial troughs, truncated spurs, hanging valleys and ribbon lakes

Glaciers, moving downhill from their source in the mountains, follow the easiest possible route which, in most cases, is an existing river valley. Unlike a river, however, the glacier often fills the whole valley and this gives it much greater erosive power. This means that, instead of having to wind around obstacles, such as interlocking spurs, the glacier is able, mainly through abrasion, to widen, deepen and straighten its valley (Figure 19.7). The result is that the characteristic V shape of a river valley in a highland area is converted into the equally characteristic U shape of a **glacial trough** (Figure 19.8). As the glacier moves downvalley it removes the ends of interlocking spurs to leave steep, cliff-like, **truncated spurs**.

Between adjacent truncated spurs are **hanging valleys**. Before the Ice Age, tributary rivers would have their confluence with the main river at the same height. During the Ice Age, the glacier in the main valley would be much larger than glaciers in the tributary valleys, and so it could erode downwards much more rapidly. When the ice melted, the tributary valleys were left 'hanging' above the main valley. Each tributary river has now to descend to the main river by a waterfall (Figure 19.9).

Many glacial troughs in highland Britain contain long, narrow, ribbon lakes (Figure 19.8). **Ribbon lakes** are partly the result of erosion when a glacier over-deepens part of its valley, perhaps in an area of softer rock or due to increased erosion after being joined by a tributary glacier. They may also be partly created by deposition of moraine across the main valley.

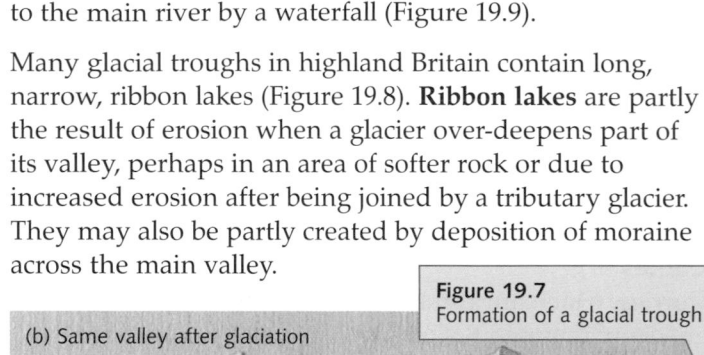

Figure 19.7
Formation of a glacial trough

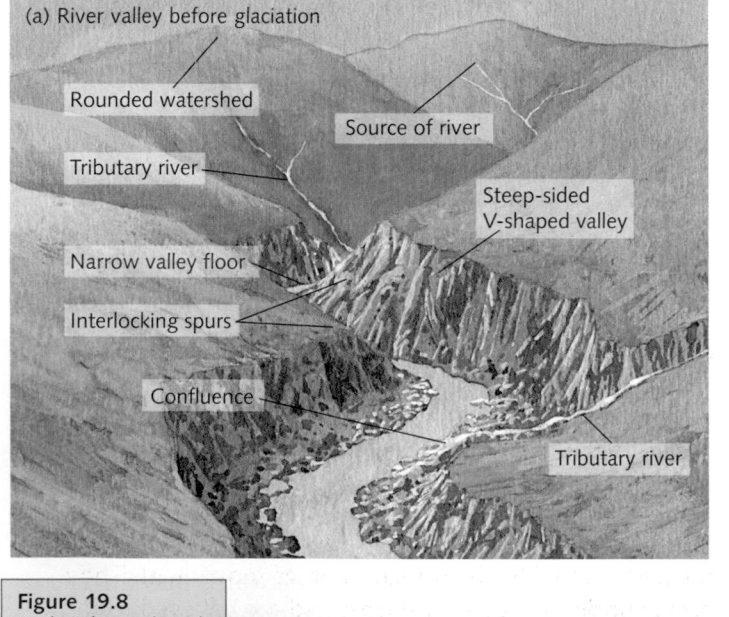

(a) River valley before glaciation

Rounded watershed

Source of river

Tributary river

Steep-sided V-shaped valley

Narrow valley floor

Interlocking spurs

Confluence

Tributary river

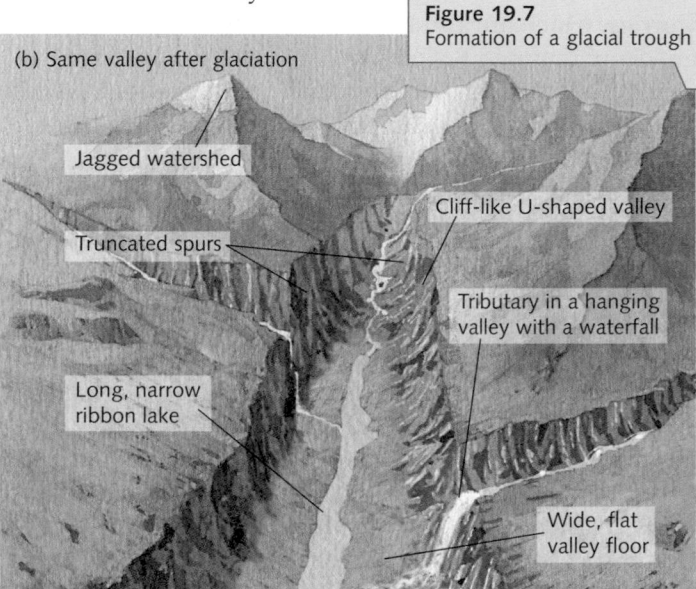

(b) Same valley after glaciation

Jagged watershed

Cliff-like U-shaped valley

Truncated spurs

Tributary in a hanging valley with a waterfall

Long, narrow ribbon lake

Wide, flat valley floor

Figure 19.8
A glacial trough with ribbon lake in Austria

Figure 19.9
Hanging valley with waterfall in California

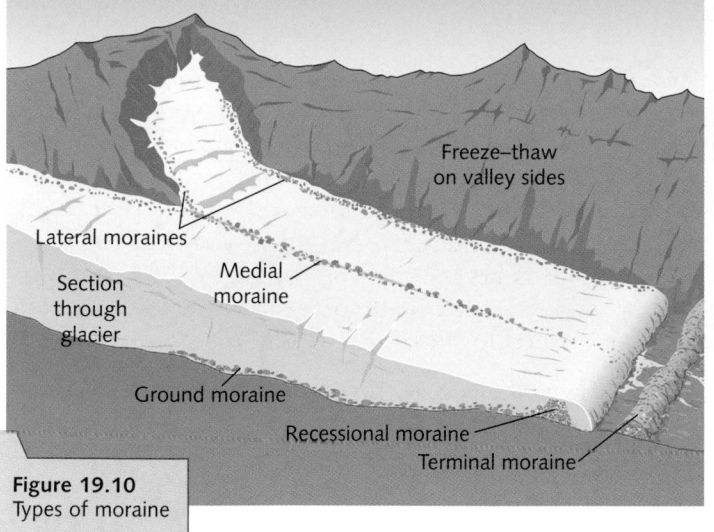

Figure 19.10
Types of moraine

Labels in figure: Freeze–thaw on valley sides; Lateral moraines; Medial moraine; Section through glacier; Ground moraine; Recessional moraine; Terminal moraine

Figure 19.11
Terminal moraine in Greenland

Transportation and deposition

Moraine is material, mainly angular rock, which is transported and later deposited by a glacier. It is deposited when there is a rise in temperature. As the glacier begins to melt, it cannot carry as much material. There are several types of moraine (Figure 19.10):

- **Lateral moraine** is material derived from freeze–thaw weathering of valley sides and which is carried at the sides of a glacier.
- **Medial moraine** is found in the centre of a glacier and results from two lateral moraines joining together.
- **Ground moraine** is material dragged underneath a glacier which, when deposited, forms the flat valley floor. Ground moraine is also referred to as **till** or **boulder clay**.

- **Terminal moraine** marks the maximum advance of a glacier. It is material deposited at the snout, or end, of a glacier (Figure 19.11). If a glacier remains stationary for a lengthy period then a sizeable mound of material, extending across the valley, can build up.
- **Recessional moraines** form behind, and parallel to, the terminal moraine. They mark interruptions in the retreat of a glacier when it remained stationary for long enough for further ridges to develop across the valley. Both terminal and recessional moraines can act as natural dams behind which ribbon lakes can form.

Glaciers can transport material many kilometres. **Erratics** are rocks and boulders carried by the ice and deposited in an area of totally different rock (Figure 19.12). Material from Norway can be found on parts of England's east coast, and Lake District rock on Anglesey.

Drumlins are smooth, elongated mounds of material formed parallel to the direction of ice movement. They often consist of stones and clay, and are believed to result from the load, carried by a glacier, becoming too heavy and being deposited. They owe their streamlined shape to later ice movement (Figure 19.13).

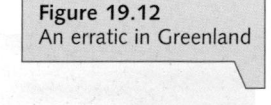

Figure 19.12
An erratic in Greenland

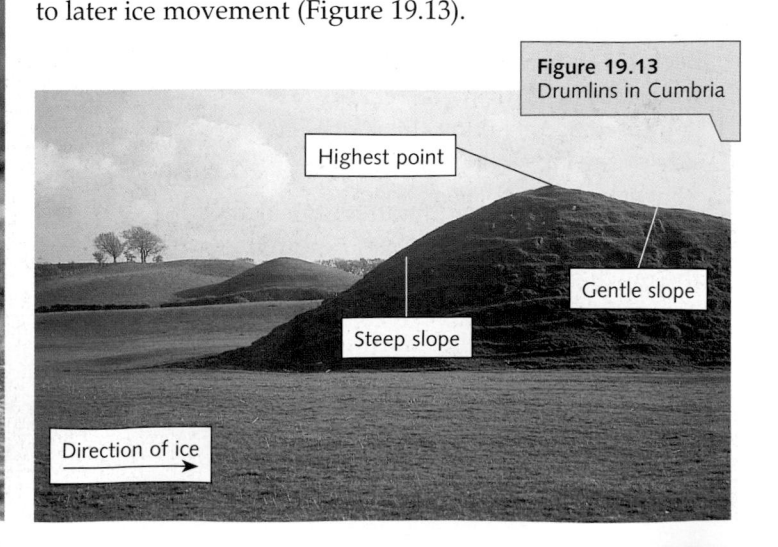

Figure 19.13
Drumlins in Cumbria

Labels in figure: Highest point; Gentle slope; Steep slope; Direction of ice

319

Glacial landforms on OS maps

Figure 19.14 shows part of Snowdonia in north Wales. You should be able to recognise:

- corries with their lakes (Glaslyn 6154 and Llyn Llydaw 6254)
- arêtes (Grib Goch 6255 and Y Lliwedd 6253) and a (partly formed) pyramidal peak (Snowdon 6054)
- glacial troughs (Llanberis 6356 and Nant Gwynant 6250) and a ribbon lake (Llyn Gwynant 6451)
- hanging valleys (the waterfalls in 6251 and 6352) and truncated spurs (6356).

The map does not show any features of glacial deposition as these are too small to be shown on a map of this scale. This does not mean, however, that there are none present – there are, for example, moraines along both sides of Llyn Llydaw (6354).

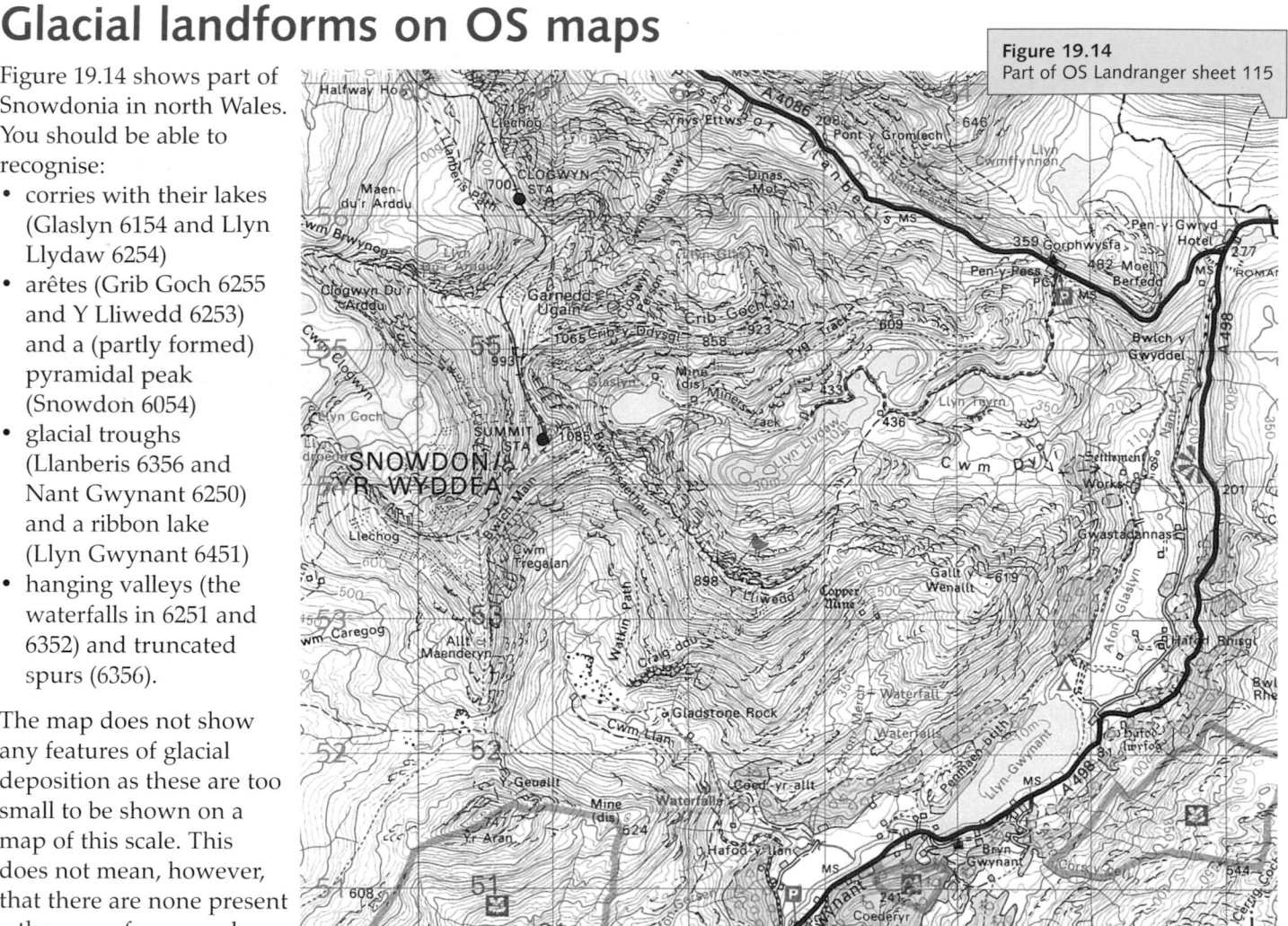

Figure 19.14
Part of OS Landranger sheet 115

Figure 19.15 shows a small area in the Lake District of north-west England. You should be able to recognise:

- corrie lakes (Blea Water 4410 and Small Water 4509) and arêtes (Riggindale Crag 4411)
- a glacial trough (containing Riggindale Beck 4511), hanging valleys (the waterfalls in 4510 and 4811) and truncated spurs (along the south-east side of the lake in 4711).

The map does not show pyramidal peaks (none were formed in this area) nor any of the much smaller features of glacial deposition. You should be aware that, had the map extended further north, it would have shown that the lake (Haweswater in 4712) is a reservoir (i.e. an artificial lake) in a glacial trough and is not a ribbon lake.

Figure 19.15
Part of OS Landranger sheet 90

Glacial highlands and human activity

Glaciated highlands have certain advantages, as well as being nearly always scenically attractive, that encourage human activity (Figure 19.16). There is, however, the danger that development for economic reasons may spoil the natural beauty of these areas.

Glacial highlands as an attraction for human activities

Farming Although highland areas in Britain have often had much of their soil removed by glaciers, have steep slopes due to glacial erosion and, because of their height, have a cold, wet climate, they tend to be ideal for sheep farming (page 100). In contrast, farming in mountainous areas of countries such as Norway and Switzerland involves a seasonal movement of animals, including cattle, from high summer pastures (which become snow-covered in winter) to low-lying winter pastures (transhumance).

Forestry Where the land has some soil, even if it is of poor quality (podsols, Figure 15.32), and the land is not too steep, forests, usually of either natural or replanted coniferous trees, may be found (page 235). The trees, when located in more accessible areas, give rise to a logging industry.

Water supply Ribbon lakes, especially when located in areas of high rainfall, form natural reservoirs (Figure 19.17), while glacial troughs with their steep sides can provide ideal sites for artificial reservoirs (e.g. Haweswater, Figures 19.15 and 13.27).

Energy resources Hanging valleys provide a natural 'head' of water needed to turn turbines in hydro-electric power stations (Figure 8.10). Exposed hillsides provide ideal sites for wind farms (Figure 8.18).

Tourism Areas with glaciers and/or winter snow often provide locations for ski-resorts (e.g. Courmayeur, page 166). Elsewhere, steep mountainsides are ideal for rock-climbing, the lower slopes for walking, and the scenic views for the passive tourist who wishes simply to enjoy the natural beauty. Ribbon lakes, such as Windermere, can be used for watersports.

Figure 19.16
Snowdon in winter

Figure 19.17
Reservoir in the Lake District

The effect of human activities on glaciated highlands

Farming Apart from modifying the natural vegetation, sheep farming and transhumance have few adverse effects.

Forestry Some groups of people feel that re-afforestation spoils the natural look of hillsides. Felling (logging), on a big scale, can leave huge scars and, if not managed carefully, can cause soil erosion (page 254).

Water supply The drowning of a glacial trough to provide a reservoir means the loss of homes and jobs for people previously living and working there.

Energy resources Hydro-electric power stations may, if they are not screened, be unsightly, as are the pylons and power lines needed to transfer the electricity to the National Grid. Wind farms are considered, by some people, to be inappropriate in areas of scenic beauty.

Tourism Winter sports resorts, with their accommodation, après-ski amenities, ski-runs, ski-lifts and approach roads, spoil the natural scenery and erode the fragile natural alpine vegetation. Skiing off-piste can also increase the risk of avalanches (Figure 19.18). Walkers can wear away footpaths (Figure 10.38), while watersports enthusiasts may pollute the ribbon lakes. As more people travel to these areas just to admire the scenery, there is an increased demand for amenities such as hotels, B&B facilities, camping and caravan sites, car parks, restaurants and gift shops (Case Study 10 – The Lake District).

Figure 19.18
Avalanche!

321

Glaciation

Key Words and Terms

a You should know the meaning of the following terms:
- glacier • moraine • freeze–thaw/frost shattering
- abrasion • plucking • transhumance
- avalanche.

b You should know the difference between:
- glacial systems and glacial budgets
- zone of accumulation and zone of ablation.

Key Ideas

You should know and understand the following:

- Highland areas that are modified by glacial processes have very distinctive landforms.
- The processes of ice erosion – freeze–thaw/frost shattering, abrasion and plucking.
- That material – moraine – is transported by ice/glaciers.
- The conditions under which ice/a glacier deposits material.
- The processes that lead to the formation of distinctive glacial features – corries (cirques/cwms), arêtes, pyramidal peaks, glacial troughs, hanging valleys, truncated spurs, ribbon lakes, moraines, erratics and drumlins.
- Characteristics of valleys changed by glacial activity.
- Recognition of glacial features on an OS map.
- The land use and resultant conflicts in glaciated uplands/highlands.
- Interaction between people and highland glacial environments.

Skills, Theories and Models

- Interpret and use diagrams
- Analyse and use photographs
- Interpret and use OS maps
- Understand the glacier system
- Appreciate the relationship between the natural environment and human activity

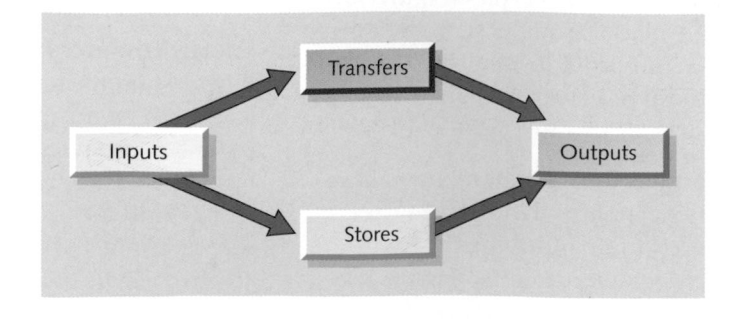

1 *(Page 316)*

a Copy and complete the diagram to show a glacier system. *(8)*

b How does the glacial budget explain why a glacier either retreats or advances? *(2)*

c i) Describe the process of freeze–thaw weathering. *(2)*

ii) Describe two processes by which a glacier can erode the land. *(4)*

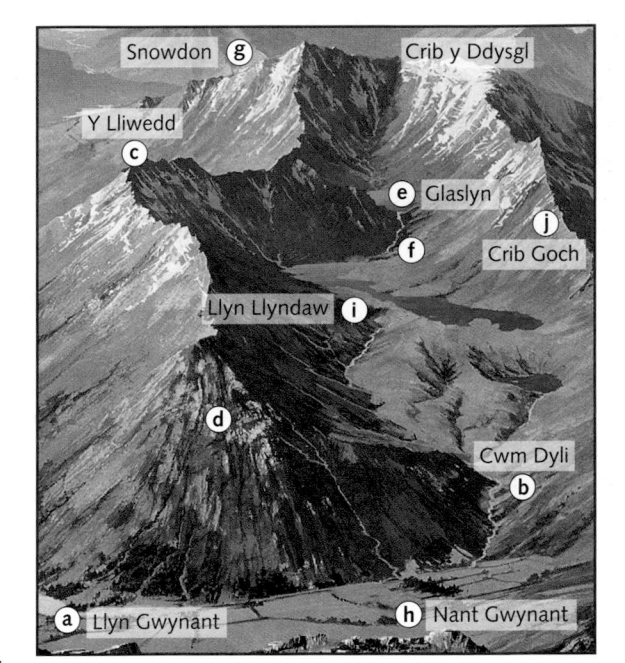

2 *(Page 317)*

Rearrange the following eleven statements to give a description of the formation of a corrie (cirque or cwm).

• ice melts to leave a corrie lake
• snow collects in north-facing hollows
• freeze–thaw still continues producing scree
• moraine left at the end of the glacier
• decrease in erosion leaves a rock lip
• freeze–thaw weathering loosens rock
• glacier moves downhill under gravity
• snow is compressed until it becomes ice
• corrie floor is overdeepened by abrasion
• ice with all air (oxygen) squeezed out turns blue
• plucking removes rocks from the backwall. *(11)*

3 *(Pages 317 to 320)*

The fieldsketch shows ten glacial landforms found in the Snowdon area of North Wales. These ten features, labelled **a** to **j**, include (but not in this order) two corries, two arêtes, a pyramidal peak, a glacial trough, two hanging valleys, a truncated spur, and a ribbon lake.

Either

a Make a list matching the landforms with their appropriate letter. *(10)*

Or

b Make a copy of the sketch and add the names of the ten glacial landforms. *(10)*

c With the help of well-labelled diagrams describe the formation of a:
• pyramidal peak   • glacial trough
• hanging valley   • terminal moraine. *(12)*

4 *(Page 321)*

With the help of the star diagram, describe how:
• a glaciated highland influences human activity
• human activity affects a glaciated highland. *(5 × 2)*

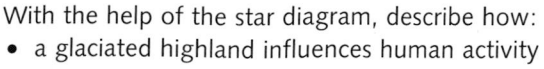

What is globalisation?

Had you seen a geography syllabus or textbook a decade ago, it is unlikely that you would have read the term 'globalisation'. Yet since that time, about when you first started school, the word has become almost overused with it being referred to seemingly daily in the media. Globalisation has a wide range of meanings, including 'transforming local, regional and national phenomena into global ones'. From a geographer's point of view, it includes any process of change that occurs on a world scale and which has worldwide effects either on people or on the environment. These processes might be:

- **physical** as illustrated by the global rise in sea-level
- **human** as in the case of trade, banking, tourism, sport and fashion
- **environmental** as with global warming.

Although the term may be new, the concept is not. For example, the Roman Empire spread across most of the then known world; Magellan was the first truly global explorer when he sailed around the world; and transnational corporations (TNCs) have, for several decades, operated as global organisations. At present, the increasingly rapid growth and effect of globalisation is probably mainly due to **space-shrinking technologies** such as satellite television, the internet, emails and the mobile phone (page 332). These innovations mean that events occurring in one part of the world can be seen across the globe as, or very soon after, they occur – events such as a car bomb explosion, an earthquake or a sporting world cup final. Likewise, events in one part of the world can spread rapidly to affect all other parts, as in the case of the present global economic recession or the threat of a swine flu pandemic. The links created by globalisation may be grouped under the terms environmental, economic, technological, social-cultural and political, although in reality it is difficult to draw borders between them (Figure 20.1).

a

b

Environmental	Economic	Technological	Social-cultural	Political
Climate change Rising sea-levels Clean water supply Health	Trade/free trade Migration (workers/refugees) Finance/banking TNCs Transport Aid	Internet Email Mobile phones Satellite TV	Sport Fashion Music Tourism	Trade groups United Nations

Notice that several of the above could be placed in more than one column.

c

d

Figure 20.1
Global links
(a) Banking: Hong Kong and Shanghai International Bank, Hong Kong
(b) Sport
(c) Transport: Terminal 5, Heathrow
(d) Technology – mobile phone

a

b

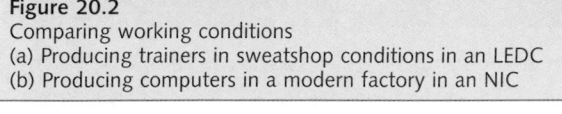

Figure 20.2
Comparing working conditions
(a) Producing trainers in sweatshop conditions in an LEDC
(b) Producing computers in a modern factory in an NIC

The effects of globalisation may be regarded as either beneficial or detrimental according to your point of view, or perhaps depending on which country or part of the world you live in.

- Some people argue that globalisation spreads wealth, knowledge and personal contacts across the world.
- Others claim that it is creating an unfair world in which rich countries and large organisations exploit the world's poorest people, which increases rather than decreases the development gap.

Figure 20.2a shows a sweatshop in South-east Asia where cheap trainers are made for the global market. To those employed, this factory may provide the sole source of regular paid employment, but to others it can be seen as an international company exploiting the workers by

paying them very little for working long hours in poor conditions. This can be compared with Figure 20.2b where those employed are highly paid and work in a more pleasant environment. Figure 20.3 compares the inside of a large luxury department store in a developed country and a small rural market in a developing country. These photographs illustrate how globalisation has increased the wealth and quality of life of some people and places, but often has only increased the gap between them and the larger number of poor people and poor countries.

This chapter is unlike all the others in this book. It looks at a series of topics, each of which, although linked under the general heading of globalisation, has been written as a self-contained issue.

a

b

Figure 20.3
Contrasts in commerce
(a) Luxury department store in an MEDC (South Korea)
(b) Village market in an LEDC (South Africa)

Millennium Development Goals

At the Millennium Summit in 2000, world leaders committed their countries to a new global partnership. They did this by setting up a series of targets, which have become known as the **Millennium Development Goals (MDGs)**, aimed at reducing **extreme poverty**. Extreme poverty has been defined by the UN as when people 'lack the basic human rights of health, diet, shelter, education and security' (Figure 20.4) and is measured by having to live on $1 or less per day. The UN claimed that at the beginning of this century, nearly 1 billion people – about one in every six of the world's population – lived in extreme poverty. Figure 20.5 shows what the MDGs have set out to achieve by the year 2015.

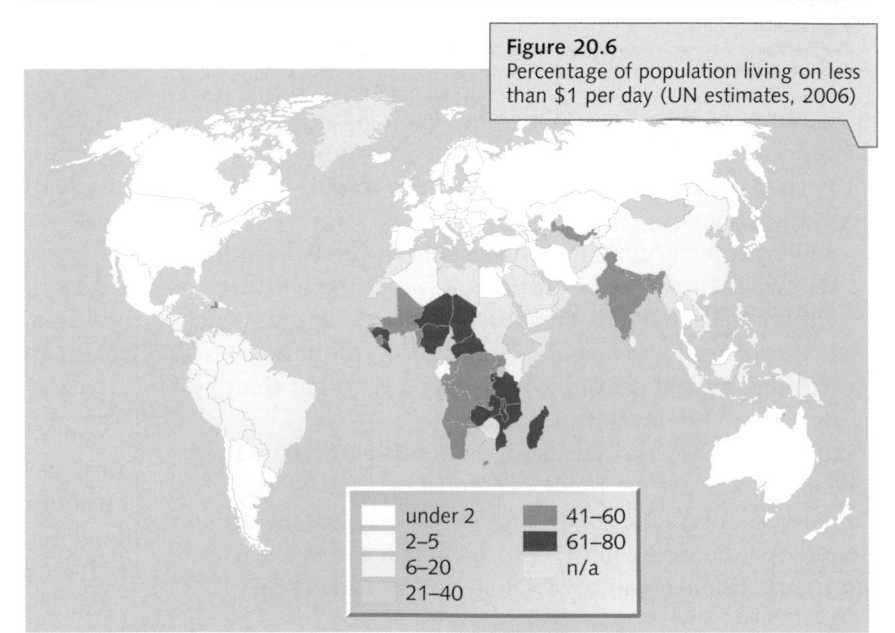

Figure 20.4
Living in extreme poverty

Millennium Development Goals
• To halve the number of people: – living on less than $1 per day – without access to safe water and basic sanitation – suffering from hunger
• To reduce: – by two-thirds the under-5s mortality rate – by three-quarters maternal mortality
• To halt and to begin to reverse the spread of HIV/AIDS
• To improve the availability of adequate housing
• To eliminate gender disparity in education

Figure 20.5
MDGs and basic human rights

A report by the MDGs team in 2006 claimed that significant progress had been made in several parts of the world. The number living in extreme poverty had declined by 13 per cent (130 million), average incomes and life expectancy had both risen, infant mortality had fallen, and more people had access to clean water and improved sanitation (up 8 per cent and 15 per cent respectively). However, the team also pointed out large disparities between people in countries in sub-Saharan Africa, where poverty was often still increasing (Figure 20.6); and between the richer urban areas and poorer rural areas. People in both the less well-off places are still caught up in the so-called **cycle of poverty** (Figure 20.7).

Figure 20.6
Percentage of population living on less than $1 per day (UN estimates, 2006)

under 2	41–60
2–5	61–80
6–20	n/a
21–40	

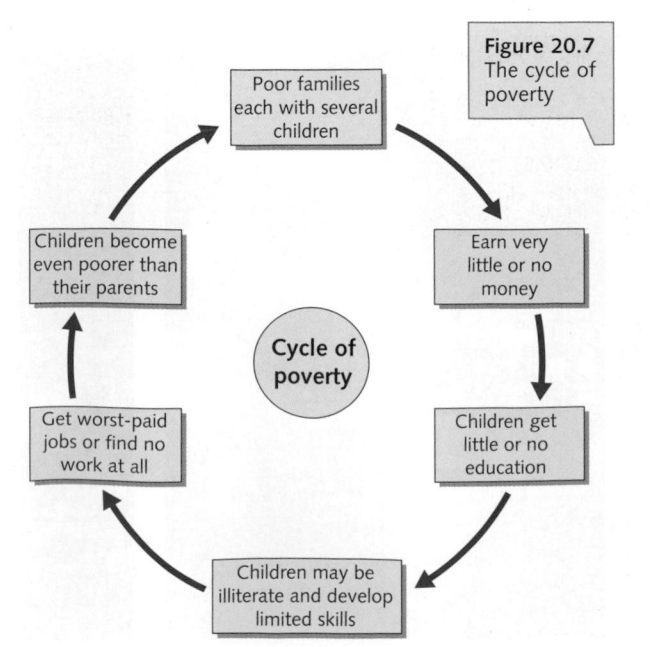

Figure 20.7
The cycle of poverty

Poor families each with several children → Earn very little or no money → Children get little or no education → Children may be illiterate and develop limited skills → Get worst-paid jobs or find no work at all → Children become even poorer than their parents →

Cycle of poverty

MDGs – health

Good health, one of the MDGs, is likely to be denied to people living in extreme poverty. This may be due to the presence of disease and hunger, a lack of clean water and basic sanitation, and insufficient money to provide hospitals, doctors, nurses and medical supplies. Health is linked to economic development, with several measures of development (Figure 11.3) being birth rate, the total fertility rate, infant mortality and life expectancy, to which can be added an adequate and balanced diet.

Globalisation has increased the speed at which disease may spread around the world. Examples include the threat of an influenza **pandemic** (meaning a worldwide epidemic), such as the bird flu scare in 2008 and the spread of swine flu in 2009. These, fortunately, have not had the same devastating effect as has HIV/AIDS, a disease first written about in 1981 (it had been identified in the 1970s). AIDS (acquired immune deficiency syndrome) has become pandemic since the 1980s and is one of the greatest threats to human health. HIV (human immunodeficiency virus) can be passed from person to person in one of three ways: by the exchange of body fluids during sexual intercourse, through infected blood after sharing needles/syringes or contaminated transfusions, and from mother to child during pregnancy. HIV/AIDS reverses the life expectancy gains, limits the ability to work and therefore productivity, decimates the workforce and increases poverty which, together, threaten the realisation of the MDGs.

In 2007, UNAID and WHO (the World Health Organization) published a report that included both disheartening and encouraging signs (Figure 20.8).

Continuing problems
All countries across the world are affected but it is most prevalent in sub-Saharan Africa where two-thirds of all cases are found (Figure 20.9).
Life expectancy, especially in the worst-affected countries, is continuing to fall (down to 41 in Zambia).
Some countries have developed a very uneven population structure – compare the population structure in Figure 20.10 with that of a normal LEDC (Figure 1.9).
Still no known cure

Encouraging signs
The number of people with HIV, many of whom were born with it, has fallen from a peak of 38 million in 2005 to 33.2 million in 2007.
The number of new infections has fallen from just over 3 million in the late 1990s to 2.5 million in 2007 (this is still 6800 new infections each day).
Deaths from AIDS fell to 2.1 million in 2007 (an average of 5700 each day).
Recently, antiretroviral therapy is beginning to prolong life.

Figure 20.8
HIV/AIDS: 2007 UNAID/WHO report

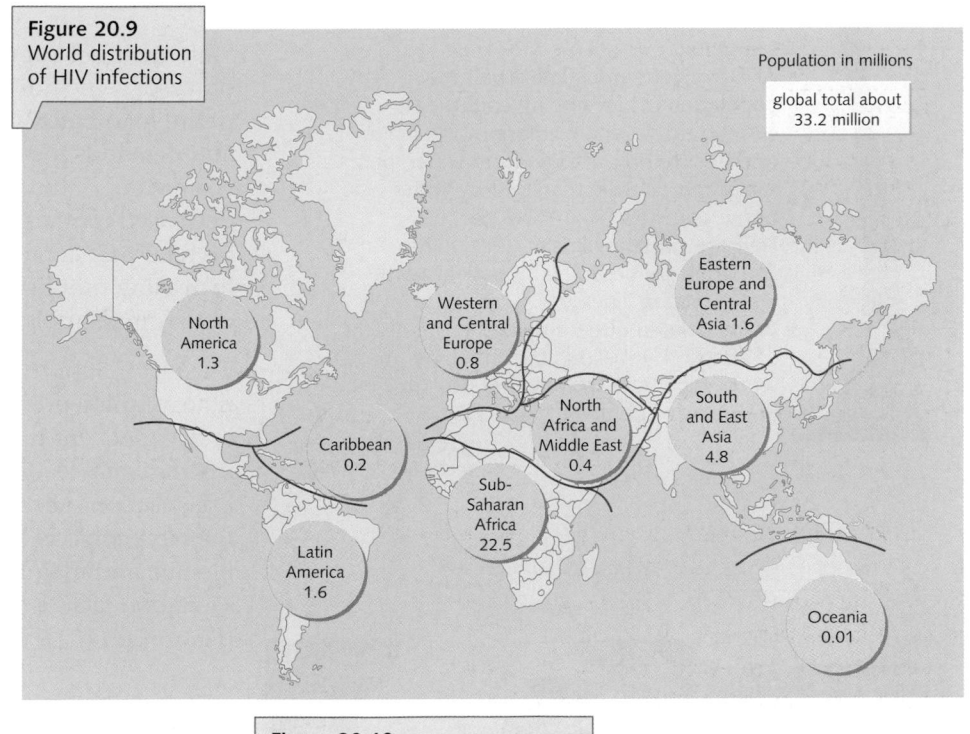

Figure 20.9
World distribution of HIV infections

Population in millions

global total about 33.2 million

North America 1.3

Western and Central Europe 0.8

Eastern Europe and Central Asia 1.6

Caribbean 0.2

North Africa and Middle East 0.4

South and East Asia 4.8

Sub-Saharan Africa 22.5

Latin America 1.6

Oceania 0.01

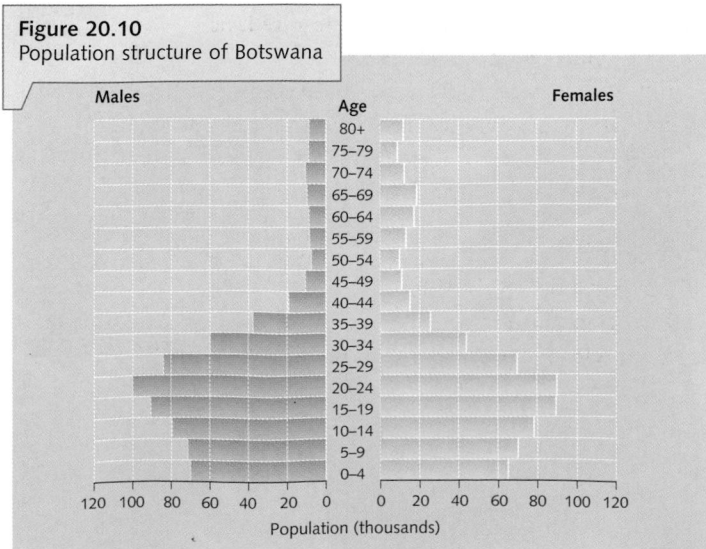

Figure 20.10
Population structure of Botswana

Males — Age — Females

80+
75–79
70–74
65–69
60–64
55–59
50–54
45–49
40–44
35–39
30–34
25–29
20–24
15–19
10–14
5–9
0–4

120 100 80 60 40 20 0   0 20 40 60 80 100 120

Population (thousands)

MDGs – clean water

The amount of fresh water in the world is limited. The human race, together with the other species that share the planet, cannot expect an infinite supply. Only 2.5 per cent of the world's water is not salty and two-thirds of that is stored in icecaps and glaciers. Yet latest estimates suggest that, over the next two decades, the demand for water will increase by 40 per cent, which would mean the average available supply per person will drop by one-third. Over 200 scientists in 50 countries have identified water shortage as the most worrying environmental problem, after global warming, to be faced in this century.

UN Report 2008

884 million people – one in eight of the world's population – do not have access to safe water (Figure 13.35).

- 2.5 billion – more than two in five of the world's population – do not have adequate sanitation.
- At any given time, almost half the population of developing countries is suffering from one or more of the main diseases that result from dirty water, poor sanitation or are transferred by water-borne parasites (Figure 20.12).
- Half the hospital beds in developing countries are occupied by patients with water-related illnesses.
- Water-related disease results in the deaths of almost 1.4 million children across the world each year, including 4000 a day in developing countries.

Figure 20.11
UN Report 2008

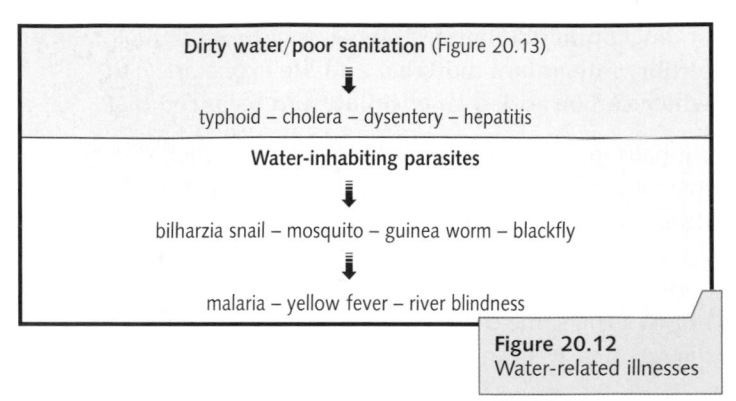

Dirty water/poor sanitation (Figure 20.13)
⬇
typhoid – cholera – dysentery – hepatitis

Water-inhabiting parasites
⬇
bilharzia snail – mosquito – guinea worm – blackfly
⬇
malaria – yellow fever – river blindness

Figure 20.12
Water-related illnesses

Water shortages are expected to increase for a combination of reasons, including world population growth and its desire for a better quality of life, increased water pollution and demands for irrigation (agriculture already uses 70 per cent of the water) and, in some parts of the world, decreased rainfall resulting from climate change. The UN Secretary-General has expressed concern over the increasing number of global conflicts inflamed by disputes over water supply.

The fact that in the first six years of the MDGs 8 per cent more of the world's population had access to clean water and 15 per cent to improved sanitation was not due to international leadership (page 326) but to charitable organisations such as WaterAid. These organisations aim to help people in some of the poorest countries to set up, operate and maintain their own safe domestic water and sanitation facilities and to learn safe hygiene practices (Figure 20.14).

Figure 20.13
Collecting water in Kolkata

Figure 20.14
A WaterAid project in Africa

MDGs – availability of food

Hunger has been defined as 'scarcity of food in a country leading to the painful sensation experienced by its inhabitants'. This is in contrast to malnutrition (page 110), which indicates a lack of some or all of the nutritional elements necessary for a healthy life. There are two types of malnutrition:

1 Protein-energy malnutrition, which is the more serious, where there is insufficient protein (e.g. from meat and milk) to provide the energy, measured in calories (Figure 7.35), needed for a healthy, active life.

2 Micronutrient (vitamin and mineral) deficiency, e.g. a lack of:
   • vitamin A, which reduces the body's resistance to disease and retards growth in children
   • iron, which can cause anaemia
   • iodine, which can endanger children's mental health and, in pregnancy, result in stillbirths.

Malnutrition is believed to be the underlying cause of almost half of all child deaths worldwide each year.

The main cause of hunger today is poverty, followed by conflict. The world still manages to grow sufficient to feed every person but far too many either cannot afford to buy it – the 850 million who have to live on $1 a day – or are unable to access what food is available (compare the maps in Figures 20.16 and 20.6). In 1996, the FAO set a target of reducing by half the then 824 million people who suffered from malnutrition in developing countries, by 2015. Their most recent estimate, in 2006, suggests that the number has actually risen (to 854 million) and that 75 million people now rely entirely on food aid organised by the United Nations (Figure 20.15).

Rising food prices

After several decades of relative stability, food prices have risen sharply with, as one example, wheat being double the price in 2008 that it had been 12 months earlier. These price rises have been attributed to a variety of causes:

Figure 20.15
Distribution of food aid in Somalia

• the continuing growth in the world's population and in the number of mouths to be fed
• the economic emergence of China and India where greater affluence has led to a rising demand for meat and a wider range of cereals
• more land and cereal crops being used to produce biofuels, i.e. to fuel vehicles not feed people (page 125)
• desertification (Case Study 15), which is increasing in China; several years of drought which has affected Australia's croplands; and climatic change which is resulting in less rain falling in parts of sub-Saharan Africa
• developed countries, and especially supermarket chains located there, encouraging developing countries to grow crops for export rather than to feed themselves (pages 104 and 105)
• the reduction in subsidies paid to farmers in North America and the EU which means less food is being grown and held in storage (page 107).

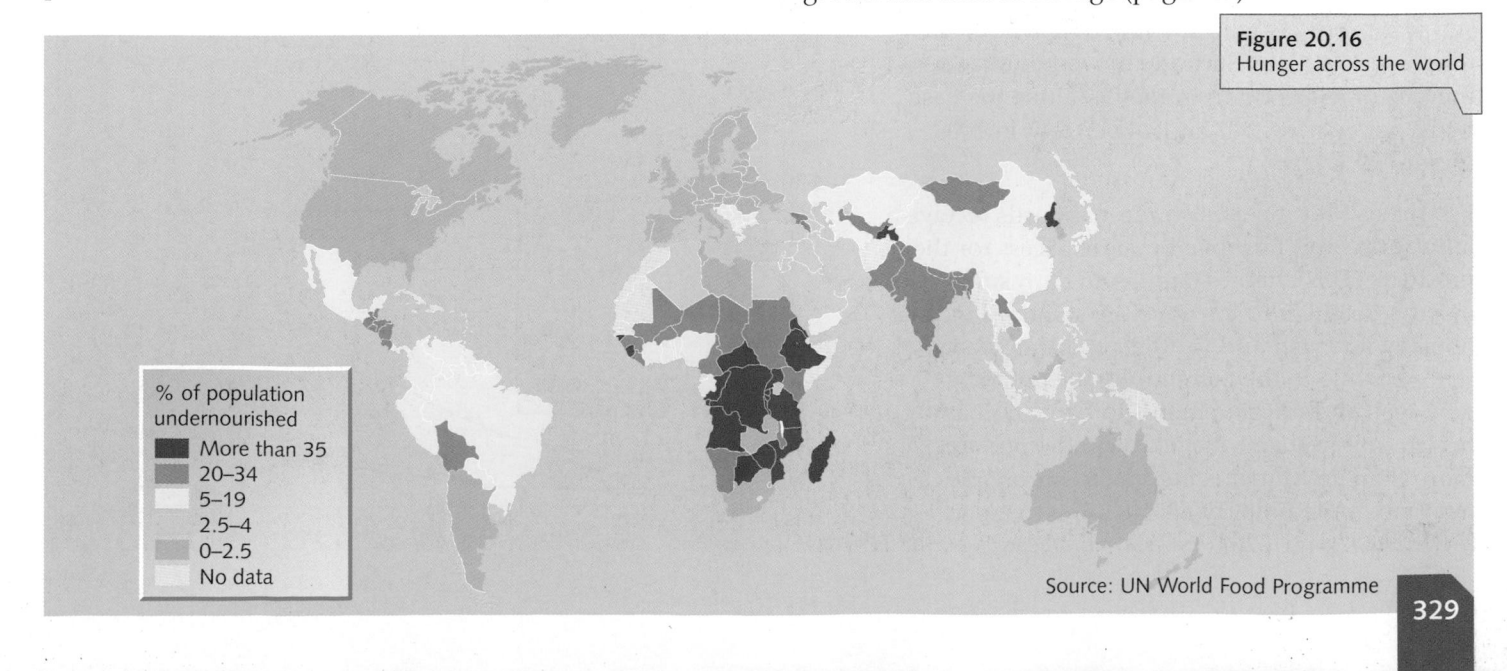

Figure 20.16
Hunger across the world

% of population undernourished
- More than 35
- 20–34
- 5–19
- 2.5–4
- 0–2.5
- No data

Source: UN World Food Programme

Global linkages – fair trade

World trade is not fair, with farmers, workers and countries at the beginning of the chain rarely getting a fair share of its benefits (page 331). As described on pages 186–7, developing countries have to rely on the export of low-value primary goods such as foodstuffs and minerals, whereas the developed countries are able to process these materials and sell them at a much higher price. Added to this, it is often the developed countries that control the prices of the primary products, keeping them as low as possible for their own benefit, and they may also, in order to protect their own industries and workers, impose quotas and restrictions on goods made more cheaply in developing countries.

Fairtrade is an international movement that aims to give better prices, improved working conditions, local sustainability and fairer terms of trade for farmers and workers in developing countries. By requiring large companies to pay a sustainable fair price, Fairtrade tries to overcome the injustices of conventional trade which traditionally discriminates against the poorest and weakest producers. It sets out to help these underprivileged producers – many of whom have to live on under $1 a day (page 326) – move towards self-sufficiency and sustainability. It also helps to improve both social and environmental standards (Figure 20.18). An increasing number of transnational corporations, including Nestlé (coffee), Cadbury (chocolate) and Tate & Lyle (sugar) together with superstores such as Asda, Sainsbury's, Tesco and Marks and Spencer, are being encouraged by shoppers to stock and support products with the FAIRTRADE Mark (Figure 20.17). The Fairtrade movement has seen the sale of its products in the UK alone increase by 72 per cent in 2007 and 44 per cent in 2008 (Figure 20.19).

Producers that meet Fairtrade standards receive a fairer price and improved market access for their products. This enables families to earn sufficient income for them to improve the standard of their housing and diet and for their children to stay at school, while to the community it can mean a new well and improvements to education and health care in their village. Fairtrade, perhaps more than anything, is helping to lift people out of extreme poverty and helping to meet the MDGs (page 326).

Figure 20.17
Fairtrade products

Figure 20.18
A Fairtrade farmer

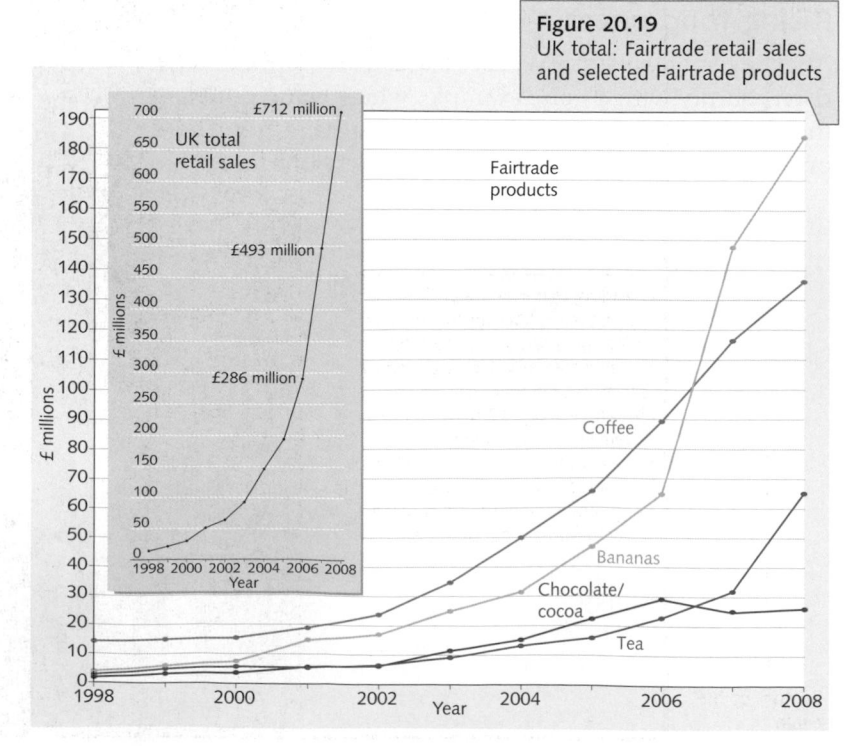

Figure 20.19
UK total: Fairtrade retail sales and selected Fairtrade products

Global linkages – industry

The **commodity chain**, also now referred to as the **global value chain**, is a connected series of activities that sees a product from its initial concept (design) through production (processing) to distribution (retailing). Initially all these stages would have been done within one country, e.g. shipbuilding in the UK in the early twentieth century (page 138). Later, with the growth of TNCs (page 144), the commodity chain was extended globally as the parent company located factories and retail outlets in different countries. This was done so that each stage could be located in a country that offered either the lowest costs (e.g. land values and labour), the greatest technological expertise (e.g. design) or access to capital (banks).

The fashion industry, which includes the production of training shoes and sportswear, is one that has been revolutionised by globalisation. As Figure 20.20 shows, farmers producing the raw materials are at the bottom of the chain, followed, in ascending order, by workers in developing countries making the product, and customers, designers and managers in the NICs and the more economically developed countries.

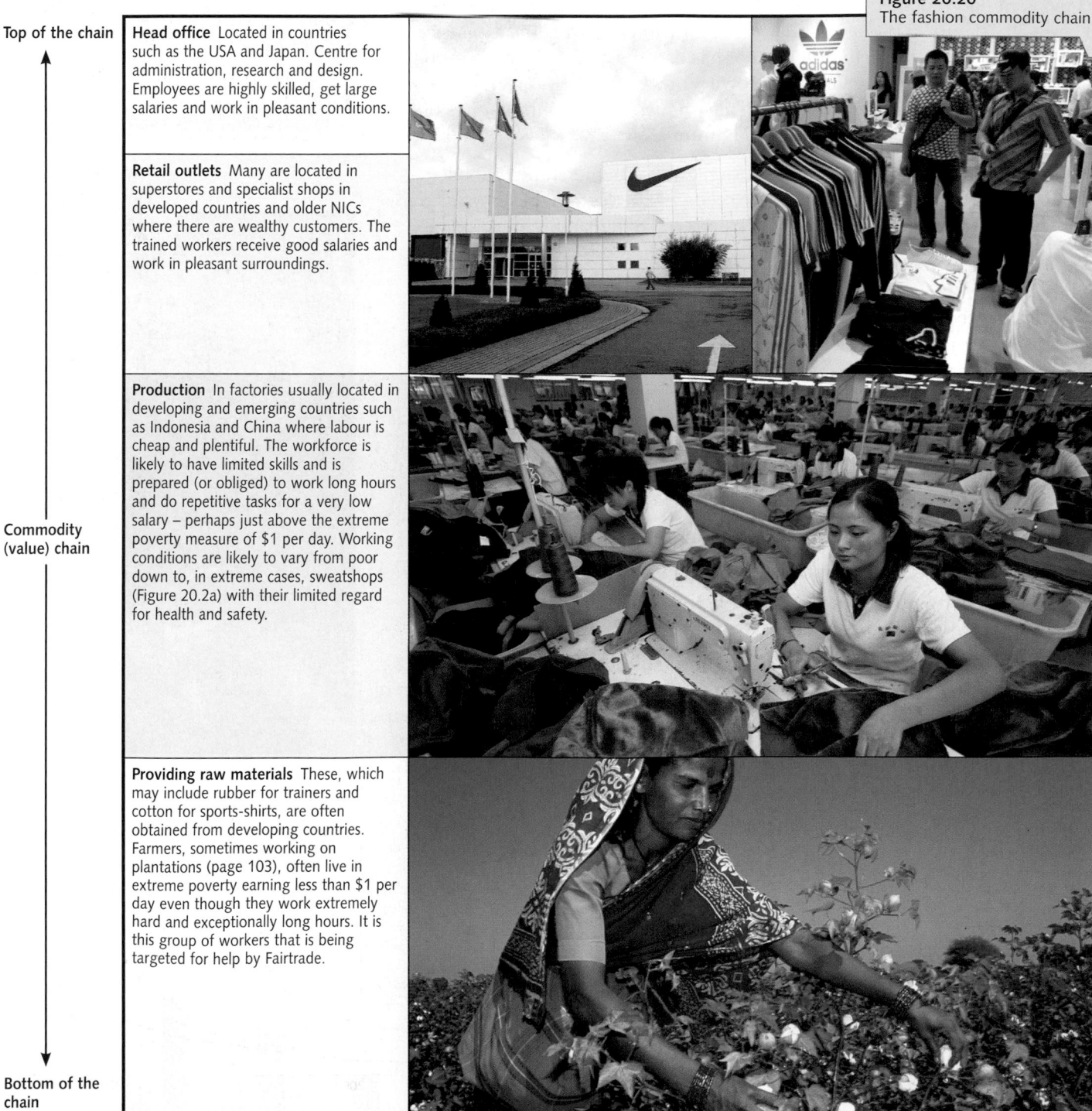

Figure 20.20
The fashion commodity chain

Top of the chain

Head office Located in countries such as the USA and Japan. Centre for administration, research and design. Employees are highly skilled, get large salaries and work in pleasant conditions.

Retail outlets Many are located in superstores and specialist shops in developed countries and older NICs where there are wealthy customers. The trained workers receive good salaries and work in pleasant surroundings.

Commodity (value) chain

Production In factories usually located in developing and emerging countries such as Indonesia and China where labour is cheap and plentiful. The workforce is likely to have limited skills and is prepared (or obliged) to work long hours and do repetitive tasks for a very low salary – perhaps just above the extreme poverty measure of $1 per day. Working conditions are likely to vary from poor down to, in extreme cases, sweatshops (Figure 20.2a) with their limited regard for health and safety.

Providing raw materials These, which may include rubber for trainers and cotton for sports-shirts, are often obtained from developing countries. Farmers, sometimes working on plantations (page 103), often live in extreme poverty earning less than $1 per day even though they work extremely hard and exceptionally long hours. It is this group of workers that is being targeted for help by Fairtrade.

Bottom of the chain

Global linkages – space-shrinking technologies

People woke up on the morning of the Queen's coronation in 1953 to hear that Edmund Hillary and Sherpa Tenzing had become the first people to climb Mount Everest. They had, in fact, reached the summit several days before, but it had taken that long for them to return to base camp and for news of their ascent to reach London. How different when, in 2009, almost 56 years later to the day, Sir Ranulph Fiennes sent a live commentary and pictures of his Everest climb to people across the world as he approached and reached the summit (Figure 20.21). Technology has changed!

Since the mid-1990s, the telecommunications/ICT sector has experienced major advances. In 1995, less than 3 per cent of the world's population owned a mobile phone and only 1.5 per cent had access to the internet. By 2007, over 53 per cent had mobiles sending, collectively, up to 300 000 texts each minute (Figure 20.22) and 23 per cent had access to the internet. It is the expansion of these **space-shrinking technologies** that has become a major feature in the growth and spread of globalisation. However, as with other aspects of development, the increasing use of mobile phones and the internet has not been spread evenly across the world (Figure 20.23).

Mobile cellular phones Latest figures suggest that ownership of mobile phones in some developed countries exceeds 100 per cent, which means that there is more than one phone for every person. Ownership in developing countries is now also increasing rapidly mainly due to the higher level of competition between mobile phone manufacturers and a resulting drop in their prices. In 2007, more than one out of every four African, one in three Asian and one in two globally (over 3 billion people) had mobiles (Figure 20.23).

The internet A much wider digital divide still exists, and is likely to continue, in the use of the internet and especially broadband. Whereas in 2007 63 per cent of people living in developed countries could go on line, the figure was much lower in developing countries, especially in Africa where the average was only 5 per cent. Together, this means only 23 per cent globally (about 1.4 billion people) use the internet. This difference is partly due to discrepancies in the **international internet bandwith** (the speed by which websites in other countries can be accessed), the high cost of its installation (developing countries have to pay the full cost of linking to a hub in a developed country), a lower level of literacy (Figure 11.5), and a lack of electricity.

It is the ICT 'revolution', with its constant and rapid flow of data and finance, that arguably has been the single most important process in the speeding up of globalisation, a process that has occurred at the expense of national economies. Space-shrinking technologies have also allowed industries and services, from large-scale TNCs and international banks down to self-employed individuals, a freer choice of location for their site or place of work.

Figure 20.21
Ranulph Fiennes – the first man to cross both the world's ice caps and, at 65, the oldest Briton to climb Mount Everest

Figure 20.22
Mobile phones are part of the ICT 'revolution'

Figure 20.23
Growth of mobile phones and access to the internet, 1995–2007

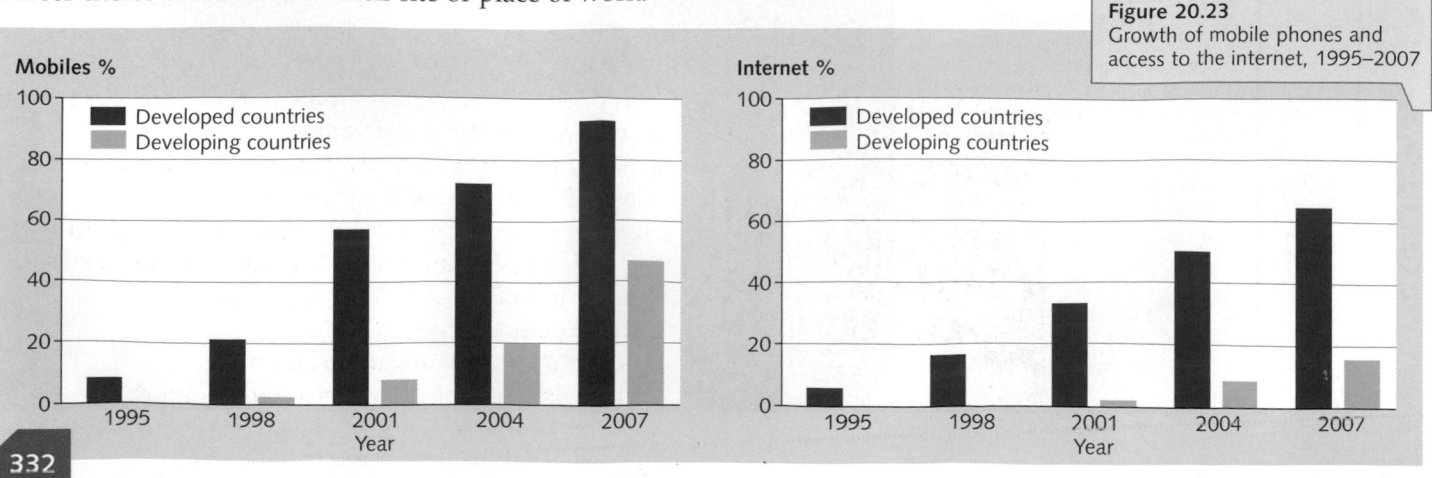

Global linkages – call centres

The introduction of space-shrinking technologies and the globalisation of industry and services has led to the growth of international **call centres**. A call centre is a centralised office used for the purpose of receiving and transmitting large volumes of requests by telephone. In most cases it is operated by a large company to receive requests from its customers, both real and potential, and to give data and after-sales service in return. They represent a company–customer relationship which can cover a wide range of support services, such as:

- sales, marketing and market research
- after-sales advice and product services
- technical support, especially in the use of computers and software
- claims enquiries, either for faulty products or on house and car insurance
- advance seat reservations on trains and planes or for concerts and the cinema
- banking and utility services such as gas, water and electricity
- mail-order catalogue retailing.

Call centres are used, therefore, by a wide range of firms including American Express, Bank of America, Swissair, British Telecom (BT), Dell, First Direct and Lloyds TSB Insurance.

Figure 20.24
A call centre

A call centre is usually operated within a large open-planned workspace. Each employee has his or her own work station with its computer, telephone head-set connected to a telecom switch, and one or more supervisor stations (Figure 20.24). The centre may work independently or be linked to a larger network.

Call centres form a part of the global linkage chain (e.g. pre-sales and after-sales) described on page 331 and, as such, seek to locate in low-cost locations. This has meant that many companies based in America and the UK have, since the early 2000s, relocated to places such as India where labour is much cheaper. At that time, an experienced IT employee could earn up to the equivalent of £60 000 a year in the USA, £45 000 in the UK and £15 000 in India. In the last decade, globalisation has seen India, and cities like Bangalore and Hyderabad in particular, specialising to become a world hub in this sector of business, which is the final link in the commodity chain. India has become a prime location for a number of reasons:

- lower labour costs, enabling the parent company to stay competitive in the global market
- a large number of university-trained ICT graduates
- a large English-speaking population
- employees prepared to work through the night to give 24-hour support and to fit with different time zones in the EU and North America (Figure 20.25)
- a stable democracy
- modern infrastructure and technology
- both national and state government support.

Figure 20.25
Call centres in India

Backwards in time

Los Angeles	New York	London	India
10 am	1 pm	6 pm	Midnight

- Mumbai
- Hyderabad
- Bangalore
- Chennai

International co-operation

Concern over the Earth's environment has previously been expressed by both individuals and conservation groups. Into the twenty-first century, it has become increasingly obvious that when it comes to protecting our planet, it needs the good will and co-operation of all countries.

Global warming and climate change

Under the Kyoto Protocol of 1992 – which took until 2005 to come into force in its entirety – industrialised countries were meant to cut their greenhouse gas emissions by approximately 5.2 per cent. Since then, total global emissions have in fact soared, mainly due to the emergence of China and India and the refusal of countries such as the USA to sign the agreement. There is, as with development, a wide global gap between the developed countries with their high emissions and developing countries with low emissions.

Under the Protocol, countries have been allocated a certain number of **carbon credits**, with each credit being worth 1 tonne of carbon dioxide or the equivalent of another greenhouse gas. A country can sell any surplus allowances that it has. This could be either an industrialised country that reduces the amount of carbon it creates, or a developing country that is unlikely to reach the limit of credits it is allowed (Figure 20.26a). This so-called **carbon trading** aims to encourage industrialised countries to produce less carbon while at the same time providing a welcome source of income for the poorest countries. The danger is that the wealthy industrialised countries will buy credits from the poorer countries and so see less need to reduce their own carbon emissions. If it is to work, carbon trading needs the global agreement of some 200 countries which at present have a wide divergence of interests (Figure 20.26b).

Figure 20.26
Carbon trading
(a) A hypothetical example of carbon trading
(b) Problems in reaching an international agreement

a

A power company in **Brazil** decides to use biomass instead of coal. This reduces its carbon emissions by 100 000 tonnes, for which it earns 100 000 credits.	The **UK** needs to reduce its carbon emissions by 1 million tonnes. If it buys the 100 000 credits from Brazil then it will only have to cut its emissions by 900 000 tonnes.

b

USA	Fears that reducing its emissions would mean job losses and a lower standard of living.
EU	Aims to reduce emissions by 30 per cent but is finding this difficult to achieve.
China	Needs extra thermal energy to create jobs.
India	Needs extra thermal energy to improve its standard of living.
Developing countries	They have neither the money nor the technology, nor do they see it as their problem.

Ecological footprint

The **ecological footprint** is a measure of the impact of people's way of life on planet Earth. It calculates how much productive land and sea a human population needs in order to generate the resources it consumes in terms of food, energy, water and raw materials in our everyday life. It also calculates how long it takes to absorb and render harmless the wastes that humanity creates, or for the ecological balance to renew itself.

Figure 20.27 shows how the ratio between human demands and nature's biocapacity has changed over time. The biocapacity, expressed in terms of 'Number of planet Earths' is always 1.0 (the horizontal line). The graph shows that whereas in net terms humanity only used about half (0.5) the planet's capacity in 1961, by 2003 this had increased to 1.25. This global deficit of 0.25 represents the world's **ecological overshoot** and means that each year humanity's ecological footprint is 25 per cent more than the planet can regenerate. Expressed another way, it now takes the Earth one year and three months (i.e. 1.25 years) to replace the resources that people use and waste in a single year. By measuring the ecological footprint of a population (a person, a city, a country or the world) we can assess our overshoot and, therefore, we should be able to manage the Earth's resources more carefully.

The biocapacity for the world is now said to be 2.26 ha per person. Figure 20.28 shows how selected countries have either a surplus or a deficit.

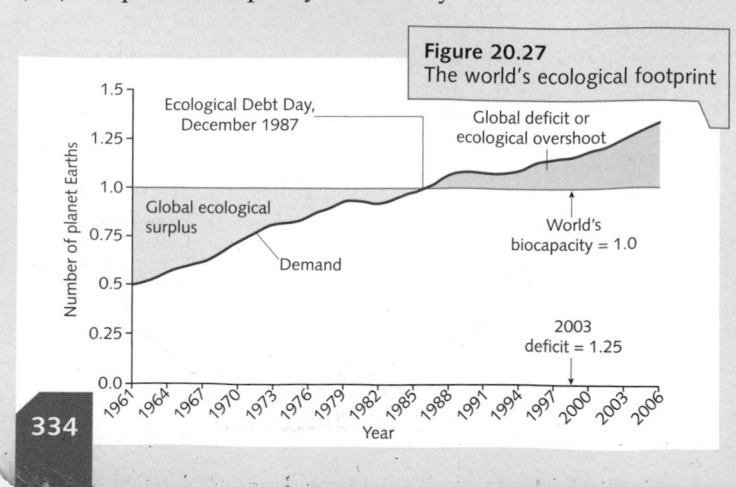

Figure 20.27
The world's ecological footprint

Figure 20.28
The ecological footprint

Surplus		Deficit	
1 Gabon	17.8	UAE	−11.0
2 Bolivia	13.7	Kuwait	−7.0
3 New Zealand	9.0	USA	−4.8
4 Mongolia	8.7	Belgium	−4.4
5 Brazil	7.8	Israel	−4.2
6 Congo	7.2	UK	−4.0
7 Canada	6.9	Saudi Arabia	−3.7
8 Australia	5.9	Japan	−3.6

Other selected countries: Germany −2.4, China −0.9, India −0.4, Kenya and Bangladesh −0.2, Ghana +3, Malaysia +1.5, Korea, Sweden and Spain each +3.5

Globalisation

Key Words and Terms

a You should know the meaning of the following terms:
- globalisation • space-shrinking technologies • Millennium Development Goals (MDGs)
- basic human rights • extreme poverty • the cycle of poverty • HIV/AIDS
- water-related illnesses • Fairtrade • global linkages
- carbon credit trading • the global commodity/value chain • sweatshops
- the ICT revolution • call centres • carbon trading • number of planet Earths
- biocapacity.

b You should know the difference between:
- epidemic and pandemic
- hunger and malnutrition
- carbon credits and ecological footprint.

Key Ideas

You should know and understand the following:
- Globalisation refers to processes that affect people and the environment across the world.
- Globalisation has led to the development of manufacturing and service industries across the world.
- The Millennium Development Goals are a group of aims created to reduce world poverty.
- The need to reduce the number of people living on $1 or less per day.
- The need to halt and reverse the global spread of HIV/AIDS.
- Globalisation increases the risk of a global pandemic.
- The need to reduce the number of people without clean water or basic sanitation.
- Differences in the availability and cleanliness of water supplies.
- The need to reduce the number of people suffering from hunger and malnutrition.
- Differences in the quality and quantity of food supplies between MEDCs and LEDCs.
- The contribution of Fairtrade in reducing the imbalance in world trade.
- The effects of the globalisation of industry on labour costs and people's standard of living.
- The development and importance of industrial linkages.
- The effect of developments in ICT and space-shrinking technologies.
- The global expansion of call centres.
- The need for international co-operation to reduce the effects of global warming/climate change and develop a global ecological footprint.

Skills, Theories and Models

- Interpret and use diagrams and maps
- Analyse and use photographs
- Interpret and use graphs
- Appreciate the benefits and problems of globalisation
- Appreciate the need to implement the Millennium Development Goals
- The cycle of poverty
- Understand how global linkages can bring together people and countries from across the world
- Identify inequalities in development
- Appreciate other people's values and attitudes

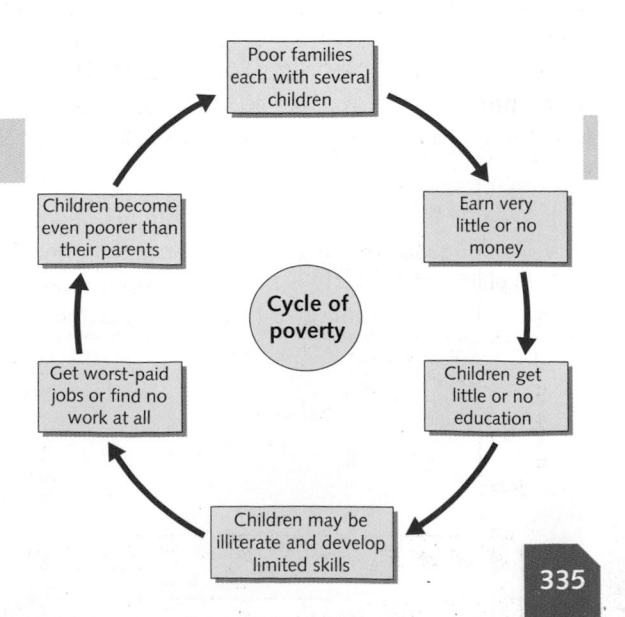

Poor families each with several children → Earn very little or no money → Children get little or no education → Children may be illiterate and develop limited skills → Get worst-paid jobs or find no work at all → Children become even poorer than their parents

Cycle of poverty

1 (Pages 324 and 325)

a What do you understand by the term 'globalisation'? *(1)*

b i) From the lists in Figure 20.1, choose one example from each of the groups headed:
  - Environmental   • Economic   • Technological
  - Social-cultural   • Political.

For each example, describe how it has been affected by globalisation. *(5)*

ii) Which ones do you think have had the greatest impact on people? *(2)*

c i) How has globalisation improved some people's standard of living? *(2)*

ii) In what ways has it widened the gap between developed and developing countries? *(2)*

Globalisation is ...

2 (Page 326)

a i) What are the Millennium Development Goals? *(2)*

ii) Why were they needed? *(1)*

iii) Using the diagram, what changes does the UN hope that the Millennium Development Goals will bring between 2000 and 2015? *(8)*

b i) What is the 'cycle of poverty'? *(2)*

ii) Why is it often difficult for poor people to break out of this cycle? *(2)*

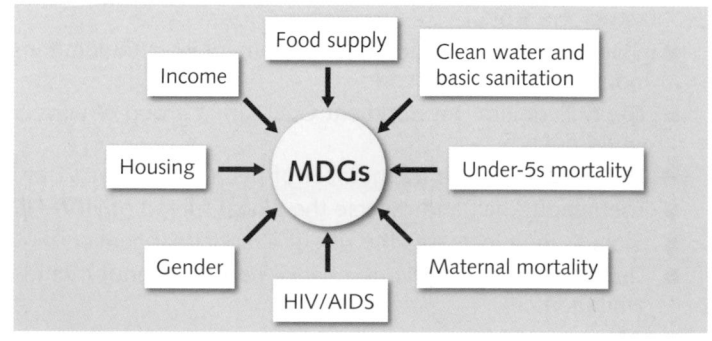

3 (Pages 327 to 329)

a i) What is HIV/AIDS? *(1)*

ii) Give three ways by which this disease might be spread. *(3)*

iii) What effects can it have on people? *(2)*

iv) Give two continuing problems of the disease. *(2)*

v) Give two encouraging signs concerning the disease. *(2)*

b i) Draw a star diagram to show four effects of dirty water. *(4)*

ii) Name two diseases that result from dirty water. *(2)*

iii) Name two diseases caused by parasites that live in water. *(2)*

iv) How are certain charitable organisations trying to overcome problems caused by dirty water and poor sanitation? *(3)*

v) Name one of these charitable organisations. *(1)*

c i) What is the difference between hunger and malnutrition? *(2)*

ii) Describe two ways by which malnutrition prevents people having a healthy life. *(2)*

iii) Explain why poverty is the main cause of hunger. *(2)*

iv) Using the diagram, give five reasons for the recent rise in global food prices. *(5)*

4 *(Pages 330 to 333)*

a i) What is Fairtrade? (1)

  ii) Name five Fairtrade products likely to be found in your local supermarket.
For each one, name a country where it may have been grown. (5)

  iii) Why is Fairtrade so important to people living in the poorest countries? (2)

  iv) How are people in those countries likely to benefit from Fairtrade? (3)

b Make a larger copy of the diagram below.

  i) Along the top row, put the following in the order in which they appear
in the commodity chain:

    • Retail   • Assembly   • Producing raw materials

    • After-sales and future designs   • Simple processing. (5)

  ii) For each of these stages:

    • in row 2, name a country or region where it occurs (5)

    • in rows 3 and 4, describe differences in wages earned and in working conditions. (5 × 2)

c i) What is meant by the term 'space-shrinking technologies'? (2)

  ii) Give three examples of space-shrinking technologies. (3)

  iii) How have space-shrinking technologies benefited MEDCs more than LEDCs? (2)

d i) What is a call centre? (1)

  ii) What types of service industry do they serve? (2)

  iii) Give four reasons why India has become a major hub for call centres. (4)

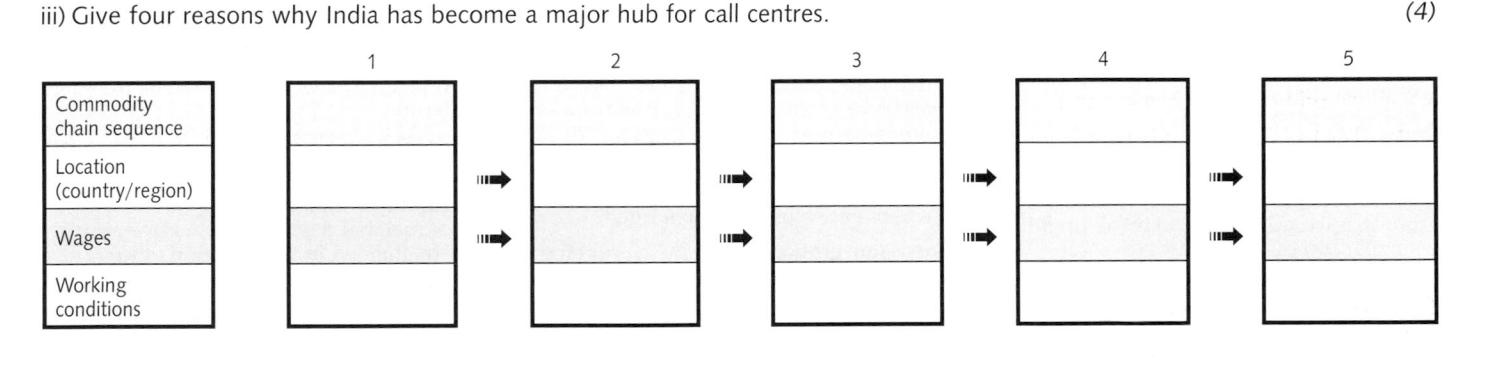

5 *(Page 334)*

a i) Why is international agreement necessary if the
emissions of greenhouse gases are to be reduced? (2)

  ii) What is the Kyoto Protocol? (1)

  iii) What is meant by the terms 'carbon credit'
and 'carbon trading'? (2)

b Refer to the map.

  i) Name countries A, B and C. Why has each
one so far failed to sign the Kyoto Agreement? (3 × 2)

  ii) Name the group of countries labelled D.
How successful have they been in reducing
carbon emissions? (2)

  iii) Why do developing countries feel carbon
emissions are not their problem? (1)

  iv) Give two reasons why developing countries might
find it difficult to reduce their carbon emissions. (2)

  v) What is meant by the term 'ecological footprint'? (1)

  vi) According to Figure 20.27, what was the
importance of December 1987? (1)

Mainly developing
countries

abrasion erosion caused by the rubbing and scouring effect of material carried by rivers, glaciers, waves and the wind.

acid rain rainwater containing chemicals that result from the burning of fossil fuels.

age–gender structure see population pyramids/structure.

ageing population the increase in the number of elderly people.

aid the giving of resources by one country, or an organisation, to another country.

air mass a large mass of air with uniform temperatures and humidity.

anticyclone an area of high pressure usually associated with settled weather.

appropriate technology an alternative technology suited to the needs, skills, knowledge and wealth of local people in the environment in which they live.

arch a coastal feature formed when waves erode through a small headland.

arête a narrow, knife-edged ridge formed by glacial erosion when two adjacent corries erode towards each other.

assisted area a region which has, based on EU guidelines, either a very low standard of living or high unemployment.

asylum seeker a person forced to leave their home country, where their life might be in danger, to seek safety in another country.

attrition erosion caused when rocks and boulders, transported by rivers and waves, bump into each other and break up into smaller pieces.

biofuels energy obtained from biomass, e.g. from vegetation, crops or dung.

biomass the total amount of organic matter contained by plants and animals in a given area.

biome an ecosystem, such as the tropical rainforest, occurring at a global scale.

birth rate the number of live births per 1000 people per year.

brownfield site disused and derelict land in an urban area that is available for redevelopment.

Burgess model an urban land use model showing five concentric zones, based on age of houses and wealth of their inhabitants.

call centre a centralised office from which companies and customers across the world are linked by telephone.

carbonation the process of chemical weathering by which carbonic acid in rainwater (carbon dioxide in solution) dissolves rocks containing limestone.

carbon credits/trading countries are allocated a quota of credits based on the amount of greenhouse gases, mainly carbon dioxide, that they emit. Those that reduce their emissions, or produce less than their quota, may sell surplus credits to countries that exceed their quota.

central business district (CBD) the commercial and business centre of a town or city where land values are highest.

central place any settlement that provides goods and services for smaller neighbouring settlements.

chemical weathering the decomposition of rock caused by a chemical change.

climate the average weather conditions for a place taken over a period of time.

coastal management the protection of the coastline and its wildlife, the improvement in the quality of life of local residents and the provision of recreational facilities for visitors.

collision margin a boundary between two plates moving together where, as both consist of continental crust, fold mountains form.

Common Agricultural Policy (CAP) the system of agricultural support adopted by the European Union for member states.

community/value added chain a connected series of activities that sees a product from its initial design through various stages of production to its sales and after-sales, with each activity seeing an increase in value.

commuting the process by which people living in one place, travel to another place for work.

conservative margin a boundary between two plates that are sliding past each other and where crust is being neither formed nor destroyed.

constructive margin a boundary between two plates that are moving apart and where new crust is being formed.

corrasion erosion caused by the rubbing and scouring effect of material carried by rivers, glaciers, waves and the wind.

corrie (cirque) a deep, steep-sided, rounded or semi-circular hollow, often with a lake, found in glaciated highlands.

corrosion erosion caused by acids in rivers and waves dissolving rocks by chemical action.

counterurbanisation the movement of people and employment away from large cities to smaller settlements within more rural areas.

crust the thin outer layer of the Earth.

death rate the number of deaths per 1000 people per year.

deforestation the complete clearance of forested land.

delta an area of silt deposited by a river where it enters the sea or a lake.

demographic transition model a model that tries to show how changes in birth and death rates over a period of time may be related to different stages of economic development.

dependency ratio the ratio between those in the non-economically active age group (taken to be children under 15 and adults over 65) and those in the economically active age group (15 to 64).

deposition the laying down of material previously transported by mass movement, water, glaciers, waves and the wind.

depression an area of low pressure in temperate latitudes usually associated with cloud, rain and strong winds.

desertification the turning of land, often through physical processes and human mismanagement, into desert.

destructive margin a boundary between two plates that are moving together and where one, consisting of oceanic crust, is forced downwards and destroyed.

development the level of economic growth of a country or region and the processes of change taking place within it.

dew point the temperature at which air becomes saturated and water vapour within it condenses to form water droplets.

diet the amount (measured in calories) and quality (the balance between proteins, carbohydrates and vitamins) of food needed to keep a person healthy and active.

discharge the volume of water in a river at a given time, usually measured in cumecs (cubic metres per second).

drainage basin the area of land drained by a main river and its tributaries.

drainage density the total length of all the streams and rivers in a drainage basin divided by the total area of the drainage basin.

drought a prolonged period of dry weather leading to a water shortage.

drumlin a smooth, elongated mound of material deposited by a glacier and streamlined in the direction of ice movement.

dry valley a valley, usually on limestone or chalk, originally formed by rivers which now only flow under extreme wet conditions.

earthquake a sudden movement within the Earth's crust, usually close to a plate boundary.

ecological footprint a measure of the time it takes to generate the resources consumed by the human population in its daily life and to render useless the waste it creates.

ecosystem a natural system in which plants (flora) and animals (fauna) interact with each other and the non-living environment.

ecotourism a sustainable form of 'green' tourism aimed, unlike mass tourism, at protecting the environment and local cultures.

emerging country a developing country, notably China and India, that is now experiencing rapid industrialisation and economic growth.

emigrant a person who leaves one country to live in another country.

employment structure the division of jobs into, traditionally, primary, secondary and tertiary sectors, together with, more recently, the quaternary sector.

energy flows the transfer of energy through an ecosystem by means of a food chain (food web).

environment the surroundings in which plants, animals and people live.

Environment Agency the amalgamation (1996) of several organisations, including the National Rivers Authority, to secure the protection of water, land and air in the UK.

Environmental Stewardship Scheme (ESS) a farming–environment partnership that encourages farmers, through subsidies, to conserve wildlife, enhance the landscape and protect natural resources.

epicentre the place on the Earth's surface immediately above the focus of an earthquake.

erosion the wearing away of the land by material carried by rivers, glaciers, waves and the wind.

erratic a large boulder transported by ice and deposited in an area of totally different rock.

ethnic group a group of people with common characteristics related to race, nationality, language, religion or culture.

eutrophication the process by which fertiliser (mainly nitrate) causes, on reaching rivers and lakes, rapid algae and plant growth and, subsequently, the depletion of oxygen available for fish.

evapotranspiration the loss of moisture from water surfaces and the soil (evaporation) and vegetation (transpiration).

exfoliation a physical weathering process by which, due to extreme changes in temperature, the surface layers of exposed rock peel away.

exports goods transported by one country for sale in a different country.

extreme poverty when people, lacking the basic human rights of health, food, shelter, education and security, have to live on $1 or less a day.

Fairtrade an attempt to alleviate poverty and, through trade, to generate sustainable development amongst disadvantaged farmers in some of the world's poorest countries.

favela a Brazilian term for an informal, shanty-type settlement.

fiord a long, narrow, steep-sided inlet formed by glaciers and later drowned by a rise in sea-level.

flood a period of either high river discharge (when a river overflows its banks) or, along the coast, an extremely high tide or storm surge.

floodplain the wide, flat valley floor of a river where silt is deposited during times of flood.

fold mountains a range of high mountains formed by two plates moving together at a collision margin.

food chain/web the transfer of energy through an ecosystem from primary producers to consumers and decomposers.

footloose industry an industry which, as it is not tied to raw materials, has a free choice of location.

formal and informal sector the difference between employment controlled by large companies and the government, and employment dependent upon the initiative of individuals.

fossil fuels non-renewable forms of energy which, when used, release carbon.

fragile environment an environment which, if not carefully managed, may be irretrievably damaged.

freeze–thaw weathering/frost shattering a process of physical weathering by which rock disintegrates due to water in cracks repeatedly freezing and thawing.

front the boundary between two air masses which have different temperature and humidity characteristics.

genetically modified (GM) crops inserting genes from one plant species into another to increase yields and disease-resistance in crops.

geothermal energy a renewable resource of energy using heated rock within the Earth's crust to produce steam and generate energy.

glacial trough a steep-sided, flat-floored glaciated valley with a characteristic U-shape.

globalisation any process of change operating at a world scale and having worldwide effects. It can be physical (sea-level change), human (banking, trade) or both (global warming).

global linkages the connection between firms located in different countries involved in the same line of production or services.

global warming the increase in the world's average temperature, believed to result from the release of carbon dioxide and other gases into the atmosphere by the burning of fossil fuels.

green belt an area of land around a large urban area where the development of housing and industry is severely restricted and the countryside is protected for farming and recreation.

greenfield site an area of countryside/open space that has not yet been built upon.

greenhouse effect the warming of the Earth's atmosphere by the release of gases that include carbon dioxide and methane.

Green Revolution the introduction of high-yielding varieties (HYVs) of cereals (rice and wheat) into less economically developed countries.

Gross Domestic Product (GDP) per capita the total value of goods produced and services provided by a country in a year, divided by the total number of people living in that country.

groundwater water stored underground in permeable rocks.

groyne an artificial structure running out to sea to limit longshore drift.

hanging valley a tributary valley left high above the main valley as its glacier was unable to erode downwards as quickly as the larger glacier in the main valley, and whose river now descends as a waterfall.

hierarchy a ranking of settlements or shopping centres according to their size or the services they provide.

high-tech industry an industry using advanced information technology and/or processes involving micro-electronics.

HIV/AIDS AIDS is a death-threatening disease that has spread around most of the world. It starts with victims contracting HIV which progressively reduces the human body's ability to combat disease.

honeypot a place of attractive scenery or historic interest which attracts tourists in large numbers.

Hoyt model an urban land use model showing wedges (sectors), based on main transport routes and social groupings.

Human Development Index (HDI) a social welfare index, adopted by the United Nations as a measure of development, based on life expectancy (health), adult literacy (education) and real GDP per capita (economic).

humus organic material found in soil derived from the decomposition of vegetation, dead organisms and animal excreta.

hurricane *see tropical cyclone.*

hydraulic action erosion caused by the sheer force of water breaking off small pieces of rock.

hydrograph a graph showing changes in the discharge of a river over a period of time.

hydrological cycle the continuous recycling of water between the sea, air and land.

igneous rock a rock formed by volcanic activity, either by magma cooling within the Earth's crust or lava at the surface.

illegal immigrant a person who enters a country without permission or any legal documents.

immigrant a person who arrives in a country with the intention of living there.

impermeable rock a rock that does not let water pass through it.

imports goods bought by a country from another country.

infant mortality the average number of deaths of children under 1 year of age per 1000 live births.

inner city the part of an urban area next to the city centre that was characterised by older housing and industry.

interdependence the reliance of a country, especially in terms of trade and aid, on other countries in the global economy.

interlocking spur one of a series of spurs that project alternately from the sides of a V-shaped river valley.

island arc a curving line of islands formed by volcanic activity at a destructive plate boundary.

lag time the period of time between peak rainfall and peak river discharge.

lava molten rock (magma) ejected onto the Earth's surface by volcanic activity.

leaching the downward movement, and often loss, of nutrients (minerals) in solution in the soil.

less economically developed countries (LEDCs) countries that tend to have a low income and standard of living.

levée (dyke) an artificial embankment built to prevent flooding by a river or the sea.

life expectancy the average number of years a person born in a particular country might be expected to live.

limestone pavement an exposed, horizontal area of limestone containing clints and grykes.

literacy rate the proportion of the total population able to read and write.

longshore drift the movement of material along a coast by breaking waves.

magma molten rock occurring beneath the Earth's crust.

malnutrition ill-health caused by a diet deficiency, either in amount (quantity) or balance (quality).

managed coastal retreat the building of new sea defences inland from the present coastline.

mantle that part of the Earth's structure between the crust and the core.

mass movement the downhill movement of weathered material under gravity.

meander the winding course of a river.

metamorphic rock a rock that has been altered by extremes of heat and pressure.

migration the movement of people (and animals/birds) either within a country or between countries, either voluntary or forced.

Millennium Development Goals (MDGs) a group of targets set at a global scale with the aim of reducing extreme poverty.

model a theoretical representation of the real world in which detail and scale are simplified in order to help explain the reality.

moraine material, usually angular, that is transported and later deposited by a glacier.

more economically developed countries (MEDCs) countries that have a high income and standard of living.

mudflow a rapid form of mass movement consisting mainly of mud and water.

National Park an area set aside for the protection of its scenery, vegetation and wildlife, so that it may be enjoyed by people living and working there at present, by visitors, and by future generations.

natural increase the growth in population resulting from an excess of births over deaths.

New Commonwealth immigrant a person born in a former British colony in Africa, Asia and the Caribbean (as opposed to the Old Commonwealth countries of Canada, Australia and New Zealand) and who has moved to Britain to live.

newly industrialised country (NIC) a country, mainly in the Pacific Rim of Asia, that has undergone rapid and successful industrialisation since the early 1980s.

non-renewable resource a finite resource, such as a fossil fuel or a mineral, which, once used, cannot be replaced.

nutrient cycle the process by which minerals necessary for plant growth are taken up from the soil, and returned when plants shed their leaves or vegetation dies.

organic farming a sustainable type of farming that does not use chemical fertiliser or pesticide but uses animal and green manures instead.

overcultivation the exhaustion of the soil by growing crops, especially the same crop, on the same piece of land year after year.

overgrazing the destruction of the protective vegetation cover by having too many animals grazing on it.

overpopulation when the number of people living in an area exceeds the amount of resources available to them.

ox-bow lake a crescent-shaped lake formed after a river cuts through the neck of, and later abandons, a former meander.

pandemic a worldwide disease epidemic.

permeable rock a rock that allows water to pass through it.

physical weathering the disintegration of rock by mechanical processes without any chemical changes within the rock.

plantation agriculture a type of tropical farming where one crop is grown, usually for export by a transnational corporation, on a large, efficiently run estate.

plate margin the boundary between two plates which may be moving towards, away from or sideways past each other.

plate tectonics the theory that the surface of the Earth is divided into a series of plates, consisting of continental and oceanic crust.

plucking a process of glacial erosion by which ice freezes onto weathered rock and, as it moves, pulls pieces of rock with it.

population density the number of people living within a given area (usually a square kilometre).

population distribution the spread of people across the Earth's surface.

population pyramids/structure the proportion of males and females within selected age groups, usually shown as a pyramid.

porous rock a rock containing tiny pores through which water can either pass or be stored.

precipitation that part of the hydrological cycle where atmospheric moisture is deposited at the Earth's surface as rain, hail, snow, sleet, dew, frost or fog.

prevailing wind the direction from which the wind usually blows.

primary industry an industry, such as farming, fishing, forestry and mining, that extracts raw materials directly from the land or sea.

pyramidal peak (horn) a triangular-shaped mountain formed by three or more corries cutting backwards, and with arêtes radiating from the central peak.

quality of life the satisfaction of people with their environment and way of life.

quaternary industry an industry, such as micro-electronics, that provides information and expertise.

quota a limit imposed on the quantity of goods produced or purchased.

range of goods the maximum distance that people are prepared to travel for a specific service.

refugees people forced to move from an area where they lived, and made homeless.

renewable resource a sustainable resource, such as solar energy or water power, which can be used over and over again.

resource a feature of the environment that is needed and used by people.

retailing the sale of goods, usually in shops, to the general public.

ria a river valley drowned by a rise in sea-level.

ribbon lake a long, narrow lake found on the floor of a glaciated valley.

Richter scale the scale used to measure the magnitude of earthquakes.

runoff the surface discharge of water derived mainly from excessive rainfall or melting snow.

rural–urban fringe a zone of transition between the built-up area and the countryside, where there is often competition for land use.

rural–urban migration the movement of people from the countryside to towns and cities where they wish to live permanently.

saltation a process of transportation by rivers in which small particles bounce along the bed in a 'leap-frog' movement.

satellite image a photograph, sometimes using false colours, taken from space and sent back to Earth.

science park/city an estate, often with an edge-of-city location, or a newly planned city, with high-tech industries and a university link.

sea-floor spreading a process where crustal plates move apart at constructive margins, allowing magma to rise to the surface.

secondary industry an industry that processes or manufactures primary raw materials (such as steelmaking), assembles parts made by other industries (such as cars) or is part of the construction industry.

sedimentary rock a rock that has been laid down in layers, often as sediment derived from the erosion and transport of older rocks.

self-help housing scheme where groups of people, especially in developing countries, are encouraged to build their own homes using materials provided by the local authority.

service industry see tertiary industry.

set-aside land land for which farmers were paid a subsidy if it was taken out of production.

settlement function the main activity, usually economic or social, of a place.

settlement pattern the shape and spacings of individual settlements, usually dispersed, nucleated or linear.

shanty town an area of poor-quality housing, lacking in amenities such as water supply, sewerage and electricity, which often develops spontaneously and illegally (as a squatter settlement) in a city in a developing country.

shifting cultivation a type of farming where people move every few years as the soil becomes exhausted.

site the actual place where a settlement (or farm/factory) is located.

situation the location of a settlement in relation to places (physical and human) surrounding it.

soil the thin, loose, surface layer of the Earth which provides a habitat for plants and which consists of weathered rock, water, gases (air), living organisms (biota) and decayed plant matter (humus).

soil creep the slowest type of downhill movement (mass movement) of soil due to gravity.

soil erosion the wearing away and loss of soil due to the action of rain, running water and strong winds, often accelerated by human activity.

soil profile a vertical section of soil showing its different layers (horizons).

solar energy the prime source of energy on Earth, taken into the food chain by photosynthesis in plants, or used by people as a source of electricity.

solution a type of chemical weathering in which water dissolves minerals in rocks.

space-shrinking technologies ICT innovations such as satellite television, the internet and mobile phones, which have speeded up the process of globalisation.

sphere of influence the area served by a settlement, shop or service.

spit a long, narrow accumulation of sand or shingle formed by longshore drift, with one end attached to the land and the other, projecting out to sea, often with a curved (hooked) end.

spring the emergence of underground water which often provided an ideal site for early settlement.

stack an isolated piece of rock detached from the mainland by wave erosion.

stalactite/stalagmite formed by water containing calcium carbonate in solution, evaporating in limestone caverns to leave an icicle-shaped feature hanging from the roof (stalactite) or a more rounded feature on the floor (stalagmite).

standard of living the degree of material comfort (i.e. wealth) enjoyed by an individual, group or country.

storm surge a rapid rise in sea-level caused by storms, especially tropical cyclones, forcing water into a narrowing sea area.

subduction zone occurs at a destructive plate margin where oceanic crust, moving towards continental crust, is forced downwards into the mantle and destroyed.

subsidies grants of money made by governments to maintain the price of a specific industrial or agricultural product, e.g. milk.

subsistence farming where all farm produce is needed by the farmer's family or village, and where there is no surplus for sale.

suburb the outer, usually residential, area of a town or city.

suburbanised village a village that has increasingly adopted some of the characteristics (new housing estates, more services) of urban areas.

suspension a process of transportation by rivers in which material is picked up and carried along within the water itself.

sustainable development a way of improving people's standard of living and quality of life without wasting resources or harming the environment.

swallow hole/sink a hole in the surface of a limestone area, usually formed by solution, down which a river may disappear.

sweatshop a factory found in a developing country where people have to work long hours in poor conditions and for low wages.

synoptic chart a map showing the state of the weather at a given time.

tariff a customs duty charged on goods imported into a country.

tectonics see plate tectonics.

tertiary industry an occupation, such as health, education, transport and retailing, that provides a service for people.

threshold population the minimum number of people needed to ensure that a specific service (shop, school, hospital) will be able to operate economically.

tor a small rocky outcrop that is the characteristic feature of granite areas.

total fertility rate (TFR) the average number of children a woman is likely to have if she lives through her child-bearing years.

traction a process of transportation by rivers in which material is rolled along the bed.

trade the movement and sale of goods from one country (the producer/exporter) to another country (the consumer/importer).

trade balance the difference between the value of imports and of exports of a country.

trading group/bloc a group of countries that have joined together for trading purposes.

transnational corporation a company which, by having factories and offices in several countries, is global in that it operates across national boundaries.

transpiration the loss of moisture from vegetation into the atmosphere.

transportation the movement of material by rivers, glaciers, waves and the wind.

tropical cyclone a severe tropical storm, characterised by low pressure, heavy rainfall and winds of extreme strength, which is capable of causing widespread damage and loss of life.

truncated spur a former interlocking spur in a pre-glacial V-shaped valley which, during a later period of glaciation, had its end removed by a glacier.

urban development corporation (UDC) created by the UK government to promote new industrial, housing and community developments in an urban area with large amounts of derelict land and buildings.

urbanisation the increase in the proportion of people living in towns and cities.

urban redevelopment the total clearance of parts of old inner city areas and starting afresh with new houses, especially high-rise flats.

urban regeneration/renewal the improvement of old houses and the addition of amenities in an attempt to bring new life to old inner city areas.

urban sprawl the unplanned, uncontrolled growth of urban areas into the surrounding countryside.

volcano a mountain or hill, often cone-shaped, through which lava, ash and gases may be ejected at irregular intervals.

V-shaped valley a narrow, steep-sided valley formed by the rapid vertical erosion of a river.

water cycle see hydrological cycle.

waterfall a vertical, or near vertical, drop of water resulting from a sudden change in the gradient of a river.

watershed a ridge of high land that forms the boundary between two adjacent drainage basins.

water table the upper limit of the zone of saturation found in a porous or permeable rock, or soil.

wave-cut notch an indentation at the foot of a cliff caused by wave erosion.

wave-cut platform a gently sloping, rocky platform found at the foot of a retreating cliff and exposed at low tide.

waves formed when wind blows over areas of water; in the case of the sea, the waves may be either constructive or destructive as they break on reaching the land.

weather the hour-to-hour, day-to-day state of the atmosphere in relation to temperature, sunshine, precipitation and wind.

weathering the breakdown of rocks in situ by either mechanical processes (physical weathering) or chemical changes (chemical weathering).

Location map: Places and Case Studies

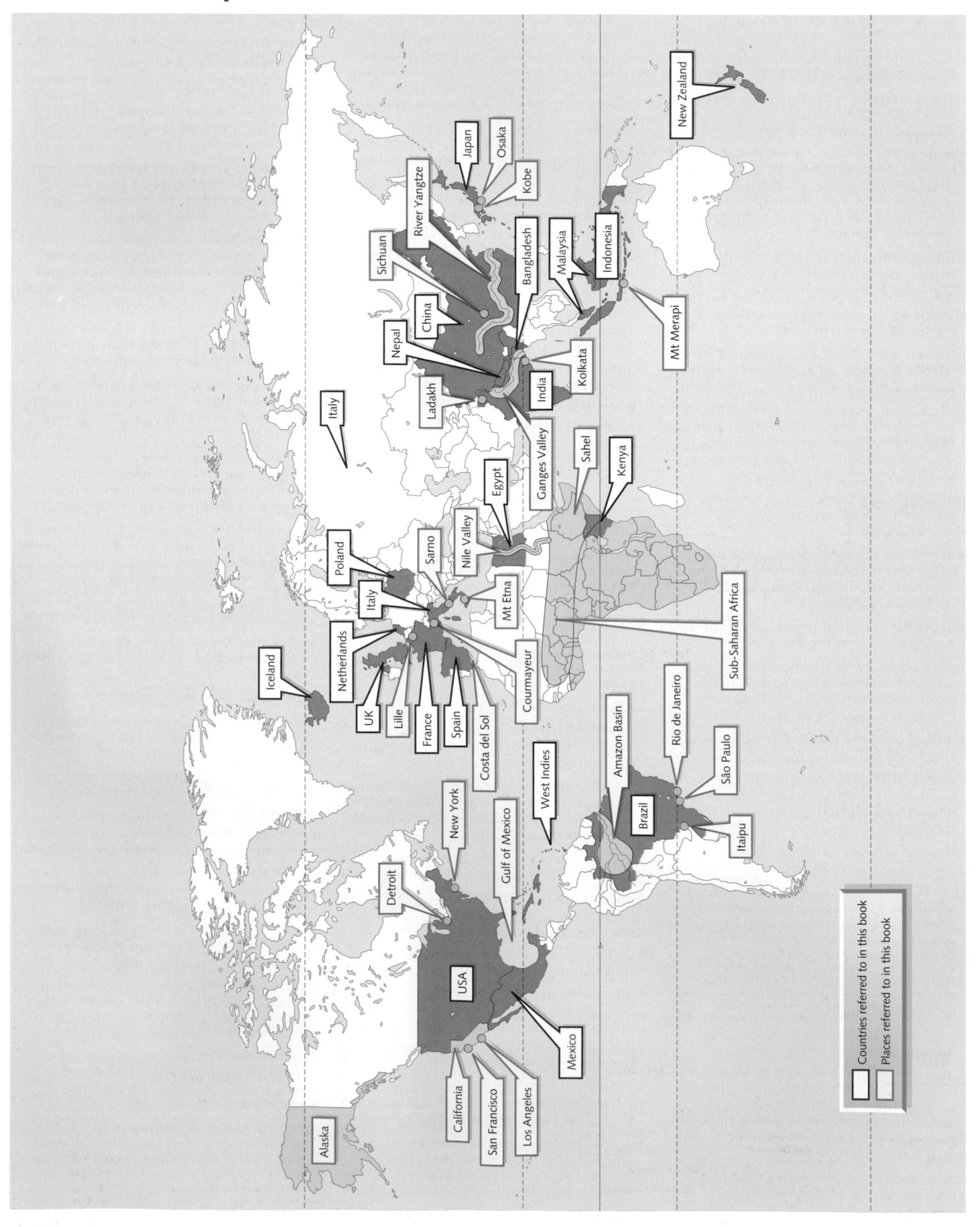

Japan
Osaka
Kobe
New Zealand
River Yangtze
Sichuan
Bangladesh
Malaysia
Indonesia
China
Nepal
Kolkata
Mt Merapi
Ladakh
India
Italy
Ganges Valley
Sahel
Kenya
Egypt
Nile Valley
Poland
Sarno
Sub-Saharan Africa
Italy
Mt Etna
Netherlands
Courmayeur
Iceland
UK
Lille
France
Spain
Costa del Sol
Amazon Basin
Rio de Janeiro
São Paulo
West Indies
Brazil
Itaipu
New York
Gulf of Mexico
Detroit
USA
Mexico
California
San Francisco
Los Angeles
Alaska

Countries referred to in this book
Places referred to in this book

Key to 1:50 000 OS Landranger map sheets

ROADS AND PATHS

Not necessarily rights of way

Service area — Junction number **1** — Elevated
M 1
Motorway (dual carriageway)

Unfenced — Footbridge
A 470 — Dual carriageway
Primary route

A 493
Main road

B 4518
Primary route/main road under construction

A 855 — Bridge — B 885
Secondary road

Narrow road with passing places

Road generally more than 4m wide

Road generally less than 4m wide

Path / other road, drive or track

Gradient: 20% (1 in 5) and steeper,
14% (1 in 7) to 20% (1 in 5)

Gates / road tunnel

Ferry P — Ferry V
Ferry (passenger) / ferry (vehicle)

RAILWAYS

Track multiple or single

Siding

Tunnel

Light rapid transit system, narrow gauge or tramway

Bridges / footbridge

LC Level crossing

a Viaduct

Station, (a) principal

Light rapid transit system station

WATER FEATURES

Marsh or salting

Slopes — Cliff — Shingle

Towpath — Lock

Aqueduct — Canal — Ford — Beacon — Flat rock — Lighthouse (in use)

Weir — Sand — Lighthouse (disused)

Lake — Footbridge — Bridge — Normal tidal limit — Dunes — Low water mark

Mud

High water mark

Canal (dry)

HEIGHTS

1 metre = 3.2808 feet

Contours are at 10 metres vertical interval

144 Heights are to the nearest metre above mean sea level

Heights shown close to a triangulation pillar refer to the ground at the base of the pillar and not necessarily to the summit.

ROCK FEATURES

Outcrop — Cliff — 650 — 600 — Scree

PUBLIC RIGHTS OF WAY

- - - - - - - - - - Footpath

– – – – – – – – Bridleway

–·–·–·–·– Restricted byway (not for use by mechanically propelled vehicles)

–+–+–+–+– Byway open to all traffic

The symbols show the defined route so far as the scale of mapping will allow.
The representation on this map of any other road, track or path is no evidence of the existence of a right of way

Danger Area Firing and Test Ranges in the area. Danger! Observe warning notices.

OTHER PUBLIC ACCESS

• • • • Other route with public access (not normally shown in urban areas). Alignments are based on the best information available. These routes are not shown on maps of Scotland.

◆ ◆ National Trail, European Long Distance Route, Long Distance Route, selected Recreational Routes

● ● National/Regional Cycle Network

— — Traffic-free cycle route

4 8 National/Regional Cycle Network number

BOUNDARIES

–+– – –+– – National

–·–+–·–+–·– District

— — — County, Unitary Authority, Metropolitan District or London Borough

National Park

ARCHAEOLOGICAL AND HISTORICAL INFORMATION

✛ Site of antiquity

⚔ Battlefield (with date)

☆ ···· Visible earthwork

VILLA Roman

Castle Non-Roman

LAND FEATURES

Electricity transmission line (pylons shown at standard spacing)

> --> --> Pipe line (arrow indicates direction of flow)

ruin Buildings

Important building (selected)

Bus or coach station

Place of worship { with tower / with spire, minaret or dome / without such additions }

Glass structure

(H) Heliport

△ Triangulation pillar

Mast

Wind pump/wind turbine

Windmill with or without sails

Graticule intersection at 5' intervals

Cutting, embankment

Quarry

Landfill site or slag/spoil heap

Coniferous wood

Non-coniferous wood

Mixed wood

Orchard

Park or ornamental ground

Forestry Commission access land

National Trust-always open

National Trust-limited access, observe local signs

National Trust for Scotland

TOURIST INFORMATION

Camp site/caravan site

Garden

Golf course or links

i *i* Information centre, all year / seasonal

Nature reserve

P P&R Parking / Park and ride, all year / seasonal

Picnic site

Selected places of tourist interest

C C Telephone, public / roadside assistance

Viewpoint

V Visitor centre

! Walks / Trails

▲ Youth hostel

⊗ World Heritage site/area

Recreation / leisure / sports centre

ABBREVIATIONS

CH	Clubhouse	CG	Cattle grid
PH	Public house	P	Post office
PC	Public convenience (in rural area)	MP	Milepost
TH	Town Hall, Guildhall or equivalent	MS	Milestone

INDEX   Bold type indicates geographical locations